Outstanding Contributions to Logic

Volume 29

Outstanding Contributions to Logic puts focus on important advances in modern logical research. Each volume is devoted to a major contribution by an eminent logician. The series will cover contributions to logic broadly conceived, including philosophical and mathematical logic, logic in computer science, and the application of logic in linguistics, economics, psychology, and other specialized areas of study.

A typical volume of Outstanding Contributions to Logic contains:

- A short scientific autobiography by the logician to whom the volume is devoted
- The volume editor's introduction. This is a survey that puts the logician's contributions in context, discusses its importance and shows how it connects with related work by other scholars
- The main part of the book will consist of a series of chapters by different scholars that analyze, develop or constructively criticize the logician's work
- Response to the comments, by the logician to whom the volume is devoted
- A bibliography of the logician's publications.

Outstanding Contributions to Logic is published by Springer as part of the Studia Logica Library.

This book series, is also a sister series to Trends in Logic and Logic in Asia: Studia Logica Library. All books are published simultaneously in print and online. This book series is indexed in SCOPUS.

Proposals for new volumes are welcome. They should be sent to the editor-in-chief sven-ove.hansson@abe.kth.se.

Thomas Piecha · Kai F. Wehmeier
Editors

Peter Schroeder-Heister on Proof-Theoretic Semantics

Springer

Editors
Thomas Piecha
University of Tübingen
Tübingen, Baden-Württemberg, Germany

Kai F. Wehmeier
University of California
Irvine, CA, USA

ISSN 2211-2758 ISSN 2211-2766 (electronic)
Outstanding Contributions to Logic
ISBN 978-3-031-50980-3 ISBN 978-3-031-50981-0 (eBook)
https://doi.org/10.1007/978-3-031-50981-0

Cover photograph by Christoph Jäckle, 2016.

This Springer imprint is published by the registered company Springer Nature Switzerland AG
The registered company address is: Gewerbestrasse 11, 6330 Cham, Switzerland

Paper in this product is recyclable.

Preface

Peter Schroeder-Heister began his academic career in 1971 as a student of mathematics and catholic theology — the latter of which would soon be replaced by philosophy — at the University of Bonn. There he was almost immediately drawn into mathematical logic by way of Gisbert Hasenjaeger's *Seminar für Logik und Grundlagenforschung*. After passing his *Staatsexamen* in philosophy and mathematics in 1977, Peter moved to the then still very new University of Konstanz, where he wrote his doctoral dissertation while editing, with Gereon Wolters, the academic estate of Hugo Dingler, and later working on Jürgen Mittelstraß's *Enzyklopädie Philosophie und Wissenschaftstheorie*. He obtained his Ph.D. in 1981 from the University of Bonn, Hasenjaeger serving as his *de iure* and Dag Prawitz as his *de facto* thesis adviser. Peter remained in Konstanz as Mittelstraß's *Assistent* until 1989, obtaining his habilitation on the way in 1987, before being appointed to a professorship in logic and philosophy of language at the University of Tübingen, where he would spend the remainder of his academic career, despite prestigious offers elsewhere, including from the University of Salzburg in 2000.

This volume reflects but one of Peter's research interests, though arguably his principal one and the one through which he has made the most significant impact: proof-theoretic semantics. It thus seems imperative to mention, in this preface, some of the other areas of intellectual endeavor on which he has left a mark, which include the history of logic (with publications on Dingler, Hertz, Frege, and Popper, not to mention Aristotle and Leibniz, but also as a long-time member of the editorial board for *History and Philosophy of Logic*), logic programming, Martin-Löf type theory, the philosophy of science, substructural and relevant logic, and empirical psychology.

Finally, the editors would each like to say just a few personal words about their relationship with Peter. TP met Peter during his studies at the University of Tübingen, where he participated with great joy in his very instructive courses in computer science and philosophy, first as a student and soon after also as a teaching assistant. He continued to work in Peter's group for many years and benefited enormously from Peter's constant support and the stimulating work environment, which, in particular, enabled TP to contribute to several international research projects, conferences organized in Tübingen, and joint publications with Peter. KW originally knew of

Peter as a scholar of Frege, through his articles on the permutation argument in §10 of *Grundgesetze* and on Frege's anticipation of propositional resolution, and had corresponded with him about this work before ever meeting him. In 2001, Peter literally rescued KW from having to leave academia by offering him a temporary position in Tübingen as his *wissenschaftlicher Mitarbeiter*. After one immensely enjoyable and productive year in Tübingen, KW moved on, with much help, encouragement, and both moral and material support from Peter, to more permanent employment with the University of California.

It was thus with great pleasure that we undertook the editing of this volume, and we are delighted at having this opportunity to express our appreciation of and gratitude to Peter. It only remains for us to thank the authors for their inspiring contributions.

Tübingen/Irvine, March 2, 2023

Thomas Piecha
Kai F. Wehmeier

Contents

Proof-Theoretic Semantics: An Autobiographical Survey 1
Peter Schroeder-Heister

***Grundlagen der Arithmetik*, §17: Part 1. Frege's Anticipation of the Deduction Theorem** 53
Göran Sundholm

Frege's Class Theory and the Logic of Sets 85
Neil Tennant

The Validity of Inference and Argument 135
Dag Prawitz

Kolmogorov and the General Theory of Problems 161
Wagner de Campos Sanz

Disjunctive Syllogism without *Ex falso* 193
Luiz Carlos Pereira, Edward Hermann Haeusler and Victor Nascimento

The Logicality of Equality 211
Andrzej Indrzejczak

Eight Rules for Implication Elimination 239
Michael Arndt

Focusing Gentzen's LK Proof System 275
Chuck Liang and Dale Miller

Intensional Harmony as Isomorphism 315
Paolo Pistone and Luca Tranchini

A Note on Synonymy in Proof-Theoretic Semantics 339
Heinrich Wansing

Paradoxes, Intuitionism, and Proof-Theoretic Semantics 363
Reinhard Kahle and Paulo Guilherme Santos

On the Structure of Proofs 375
Lars Hallnäs

Truth-Value Constants in Multi-Valued Logics 391
Nissim Francez and Michael Kaminski

Counterfactual Assumptions and Counterfactual Implications 399
Bartosz Więckowski

Some Set-Theoretic Reduction Principles 425
Michael Bärtschi and Gerhard Jäger

Comments on the Contributions 443
Peter Schroeder-Heister

Index 457

List of Contributors

Michael Arndt
Department of Computer Science, University of Tübingen, Germany,
e-mail: arndt@cs.uni-tuebingen.de

Michael Bärtschi
Institute of Computer Science, University of Bern, Switzerland,
e-mail: michael.baertschi@inf.unibe.ch

Wagner de Campos Sanz
Faculdade de Filosofia, Universidade Federal de Goiás, Goiânia, Brazil,
e-mail: wsanz@ufg.br

Nissim Francez
Technion – Israel Institute of Technology, Haifa, Israel,
e-mail: francez@cs.technion.ac.il

Edward Hermann Haeusler
Informatica, PUC-Rio, Rio de Janeiro, Brazil, e-mail: hermann@inf.puc-rio.br

Lars Hallnäs
The Swedish School of Textiles, University of Borås, Sweden,
e-mail: lars.hallnas@hb.se

Andrzej Indrzejczak
Department of Logic, University of Łódź, Poland,
e-mail: andrzej.indrzejczak@filhist.uni.lodz.pl

Gerhard Jäger
Institute of Computer Science, University of Bern, Switzerland,
e-mail: gerhard.jaeger@inf.unibe.ch

Reinhard Kahle
Carl Friedrich von Weizsäcker Center, University of Tübingen, Germany, and CMA, FCT, Universidade Nova de Lisboa, Portugal,
e-mail: reinhard.kahle@uni-tuebingen.de

Michael Kaminski
Technion – Israel Institute of Technology, Haifa, Israel,
e-mail: kaminski@cs.technion.ac.il

Chuck Liang
Department of Computer Science, Hofstra University, Hempstead, NY, United States of America, e-mail: cscccl@hofstra.edu

Dale Miller
Inria & LIX/Ecole Polytechnique, Palaiseau, France, e-mail: dale.miller@inria.fr

Victor Nascimento
Filosofia, UERJ, Rio de Janeiro, Brazil, e-mail: victorluisbn@gmail.com

Luiz Carlos Pereira
Filosofia, PUC-Rio/UERJ/CNPq, Rio de Janeiro, Brazil, e-mail: luiz@inf.puc-rio.br

Paolo Pistone
Department of Computer Science and Engineering, University of Bologna, Italy,
e-mail: paolo.pistone@uniroma3.it

Dag Prawitz
University of Stockholm, Sweden, e-mail: dag.prawitz@philosophy.su.se

Paulo Guilherme Santos
Centro de Matemática e Aplicações, NOVA School of Science and Technology, Universidade Nova de Lisboa, Portugal, e-mail: pgd.santos@campus.fct.unl.pt

Peter Schroeder-Heister
Department of Computer Science, University of Tübingen, Germany,
e-mail: psh@uni-tuebingen.de

Göran Sundholm
Institute for Philosophy, Leiden University, The Netherlands,
e-mail: goran.sundholm@gmail.com

Neil Tennant
Department of Philosophy, The Ohio State University, Columbus, OH, United States of America, e-mail: tennant.9@osu.edu

Luca Tranchini
Department of Computer Science, University of Tübingen, Germany,
e-mail: luca.tranchini@gmail.com

Heinrich Wansing
Department of Philosophy I, Ruhr University Bochum, Germany,
e-mail: Heinrich.Wansing@rub.de

Bartosz Więckowski
Institut für Philosophie, Goethe-Universität Frankfurt am Main, Germany,
e-mail: wieckowski@em.uni-frankfurt.de

Proof-Theoretic Semantics: An Autobiographical Survey

Peter Schroeder-Heister

Abstract In this autobiographical sketch, which is followed by a bibliography of my writings, I try to relate my intellectual development to problems, ideas and results in proof-theoretic semantics on which I have worked and to which I have contributed.

1 The term: Proof-theoretic semantics

Proof-theoretic semantics, from various perspectives, has been a predominant occupation for me since my doctorate. The field itself was already in existence prior to any contribution of mine. It was created by Gerhard Gentzen by designing a logical calculus representing the "natural" way of deductive reasoning, and by claiming that its special features — in particular its way of handling assumptions, and its classification of inference rules into rules for the introduction and the elimination of logical symbols — give certain rules a 'definitional' status, thus equipping logical symbols with their meaning (Gentzen, 1935). It was only consequential that the German logician Franz von Kutschera called this approach "Gentzen semantics" (von Kutschera, 1968) and thus created a term in analogy to "Tarski semantics" which is often used for the dominant approach to denotational semantics established by Alfred Tarski. Similarly, I came up with the term "proof-theoretic semantics" as a systematic term in analogy to "model-theoretic semantics", which emanated from what Tarski had put forward.[1] In 1985, I first mentioned it as the title of a planned book in a letter to Dag Prawitz. Later I used the term in lectures I gave in Stockholm. It appeared in

Peter Schroeder-Heister
Department of Computer Science, University of Tübingen, Germany, e-mail: psh@uni-tuebingen.de

[1] For further reflections on the idea of proof-theoretic semantics see my *Stanford Encyclopedia of Philosophy* entry (E2012c) and, from slightly different perspectives, Wansing (2000) and Francez (2015).

T. Piecha and K. F. Wehmeier (eds.), *Peter Schroeder-Heister on Proof-Theoretic Semantics*, Outstanding Contributions to Logic 29,
https://doi.org/10.1007/978-3-031-50981-0_1

print, I think, first in 1991 in an abstract (A1991e)[2]. Nowadays it seems a 'natural' term, which has become a standard designation for a certain field of study and is often used without much consideration of its conceptual origin; something it shares with Robert Brandom's term "inferentialism" (Brandom, 1994, 2000). It suggests that we are talking about a semantics that is based on, or crucially uses the notion of "proof" or "proof theory".

2 Academic roots: Hasenjaeger's institute

I discovered logic relatively early. At the University of Bonn, where I started studying in 1971, I came into contact with the *Institute for Logic and Foundational Research*[3] during my first year of study, even though it had nothing to do with my official subjects, which at that time were Catholic theology and mathematics. The head of this institute was Gisbert Hasenjaeger (M2006c, Wirth, 2021), who, in the 1940s and 1950s, had studied, worked and published with Heinrich Scholz in Münster. He had modelled his institute on Scholz's "Institute for Mathematical Logic and Foundational Research", which was the first institute of its kind in Germany. Hasenjaeger — who, incidentally, was a sort of counterpart to Turing[4] — had a very liberal attitude towards what logic should be, so he did not follow a particular strand of logic exclusively. This attitude also applied to those whom he admitted to his group's seminars. He did not mind that I, as an undergraduate, attended seminars which were officially intended for graduates only. As early as late 1972, for a seminar directed by Wolfram Schwabhäuser in my third semester of study, I quite enthusiastically produced a detailed exposition of Gödel's incompleteness theorems. In that essay of nearly 50 pages I even included Y. Matiyasevich's result from two years prior that every recursively enumerable predicate is diophantine, from which the (negative) solution of Hilbert's tenth problem directly follows. Alexander Prestel, who later, as a professor in Konstanz, was to become instrumental for my career, had just completed his Habilitation in Bonn and, following his inaugural lecture on this topic, left me his notes. I had become acquainted with the issues of incompleteness and undecidability through Wolfgang Stegmüller's excellent book of 1959, on which I must have hit during my first semester of study, or even earlier. I remember Hasenjaeger, during one of my presentations, correcting my pronunciation of the name "Kleene", which I pronounced like the word "clean", as I had not listened to any logic lectures before and knew the name only from printed

[2] References of the form A⟨year⟩, P⟨year⟩, C⟨year⟩, D⟨year⟩, R⟨year⟩, E⟨year⟩, M⟨year⟩ refer to the corresponding sections of the bibliography at the end of this article.

[3] "Seminar für Logik und Grundlagenforschung"

[4] After being severely wounded in the war, he was, as a gifted mathematician (who had not yet started university studies), appointed to the cryptology department of the German military and tasked with testing the cryptological reliability of the Enigma machine. He was not able to discover those weaknesses that allowed Alan Turing and Gordon Welchman (based on ideas of Marian Rejewski and other Polish cryptologists) to decrypt the code it generated — "fortunately", as he said after the war, as otherwise the war might have lasted even longer.

sources. In 1973, still in my second year at Bonn, I wrote another essay on Gödel's functional interpretation of arithmetic (the *Dialectica* interpretation), which meant that by this point I had built up a significant background in logic. However, I regret that I did not take the chance of learning set theory or model theory in more depth. I would have had a perfect chance to do so, as besides Prestel, Keith Devlin, Ronald Björn Jensen, Sabine Koppelberg and Wolfram Schwabhäuser were all at the institute at the time. In any case, there was a spirit at the department that motivated young researchers like myself.

There was also funding available for students to attend meetings in mathematical logic; much of it from the Volkswagen Foundation. Already in early April of 1973, in my second undergraduate year, I had the chance to visit such a meeting in Tübingen, organized by my future colleague Walter Felscher, who had just become a professor of mathematical logic there. Solomon Feferman gave a two-day course on advanced proof theory, and I was able to meet quite a prominent (or soon-to-be prominent) section of the German logic scene including Wilfried Buchholz, Justus Diller, Ulrich Felgner, Wolfgang Maas, Gert Müller, Helmut Pfeiffer, Wolfram Pohlers, Kurt Schütte and Helmut Schwichtenberg. Very impressive was a week in January 1974 at the Mathematical Research Institute in Oberwolfach, where I had been invited to a meeting on set theory and model theory led by Felgner (another future Tübingen colleague), where many outstanding German logicians were present. Of the other conferences or schools that I was able to attend as an undergraduate, I remember one in Münster on intuitionistic logic with courses given by Anne Troelstra about choice sequences and Dirk van Dalen about intuitionistic logic in general.

With other students at the Hasenjaeger institute, notably Benedikt Peppinghaus and Hans Leiß, we embarked on all kinds of interesting ventures. Benedikt, for example, invited Eduard Wette, who was considered by many an enfant terrible of mathematical logic[5], to present his purported system-internal consistency proof of arithmetic which, by Gödel's second incompleteness theorem, would prove the inconsistency of arithmetic. I found this extremely stimulating, even though there were gaps in the proof he presented (as one might have expected). Again, it speaks for Hasenjaeger's liberal attitude that he permitted such things to take place. It is obvious that all this could pull somebody like me in the direction of proof theory and constructive logic. However, logic was only a part of my student life in Bonn between 1971 and 1977 (see Section 4), and I did not acquire any university degree in logic until my doctorate.

3 Family roots: Upbringing and school

I do not have an academic family background. My mother came from the household of a primary school teacher and was very well-educated despite the fact that due to the difficult circumstances of her time (including gender-based disadvantages) and family

[5] For a fair characterization of Wette see Paul Bernays's letter to Kurt Gödel of 24 January 1975 (Gödel, 2003).

obligations she was unable to attend high school. My father, who, like my mother, was well-educated and had a talent for mathematics, ran a business that produced stationery (exercise books, writing pads etc.), which he had taken over from his father. This prevented him from going to university after returning from America, where he had been a prisoner of war. Despite it not being his first-choice career, he was a very successful businessman. When he sold the company in the 1980s, he was able to establish, from the profits, a charitable foundation that built and ran an innovative local carehome. In our town of Düren[6], where I was born on 2 March 1953 as the eldest of four children, I attended the "humanist" high school with its emphasis on Greek and Latin, at the expense of modern languages and natural sciences, though provided with an excellent education in mathematics.

At school, I was always good at mathematics, and it was always assumed I would study mathematics at university (which I did). I otherwise became interested in philosophical and theological questions when I was fifteen or so, partly through the influence of a Catholic youth organization and its local leaders[7], who were pretty left-wing (it was 1968, after all). Through a friend — Lothar Stresius, who was four years older, a theology student at Bonn and Tübingen[8], and in many respects a role model — I had the chance, while still a high school student, to attend lectures in Tübingen, I think in 1970, by many famous German theologians of the time. These included Eberhard Jüngel (20 years later my colleague in the Department of Philosophy), Ernst Käsemann, Walter Kasper (today cardinal of the Roman Curia), Hans Küng and Jürgen Moltmann (Joseph Ratzinger — later pope Benedict XVI — had left Tübingen the year before). All this impressed me so much that, after my high school diploma in 1971, I enrolled in Bonn in Catholic theology, in addition to mathematics. After one year I replaced theology with philosophy, which I had been studying as part of the theology course, and which also interested me deeply.

4 Bonn 1971–1977: Undergraduate study

Philosophy and mathematics. My interest in philosophy dates back to when I was 16 or 17 and read, in addition to the fashionable philosophical literature of the time (Frankfurt school, especially Adorno and Horkheimer), some Popper (my earliest copy of the *Logik der Forschung* dates from 1970) and also some Kant. I was introduced to the latter by my excellent high school philosophy teacher Hubert Fackeldey, who later became a professor in Cologne and made significant contributions to philosophical aspects of deontic logic.

Philosophy in Bonn was a traditional department covering the whole history of the discipline from antiquity to modern times, where modern times meant essentially

[6] To mention its mathematical connection: Peter Gustav Lejeune Dirichlet (1805–1859), who propagated the notion of a mathematical function as an abstract (that is, operation-independent) mapping, was born there.

[7] Especially Paul Georg Meyer, who later became a linguistics professor at RWTH Aachen.

[8] Later, after a Ph.D. on Adorno, he became high school principal.

hermeneutics and Heidegger, but also some Wittgenstein and analytical philosophy. Even the normally underrepresented (if present at all) medieval philosophy was very strong. Most impressive in philosophy to me was the approach of the Kantian Gerold Prauss, who was an associate professor who later went to Cologne and then to Freiburg. He was working on a novel interpretation of Kant's concept of "thing-in-itself" and had just completed a book manuscript on it (Prauss, 1974). Although he did not know me he had so much trust in me that he lent me a carbon copy of it (not a photocopy — this technique was just about to enter wider use), and at some stage, because he unexpectedly needed it for the publisher, had to find out where I was living to have it returned.

Prauss later also supervised the philosophy part of my university degree, which I completed in 1977, writing a thesis on the concept of truth in natural languages (originally I had even considered writing something about Heidegger). Unlike its title might suggest from the modern point of view, it was a piece in philosophy of language in the Kantian spirit, which had nothing to do with formal natural language semantics. My final university degree was actually a combined teachers' degree in philosophy and mathematics. Like most humanities students at the time, I chose the teachers' degree simply because this gave one the possibility to enter the high school teaching profession if anything went wrong with the Ph.D.

Through Prauss I also got in contact with Günter Buhl, who had written an excellent dissertation on consequence and grounding in Bolzano (Buhl, 1961), the outstanding quality of which I only appreciated much later when working in proof-theoretic semantics, as he was the first to fully recognize the proof-theoretic aspects in Bolzano's work. Buhl taught elementary logic in the philosophy department using a system of natural deduction. He hired me as a student assistant, and I was able to give tutorials in formal logic, an experience that shaped all my later teaching of logic to philosophers.

I spent my student time until graduation essentially in Bonn, but also as a "guest student" in Cologne and Aachen. The initial choice of Aachen was chiefly due to the fact that my girlfriend Gabi was a student there. In Aachen, I attended seminars run by Christian Thiel, who later became the successor to Paul Lorenzen in Erlangen. I learned a lot from him. He drew my attention to Lorenzen's *Operative Logic* of 1955, which from today's perspective must be considered one of the milestones of proof-theoretic semantics[9], even though Lorenzen himself abandoned his operative approach in favour of dialogical semantics in the 1970s, something also appreciated by Thiel. Dialogical logic was a topic I became strongly interested in myself — much later we even had a research project on it (see Section 11). Outstanding was, of course, Thiel's knowledge and mastership as far as Gottlob Frege's work was concerned. In the German-speaking world he had initiated, through his doctoral dissertation on sense and denotation in Frege's logic (Thiel, 1965), the revival of interest in Frege's work and played the role which in the English-speaking world was taken by Michael Dummett and his monograph on Frege's philosophy of language (Dummett, 1973).

---

[9] See my later papers A2007b, A2007c, A2008a.

Music and musicology. I had a modest background in musical performance (piano) and therefore a corresponding interest in music and its theory. For three semesters I studied musicology in Bonn as an additional subject, which was possible at that time. Of the many courses I took, most stimulating were theories of harmony, which was only natural given my mathematical inclinations. A lasting impression on me was made by Martin Vogel, who had developed a systematic approach to the topic, which very much differed from the historical attitude typical for German musicology. Together with a small group of students, some of whom came from mathematics, he developed a theory of harmony based on pure tuning, mathematically modelled as (an extension of) Leonhard Euler's web of tones and based on approaches by the physicists Hermann von Helmholtz and Arthur von Oettingen (Vogel, 1975). Had I continued musicology, I probably would have followed this path of study myself, in particular as my later occupation with computer science would have presented the right environment and interesting practical applications for it.

Private life. The summer 1974 changed my life, when I met Gabi (Gabriele Heister), who was to become my wife five years later. Gabi was studying in Aachen for two degrees at the same time, one in psychology and one joint degree in German and education. I met her on a holiday trip to the Netherlands in a larger group of students. Apart from sharing my interest in music (she had at one point trained as a soprano), she also influenced my general scientific views. Whereas I was very much a humanities scholar (including logic and mathematics), she had a thorough grasp of empirical research. Moreover, through experimental psychology she was concerned with statistical analysis of data, which was completely new to me, even though I knew some probability theory and mathematical statistics. Discussing these topics with Gabi broadened my intellectual views significantly. Drawing up hypotheses, looking at the outcome of experiments, and using statistical methods in the evaluation of data — the absolutely normal approach in the sciences — was something I was not used to.

5 Konstanz 1978–1981: Encyclopedia and doctorate

What next? When I had completed my teacher's degree in May 1977, the question was: what to do now. I wanted to do a doctorate but I was not keen to continue with the topic of my undergraduate thesis, and apart from that there was no job in sight to finance it. As it turned out, Gabi, who had completed her diploma in psychology around the same time, had been offered several positions, including one at the University of Konstanz for the year. With no other options for me in sight, we decided to go. In November 1977 we found a flat in Allensbach near Konstanz, and in December we moved. For Gabi the working conditions were less than ideal, to put it mildly. For me, the move turned out to be one of the best decisions of my life. There was no job initially, but that was soon to change.

Gereon Wolters, the Dingler Nachlass, and the Encyclopedia. The opportunity to start an academic career in Konstanz I owe first to Gereon Wolters and then to

Jürgen Mittelstraß. In early 1978, I walked into Wolters's office. He had completed his Ph.D. the year before and was affiliated with Mittelstraß, who held one of Konstanz's three chairs in philosophy. It turned out that Wolters had just successfully applied for a grant to make the Nachlass (literary estate) of the philosopher Hugo Dingler accessible. There was a part-time position available for 10 months on this project with no candidate yet nominated. Wolters hired me on the spot. He even gave me an office of my own, the window of which offered a magnificent view over Lake Constance and the Isle of Mainau. As the university was still under development, it could be relatively generous with respect to office space. My first publication resulted from this job: a bibliography of the works of Hugo Dingler (A1981a). It was a piece of meticulous work, of which I am still proud.

Through Wolters, I joined the team assisting Jürgen Mittelstraß in his editorial work on the *Encyclopedia of Philosophy and Philosophy of Science* (1980–2018), a voluminous work that he had started some years ago, with the first volume covering the letters A–G due to be completed in spring 1979. A logician with interests in adjacent fields was a very welcome addition to the team. When in 1979 there was a two-year position available, I received a part-time appointment to work on the Encyclopedia, being free to work on my dissertation during the rest of my time. Over the years (the eighth and final volume of the second edition appeared in 2018, the year before my retirement) I wrote around 200 articles for it (E1980–2018); tiny ones such as *a* (the letter mnemotechnically representing universal affirmative judgements in traditional syllogistics), but also longer ones such as *Popper, Karl Raimund* (E1995a). We also had a lot of fun creating a number of fictitious entries ('Nihilartikel', 'mountweazels') — a long tradition in encyclopedic works, which reached a new level of sophistication under Mittelstraß.

Linguistics and mathematics. In Bonn I had come into contact with logic almost exclusively from the mathematical perspective. As to philosophical logic, I was only acquainted with topics relating to the foundations of mathematics, such as the dispute about classical versus intuitionistic logic or the discussion of the logical and set-theoretical paradoxes. I was not aware of extra-mathematical applications of logic.

Making interdisciplinary contacts was very easy in Konstanz. All disciplines were under the same roof in one huge building complex, where one would frequently run into people from other departments in the corridor, in the cafeteria and in the large entrance lounge of the compound. People in general linguistics were natural partners to whom one would speak. In Konstanz, general linguistics had a strong leaning towards logic-based semantics. It was through this that I engaged more seriously with Montague grammar[10], higher-level linguistic type theory and similar issues. I discovered that general linguists were all doing logic in a very original fashion and at an advanced level that went beyond the sort of mathematical logic I was used to. This was essentially due to the intensional phenomena one has to deal with in linguistics and that do not play a role in mathematics, which is essentially (though not throughout) extensional. My main contacts on the professorial side were Urs Egli and

[10] Had I followed Hans Leiß's initiative, I would have studied this subject already together with him in Bonn.

Arnim von Stechow (who later became my colleague in Tübingen), and at the level of the postdocs Thomas Ede Zimmermann (who was later in Tübingen and Stuttgart, and then became professor in Frankfurt) and Wolfgang Sternefeld (later a colleague in Tübingen). I learned a great deal that I would probably never have encountered at a less communicative university with a more conventional department structure.

Konstanz had a strong mathematics department, and also a strong group in mathematical logic, headed by Alexander Prestel, whom I knew from Bonn. Prestel was a set theorist and model theorist and held a weekly seminar which I regularly attended. A special role was played by Ulf Friedrichsdorf, a permanent member of Prestel's group. He was a hardcore mathematical logician, but with an incredibly wide range of interests — he co-authored textbooks on model theory and set theory with Prestel — which extended in particular into philosophical logic and into linguistics, exemplified, for example, by his brilliant textbook on classical and intensional logics (Friedrichsdorf, 1992), which even covered the arithmetical completeness of derivability logic proved by Solovay some time ago. I have frequently used it for teaching. In Konstanz I realized for the first time what a collaboration between philosophers, mathematicians and linguists could achieve.[11]

Teaching. From 1981 to 1989 I held full-time positions in Konstanz, made possible by Jürgen Mittelstraß. This meant that I was obliged to teach, and in Konstanz I was often able to choose which subjects to teach, which elsewhere would not have been possible at this early stage of my career. I taught logic courses, but also other topics including Frege, Husserl, Carnap, Popper, ontology, epistemology, general philosophy of science, and later also logic programming and automated theorem proving etc. I learned a lot from teaching, and I had quite a number of excellent students. As mentioned above, the architecture of the university encouraged interdisciplinary activities, and there were always students from mathematics, linguistics and other subjects in my courses.

Marriage and children. Gabi and I got married in December 1979. This is how I arrived at my double-barrelled surname: "Schroeder-Heister", adding Gabi's surname to my birth name, "Schroeder". As the ability of husbands to take their wives' names had only just been introduced in Germany, the administration was not used to it. The registrar visited us at home after the ceremony to fill in critical signatures they had forgotten at the ceremony. Our children were born a couple of years later. Paula in 1984, and Justin in 1987; both in Münsterlingen, on the Swiss side of the lake. This did not give them Swiss citizenship, however: it is not that easy to become Swiss.

A Ph.D. without a supervisor. I took the risk of doing my Ph.D. without a formal supervisor. There was no "natural" supervisor in Konstanz for my field of interest. In fact, for a long time I did not know what to do contentwise. What would become the final topic of my thesis only emerged about twelve to eighteen months before I submitted it. Until then, it was not even clear whether it would be mathematical or philosophical logic, and how strong its philosophical component would be.

I was working in all sorts of directions, many of which I no longer remember.

[11] For the history of logic at Konstanz University see Buldt (2022).

What I do remember is that I somehow hit onto von Kutschera's 1968 article, in which he coined the term "Gentzen semantics" (see Section 1). There he showed, for intuitionistic propositional logic, the expressive completeness of the four standard connectives conjunction, disjunction, implication and absurdity, corresponding to the classical idea of functional completeness. This was established relative to a general schema of rules characterizing arbitrary n-ary connectives. Crucial was the idea of introducing some iteration of a sort of structural implication leading to what he called "S-formulas", which played a critical role in the interpretation of implication as a connective. I also at the time hit upon Prawitz's 1979 article, which had the same aim of proving expressive completeness. Whereas von Kutschera worked in a sequent-style framework, Prawitz used the apparatus of natural deduction that he had put forward in his monograph (Prawitz, 1965). It turned out that Prawitz's paper contained an error, as he could not give a proper schema for implication without presupposing it. My idea was to combine von Kutschera's idea of S-formulas with Prawitz's natural deduction schema. To achieve this, I developed a system of rules of higher levels, that is, of rules depending on rules, and generalized elimination rules using this tool. This provided a general framework to deal with arbitrary connectives. The introduction and elimination rules for logical connectives were understood as telling that connectives express a system of inference rules, which gave the connectives a semantics in terms of rules, thus a sort of proof-theoretic semantics (although I did not yet use this term in my thesis). In the second part of the thesis, following some ideas of von Kutschera (1969) I extended this approach to a system with an operation of denial and corresponding refutation rules, something which nowadays would be called bilateralist, but which I never published afterwards. I developed this very quickly, and my thesis was essentially written in the first half of 1980.

How to proceed formally towards a Ph.D.? The thesis was very much done in the spirit of Dag Prawitz, whose works I admired, in particular his combination of formal investigations with philosophical interpretation. However, I did not know him personally, and had not seen him at any conferences. In fact, Prawitz to me was somebody of such high status that I did not dare approach him. Help came from Alexander Prestel. He encouraged me to contact Prawitz after all, and actually sent him a photocopy of my thesis draft recommending me to him. With Prestel I also discussed the idea of finally submitting the thesis in Bonn with Hasenjaeger as examiner, even though he had not been involved with my thesis at all. From his time in Bonn, Prestel knew Hasenjaeger very well and was aware that I had left a good impression on him years ago. The arrangement with Hasenjaeger finally worked out. After seeing my draft, and in light of my earlier achievements as an undergraduate student, he accepted me. The final plan was, in addition to Hasenjaeger as the first examiner, to get Prawitz appointed as external examiner, which the philosophical faculty agreed to.

The only problem was that I had no reply from Prawitz, who at the time was on a research stay in Italy. Per Martin-Löf, who knew from Prawitz about my thesis draft supported me and encouraged him to respond to my request. I met Martin-Löf for the first time in Munich in early 1981 at a meeting organized by Helmut Schwichtenberg, where he presented his constructive type theory, and where Peter Aczel presented his version of intuitionistic set theory. This meeting impressed me enormously, in particular the way in which Martin-Löf presented his system. It was very philosophical; he argued for the rules he was giving from first principles.

Suddenly, in mid-January 1981, after having heard nothing from Prawitz for months, there came a 25-page handwritten letter, in which he discussed my thesis draft in detail. This was extremely helpful, in particular as he pointed out to me certain problems associated with introducing rules as structural entities, as this essentially meant the duplication of implication. However, in principle he was quite happy with the approach. This allowed me to revise and finish the thesis very quickly and to formally submit it, so that in May 1981 my oral doctoral exam ("Rigorosum") took place in the form of three examinations on one day, in logic as the main subject, and philosophy and musicology as minor subjects. Due to the existence of Hasenjaeger's institute (Section 2) the formal degree subject of the doctorate was "Logic and Foundational Research" (rather than "Philosophy"). I only met Prawitz personally the year after. Thus, he had been the hidden supervisor of my thesis.

Part of my thesis (A1981b), which itself was written in German, appeared in English in revised form under the title "A natural extension of natural deduction" (A1984a) in the *Journal of Symbolic Logic*[12]. The title goes back to comments made by Göran Sundholm, who acted as a reviewer for the *JSL*[13]. It was to become my most-cited paper. Perhaps this confirms the view that for normal researchers, the work for their doctorate is often the most substantial work they ever do, with all the rest being incremental extensions of it. Somebody like Gödel had several different and independent fundamental ideas in his lifetime, but most people are no Gödel.

6 Konstanz 1981–1989: Postdoctoral research and Habilitation

Being free after achieving the Ph.D., I continued my research in various ways on topics that occupy me still today.

Semantical completeness. My dissertation gave a proof-theoretic semantics of propositional logics and proved the *expressive completeness* of connectives with respect to this semantics. What I tried in the sequel was to prove Prawitz's completeness conjecture of 1971 (Prawitz, 1971, p. 257), namely that the formalism of intuitionistic logic is *semantically complete* with respect to a certain sort of proof-theoretic semantics. I published two papers on this issue (A1983c, A1985b) with practically no feedback. Only some 35 years later I came back to these issues in joint work with Thomas Piecha, where we showed the limitations of the goal to prove completeness — see below (Section 10). Today, on the background of the modern discussion (see Piecha, 2016), this topic has regained some interest through work by Tor Sandqvist (2015), David Pym and others (2022).

[12] A transfer to quantifier logic appeared in A1984c.

[13] The original paper submitted to the *JSL*, which differs substantially from the final printed version, is still worth reading. It is attached to the online version of my thesis A1981b.

Popper. One of the outstanding experiences of my intellectual life was meeting with Karl Popper. I had already once been in contact with him when I asked him for possible information about Kurt Grelling, on whom in 1980 I had to write an article for the *Encyclopedia*. I was extremely impressed that he replied immediately to my request with a letter in his beautiful handwriting. Shortly after my doctorate, I came, more or less accidentally, across his papers on deductive logic that he had published in the late 1940s after finishing his *Open Society* (Popper, 1945). These papers were very different from the topics that one would expect from Popper. They had received a mixed reception in the logic community. Even though Popper was one of the leading philosophers of the twentieth century, his contribution to deductive logic remained largely unknown.

I immediately saw the significance of these papers for proof-theoretic semantics. They attempted to provide an inferentialist foundation for logic. What I discovered was that Popper's papers provided an original contribution towards the problem of the logicality of sentence operations. I tried to work out this idea in a long paper, which appeared in *History and Philosophy of Logic* (A1984d). A later paper in the *Popper Centenary* volume (A2006b) attempted to change the ignorance regarding Popper's logical theory, as did papers by my students and colleagues David Binder and Thomas Piecha a decade later (2017, 2021). Recently, the three of us completed a volume containing Popper's logical papers as well as manuscripts from his Nachlass (estate) on these topics, including exchanges of letters related to logic, as well as a detailed introduction (C2022b, A2022a)[14]. We look forward to seeing the reaction from the logic community to Popper's ideas.

In July 1982, I sent the first version of my manuscript (of A1984d) to Popper. I immediately received a reply, followed by another one with more detailed comments a few days later. This extended into an exchange of letters (M2022). Popper started this exchange with me, somebody who was a complete no-name to him, simply because he was interested in the topic and in what I had done on it. Some time later Gabi and I met Popper in person. We were staying in Edinburgh, where I had a fellowship at the Institute for Advanced Studies in the Humanities for three months, and Popper was the keynote speaker at a meeting in Leicester in spring 1983 that we attended. He greeted us there, and I had the chance to speak with him about the paper. Soon after we had another exchange of letters on inductive probability and his joint paper with David Miller (Popper and Miller, 1983)[15]. Years later, after a lecture in St. Gallen in June 1989, with his characteristic sense of humour, he greeted me with "Ach, Herr Schroeder, ich hätte Sie fast nicht erkannt, Sie sind aber dick geworden" ("Ah, Mr Schroeder, you have put on weight — I almost didn't recognize you").

My interest in the foundations of probability theory and the notion of randomness that I had maintained since my mathematical studies in Bonn was also, in part, due to Popper. The chapter on probability in the *Logik der Forschung* presented a frequentist definition of probability which was highly original and, being a definition of randomness for finite sequences, anticipated certain aspects of A. N. Kolmogorov's

[14] For a summary of (our view of) Popper's ideas on deduction see Piecha (2023).

[15] In Popper and Miller (1987), they explicitly acknowledged comments of mine on the possible dualization of their argument, which was also quite rewarding at this stage of my career.

later definition in terms of algorithmic complexity. I studied some of these theories, in particular by R. von Mises, A. Wald, Martin-Löf, C. P. Schnorr and Kolmogorov, but have never been able to turn this into a research topic — too many excellent researchers were already involved in it, and I was fully occupied with other issues. However, a presentation of Popper's approach in a collection of essays on the *Logik der Forschung* resulted from it (A1998b).

Psychology. Gabi completed her Ph.D. in experimental neuropsychology in 1985. Through following Gabi's work, I found this type of research more and more interesting and gradually started looking into it myself. Her experiments were based on reaction times upon tachistoscopically presented stimuli. She realized that work in general psychology in the area of perception played a significant role in the effects observed, and her work shifted somewhat towards that realm. She became interested in spatial stimulus-response compatibility: the phenomenon that in most situations spatial features of a stimulus correspond to spacial features of the response in terms of shorter reaction times. This finding allows certain insights into the way spatial information is processed and how cognitive processing is organized in the brain. My contribution went slightly (but only slightly) beyond statistical evaluation and towards the discussion and the theoretical modelling of results, which is reflected in the fact that I made first author in one of around ten papers with which I was involved (P1988).[16] This work absorbed me for quite some time, as it was so interesting, even though it distracted me from proof-theoretic semantics. It fitted my interest in cognitive science, nevertheless.

Philosophy of science. The 1970s and 1980s were the heydays of philosophy of science, in which general topics such as the status of theoretical terms and grand themes such as normal versus revolutionary science were discussed. A particularly important topic was the thesis of the incommensurability of theoretical terms in the case of revolutionary science, put forward by T. S. Kuhn. In Germany J. D. Sneed's model-theoretic unterstanding of scientific reasoning was prominent, especially through Stegmüller's (1979) propagation of it. Every logically inclined philosopher would have studied these approaches at the time. In my case this led to a paper in *Philosophy of Science* (A1989) which argued, by using model-theoretic means, that reducibility and incommensurability of theories are not necessarily incompatible. It was to become my only model-theoretic publication and used the expertise of my co-author Frank Schaefer, who was studying model theory with Prestel.

Frege. Something that has always intrigued me was the philosophy and logic of Gottlob Frege. I had come across it at a very early stage of my studies — my copy of the *Grundgesetze der Arithmetik* dates from 1971, my first semester of study. However, practically all interesting questions I have dealt with in various papers were triggered by remarks or exchanges with Christian Thiel, partly by remarks in discussions with him, partly by his many publications on Frege and Fregean themes.

The first paper of this kind was on Section 10 of the *Grundgesetze* (A1987a),

[16] See subsection P of the bibliography. Sadly, our co-author of many papers, Walter Ehrenstein, died in 2009, aged only 58 (Paramei, 2009).

which I also presentend at a Frege conference in Schwerin in 1984 (A1984b). At this conference I had the chance to meet almost the whole logic community of the German Democratic Republic (GDR), but also many people from elsewhere. Doing logic in the GDR was a way to avoid the ideological pressure exerted on other branches of philosophy. Consequently, in communist East Germany, logic was better represented within philosophy than in West Germany. In the paper I tried to reconstruct Frege's arguments of whether and which courses-of-values could be chosen to function as truth values, which is fundamental for the ontology of the *Grundgesetze*. At that time I was quite satisfied with the paper, and later even more when it occurred amongst the very few quotations Dummett made in his book on Frege's philosophy of mathematics (Dummett, 1991). Initiated by Kai Wehmeier, the topic of this paper later received a new rethinking in a joint paper (A2005), which we then dedicated to Thiel.

When looking at Frege's notation and terminology in the *Grundgesetze* I realized that in a certain way this resembles very much the structure of Gentzen sequents and the terminology applied there. I wrote a paper on this but realized just when it was finished that von Kutschera, to whom I had sent it, had independently had very similar ideas[17]. However, for another paper on Frege I claim full originality (A1997). In it I interpreted his propositional calculus from the *Grundgesetze*, which he was applying to solve a problem discussed by G. Boole, E. Schröder, W. Wundt and H. Lotze, as a system of propositional resolution. I found it very intriguing that systems with the cut rule (plus substitution) as the only inference rule, which have become prominent in automated theorem proving, could be traced back to Frege, at least for the propositional fragment. Needless to say Frege has been a frequent topic of my teaching.

Historical foundations of logic. Frege is the greatest figure as far as the foundations of modern mathematical and philosophical logic are concerned. Two other great figures in the (occidental) history of logic before were Aristotle and Leibniz. I was interested in both of them and have taught and written on both. On Leibniz I wrote a paper with Mittelstraß concerning his arithmetical calculus (A1986), towards which I had been directed by Thiel, and which can be seen as a model-theoretic semantics in terms of pairs of coprime natural numbers. This is closely related to Leibniz's programme of a *Characteristica Universalis* that would code concepts by arithmetical means and allow one to decide the validity of inferences by calculation, sometimes seen as the precursor of modern AI. The paper also contains remarks on the concept of probability in Leibniz as the degree of possibility. On Aristotle I have a paper (A2008), with Mittelstraß also, which discusses, in a certain context, his modal syllogistics and also the significance of the fourth figure of traditional syllogistics (not yet present in Aristotle). I emphasized in particular a result by Daniel Merrill (whom I knew already from my work on Popper's logic), which showed that in the presence of obversion, every syllogism (even an invalid one) can be reduced to the fourth figure, which is not possible for the other figures (to which, as Leibniz was already aware, every valid syllogism can be reduced). I also studied the work of Paul Hertz, a predecessor of

[17] Von Kutschera (1996). Though presenting them on various occasions (e.g. A1999), I published my results only much later (A2014a), of course with proper acknowledgement to von Kutschera.

Gentzen, who developed a system of purely structural reasoning, whose inference steps can be viewed as applications of the resolution rule, and whose proofs can be put into certain normal forms (A2002b). This is highly significant for proof-theoretic semantics, as this system is the starting point of Gentzen's research and the topic of his first publication (1933). His iteration of structural implications could remind one of my own higher-level rules. If I remember correctly, again it was Thiel who had drawn my attention to Hertz. Michael Arndt (D2008) continued and further advanced this work on Hertz.

Logic programming and computer science. Around the time of my doctorate computing took off for the general public. My doctoral thesis was still written using a typewriter with the formulas inserted by hand and larger corrections made with scissors and sticky tape, thus by cut and paste in the literal sense. I had already come across advanced computing through the linguists, who owned a Lisp machine. I do not remember which brand it was. In any case, it was a revelation to see the graphics display operated with a mouse, and the large 8-inch floppies to store programs and data. Through them I also acquired the reference to PROLOG and logic programming, and, following that, I hit on J. W. Lloyd's 1984 book on its foundations, which had just come out.

After reading Lloyd's book, I realized that logic programming, when understood proof-theoretically, had very much to do with what I myself was doing. In particular, my own framework with rules as assumptions offered not only a neat interpretation of logic programming, but even allowed for an extension of logic programming using embedded implications. These things were in the air, with people such as Dov Gabbay and Uwe Reyle (1984) as well as Dale Miller (1986) doing similar things.

Beginning with Schwichtenberg's Munich conference of 1981 (see Section 5) and through my contacts to Stockholm, I also became involved with the type-theory oriented programming community, which did not have much overlap with the logic programming community, apart from people like myself, Lars Hallnäs and Dale Miller. This community was much nearer to functional programming which has often tended to distance itself from other paradigms. People in Edinburgh such as Gordon Plotkin and the *Logical Framework* group realized that my higher-level rules and the generalized elimination rules for logical operators seemed to be of considerable computational use for the implementation of logic systems, to the effect that I was invited to conferences there with many prominent computer scientists and computational logicians present. My contribution to Gérard Huet's and Plotkin's volume on logical frameworks (A1991d), in which, by reference to substructural logics, I gave the general elimination inferences the status of a kind of structural rule, resulted from this. I also made contact with colleagues in Cambridge; I remember talks in Larry Paulson's and in Martin Hyland's seminar[18]. My strong interest in Martin-Löf type theory and the fact that I could apply my methodology to it (A1989a), kept me in lasting contact with the Programming Methodology Group in Gothenburg, quite independent of the very many and regular encounters and personal discussions with Per Martin-Löf himself over the decades.

[18] And in Paulson's office a young French postdoc, Thierry Coquand.

All this coincided with me developing an interest in cognitive science and artificial intelligence, not only from the point of view of rule-based AI (today the "old" AI), but also from the viewpoint of neural networks, which were enthusiastically discussed at the time. From a conference in Berne, which I attended together with Gabi in the early 1980s or perhaps even before, I still remember a presentation by Allen Newell, where he demonstrated one of the big expert systems of the time in which he was involved, I think it was MYCIN. If I remember correctly, Herbert Simon was there, too (perhaps also Marvin Minsky, but my memory might be deceiving me). Later in the 1980s I attended a lecture by David Rumelhart in Paul Feyerabend's colloquium at the ETH Zurich, where he described his backpropagation algorithm for neural networks.

Habilitation. From October 1983 onwards, I had an assistantship with Mittelstraß[19] with much time for research. As my professional duty apart from teaching, I worked on his Encyclopedia. It was also expected that within the time period of this assistantship — six years — I would do my Habilitation, as normally required for the position of a professor in Germany. In the years after my doctorate, I wrote quite a number of articles on the topics mentioned above (Section 6), but no single coherent piece of work as demanded by the Philosophical Faculty as a Habilitationsschrift. In 1987 I decided to put all non-historic logic materials together, essentially around the theme of higher-level rules, which was already the topic of my doctoral thesis, but now extended in various directions. I included a chapter on relating my framework to the sequent calculus rather than natural deduction, to the bunch-based system of relevant logic (see below, Section 7), to logic programming and also to the framework of Martin-Löf type theory. Putting my family under enormous stress — Gabi was working as a postdoc in Zurich and our second child was born in March 1987 — I wrote it in a couple of months and submitted it in late summer, shortly before we left for a stay in Sweden from September 1987 until January 1988. The thesis (A1987c) was relatively short, but satisfied the requirements.[20] Prawitz was one of the reviewers, and I was later told that Prestel, who had attended the faculty meeting as an external expert, argued: "If Prawitz says yes, the faculty can't say no". In the obligatory colloquium with the Philosophical Faculty I spoke about "What is probability?" — the topic had to be different from that of the thesis and was selected from three themes submitted by the candidate. I talked about probability in a very general way to the humanities scholars in the audience. Because musicology was not represented as a subject at Konstanz, I even dared to say something about the Tristan chord and its relative frequency in music history, mentioning its occurrence in Beethoven's op. 31,3 piano sonata which I had learned about in Bonn with Martin Vogel: something that would have been a no-go for me if musicology professors had been there.

There was a formal inaugural lecture shortly afterwards. I still have the handwritten slides for it. I talked about logic and its future with quite some emphasis on issues of computer science and of artificial intelligence I had become interested in (including

---

[19] Assistant professor without tenure would be the closest American analogue.

[20] It circulated as a manuscript and was later made available on my website. A publication as a book in the Bibliopolis series "Studies in Proof Theory" was recommended by Dag Prawitz, but did not materialize.

nonmonotonic reasoning, polymorphic typing and hardware verification). So my general intellectual attitudes towards the end of the 1980s, when I moved to Tübingen, was a compound of proof-theoretic semantics, foundations of computer science, logic programming, general psychology and cognitive science.

7 Konstanz 1981–1989: Start of long-term collaborations and long-term friendships

At the time of my Ph.D. and shortly after I made scientific contacts and established personal friendships that lasted for the whole of my career and shaped my research topics.

Lars Hallnäs, definitional reflection and extensions of logic programming. Extending logic programming by means of implications in the bodies of clauses was a natural idea which fitted well with my approach of higher-level rules developed in my thesis. However, Lars Hallnäs had an idea which went much further, namely using the schema for general elimination rules as a schema for the inversion of systems of arbitrary, not necessarily logical rules, in particular rules of a logic program. He called it the principle of *definitional reflection* as one reflects on the given definition as a whole in order to invert it, the approach itself being called "definitional reasoning". Hallnäs's idea was to incorporate it into logic programming by means of a logic programming language which allowed one to evaluate goals according to this rule. In the 1990s, with his collaborators at the Swedish Institute of Computer Science (SICS), which was very much involved in implementing PROLOG, he developed such a system. We published the idea of proof-theoretic extensions of logic programming including definitional reflection in a two-part article at the end of the 1980s (A1990a, A1990b). From that the idea emerged to set up a series of conferences on *Extensions of Logic Programming (ELP)*. This worked out very well. We initiated and partly organized five conferences with corresponding proceedings: ELP1989 in Tübingen (C1991b), ELP1991 in Stockholm (C1992), ELP1992 in Bologna (C1993), ELP1993 in St. Andrews (C1994) and ELP1996 in Leipzig (C1996).

From the beginning of the 1990s Hallnäs and I have put more emphasis on definitional reasoning as a foundational approach that goes way beyond logic programming and is a general reasoning principle, as originally intended by Hallnäs (1991, 2006) (see my A1993, A1994b). Since it can be applied to any system of definitional rules, thus also to the rule defining p in terms of *not-p*, it has applications to paradoxical reasoning, and to any kind of non-wellfounded definition. This is why Hallnäs spoke of "partial" inductive definitions. This had implications for my later work on paradoxes. Hallnäs had done his Ph.D. with Prawitz in 1983 on normalization in set theory (Hallnäs, 1983), in which he had been strongly involved in the proof-theoretic treatment of paradoxes. Our work on definitional reflection quickly grew into a friendship between us and our families, with many short and long visits to each other's homes. Lars is also a composer by training[21], and his present for my 60th birthday was a composition for organ.

Unfortunately, we were not very successful with these ideas about definitional reflection. This is partly due to the fact that there is an obsession in philosophical proof-theoretic semantics to deal with logical constants, which are particularly well-behaved. Many do not realize that an inductive definition is like a set of introduction rules, a fact well established in mathematical proof theory. Dale Miller and his group were essentially the only ones beyond groups in Sweden to appreciate this approach. Had there been more resonance, we would have pursued it further into more advanced directions including definitional reflection for definitions of functions and functionals. There is some work by Hallnäs in this direction (see his contribution to this volume, Hallnäs 2023), which presents a good starting point for such investigations.[22]

Kosta Došen and logical constants. I met Kosta Došen at the logic colloquium in Florence in 1982. He had recently finished a D.Phil. thesis on logical constants with Michael Dummett and Dana Scott at Oxford, where he developed a theory of logicality based on an idea of iterated sequents (today one would speak of "hypersequents") and rules which could be read both downwards and upwards and which he called double-line rules. Later, through his publication on logical constants as "punctuation marks" (Došen, 1989), this approach became quite popular and widely discussed. I approached him personally in Florence because I was attracted by the abstract of his talk, where I saw immediate similarities to my idea of rules of higher levels. There was a certain difference, as his higher-level sequents essentially served for the interpretation of modal connectives whereas mine served for the interpretation of implication, for which he did not need any higher-order entitities. When we met, we got on well together and stayed friends for life. In Konstanz I worked with him on the notion of conservativeness and uniqueness of logical operators. In our joint papers A1985 and A1988 written in Konstanz in the mid-1980s we showed the duality of these notions and broke down the uniqueness problem to the interaction of two consequence relations. Later we initiated the topic of substructural logic (see below, Section 8). At the beginning of the 1990s he and his wife stayed with us in Tübingen for some time as he could not return to Belgrade during the Yugoslav war. In the middle of the 1990s, Kosta Došen turned towards category theory, which I did not find so interesting at the time. He became one of the leading figures in categorial proof theory, building very much on the work of Joachim Lambek. This was roughly at the time when he was giving up a full professorship in Toulouse in 1998, which he had held since 1994 after a two-year stay in Montpellier, that is, from the breakup of Yugoslavia onwards, to take up a professorship in Belgrade, after his arrival spending nights in bomb shelters during the NATO air raids of 1999. Later on, towards the 2010s, our contacts became more frequent once again, as we realized that his categorial approach to logic and my idea of the primacy of the hypothetical over the categorical converged. Sadly, in spite of an innovative cancer treatment in

[21] For his musical biography see https://quatuorbozzini.ca/en/artiste/hallnas_la.

[22] Cp. our notes presented to Dag Prawitz on his 80th birthday (M2016).

Germany the year before, he died in 2017, aged only 63. I delivered a eulogy at his funeral (M2022c).

Neil Tennant and the Scottish connection. I first got to know Neil Tennant through a paper in the *Journal of Philosophical Logic* (Tennant, 1980), in which, among other issues, he claimed to have given a proof of normalization for full classical logic, something which was correctly proved by Gunnar Stålmarck (1991)[23] some time later. I pointed Tennant to the deficiencies in his proof, and he invited me to a short stay at Stirling (Scotland) in 1982, where he had just started an appointment as a professor. It was there that I gave my first talk abroad in English. All presentations I had given before had been in German — times were different from today, where English has become the standard language of communication even in my Tübingen department. A follow-up stay — again with Gabi — took place in Winter/Spring 1983 when I was fellow of the Institute for Advanced Studies in the Humanities in Edinburgh, which was very productive.

Through Neil Tennant we met Stephen Read, and through him Roy Dyckhoff. This lead to an enjoyable and productive several-month stay in St. Andrews in spring 1985. Through Read I became acquainted with "standard" relevance logic — Tennant's deviating approach, from which his later *Core Logic* (Tennant, 2017) evolved, was already known to me. Particularly interesting was the proof-theoretic approach distinguishing different conjunctions at the structural level, something that Read and John Slaney had developed relying on earlier work by Michael Dunn. It was based on so-called "bunches" as a specific sort of structural entities and made it easy to formulate general rules for certain intensional connectives. I was quite enthusiastic about this topic and included it later in my Habilitation thesis (see above, Section 6).

Roy Dyckhoff, who had come from category theory and was now working in logic in computer science, I met originally as an attendee of the philosophical seminars in St. Andrews. As his work was so closely related to mine we collaborated intensively over the years. He was also somebody working on a contraction-free calculus for intuitionistic propositional logic independent and in parallel to the results in the Ph.D. thesis of Jörg Hudelmaier, who was a member of my group in Tübingen in the 1990s. Roy Dyckhoff was the first to propose general elimination rules without higher levels ('flat' general elimination rules, see below, Section 10). He died in 2018, aged 70. At his memorial service at St. Salvator's Chapel of the University of St. Andrews the bells were rung, for which he himself, as a passionate bell-ringer and theoretician of bell-ringing (see Dyckhoff, 2018), had raised funds.

8 Tübingen 1989–1997: Logic, philosophy, computer science

Application and appointment. After unsuccessful applications elsewhere, I became professor for logic and philosophy of language at the University of Tübingen in 1989. The support of Franz Guenthner, who was professor of general and computational

[23] I was an examiner of Stålmarck's Stockholm M.A. thesis which resulted in this article.

linguistics in Tübingen, was decisive both for the installation of this professorship and for my appointment. He was the editor with Dov Gabbay of the *Handbook of Philosophical Logic* and had first taken notice of my work through Göran Sundholm's *Handbook* conribution on proofs and meaning (Sundholm, 1986). As mentioned above in Section 5, Göran had been a reviewer for the *JSL* of my thesis publication. My application lecture was on "Logic and Cognition" — I tried to give an impression of my inclination towards cognitive science going beyond logic in the narrower sense.

Institutional struggles, offer from Berlin. In the fall of 1989 I started to teach in Tübingen. This was a wild time politically. I remember seeing the pictures of the collapse of the Berlin wall on TV during a stay in Tübingen. We were extremely lucky to buy a house in Tübingen at a very reasonable price in January 1990 that we rebuilt and refurbished, so that we could move into it at the end of the summer of 1990, just in time for school and the academic year. For me the first years in Tübingen were accompanied by certain struggles due to institutional issues between philosophy, computer science and linguistics. With the backing of the computer algebraist Rüdiger Loos, this was finally solved for me by moving into the newly founded department of computer science, with philosophy as a second affiliation, so that I had a joint appointment in computer science and philosophy, which for me and my range of interests was ideal. In 1991 I took my office in the computer science building, a former rehabilitation hospital with a beautiful view over the hills nearby and have stayed there until today.

Shortly after starting the Tübingen position I received an offer of a professorship in logic at the Free University Berlin. David Pearce, who at that time was working at Berlin, did all he could to get me there even though he was a strong candidate for the position himself. However, in the end I decided to stay in Tübingen where I had just started and our family had just settled.

Substructural logics. Apart from "proof-theoretic semantics" I am happy to have been involved in the coining of another term: "substructural logic" or "substructural logics". This term made it into the Mathematics Subject Classification, its current codes being 03B47 and 03F52 (American Mathematical Society, 2020). The situation unfolded as follows. I had just started my professorship in Tübingen in autumn 1989, while Kosta Došen had received a Humboldt grant for 1990–1991 and was staying in Konstanz. Franz Guenthner was in the possession of funds that he offered to me to spend on promising conferences. The first one in December 1989 was on *Extensions of Logic Programming* — see above (Section 7); the second one in October 1990 was organized together with Kosta Došen and originally entitled *Logics with Restricted Structural Rules*. We managed to invite top, or better the top, people in the respective fields. Our speakers were J. Lambek, S. V. Soloviev and J. van Benthem for the Lambek calculus, R. K. Meyer and J. M. Dunn for relevant logic, J.-Y. Girard, G. Sambin and A. Scedrov for linear logic, and V. Grishin and H. Ono for BCK logic. One or two days before the conference started Kosta rang me up: "*Logics with Restricted Structural Rules* is too clumsy, we must find a term that catches. I propose *Substructural Logics*. According to the Oxford English Dictionary it means 'relating to a substructure', that is, to the foundation of something." Initially, I was opposed to this, as it sounded like "subculture" to me, but eventually I gave in. The volume that resulted from the conference then bore this title (C1993). While the title of the book worked well, publication came with some unexpected problems. We had proofread everything except the title page and cover, and when the book came out, the editors appeared in non-alphabetical order: Schroeder-Heister & Došen. Because it was me who corresponded with Oxford University Press, they had assumed I was designated first editor and changed our specified author order! Cover disaster aside, the project was a great success. A vast amount of literature on "substructural" logics has appeared in the meantime, including two textbooks.

Logic and cognition. Even though the formal denomination of the professorship in Tübingen was "Logic and Philosophy of Language", and even though I was a full-fledged logician, my interests had shifted in the time after my Ph.D. to many other subjects, and in particular to the cognitive realm. There had been various influences on me, and my involvement in cognitive psychology through Gabi certainly played a major role. Most interesting to me was that people started to see connections between logical or symbolic reasoning in the narrower sense and more general reasoning models developed in the cognitive sciences. Moreover, my position was connected to the "Institute for Natural Language Systems"[24] founded by Franz Guenthner, with its associated curriculum and degree in General Linguistics with Psychology and Computer Science. This was a four-year M.A. course in cognitive science with emphasis on (both general and computational) linguistics, with computer science and psychology as obligatory minors. It had absolutely brilliant (individually selected) students from different backgrounds who had all sorts of interests besides their main fields of study. I remember, for example, a very stimulating talk by Joseph Weizenbaum whom they invited to discuss the social impact of recent developments in information technology and artificial intelligence. They were a pleasure to teach — an experience that I enjoyed already two years prior, as a substitute professor.

This suited me extremely well, and the titles of the courses I initially taught in Tübingen were all, in a sense, related to the cognitive field. Almost immediately after my start in Tübingen, in the autumn of 1990, I led seminars on connectionist modelling. My philosophy colleague Rüdiger Bubner complained when I made an announcement using the title "PDP", which sounded to him like the name of a political party in the GDR (it ceased to exist shortly afterwards in October 1990). He was right, of course, so I named it "Connectionism" and discussed the topic using its full name: "Parallel Distributed Processing", the title of the collection by David Rumelhart and James McClelland (1988) that was being widely debated at the time. I also taught philosophical foundations of cognitive science, philosophical critiques of AI, causality, connectionist reasoning systems and the like; many issues that would fit very well into the current landscape, where AI is the big thing, including in philosophical discussions.

A great success was a four-year research grant on the cognitive side of deduction as part of a programme called "Cognition and the Brain" by the German national science foundation (DFG) at the beginning of the 1990s together with a simultaneous grant

[24] "Seminar für natürlichsprachliche Systeme"

on efficient Gentzen systems within the DFG programme "Deduction". I was able to hire as postdocs Venkat Ajjanagadde, who had just finished a much quoted paper with L. Shastri on the connectionist representation of rules (later published as Shastri & Ajjanagadde 1993) and Seppo Keronen, who had done his Ph.D. with R. Stanton on "Computational Natural Deduction" (D1991); and, as a doctoral student, Uwe Oestermeier, who finished with a thesis on pictorial and logical thinking, which won a dissertation prize at the university (D1993).

What I was interested in, with respect to cognitive science, was the establishment of a proper combination of symbolic and connectionist modelling of logical reasoning, something that would be at the cutting edge of science nowadays. Had I pursued that further, my research could play a role in today's discussions, but this would have been at the expense of proof-theoretic semantics.

My interests in connectionism and AI were essentially from the philosophical side. At the same time I did a lot of teaching in theoretical computer science, as in the beginning of the 1990s the chair devoted to that field was not yet filled. I taught both advanced topics such as denotational semantics of programming languages, but also elementary courses on formal languages and computatability with large audiences. This was very much in accordance with my interests in logic programming as well as in type-theoretic approaches and thus advanced functional programming.

At the beginning of the 1990s I also had Jörg Hudelmaier working with me, who had done excellent work on the complexity of intuitionistic theorem proving, and with whom I published a paper on the non-commutative Lambek calculus (A1995) demonstrating certain limits for cut elimination. Ernst Zimmermann, whose Ph.D. thesis of 1995 on the philosophy of modal logic (D1995) I co-examined, has been present in my group for the past 20 years.

In Tübingen I had two collegues, on whom I relied heavily. On the philosophy side this was Walter Hoering, who was the logic professor in philosophy until he retired in 1998 (M1998a, M2019a) and on the informatics side this was Rüdiger Loos, who was instrumental for my affiliation to computer science (see above) and without whom computer science would not exist in Tübingen in its current, independent, and strong form.

Large public lecture series on "Music in the Sciences" and "Music and Informatics" (organized together with Loos), on "Thinking and Calculating", and a full public lecture series on the work of Karl Popper immediately after his death in 1995 (delivered with the philosopher of science Herbert Keuth), which were all very well received, serve to show that at the time, proof-theoretic semantics was not yet my key occupation.

9 London 1997–2000: Back to my roots

Sabbaticals and research stays. Longer stays at other institutes and with other colleagues, especially abroad, are normally highlights of a researcher's career. I only mention the longest one, in London, from 1997 to 2000 (others took place in

Edinburgh, Stockholm, Berne, Paris and Oxford). Initially, I had only a sabbatical in the winter term 1997–1998, for which I had an invitation from Dov Gabbay to Imperial College London. We went to England as a family and rented a house in Kingston-upon-Thames, not far from the German School. We extended our stay, and in the end remained there until summer 2000, that is, for a total of three years. From the summer semester 1998 on, I commuted between Tübingen and London, flying to Tübingen for three days a week during the semester and spending the university vacations in London. This was possible at the time without impairing my duties in Tübingen. I still conducted the full teaching programme of nine hours per week of classroom teaching, split between philosophy and computer science. Besides Imperial College, I was, through Edmund Robinson, also affiliated with Queen Mary and Westfield College (QMW, now Queen Mary, University of London). While working there in computer science, I met David Pym with whom I still collaborate today. On a personal level, staying in London was one of the best decisions we made in our lives, in particular as far as our children were concerned. It was a great experience that shaped their future. They both went on to study and live in England.

Logic in philosophy. In a sense the time in London marks my returning from a dominant interest in cognitive science (still with an emphasis on logic) which I had retained for more than a decade, to logic and in particular philosophical logic and philosophy of logic. These were my roots when I started studying in Bonn and also my occupation in my doctorate. Even the computer science aspects of my work I now started seeing from the perspective of philosophical (rather than mathematical) logic. For example, the inversion rules in extended logic programming languages could be seen as philosophically motivated inversion principles in a proof-theoretic semantics of logical and other signs. This did not mean that I gave up cognitive science or computer science, but rather considered it as a background and application for more philosophical theories. What contributed to it was the joint Tübingen-Konstanz research group *Logic in Philosophy* which I initiated, and which I applied for together with Wolfgang Spohn (who had just taken up a professorship in Konstanz) in 1996 with many brilliant philosophical projects both in Tübingen and Konstanz (C2005). The project started exactly when we went to London, so the Tübingen part was partly directed from London. Through this I collaborated with Patrizio Contu and Reinhard Kahle. With Patrizio I published my first paper on hypothetical versus categorical reasoning (A2005), with its philosophical claim that the hypothetical concept of consequence should be given conceptual priority over the categorical concepts of truth or provability.

The Salzburg offer. In the year 2000 the University of Salzburg offered me a chair in philosophy — the successorship to Paul Weingartner, who had just retired. This was quite unexpected. I almost decided to accept it; we even registered our children at local schools. However, it finally failed for administrative reasons, which had essentially to do with the German pension system and the losses I would have incured when moving to Austria, even though it was within the European Union. Turning it down was a hard decision as there was only little that could be gained in Tübingen; for example, the fact that my position in Tübingen was a second-class professorship, not a chair, was non-negotiable. Even though I declined it, the Salzburg offer contributed to my intellectual shift back towards philosophy.

10 Tübingen 2000–2019: Topics in proof-theoretic semantics

During my time in Tübingen, and in particular after the stay in London, which fixed, so to speak, the topic of proof-theoretic semantics as the key topic of my group, I have worked on various aspects of proof-theoretic semantics. There was no linear order. I tried to identify critical topics and find solutions to certain problems, though many of them are still open (A2016a). I shall just mention a few of them.

Definitional reflection, paradoxes and intensional proof-theoretic semantics. I have already mentioned Lars Hallnäs's idea of definitional reflection and the idea of definitional reasoning, which my mind always kept returning to. It is clear that definitional reasoning constitutes a kind of intensional proof-theoretic semantics, as definitions are intensional entities. It depends on the sort of definitions whether the system obtained is well-behaved. Hallnäs called this well-behaviour, where we have cut elimination and the like "total". Changing the definition, that is, changing the meaning of terms, changed the global behaviour of the system, so there was a strong sense of non-locality. The standard example was always the definition of *p* by *not-p*, which should be allowed (as in logic programming), but which globally leads to systems without normalization or cut elimination and which can be seen as the skeleton structure of the various logical, semantical and mathematical paradoxes. One follow-up question is then under which conditions definitions are "well-behaved" and are in this sense extensional entities. In several papers I showed that this strongly depends on the structural rules available, whereby these structural rules interact (A1992a, A2016b), including the seemingly trivial structural identity axiom ("*A* entails *A*"). I could also show that, in order to avoid paradoxes, a very limited restriction of the rule of contraction is sufficient, in contradistinction to abandoning this rule altogether (A2012b), based on an intensional distinction of the ways a formula is 'given' to us — either through a definition or without any specification (A2022b). I even speculated on the idea of a 'free' type theory, in which the application of rules depended on the evaluation of certain definitional terms (A2012d).

A purely intensional paradox was discovered by Jan Ekman in a thesis he wrote supervised by Hallnäs (Ekman, 1994). By translating set-theoretic paradoxes into propositional logic he could show that under certain assumptions derivations in propositional logic cannot be normalized. This has fascinated me ever since its appearance, and I often mentioned it in talks. I knew Ekman from the time he was still a doctoral student, and would have liked to employ him on a post in the research group *Logic in Philosophy*. However, he had already accepted a permanent position in industry at that point. Only in recent years did I manage to write two papers on the subject, together with Luca Tranchini (A2017, A2021). By that time Luca had embarked on a different notion of intensional proof-theoretic semantics, which is concerned with the identity of proofs, and an objection we made was that from this point of view Ekman's translation has certain deficiencies. This is closely related to the issue of harmony, and Luca is pursuing this strand of research.

In fact, I came across the issue of identity of proofs when I was external examiner ("opponent") at the doctoral defence of Prawitz's student Filip Widebäck in 2001, who had written on exactly that topic (D2001). In preparing my talk I had noticed that Widebäck presented results that had been obtained independently by Kosta Došen at the same time. Only much later I realized the significance of these issues for the sort of proof-theoretic semantics I was advocating. Incidentally, a thesis defence in Sweden, with an external opponent who first presents the thesis (to a potentially large audience), and then publicly questions the candidate about it, is a great experience in its own right, even if it costs the opponent a week of preparation as he needs to read the thesis really carefully. Before I had acted as opponent in Sweden once already, for Lars-Henrik Eriksson in computer science, who had worked on extensions of definitional reflection (D1993).

Direct negation, Béziau's square, and bilateralism. In the still unpublished part of my doctoral dissertation, I had already discussed an extended natural deduction system with a 'structural' kind of negation — structural, because it is built into the deduction machinery and not a logical constant. Such negations, also called "denial", sometimes "strong negation", have been considered for a long time, not only in philosophy but also in computer science. I discovered that the idea of definitional reflection gives rise to a slightly different sort of denial, as it allows a distinction between a direct denial based on definitional clauses, and an indirect denial induced by definitional reflection. The latter resembles, in a rough way, what in logic programming is called negation by failure. Correspondingly, we obtain two sorts of assertion, namely direct assertion (by definition) and indirect assertion (by failure). I presented some initial ideas at the first Uni-Log (Universal Logic) conference organized by Jean-Yves Béziau in Montreux in 2005. This was an excellent meeting with, for example, Saul Kripke present (as a so-called "secret speaker", who was not announced beforehand).

It turned out that this idea with two sorts of assertions and two sorts of denials could be put into the structure of a square, so I presented it in 2007 at the first conference on the square of opposition by Béziau, again in Montreux. Initially it seemed to me to be a crazy idea to devote a conference to this topic, and later a whole series of conferences, but it turned out that there were sufficient serious ideas in it to compensate for the stranger aspects. The conference proceedings only appeared in 2012. I still consider my contribution there (A2012a) significant, although it had no impact whatsoever. When I was invited for the third congress on the square in 2012 in Beirut, I presented an even further developed variant of my approach, calling it a "calculus of squares". I arranged the two assertions (or positions) and the two denials (or negations) at the corners of a square, proposed a sequent calculus for such entities and discussed major theorems (cut elimination etc.) for such a system. The conference was an impressive event at the beautifully located American University. People in Beirut tried to live their lives normally despite the Syrian war having started the year before. To us as visitors at least, it was not much felt there yet. I have not continued this line of research since, partly because of lack of resonance, partly because I have been engaged in other issues of proof-theoretic semantics. I think it has a great potential. It would fit very well into today's discussion of what is called "bilateralism".

Béziau also invited me to give a course on proof-theoretic semantics in 2007 at the second Uni-Log conference in Xi'an (China), which took place after the LMPS congress in Beijing. Even though, or perhaps because I essentially talked to colleagues rather than to students, it was extremely instructive to me. I later gave a related course at the ESSLLI conference in Bordeaux in 2009, this time to students (A2009a). I had already given a course with Lars Hallnäs at ESSLLI 1993 in Lisbon and have always considered the ESSLLI summer schools a great success story. Thomas Piecha and myself also organized a workshop at the sixth Uni-Log in Vichy in 2018 (C2018b). Jean-Yves Béziau, through marketing projects in a way one is not normally used to in logic, contributes to keeping the machinery of research interaction running. I myself have strongly profited from this, as I would probably not have developed certain ideas otherwise. The fact that we now have a *World Logic Day* (14 January) approved by UNESCO and celebrated even during the Covid pandemic as a massive collection of online events, is also the result of Béziau's initiative.

Assumption-conclusion bilateralism, hypothetical reasoning and the dogma of standard semantics. As far as bilateralism is concerned, I have always understood it in the sense that reasoning should not only be one-sided, from unspecific assumptions towards a specific conclusion, but two-sided in the sense that in the course of reasoning, both the assumptions and the conclusion can be modified according to semantic rules. This is in certain ways related to negation-bilateralism, but rests on different intuitions, as no negation is involved in the first place. I have argued in favour of this sort of bilateralism[25] in the context of criticising standard semantics. By standard semantics I mean both model-theoretic semantics and most intuitionistic semantics such as BHK, realizability, validity-based semantics in the sense of Prawitz and Dummett etc. All these semantics start using a categorical concept such as truth or validity. Then a hypothetical concept of consequence is defined as the transmission of this categorical concept under all circumstances. Assumptions are nothing but placeholders for categorical entities (truths or valid proofs) — the specific semantics only applies to conclusions. This motivates the emphasis on introduction rules in standard proof-theoretic semantics. I called this a *dogma*, as there is no real reason for choosing this sort of approach. Why not choose a hypothetical concept of consequence as a basis and then derive the categorical from the hypothetical rather than vice versa?

There is a formal model of such an approach in the form of the sequent calculus, where one can modify both the left and right side by independent rules. There is also categorial proof theory which is essentially based on such an approch, as one deals

[25] I spoke of "bidirectionality" instead of "bilateralism". Perhaps this is a preferable term to avoid terminological confusions.

with hypothetical entities (arrows) from the beginning.[26] However, this has never been developed into a philosophical foundational theory.

I developed this idea together with Patrizio Contu in our research group "Logic in Philosophy" in the late 1990, who very strongly insisted on it (A2005), and have 'preached' it ever since at many conferences and many occasions, with limited success. Certainly, my arguing was more programme than execution, but the target was clear: developing a semantics of hypothetical judgements as primordial entities, that is, whithout presupposing categorical judgements. I remember presenting it at the conference of the Society for Analytical Philosophy in Bielefeld in 2004 (A2004), the Logica conferences in Hejnice in 2007 and 2008 (A2008b, A2009b), at a conference at the Swedish Collegium for Advanced Study (SCAS) in Uppsala in 2010 (A2012e), where a great deal of the proof-theoretic semantics community was present, on several occasions in Kosta Došen's colloquium in Belgrade, as well as in a manuscript for a special issue of *Erkenntnis* that never saw publication (M2008c). Perhaps there will at some point be a student prepared to work out this concept in detail, both on the classical and intuitionistic side.

Harmony. Harmony — a term introduced in this context by Dummett (1973, pp. 396f.) — is considered one of the fundamental notions of proof-theoretic semantics. The conditions for asserting a sentence should match the conclusions that can be drawn from it. What it should mean in detail is still a matter of discussion. In my thesis I tried to achieve harmony by generalized elimination inferences for logical constants which guarantee it due to their syntactic form. To enable this I had to introduce higher-level rules, as otherwise implicational connectives could not be interpreted. They are also a crucial tool to interpret certain connectives, for example, of relevance logic such as fusion (A1987c). Now there is also another variant of generalized elimination rules put forward by Dyckhoff, Edgar López-Escobar, Tennant and Jan von Plato, which are of an elementary level (called "parallelized" by Tennant) and are a natural deduction translation of Gentzen's sequent calculus rules (see my A2014b, and above, Section 7). Their adherents have always preferred these 'flat' rules and even claimed that they provide a framework as powerful as mine to generate harmonious rules. It bothered me for many years that I did not have an easy argument at hand that certain constants could not be expressed using flat generalized elimination rules. This changed when I met Grigory Olkhovikov. He was a student of Grigori Mints, and Mints recommended him to me at a conference in Bochum in 2012. When I posed the problem to him — translated into a problem in intuitionistic propositional logic — he came up with a solution, which resulted in a joint paper in the *Review of Symbolic Logic* (A2014a, with an extension to arbitrary levels in A2014b). This settled the issue, although incorrect claims continue to be propagated.

However, this was still a limited notion of harmony, as it only laid down appropriate elimination rules given certain introduction rules (something which, by the way, also could be inverted; A2015c). A more universal notion would tell when a given pair

[26] Note that here the term "categorial" refers to mathematical category theory, whereas "categorical" is used in the traditional philosophical sense of denoting a categorical in contradistinction to a hypothetical judgement.

of introduction and elimination rules is in harmony. I made such a proposal in my contribution to the "Outstanding Contributions to Logic" volume in honour of Dag Prawitz (A2015a) and a subsequent paper in *Studia Logica* (A2014d), where I gave second-order translations of the introduction and elimination meanings of connectives and declared them to be in harmony when these translations were equivalent. This was a conceptual achievement, I think, even though it is 'only' an extensional notion of harmony based on deductive equivalence. Luca Tranchini is developing this approach further in the direction of an 'intensional' notion taking proof identity into account. (I made some remarks on this issue in A2016a.) He elaborated on this idea partly through interaction with Kosta Došen and his work, with whom we were quite intensively collaborating in the 2010s, and also worked with Paolo Pistone, who had done his Ph.D. with Girard and was in Tübingen as a postdoc in 2018–19.[27]

Incompleteness and atomic systems: Beyond (intuitionistic) logic. As mentioned above (Section 6), the idea that proof-theoretic semantics could in a sense justify intuitionistic logic, in that the latter is complete with respect to the former, has fascinated me ever since my Ph.D. thesis. In the past fifteen years, Thomas Piecha and myself have essentially come to a negative conclusion. This was considerably stimulated by exchanges (including mutual visits) with Wagner de Campos Sanz and also with Tor Sandqvist, both of whom had the idea that certain only classically valid formulas are verified by proof-theoretic semantics. Several papers with de Campos Sanz and Piecha (A2014, A2015) resulted from that, and in the end Piecha and I found a counterexample free from deficiencies, depending only on a few plausible assumptions about the proof-theoretic semantics used (A2019c). This is, of course, a somewhat negative result, and not what I had expected 35 years ago, but perhaps it is a prejudice to think that intuitionistic logic is *the* proper logic of constructive or operational reasoning. As a desideratum there remains, of course, to describe this proper logic, if there is (a finitely axiomatizable) one.

Closely interconnected with the completeness problem is the question of the atomic base of proof-theoretic semantics. In model-theoretic semantics it is structures that tell us which atomic sentences are true and which are false. In the proof-theoretic case one considers atomic deduction systems instead. However, what kind of such systems are admitted strongly influences the logic one obtains, in particular if in atomic systems not only production rules but more general rule concepts including definitional reflection are allowed, as joint work with Piecha shows (A2016b, A2017). Quite independent of any logic built on top of atomic systems, these systems in themselves represent a highly interesting topic that goes way beyond logic and impacts, for example, the theory of inductive definitions as well as rule based argumentation theory. Given the current interest in the latter, this is perhaps a future research field to invest in.

The format of deduction. One topic that interacts with almost all others is which form a logical deduction should take. Proof-theoretic semantics has a bias towards natural deduction, but the sequent calculus is another possible format which fits

[27] Cp. their joint contribution to this volume (Pistone and Tranchini, 2023).

very well in particular with approaches that want to make assumption-conclusion bilateralism explicit and allows one, for example, to establish a perfect duality between conjunction and disjunction, something on which Došen's categorial approach and Giovanni Sambin et al.'s Basic Logic (Sambin, Battilotti, and Faggian, 2000) rests. Definitional reflection works under both formats, and I have argued it has certain advantages over Sambin et al.'s approach (A2013). However, as soon as it comes to implication, not even the standard rules of the sequent calculus are exempt from criticism. I have claimed that for the introduction of implication on the left side (the assumption side) different rules might be considered, if one wants to distinguish between implications as rules and implications as links (A2011b, A2014b)[28] and thus to disentangle logical from structural features of implication. This also plays a role if one wants to give the approach of higher-level inferences rules in my doctoral thesis a sequent-style formulation, as Arnon Avron has pointed out (Avron, 1990 — incidentally the only paper referring to me in its title). In addition, there is the even more fundamental question of the role of proof search, that is, the idea that in reasoning we often start with a goal that we want to prove, and then *reduce* it to subgoals. Perhaps "reductive" reasoning in this sense should be given a more fundamental stance, something which is the case, for example, in dialogue logics. We have been involved in the latter through research grants (see Section 11), and Thomas Piecha and myself have worked on it (A2012, A2015a) — in fact, Piecha wrote his doctoral thesis on it (D2012), which represented a major advance for the field. However, I think the general problem of deduction versus reduction still needs to be settled, possibly in a much wider framework.

11 Group, grants, cooperations

In the sciences it is absolutely standard to work with a group of people with whom one can discuss problems and results, and to publish with them. In the humanities, there is still a considerable number of single researchers who reach (often outstanding) results without much interaction. I definitely wanted to have at least a small group of, say, three people to talk to. I had some money from the university to pay for one position, which I secured when I turned down the Salzburg offer. For all the rest grants were needed, both doctoral and postdoc grants.

I mentioned already the grants in cognitive science and deduction as well as the research group "Logic in Philosophy" during my first decade at Tübingen and London (Sections 8 and 9). Successful grants after 2000 were a grant from the European Science Foundation (ESF) on "Dialogical Foundations of Semantics" (2008–2012) for a joint project with groups in Lisbon (Reinhard Kahle) and Amsterdam (Benedikt Löwe) as well as a 10-year long French-German collaboration on "Hypothetical Reasoning" (2009–2019) within a joint programme of the French and German national science foundations (ANR and DFG), with the Institut d'histoire et de philosophie

[28] Cp. Michael Arndt's (2023) contribution to this volume.

des sciences et des techniques (IHPST) in Paris. All this made it possible to hire as postdocs Kai Wehmeier, who was brilliant but left after one year to become a professor at the University of California, Irvine, and Bartosz Więckowski who in his thesis had created the topic of subatomic natural deduction (D2006) — now a growing branch of proof-theoretic semantics. Later, it allowed the recruitment of Rainer Lüdecke, who wrote a thesis on the game semantics of logic programming (D2012), Harald Maurer on connectionist modelling (D2014), Tiago Rezende de Castro Alves (D2019) on identity of proofs, Hermogenes Oliveira (D2019) on proof-theoretic validity and René Gazzari on the notion of occurrence (D2020).

In addition to providing money to employ researchers, the French-German project was a very successful research cooperation through which we made many new friends. It opened up entirely new perspectives, with many exchanges and research stays in Paris, and also a number of joint conferences. I knew Michel Bourdeau and Jean Fichot, the French project leaders, from the 2007 meeting on one hundred years of intuitionism at Cerisy castle (van Atten, Boldini, Bourdeau, and Heinzmann, 2008), where I gave a talk on Lorenzen's *Operative Logic* as an approach to proof-theoretic semantics (A2008a) and met Michael Dummett again after the 1999 Tübingen conference on proof-theoretic semantics (Section 12). As the attendence of the conference exceeded the number of available single rooms, Jean Fichot and I had to share a room, which fostered our friendship. Within our own project, Jean organized a meeting at the same place in 2017.

Another collaborator over the decades was Reinhard Kahle. He worked as a postdoc in our research group "Logic in Philosophy" and has collaborated with us ever since during his long time as a professor in Lisbon. In 2018 he was appointed to the newly created Carl Friedrich von Weizsäcker chair for Philosophy and History of Science in Tübingen. It is not a dedicated logic position, but, as Reinhard is a logician, logic will be represented there (my own professorship has been discontinued as a logic position).

Most important for shaping the field of proof-theoretic semantics in Tübingen were the three members of my group who were constantly present during the last decade of my work there and whom I would name first when speaking of "my group": Michael Arndt (thesis D2008), Thomas Piecha (thesis D2012) and Luca Tranchini (thesis D2010). I benefited enormously from my interactions with them. None of the papers we wrote together could have been written by me alone, and I am sure that many of my single-authored papers would be of poorer quality.

12 The Tübingen conferences on proof-theoretic semantics

We organized three major conferences in Tübingen on proof-theoretic semantics which initially served to establish the subject under this name and then to keep its status. The first of these I organized in January 1999 at Tübingen castle together with Reinhard Kahle. We had quite prominent guests there including Dummett, Prawitz, Martin-Löf, Mints and William Tait. We had big problems obtaining a visa for Grigori

Mints, who was living in Stanford but had Russian citizenship. In the end, I made this public in the local daily newspaper. First thing in the morning on the day the paper appeared, I was rung up by Herta Däubler-Gmelin, our local Tübingen member of parliament who at the time was Minister for Justice in the federal government. She solved the problem by intervening through channels available to her, and Grigori was able to come.[29]

It took seven years for the proceedings to appear (C2006), partly due to me, as I needed quite some time for my own contribution on validity concepts in proof-theoretic semantics (A2006d). Originally I had wanted to write about definitional reflection as an alternative approach to standard proof-theoretic semantics. However, then I thought I should first elaborate the original Prawitz approach to validity, which then took the space allotted to my contribution[30]. The effect was that many considered the paper to be describing my deeply rooted proper opinion, and Prawitz was full of praise for it as presenting some of his ideas in a congenial way. I am no longer certain in what direction the 'most appropriate' approach to proof-theoretic semantics should go, and Prawitz is not absolutely certain about it either (see his contribution to this volume, Prawitz 2023), but in any case this is a paper making accessible a certain conception of proof-theoretic semantics in a thorough way. The special issue of *Synthese* probably paved the way for the term "proof-theoretic semantics". At least, it was roughly from this time on that the term took off and entered a great number of publications.

The second Tübingen conference on proof-theoretic semantics (C2016a), in March 2013, coincided with my 60th birthday. In addition to the conference, where again the big figures of the proof-theoretic semantics community were gathered, Reinhard Kahle and Thomas Piecha organized a birthday colloquium where David Pearce, Heinrich Herre, Walter Hoering, Gereon Wolters, Bartosz Więckowski, Ernst Zimmermann, Marie Duži, Pavel Materna and Gerhard Jäger spoke, all of them significant people in my intellectual development. This was a wonderful present. It also meant that we enjoyed two conference dinners, the second one being a birthday dinner hosted by me. The third Tübingen conference took place in spring 2019, one semester ahead of my retirement. It was the largest one with a lot of contributed papers and listening guests, reflecting that the subject has reached a mature state (C2019d).

In 2015, initiated by Luiz Carlos Pereira, Thomas Piecha and I organized a conference on *General Proof Theory* to celebrate the 50th anniversary of Dag Prawitz's book on natural deduction. Došen, Martin-Löf, Pereira, Schwichtenberg and Wansing, among others and in addition to Prawitz himself, all spoke. This was later published as a special issue of *Studia Logica* (C2019a). In 2001, Pereira, who had done his doctorate with Dag Prawitz around the time that Dag was acting as external referee for my doctorate, organized a conference in Rio de Janeiro celebrating Dag Prawitz's work, so the Tübingen conference was a kind of continuation of it. At the conference in Rio I presented my historical research on Hertz and on the first paper by Gentzen which was written in the spirit of Hertz's work (A2002b).

[29] As Grigori said at the dinner, it was the first time in his life that he toasted a government minister.

[30] The more inclusive and much longer first version of the paper is available as M2003b.

This conference made a lasting impression on me, not only because of the excellent presentations, but also because there was sufficient time both for socialising and for seeing the beautiful city. Besides the "standard" sightseeing places I remember well the wonderful reception by Oswaldo Chateaubriand in his house, and also the visit of Rocinha and the beautiful view we (Grigori Mints, Jan von Plato, Ernst Zimmermann and myself) had from the flat roof of a house up there.

13 Service and honours

My main service to the scientific community was my involvement in the Division of Logic, Methodology and Philosophy of Science (DLMPS, from 2015 Division of Logic, Methodology and Philosophy of Science and Technology, DLMPST). This association organizes its namesake congress every four years, dating back to 1960. I have been to several of these congresses. At the Salzburg 1983 congress I presented my first international conference paper (A1983a). For the 2011 congress, which was to take place in Nancy, the DLMPS executive committee (in particular Wilfrid Hodges, DLMPS president), supported by Gerhard Heinzmann (local organising chair) chose me as the general programme chair, which was of course a great honour. Preparing the congress was a lot of work — we could invite around 70 speakers, and accepted some 650 contributed papers (C2014, C2014–2015). The congress went quite well, in particular as the local organizers were so efficient. At the general assembly I was elected secretary general of DLMPS for the subsequent four years, and shortly afterwards I became in addition treasurer of the association, a job I kept for eight years. At the 2019 congress in Prague I retired from all DLMPST positions. All in all, this sort of activity was a great experience, but its more than ten years were enough to exhaust my administrative energy.

A great honour in my intellectual career was the doctorate *honoris causa* received from the University of Belgrade. The award ceremony took place in May 2016. It was very nice, in a beautiful Auditorium Maximum, with *Gaudeamus Igitur* being sung. The following weekend we celebrated Easter at Kosta Došen's home with some of his friends. For me it was the first year with Easter celebrated twice, this one being the orthodox feast, which was very late that year.

14 Retirement and outlook

As a professor in Germany, one can postpone one's retirement by three years beyond the retirement age, but only if the university agrees. This was not the case for me, as they needed my salary money otherwise (my pension comes from a different source). I just managed to obtain a half-year extension and retired on 1 October 2019. In November, I gave my retirement lecture on "Logic yesterday – today – tomorrow", to which plenty of former colleagues and friends came. Coincidentally, that very day the UNESCO General Assembly installed World Logic Day, something I was happy to announce. Shortly afterwards I left for a half-year research stay at the Swedish Collegium for Advanced Study (SCAS) in Uppsala, where I had intended to work with Dag Prawitz on a joint monograph. Initially it was wonderful. I had perfect working conditions, very interesting and interested colleagues from a variety of fields; Gabi and I were provided with an excellent flat where we could even accomodate our children as visitors. Unfortunately, in March the Covid-19 pandemic struck and at the urgent recommendation of the German foreign office, we broke off our stay at the beginning of April, driving 20 hours non-stop with a rental car back to Stuttgart. Now, in spring 2023, we have had three years of pandemic, a much wider scientific interaction due to videoconferencing, and much time to think about what to do next (M2022d). There are very many loose ends in the topics mentioned above (Section 10), so plenty of work and also ideas to continue. There are also other topics to consider, such as the mathematical applications of Martin-Löf's ideas and results in the context of homotopy type theory (The Univalent Foundations Program, 2013). On the other hand, given the advances proof-theoretic semantics has already made, I doubt I can make much more than incremental progress.

Nevertheless, there are two book projects that I would like to finish. The first one is the monograph with Prawitz on general proof theory. We have been planning it since Prawitz brought up this issue when we met at a conference in Dubrovnik in May 2010 (the conference was on the philosophical nature of logical consequence). Acting as his co-author is a great honour for me, and I regret very much that progress has not been as fast as envisaged. The pandemic and other matters interrupted our work in Sweden. Completing it has first priority for me. The second is a monograph on proof-theoretic semantics. This I give more time, even though I wanted to write it already four decades ago, as mentioned in the first paragraph of this autobiography. In any case, it is to rely on the book with Prawitz. While Nissim Francez's monograph (2015) covers many topics including some applications in linguistic semantics, and certainly helps to establish the discipline, I still see so many open problems in proof-theoretic semantics, in particular on the conceptual side, such that for me there is no doubt that an advanced textbook covering both philosophical and technical aspects is a desideratum[31]. I very much hope that I will continue to have the intellectual strength and good health to pursue this goal. As far as the field of proof-theoretic semantics is concerned, it is on the right track. Even though my group is not being continued in Tübingen, there are plenty of groups continuing its topics, the most outstanding one in Germany being Heinrich Wansing's in Bochum. I very much hope that in the long run proof-theoretic semantics, which originated from logic, will prove itself in a great number of applications outside logic.

Acknowledgements I would like to thank Thomas Piecha and Kai Wehmeier for having taken on the great burden of editing this volume, and, of course, the authors for their contributions. I am also

[31] Originally, my SEP entry on proof-theoretic semantics (E2012c) was considered a first step in this direction, which was actually based on a much longer draft (available online: M2011c). Given the fast development of the field and changes in my own attitudes towards it, my original conception requires further thought.

greatly indebted to my family, especially my wife Gabi and my daughter Paula, for their help with this chapter, and to Chris Jones for detailed comments on the manuscript.

References

American Mathematical Society (2020). *2020 Mathematics Subject Classification*. URL: https://mathscinet.ams.org/mathscinet/msc/msc2020.html.

Arndt, M. (2023). Eight rules for implication elimination. In: *Peter Schroeder-Heister on Proof-Theoretic Semantics (this volume)*. Ed. by T. Piecha and K. F. Wehmeier. Berlin: Springer.

Avron, A. (1990). Gentzenizing Schroeder-Heister's natural extension of natural deduction. *Notre Dame Journal of Formal Logic* 31, 127–135. DOI: 10.1305/ndjfl/1093635337.

Binder, D. and T. Piecha (2017). Popper's notion of duality and his theory of negations. *History and Philosophy of Logic* 38, 154–189. DOI: 10.1080/01445340.2016.1278517.

– (2021). Popper on quantification and identity. In: *Karl Popper's Science and Philosophy*. Ed. by Z. Parusniková and D. Merritt. Berlin: Springer, 149–169. DOI: 10.1007/978-3-030-67036-8_8.

Brandom, R. B. (1994). *Making It Explicit: Reasoning, Representing, and Discursive Commitment*. Cambridge, MA: Harvard University Press.

– (2000). *Articulating Reasons: An Introduction to Inferentialism*. Cambridge, MA: Harvard University Press.

Buhl, G. (1961). *Ableitbarkeit und Abfolge in der Wissenschaftstheorie Bolzanos*. Kantstudien Ergänzungshefte 83. Kölner Universitäts-Verlag.

Buldt, B. (2022). Logic and foundations of the exact sciences at the University of Konstanz: people & projects 1966–2021. In: *60 Jahre DVMLG*. Ed. by B. Löwe and D. Sarikaya. London: College Publications, 51–96.

Došen, K. (1989). Logical constants as punctuation marks. *Notre Dame Journal of Formal Logic* 30, 362–381. DOI: 10.1305/ndjfl/1093635154.

Dummett, M. (1973). *Frege: Philosophy of Language*. London: Duckworth.

– (1991). *Frege. Philosophy of Mathematics*. Cambridge, MA: Harvard University Press.

Dyckhoff, R. (2018). *A note on bobs-only Grandsire Triples. 2 January 2018*. St Andrews University. URL: https://www.ringing.info/bobs-only_g7.pdf.

Ekman, J. (1994). *Normal Proofs in Set Theory*. Unversity of Göteborg: Ph.D. thesis. URL: https://hdl.handle.net/2077/10743.

Francez, N. (2015). *Proof-Theoretic Semantics*. London: College Publications.

Friedrichsdorf, U. (1992). *Einführung in die klassische und intensionale Logik*. Braunschweig: Vieweg.

Gabbay, D. M. and U. Reyle (1984). N-PROLOG: an extension of PROLOG with hypothetical implications. *Journal of Logic Programming* 1, 319–355. DOI: 10.1016/0743-1066(84)90029-3.

Gentzen, G. (1933). Über die Existenz unabhängiger Axiomensysteme zu unendlichen Satzsystemen. *Mathematische Annalen* 107, 329–350. DOI: 10.1007/BF01448897. English translation in: *The Collected Papers of Gerhard Gentzen.* Ed. by M. E. Szabo. Amsterdam: North Holland 1969, 29–52.
– (1935). Untersuchungen über das logische Schließen. *Mathematische Zeitschrift* 39, 176–210 (DOI: 10.1007/BF01201353), 405–431 (DOI: 10.1007/BF01201363). English translation in: *The Collected Papers of Gerhard Gentzen.* Ed. by M. E. Szabo. Amsterdam: North Holland 1969, 68–131.
Gödel, K. (2003). *Kurt Gödel: Collected Works: Volume IV: Selected Correspondence, A–G.* Ed. by S. Feferman, J. W. Dawson, W. Goldfarb, C. Parsons, and W. Sieg. Oxford: Clarendon Press.
Hallnäs, L. (1983). *On Normalization of Proofs in Set Theory.* Unversity of Stockholm: Ph. D. thesis. Published as: *Dissertationes Mathematicae*, Vol. 261, Polska Akademia Nauk, Warszawa 1988. URL: http://pldml.icm.edu.pl/pldml/element/bwmeta1.element.zamlynska-2e4a4d2f-6a08-47b6-8e9b-1f63f3aac438/c/rm26101.pdf.
– (1991). Partial inductive definitions. *Theoretical Computer Science* 87, 115–142. DOI: 10.1016/S0304-3975(06)80007-1.
– (2006). On the proof-theoretic foundation of general definition theory. *Synthese* 148, 589–602. DOI: 10.1007/s11229-004-6291-6.
– (2023). On the structure of proofs. In: *Peter Schroeder-Heister on Proof-Theoretic Semantics (this volume).* Ed. by T. Piecha and K. F. Wehmeier. Berlin: Springer.
Lloyd, J. W. (1984). *Foundations of Logic Programming.* 2nd Edition 1987. Berlin: Springer.
Lorenzen, P. (1955). *Einführung in die operative Logik und Mathematik.* 2nd Edition 1969. Berlin: Springer.
Miller, D. (1986). A theory of modules for logic programming. In: *Proceedings of the 1986 Symposium on Logic Programming (Salt Lake City Utah).* Washington: IEEE Computer Society Press, 106–114.
Mittelstraß, J., ed. (1980–2018). *Enzyklopädie Philosophie und Wissenschaftstheorie. Vol. I–IV (1980–1996). 2nd Edition: Vol. I–VIII (2005–2018).* Stuttgart: Metzler.
Paramei, G. V. (2009). Walter H Ehrenstein (1950-2009). *Perception* 38, 784–787. DOI: 10.1068/p3805ob.
Piecha, T. (2016). Completeness in proof-theoretic semantics. In: *Advances in Proof-Theoretic Semantics.* Ed. by T. Piecha and P. Schroeder-Heister. Berlin: Springer, 231–251. DOI: 10.1007/978-3-319-22686-6_15.
– (2023). Karl Popper on deduction. In: *Perspectives on Deduction.* Ed. by A. Piccolomini d'Aragona. Berlin: Springer.
Pistone, P. and L. Tranchini (2023). Intensional harmony as isomorphism. In: *Peter Schroeder-Heister on Proof-Theoretic Semantics (this volume).* Ed. by T. Piecha and K. F. Wehmeier. Berlin: Springer.
Popper, K. and D. Miller (1983). A proof of the impossibility of inductive probability. *Nature* 302, 687–688. DOI: 10.1038/302687a0.

– (1987). Why probabilistic support is not inductive. *Philosophical Transactions of the Royal Society of London. Series A, Mathematical and Physical Sciences* 321, 569–591. DOI: 10.1098/rsta.1987.0033.

Popper, K. R. (1945). *The Open Society and Its Enemies*. London: Routledge.

Prauss, G. (1974). *Kant und das Problem der Dinge an sich*. Bonn: Bouvier.

Prawitz, D. (1965). *Natural Deduction: A Proof-Theoretical Study*. Stockholm: Almqvist & Wiksell. Reprinted Mineola, NY: Dover Publ., 2006.

– (1971). Ideas and results in proof theory. In: *Proceedings of the Second Scandinavian Logic Symposium (Oslo 1970)*. Ed. by J. E. Fenstad. Amsterdam: North-Holland, 235–308. DOI: 10.1016/S0049-237X(08)70849-8.

– (1979). Proofs and the meaning and completeness of the logical constants. In: *Essays on Mathematical and Philosophical Logic: Proceedings of the Fourth Scandinavian Logic Symposium and the First Soviet-Finnish Logic Conference, Jyväskylä, Finland, June 29 – July 6, 1976*. Ed. by J. Hintikka, I. Niiniluoto, and E. Saarinen. Dordrecht: Kluwer, 25–40 [revised German translation 'Beweise und die Bedeutung und Vollständigkeit der logischen Konstanten', *Conceptus, 16*, 1982, 31–44]. DOI: 10.1007/978-94-009-9825-4_2.

– (2023). The validity of inference and argument. In: *Peter Schroeder-Heister on Proof-Theoretic Semantics (this volume)*. Ed. by T. Piecha and K. F. Wehmeier. Berlin: Springer.

Pym, D., E. Ritter, and E. Robinson (2022). Proof-theoretic semantics in sheaves (extended abstract). In: *Scandinavian Logic Symposium (SLSS 2022). 17th – 19th June, 2022. Bergen, Norway*. URL: http://www0.cs.ucl.ac.uk/staff/D.Pym/PymRitterRobinson.pdf.

Rumelhart, D. E., J. L. McClelland, PDP Research Group, et al. (1988). *Parallel Distributed Processing*. New York: IEEE.

Sambin, G., G. Battilotti, and C. Faggian (2000). Basic Logic: reflection, symmetry, visibility. *Jounal of Symbolic Logic* 65, 979–1013. DOI: 10.2307/2586685.

Sandqvist, T. (2015). Base-extension semantics for intuitionistic sentential logic. *Logic Journal of the IGPL* 23, 719–731. DOI: 10.1093/jigpal/jzv021.

Shastri, L. and V. Ajjanagadde (1993). From simple associations to systematic reasoning: a connectionist representation of rules, variables and dynamic bindings using temporal synchrony. *Behavioral and Brain Sciences* 16, 417–451. DOI: 10.1017/S0140525X00030910.

Stålmarck, G. (1991). Normalization theorems for full first order classical natural deduction. *Journal of Symbolic Logic* 56, 129–149. DOI: 10.2307/2274910.

Stegmüller, W. (1959). *Unvollständigkeit und Unentscheidbarkeit: Die metamathematischen Resultate von Gödel, Church, Kleene, Rosser und ihre erkenntnistheoretische Bedeutung*. Wien: Springer. DOI: 10.1007/978-3-7091-3524-2.

– (1979). *The Structuralist View of Theories: A Possible Analogue of the Bourbaki Programme in Physical Science*. Berlin: Springer. DOI: 10.1007/978-3-642-95360-6.

Sundholm, G. (1986). Proof theory and meaning. In: *Handbook of Philosophical Logic, Vol. III*. Ed. by D. Gabbay and F. Guenthner. Dordrecht: Reidel, 471–506. DOI: 10.1007/978-94-017-0464-9_3.

Tennant, N. (1980). A proof-theoretic approach to entailment. *Journal of Philosophical Logic* 9, 185–209. DOI: 10.1007/BF00247748.

– (2017). *Core Logic*. Oxford University Press.

The Univalent Foundations Program (2013). *Homotopy Type Theory: Univalent Foundations of Mathematics*. Princeton: Institute for Advanced Study.

Thiel, C. (1965). *Sinn und Bedeutung in der Logik Gottlob Freges*. Meisenheim a. Gl.: Anton Hain.

van Atten, M., P. Boldini, M. Bourdeau, and G. Heinzmann, eds. (2008). *One Hundred Years of Intuitionism (1907–2007): The Cerisy Conference*. Basel: Birkhäuser. DOI: 10.1007/978-3-7643-8653-5.

Vogel, M. (1975). *Die Lehre von den Tonbeziehungen*. Bonn-Bad Godesberg: Verlag für systematische Musikwissenschaft.

von Kutschera, F. (1968). Die Vollständigkeit des Operatorensystems $\{\neg, \wedge, \vee, \supset\}$ für die intuitionistische Aussagenlogik im Rahmen der Gentzensemantik. *Archiv für mathematische Logik und Grundlagenforschung* 11, 3–16. DOI: 10.1007/BF01973339.

– (1969). Ein verallgemeinerter Widerlegungsbegriff für Gentzenkalküle. *Archiv für mathematische Logik und Grundlagenforschung* 12, 104–118. DOI: 10.1007/BF01969697.

– (1996). Frege and natural deduction. In: *Frege: Importance and Legacy*. Ed. by M. Schirn. Berlin: De Gruyter, 301–304. DOI: 10.1515/9783110813043.301.

Wansing, H. (2000). The idea of a proof-theoretic semantics and the meaning of the logical operations. *Studia Logica* 64, 3–20. DOI: 10.1023/A:1005217827758.

Wirth, C.-P. (2021). Gisbert Hasenjaeger and a most interesting unpublished draft for Hilbert and Bernays' "Grundlagen der Mathematik". *Siegener Beiträge zur Geschichte und Philosophie der Mathematik* 14, 1–57. DOI: 10.25819/ubsi/10063.

Publications by Peter Schroeder-Heister

Most publications can be downloaded from Peter Schroeder-Heister's homepage.

(A) Logic-related publications including significant abstracts

Schroeder-Heister, P. (1981a). Bibliographie Hugo Dingler (1981–1954). *Zeitschrift für philosophische Forschung* 35, 283–298. Slightly modified version in: *Der wissenschaftliche Nachlaß von Hugo Dingler (1881-1954). Verzeichnis mit einer Bibliographie der Schriften Dinglers.* Ed. by G. Wolters and P. Schroeder-Heister. DOI: 10.15496/publikation-69429.

– (1981b). *Untersuchungen zur regellogischen Deutung von Aussagenverknüpfungen.* University of Bonn: Doctoral Dissertation. Published online together with the manuscript "Sentential calculi with rules of arbitrary levels" of 1982 (Summary of Chapters 1 and 2 of the German Ph.D. thesis). DOI: 10.15496/publikation-68454.

– (1982). Logische Konstanten und Regeln. Zur Deutung von Aussagenoperatoren. *Conceptus. Zeitschrift für Philosophie* 16, 45–59. DOI: 10.15496/publikation-72304.

– (1983a). Inversion principles and the completeness of intuitionistic natural deduction systems. In: *Abstracts of the 7th International Congress of Logic, Methodology and Philosophy of Science (Salzburg 1983)*, 150–153. DOI: 10.15496/publikation-72545.

– (1983c). The completeness of intuitionistic logic with respect to a validity concept based on an inversion principle. *Journal of Philosophical Logic*, 359–377. DOI: 10.1007/BF00263483.

– (1984a). A natural extension of natural deduction. *Journal of Symbolic Logic* 49, 1284–1300. DOI: 10.2307/2274279.

– (1984b). Freges Permutationsargument: Zu §10 der "Grundgesetze der Arithmetik". In: *Frege Conference 1984. Proceedings of the International Conference Held at Schwerin (GDR), September 10-14, 1984.* Ed. by G. Wechsung. Berlin: Akademie-Verlag, 182–188. DOI: 10.15496/publikation-72305.

– (1984c). Generalized rules for quantifiers and the completeness of the intuitionistic operators $\&, \vee, \supset, \bot, \forall, \exists$. In: *Computation and Proof Theory: Proceedings of the Logic Colloquium Held in Aachen, July 18-23, 1983, Part II.* (Lecture Notes in Mathematics Vol. 1104). Ed. by M. M. Richter, E. Börger, W. Oberschelp, B. Schinzel, and W. Thomas. Berlin: Springer, 399–426. DOI: 10.1007/BFb0099494.

– (1984d). Popper's theory of deductive inference and the concept of a logical constant. *History and Philosophy of Logic* 5, 79–110. DOI: 10.1080/01445348408837064.

Došen, K. and P. Schroeder-Heister (1985). Conservativeness and uniqueness. *Theoria* 51, 159–173. DOI: 10.1111/j.1755-2567.1985.tb00093.x.

Schroeder-Heister, P. (1985a). Natural deduction calculi with rules of higher levels (Abstract). *Journal of Symbolic Logic* 50, 275–276. DOI: 10.2307/2273818.

Schroeder-Heister, P. (1985b). Proof-theoretic validity and the completeness of intuitionistic logic. In: *Foundations of Logic and Linguistics: Problems and Their Solutions*. Ed. by G. Dorn and P. Weingartner. New York: Plenum Press, 43–87. DOI: 10.1007/978-1-4899-0548-2_4.

Mittelstraß, J. and P. Schroeder-Heister (1986). Zeichen, Kalkül, Wahrscheinlichkeit. Elemente einer Mathesis Universalis bei Leibniz. In: *Pragmatik. Handbuch pragmatischen Denkens. Band I: Pragmatisches Denken von den Ursprüngen bis zum 18. Jahrhundert*. Ed. by H. Stachowiak. In collab. with C. Baldus. Hamburg: Felix Meiner Verlag, 392–414. DOI: 10.15496/publikation-72307.

Došen, K. and P. Schroeder-Heister (1987). A general interpolation and definability theorem (Abstract). *Journal of Symbolic Logic* 52, 315–316. DOI: 10.2307/2273893.

Schroeder-Heister, P. (1987a). A model-theoretic reconstruction of Frege's permutation argument. *Notre Dame Journal of Formal Logic* 28, 69–79. DOI: 10.1305/ndjfl/1093636847.

– (1987b). Judgements of higher levels in Martin-Löf's logical theory (Abstract). *Journal of Symbolic Logic* 52, 1083. DOI: 10.2307/2273851.

– (1987c). *Structural Frameworks with Higher-Level Rules*. University of Konstanz: Habilitationsschrift. DOI: 10.15496/publikation-69827.

Schroeder-Heister, P. and F. Schaefer (1987). Der strukturalistische Reduktionsbegriff und Repräsentationen von Theorien. In: *Logik, Wissenschafstheorie und Erkenntnistheorie. Akten des 11. Internationalen Wittgenstein Symposiums, 4.-13. 8. 1986, Kirchberg/Wechsel (Österreich)*. Ed. by P. Weingartner and G. Schurz. Wien: Hölder-Pichler-Tempsky, 230–233. DOI: 10.15496/publikation-72329.

Došen, K. and P. Schroeder-Heister (1988). Uniqueness, definability and interpolation. *Jounal of Symbolic Logic* 53, 554–570. DOI: 10.2307/2274524.

Hallnäs, L. and P. Schroeder-Heister (1989). Logic programming with higher-level rules (Abstract). *Journal of Symbolic Logic* 54, 656–657. DOI: 10.2307/2274896.

Schroeder-Heister, P. (1989a). Judgements of higher levels and standardized rules for logical constants in Martin-Löf's theory of logic. In: *Proceedings of the Workshop on Programming Logic*. Ed. by P. Dybjer, L. Hallnäs, B. Nordström, K. Petersson, and J. M. Smith. Göteborg: Programming Methodology Group, University of Göteborg and Chalmers University of Technology, 494–517. DOI: 10.15496/publikation-72310.

Schroeder-Heister, P. and F. Schaefer (1989). Reduction, representation and commensurability of theories. *Philosophy of Science* 56, 130–157. DOI: 10.1086/289476.

Hallnäs, L. and P. Schroeder-Heister (1990a). A proof-theoretic approach to logic programming: I. Clauses as rules. *Journal of Logic and Computation* 1, 261–283. DOI: 10.1093/logcom/1.2.261.

– (1990b). A proof-theoretic approach to logic programming: II. Programs as definitions. *Journal of Logic and Computation* 1, 635–660. DOI: 10.1093/logcom/1.5.635.

Schroeder-Heister, P. (1991a). An asymmetry between introduction and elimination inferences. In: *Abstracts of the 9th International Congress of Logic, Methodology and Philosophy of Science (Uppsala 1991)*. DOI: 10.15496/publikation-72545.

– (1991c). Hypothetical reasoning and definitional reflection in logic programming. In: *Extensions of Logic Programming: International Workshop, Tübingen, December 1989, Proceedings*. LNCS (Vol.475). Ed. by P. Schroeder-Heister. Berlin: Springer, 327–340. DOI: 10.1007/BFb0038701.
– (1991d). Structural frameworks, substructural logics, and the role of elimination inferences. In: *Logical Frameworks*. Ed. by G. Huet and G. Plotkin. Cambridge University Press, 385–403. DOI: 10.1017/CBO9780511569807.016.
– (1991e). Uniform proof-theoretic semantics for logical constants (Abstract). *Journal of Symbolic Logic* 56, 1142. DOI: 10.2307/2275092.
Hallnäs, L. and P. Schroeder-Heister (1992b). Local reflection in inductive definitions: the D-rule (Abstract). *Journal of Symbolic Logic* 57, 300. DOI: 10.2307/2275203.
Schroeder-Heister, P. (1992a). Cut elimination in logics with definitional reflection. In: *Nonclassical Logics and Information Processing: International Workshop, Berlin, November 1990, Proceedings*. Lecture Notes in Computer Science Vol. 619. Ed. by D. Pearce and H. Wansing. Berlin: Springer, 146–171. DOI: 10.1007/BFb0031929.
– (1992b). Rules of definitional reflection in logic programming. In: *Proceedings of the Workshop on Linear Logic and Logic Programming, Washington, DC, 14 November 1992 (following the 1992 Joint International Conference and Symposium on Logic Programming)*. Ed. by D. Miller. University of Pennsylvania Department of Computer and Information Science Technical Report No. MS-CIS-92-80, 45–46. DOI: 10.15496/publikation-72545. URL: https://repository.upenn.edu/cis_reports/301.
– (1993). Rules of definitional reflection. In: *Proceedings of the 8th Annual IEEE Symposium on Logic in Computer Science (Montreal 1993)*. Los Alamitos: IEEE Press, 222–232. DOI: 10.1109/LICS.1993.287585.
– (1994a). Cut elimination for logics with definitional reflection and restricted initial sequents. In: *Post-Conference Workshop on "Proof-Theoretical Extensions of Logic Programming" at the 11th International Conference on Logic Programming (ICLP'94, S. Margherita Ligure, Italy, 13–17 June 1994)*. DOI: 10.15496/publikation-72545.
– (1994b). Definitional reflection and the completion. In: *Extensions of Logic Programming: Proceedings of the 4th International Workshop, ELP '93 (St. Andrews, March/April 1993)*. Lecture Notes in Computer Science, Vol. 798. Ed. by R. Dyckhoff. Berlin: Springer, 333–347. DOI: 10.1007/3-540-58025-5_65.
Hudelmaier, J. and P. Schroeder-Heister (1995). Classical Lambek logic. In: *Theorem Proving with Analytic Tableaux and Related Methods. 4th International Workshop, TABLEAUX '95, Schloß Rheinfels, St. Goar, Germany, May 7–10, 1995. Proceedings*. Ed. by P. Baumgartner, R. Hähnle, and J. Possega. Berlin: Springer, 247–262. DOI: 10.1007/3-540-59338-1_40.
Schroeder-Heister, P. (1997). Frege and the resolution calculus. *History and Philosophy of Logic* 18, 95–108. DOI: 10.1080/01445349708837278.
– (1998b). Wahrscheinlichkeit. In: *Karl Popper: Logik der Forschung*. Ed. by H. Keuth. Klassiker Auslegen, Vol. 12, 185–213. 4th Edition 2013. Berlin: Akademie Verlag, 187–215. DOI: 10.1524/9783050050188.187.

Schroeder-Heister, P. (1999). Gentzen-style features in Frege. In: *Abstracts of the 11th International Congress of Logic, Methodology and Philosophy of Science, Cracow, Poland (August 1999)*, 449. DOI: 10.15496/publikation-72545.
– (2002a). Popper's inferential definitions of logical constants (Abstract). In: *Karl Popper 2002 – Centenary Congress (Vienna, 3-7 July 2002)*. DOI: 10.15496/publikation-69821. Full paper published as Schroeder-Heister, 2006b.
– (2002b). Resolution and the origins of structural reasoning: early proof-theoretic ideas of Hertz and Gentzen. *Bulletin of Symbolic Logic* 8, 246–265. DOI: 10.2178/bsl/1182353872. Extended version available under DOI: 10.15496/publikation-69828.
– (2003a). Definitional reflection and circular reasoning. In: *Abstracts of the 12th International Congress of Logic, Methodology and Philosophy of Science (Oviedo, Spain, August 7-13, 2003)*. Ed. by E. Álvarez, R. Bosch, and L. Villamil, 126–128. DOI: 10.15496/publikation-72545.
– (2004). On the notion of *Assumption* in logical systems. In: *Selected Papers Contributed to the Sections of GAP5, Fifth International Congress of the Society for Analytical Philosophy, Bielefeld, 22-26 September 2003*. Ed. by R. Bluhm and C. Nimtz. Paderborn: Mentis, 27–48. URL: http://www.gap5.de/proceedings/. (Abstract and Extended Summary DOI: 10.15496/publikation-72545).
– (2005a). Begründungsrationalität und Logik. In: *Homo Sapiens und Homo Faber. Epistemische und technische Rationalität in Antike und Gegenwart. Festschrift für Jürgen Mittelstraß*. Ed. by G. Wolters and M. Carrier. Berlin: De Gruyter, 285–296. DOI: 10.15496/publikation-72306. Publisher DOI: 10.1515/9783110923940.285.
– (2005c). Generalized rules, direct negation, and definitional reflection (Abstract). In: *First World Congress on Universal Logic (UNILOG 2005), Montreux, Switzerland, March 31 – April 3, 2005*. DOI: 10.15496/publikation-72545.
Schroeder-Heister, P. and P. Contu (2005). Folgerung. In: *Logik in der Philosophie*. Ed. by W. Spohn, P. Schroeder-Heister, and E. J. Olsson. Heidelberg: Synchron Wissenschaftsverlag, 247–276. DOI: 10.15496/publikation-70826.
Wehmeier, K. F. and P. Schroeder-Heister (2005). Frege's permutation argument revisited. *Synthese* 147, 43–61. DOI: 10.1007/s11229-004-6206-6.
Schroeder-Heister, P. (2006a). *Assertion* and *Denial* in proof-theoretic semantics (Abstract and Extended Summary). In: *GAP6, 6th International Congress of the Society for Analytical Philosophy, Berlin, 11-14 September 2006; UNILOG'07, 2nd World Congress and School on Universal Logic, Xi'an, 16-22 August 2007; GAP7, 7th International Congress of the Society for Analytical Philosophy, Bremen, 14-17 September 2009*. DOI: 10.15496/publikation-72545.
– (2006b). Popper's structuralist theory of logic. In: *Karl Popper: A Centenary Assessment. Vol. III: Science*. Ed. by I. Jarvie, K. Milford, and D. Miller. Aldershot: Ashgate, 17–36. DOI: 10.15496/publikation-69821.
– (2006d). Validity concepts in proof-theoretic semantics. *Synthese* 148. Special issue *Proof-Theoretic Semantics*. Ed. by R. Kahle and P. Schroeder-Heister, 525–571. DOI: 10.1007/s11229-004-6296-1.
– (2007a). Direct negation in proof-theoretic semantics and the square of opposition (Abstract). In: *First World Congress on the Square of Opposition, Montreux,*

Switzerland, 1–3 June 2007. DOI: 10.15496/publikation-72545. Full paper published as Schroeder-Heister, 2012a.

– (2007b). Generalized definitional reflection and the inversion principle. *Logica Universalis* 1, 355–376. DOI: 10.1007/s11787-007-0018-7.

– (2007c). Lorenzens operative Logik und moderne beweistheoretische Semantik. In: *Der Konstruktivismus in der Philosophie im Ausgang von Wilhelm Kamlah und Paul Lorenzen*. Ed. by J. Mittelstraß. Paderborn: Mentis, 167–196. DOI: 10.30965/9783969750582_013.

Mittelstraß, J. and P. Schroeder-Heister (2008). Nicholas Rescher on Greek philosophy and the syllogism. In: *Rescher Studies: A Collection of Essays on the Philosophical Work of Nicholas Rescher Presented to Him on the Occasion of His 80th Birthday*. Ed. by R. Almeder. Frankfurt: Ontos Verlag, 211–240. DOI: 10.15496/publikation-72303. Publisher DOI: 10.1515/9783110329094.211.

Schroeder-Heister, P. (2008a). Lorenzen's operative justification of intuitionistic logic. In: *One Hundred Years of Intuitionism (1907-2007): The Cerisy Conference*. Ed. by M. van Atten, P. Boldini, M. Bourdeau, and G. Heinzmann. Basel: Birkhäuser, 214–240 [References for whole volume: 391–416]. DOI: 10.1007/978-3-7643-8653-5_14.

– (2008b). Proof-theoretic versus model-theoretic consequence. In: *The Logica Yearbook 2007*. Ed. by M. Peliš. Prague: Filosofia, 187–200. DOI: 10.15496/publikation-70827.

– (2009a). *Proof-theoretic Semantics. Advanced Course, 21st European Summer School in Logic, Language and Information (ESSLLI), Bordeaux, 27–31 July 2009 (Abstract and Course Plan)*. DOI: 10.15496/publikation-72545.

– (2009b). Sequent calculi and bidirectional natural deduction: on the proper basis of proof-theoretic semantics. In: *The Logica Yearbook 2008*. Ed. by M. Peliš. London: College Publications, 237–251. DOI: 10.15496/publikation-70817.

– (2011a). An alternative implication-left schema for the sequent calculus (Abstract). *Bulletin of Symbolic Logic* 17, 316. DOI: 10.2178/bsl/1305810915.

– (2011b). Implications-as-rules vs. implications-as-links: an alternative implication-left schema for the sequent calculus. *Journal of Philosophical Logic* 40, 95–101. DOI: 10.1007/s10992-010-9149-z.

– (2011d). Schluß und Umkehrschluß. Ein Beitrag zur Definitionstheorie. In: *Lebenswelt und Wissenschaft. XXI. Deutscher Kongreß für Philosophie, Universität Duisburg-Essen, 15.-19.9.2008. Kolloquienbeiträge*. Ed. by C. F. Gethmann, J. C. Bottek, and S. Hiekel. (= Deutsches Jahrbuch Philosophie, Vol. 2). Hamburg: Felix Meiner, 1065–1092. DOI: 10.15496/publikation-72308.

Piecha, T. and P. Schroeder-Heister (2012). Implications as rules in dialogical semantics. In: *The Logica Yearbook 2011*. Ed. by M. Peliš and V. Punčochář. London: College Publications, 211–225. DOI: 10.15496/publikation-70828.

Schroeder-Heister, P. (2012a). Definitional reasoning in proof-theoretic semantics and the square of opposition. In: *The Square of Opposition: A General Framework for Cognition*. Ed. by J.-Y. Béziau and G. Payette. Bern: Peter Lang, 323–349. DOI: 10.15496/publikation-72333.

Schroeder-Heister, P. (2012b). Paradoxes and structural rules. In: *Insolubles and Consequences: Essays in Honour of Stephen Read*. Ed. by C. Dutilh Novaes and O. Hjortland. London: College Publications, 203–211. DOI: 10.15496/publikation-70824.

– (2012d). Proof-theoretic semantics, self-contradiction, and the format of deductive reasoning. *Topoi* 31, 77–85. DOI: 10.1007/s11245-012-9119-x.

– (2012e). The categorical and the hypothetical: a critique of some fundamental assumptions of standard semantics. *Synthese* 187. Special issue *The Philosophy of Logical Consequence and Inference*. Ed. by S. Lindström, E. Palmgren and D. Westerståhl, 925–942. DOI: 10.1007/s11229-011-9910-z.

– (2012f). What is the proper logic of consequence? In: *GAP8, 8th International Congress of the Society for Analytical Philosophy, Konstanz, 17-20 September 2012*, 40–41. DOI: 10.15496/publikation-72545. URL: http://www.gap8.de/mentis-gap8-druck-endversion.pdf.

– (2013). Definitional Reflection and Basic Logic. *Annals of Pure and Applied Logic* 164. Special issue *Articles in Honor of Giovanni Sambin's 60th Birthday*. Ed. by M.E. Maietti, E. Palmgren and M. Rathjen, 491–501. DOI: 10.1016/j.apal.2012.10.010.

de Campos Sanz, W., T. Piecha, and P. Schroeder-Heister (2014). Constructive semantics, admissibility of rules and the validity of Peirce's law. *Logic Journal of the IGPL* 22, 297–308. DOI: 10.1093/jigpal/jzt029.

Olkhovikov, G. K. and P. Schroeder-Heister (2014a). On flattening elimination rules. *Review of Symbolic Logic* 7, 60–72. DOI: 10.1017/S1755020313000385.

– (2014b). Proof-theoretic harmony and the levels of rules: general non-flattening results. In: *Second Pisa Colloquium in Logic, Language and Epistemology*. Ed. by E. Moriconi and L. Tesconi. Pisa: Edizioni ETS, 245–287. DOI: 10.15496/publikation-72309.

Schroeder-Heister, P. (2014a). Frege's sequent calculus. In: *Trends in Logic XIII: Gentzen's and Jaśkowski's Heritage – 80 Years of Natural Deduction and Sequent Calculi*. Ed. by A. Indrzejczak, J. Kaczmarek, and M. Zawidzki. Łódż University Press, 233–245. DOI: 10.15496/publikation-72324.

– (2014b). Generalized elimination inferences, higher-level rules, and the implications-as-rules interpretation of the sequent calculus. In: *Advances in Natural Deduction: A Celebration of Dag Prawitz's Work*. Ed. by L. C. Pereira, E. H. Haeusler, and V. de Paiva. Berlin: Springer, 1–29. DOI: 10.1007/978-94-007-7548-0_1.

– (2014c). Proof-theoretic harmony: the issue of propositional quantification. In: *Trends in Logic XIII: Gentzen's and Jaśkowski's Heritage – 80 Years of Natural Deduction and Sequent Calculi*. Ed. by A. Indrzejczak, J. Kaczmarek, and M. Zawidzki. Łódż University Press, 5–16. DOI: 10.15496/publikation-72327.

– (2014d). The calculus of higher-level rules, propositional quantifiers, and the foundational approach to proof-theoretic harmony. *Studia Logica* 102. Special issue *Gentzen's and Jaśkowski's Heritage: 80 Years of Natural Deduction and Sequent Calculi*. Ed. by A. Indrzejczak, 1185–1216. DOI: 10.1007/s11225-014-9562-3.

Piecha, T., W. de Campos Sanz, and P. Schroeder-Heister (2015). Failure of completeness in proof-theoretic semantics. *Journal of Philosophical Logic* 44, 321–335. DOI: 10.1007/s10992-014-9322-x.

Piecha, T. and P. Schroeder-Heister (2015a). Dialogical logic for definitional reasoning and implications as rules. In: *Dialogische Logik*. Ed. by J. Mittelstraß and C. von Bülow. Paderborn: Mentis, 91–125. DOI: 10.30965/9783897856394_009.

Schroeder-Heister, P. (2015a). Harmony in proof-theoretic semantics: a reductive analysis. In: *Dag Prawitz on Proofs and Meaning*. Ed. by H. Wansing. Berlin: Springer, 329–358. DOI: 10.1007/978-3-319-11041-7_15.

– (2015c). Proof-theoretic validity based on elimination rules. In: *Why is this a Proof? Festschrift for Luiz Carlos Pereira*. Ed. by E. H. Haeusler, W. de Campos Sanz, and B. Lopez. London: College Publications, 159–176. DOI: 10.15496/publikation-70825.

Piecha, T. and P. Schroeder-Heister (2016b). Atomic systems in proof-theoretic semantics: two approaches. In: *Epistemology, Knowledge and the Impact of Interaction*. Ed. by J. Redmond, O. P. Martins, and A. N. Fernández. Berlin: Springer, 47–62. DOI: 10.1007/978-3-319-26506-3_2.

Schroeder-Heister, P. (2016a). Open problems in proof-theoretic semantics. In: *Advances in Proof-Theoretic Semantics*. Ed. by T. Piecha and P. Schroeder-Heister. Berlin: Springer, 253–283. DOI: 10.1007/978-3-319-22686-6_16.

– (2016b). Restricting initial sequents: the trade-offs between identity, contraction and cut. In: *Advances in Proof Theory*. Ed. by R. Kahle, T. Strahm, and T. Studer. Basel: Birkhäuser, 339–351. DOI: 10.1007/978-3-319-29198-7_10.

Piecha, T. and P. Schroeder-Heister (2017). The definitional view of atomic systems in proof-theoretic semantics. In: *The Logica Yearbook 2016*. Ed. by P. Arazim and T. Lávička. London: College Publications, 185–200. DOI: 10.15496/publikation-70829.

Schroeder-Heister, P. and L. Tranchini (2017). Ekman's paradox. *Notre Dame Journal of Formal Logic* 58, 567–581. DOI: 10.1215/00294527-2017-0017.

Piecha, T. and P. Schroeder-Heister (2018a). Intuitionistic logic is not complete for standard proof-theoretic semantics (Abstract). *Bulletin of Symbolic Logic* 24, 262. DOI: 10.1017/bsl.2018.13.

Schroeder-Heister, P. (2018). Die Grundlagen mathematischer Beweise. Philosophisch-logische Überlegungen. *Mitteilungen der Mathematischen Gesellschaft in Hamburg* 38, 117–137. URL: https://mathges.hamburg/Zeitschriften.html.

Piecha, T. and P. Schroeder-Heister (2019b). General proof theory: introduction. *Studia Logica* 107. Special issue *General Proof Theory*. Ed. by T. Piecha and P. Schroeder-Heister, 1–5. DOI: 10.1007/s11225-018-9818-4.

– (2019c). Incompleteness of intuitionistic propositional logic with respect to proof-theoretic semantics. *Studia Logica* 107. Special issue *General Proof Theory*. Ed. by T. Piecha and P. Schroeder-Heister, 233–246. DOI: 10.1007/s11225-018-9823-7.

Schroeder-Heister, P. and L. Tranchini (2021). How to Ekman a Crabbé-Tennant. *Synthese* 199, 617–639. DOI: 10.1007/s11229-018-02018-3.

Binder, D., T. Piecha, and P. Schroeder-Heister (2022a). Popper's theory of deductive logic. In: *The Logical Writings of Karl Popper*. Ed. by D. Binder, T. Piecha, and P. Schroeder-Heister. Berlin: Springer, 1–79. DOI: 10.1007/978-3-030-94926-6_1.

Schroeder-Heister, P. (2022a). Axiomatic thinking, identity of proofs and the quest for an intensional proof-theoretic semantics. In: *Axiomatic Thinking I*. Ed. by F. Ferreira, R. Kahle, and G. Sommaruga. Berlin: Springer, 145–163. DOI: 10.1007/978-3-030-77657-2_8.

– (2022b). Intensional proof-theoretic semantics and the rule of contraction. In: *Logically Speaking: A Festschrift for Marie Duží*. Ed. by P. Materna and B. Jespersen. London: College Publications, 5–21.

(P) Psychological publications

Heister, G. and P. Schroeder-Heister (1984). A note on location of the decisional stages of choice reactions in the left hemisphere. *Cortex* 20, 271–275. DOI: 10.1016/S0010-9452(84)80044-1.

– (1985). S-R compatibility effect or cerebral laterality effect? Comments on a controversy. *Neuropsychologia* 23, 427–430. DOI: 10.1016/0028-3932(85)90030-2.

Heister, G., W. H. Ehrenstein, and P. Schroeder-Heister (1986a). Spatial S-R compatibility effects with unimanual two-finger choice reactions for prone and supine hand positions. *Perception & Psychophysics* 40, 271–278. DOI: 10.3758/BF03211507.

– (1986b). Spatial S-R compatibility with two-finger choice reactions (Abstract, *9th European Vision Conference, Bad Nauheim, Germany, 17–20 September 1986*). *Perception* 15, Number 1, A33. DOI: 10.15496/publikation-72545.

– (1987). Spatial S-R compatibility with unimanual two-finger choice reactions: effects of irrelevant stimulus location. *Perception & Psychophysics* 42, 195–201. DOI: 10.3758/BF03210508.

Heister, G. and P. Schroeder-Heister (1987). Evidence for stimulus-response compatibility effects in a divided visual field study of cerebral lateralization. *Acta Psychologica* 66, 127–138. DOI: 10.1016/0001-6918(87)90029-1.

Heister, G., P. Schroeder-Heister, and W. H. Ehrenstein (1987). Spatial stimulus-response (S-R) compatibility under head tilt: evidence for a factorial model (Abstract, *10th European Conference on Visual Perception, Varna, Bulgaria, 21–24 September 1987*). *Perception* 16, Number 2, A17 (p. 241). DOI: 10.15496/publikation-72545. DOI for all conference abstracts: 10.1068/p160225.

Ehrenstein, W. H., P. Schroeder-Heister, and G. Heister (1988). Spatial visuo-motor compatibility with orthogonal stimulus-response arrangement (Abstract, *11th European Conference on Visual Perception, Bristol, UK, 31 August – 3 September1988*). *Perception* 17, Number 3, A77 (p. 415). DOI: 10.15496/publikation-72545.

Schroeder-Heister, P., G. Heister, and W. H. Ehrenstein (1988). Spatial S-R compatibility under head tilt. *Acta Psychologica* 69, 35–44. DOI: 10.1016/0001-6918(88)90028-5.

Ehrenstein, W. H., G. Heister, and P. Schroeder-Heister (1989). Spatial visuomotor compatibility as a function of retinal eccentricity (Abstract, *12th European Conference on Visual Perception, Zichron Yaakov, Israel, 17 – 22 September 1989*). *Perception* 18, Number 4, A62 (p. 544). DOI: 10.15496/publikation-72545.

Ehrenstein, W. H., P. Schroeder-Heister, and G. Heister (1989). Spatial S-R compatibility with orthogonal stimulus-response relationship. *Perception & Psychophysics* 45, 215–220. DOI: 10.3758/BF03210700.

Heister, G., T. Landis, M. Regard, and P. Schroeder-Heister (1989a). Shift of functional cerebral asymmetry during the menstrual cycle. *Neuropsychologia* 27, 871–880. DOI: 10.1016/0028-3932(89)90010-9.

– (1989b). Shift of visual half-field superiority for face perception during the menstrual cycle (Abstract, *General Conference of the European Brain Behaviour Society, Zurich, Switzerland, 7–9 September 1988*). *Behavioural Brain Research* 33, 314. DOI: 10.1016/S0166-4328(89)80124-X.

Heister, G., P. Schroeder-Heister, and W. H. Ehrenstein (1990). Spatial coding and spatio-anatomical mapping: evidence for a hierarchical model of spatial stimulus-response compatibility. In: *Stimulus-Response Compatibility: An Integrated Perspective*. Ed. by R. W. Proctor and T. G. Reeve. Amsterdam: North-Holland, 117–143. DOI: 10.1016/S0166-4115(08)61220-0.

Heister, G. and P. Schroeder-Heister (1994). Spatial S-R compatibility: positional instruction vs. compatibility instruction. *Acta Psychologica* 85, 15–24. DOI: 10.1016/0001-6918(94)90017-5.

(C) Collections edited or initiated

Wolters, G. and P. Schroeder-Heister, eds. (1979). *Der wissenschaftliche Nachlaß von Hugo Dingler (1881-1954). Verzeichnis mit einer Bibliographie der Schriften Dinglers*. Universität Konstanz. Digital Edition in: Hugo Dingler. Gesammelte Werke auf CD-ROM. Ed. by U. Weiß. Karsten Worm InfoSoftWare: Berlin 2004, Appendix, Chapters 1986 and 1987. Republished online Universität Tübingen 2022. DOI: 10.15496/publikation-69429.

Schroeder-Heister, P., ed. (1990). *Proceedings of the SNS Logic Colloquium, March 1990*. SNS-Bericht 90-58. Universität Tübingen. DOI: 10.15496/publikation-69809.

– ed. (1991b). *Extensions of Logic Programming: International Workshop, Tübingen, FRG, December 8-10, 1989. Proceedings*. Berlin: Springer. DOI: 10.1007/BFb0038689.

Eriksson, L.-H., L. Hallnäs, and P. Schroeder-Heister, eds. (1992). *Extensions of Logic Programming: Second International Workshop, ELP'91, Stockholm, Sweden, January 27-29, 1991. Proceedings*. Berlin: Springer. DOI: 10.1007/BFb0013601.

Došen, K. and P. Schroeder-Heister, eds. (1993). *Substructural Logics*. Oxford University Press.

Lamma, E. and P. Mello, eds. (1993). *Extensions of Logic Programming: Third International Workshop, ELP'92, Bologna, Italy, February 26-28, 1992. Proceedings.* Berlin: Springer. DOI: 10.1007/3-540-56454-3.

Dyckhoff, R., ed. (1994). *Extensions of Logic Programming: 4th International Workshop, ELP'93, St Andrews, UK, March 29–April 1, 1993. Proceedings.* Berlin: Springer. DOI: 10.1007/3-540-58025-5.

Dyckhoff, R., H. Herre, and P. Schroeder-Heister, eds. (1996). *Extensions of Logic Programming: 5th International Workshop, ELP'96, Leipzig, Germany, March 28-30, 1996. Proceedings.* Berlin: Springer. DOI: 10.1007/3-540-60983-0.

Kahle, R., P. Schroeder-Heister, and R. Stärk, eds. (2001). *Proof Theory in Computer Science: International Seminar, PTCS 2001, Dagstuhl Castle, Germany, October 7-12, 2001. Proceedings.* Berlin: Springer. DOI: 10.1007/3-540-45504-3.

Spohn, W., P. Schroeder-Heister, and E. J. Olsson, eds. (2005). *Logik in der Philosophie.* Heidelberg: Synchron Wissenschaftsverlag.

Kahle, R. and P. Schroeder-Heister, eds. (2006). *Proof-Theoretic Semantics.* Special issue of *Synthese* (Vol. 148 No. 3). DOI: 10.1007/s11229-004-6292-5.

Bour, P. E., G. Heinzmann, W. Hodges, and P. Schroeder-Heister, eds. (2014–2015). *Logic and Philosophy of Science in Nancy (I-II). Selected Contributed Papers from the 14th International Congress of Logic, Methodology and Philosophy of Science.* (I) *Philosophia Scientiae*, Vol. 18 No. 3 (DOI: 10.4000/philosophiascientiae.957), (II) Vol. 19 No. 1 (DOI: 10.4000/philosophiascientiae.1027).

Schroeder-Heister, P., G. Heinzmann, W. Hodges, and P. E. Bour, eds. (2014). *Logic, Methodology and Philosophy of Science: Proceedings of the 14th International Congress (Nancy). Logic and Science Facing the New Technologies.* College Publications. DOI: 10.15496/publikation-73440.

Piecha, T. and P. Schroeder-Heister, eds. (2015b). *Proceedings of the Conference on Hypothetical Reasoning, 23–24 August 2014, Tübingen.* DOI: 10.15496/publikation-415.

– eds. (2016a). *Advances in Proof-Theoretic Semantics.* Berlin: Springer. DOI: 10.1007/978-3-319-22686-6.

– eds. (2016c). *General Proof Theory: Celebrating 50 Years of Dag Prawitz's "Natural Deduction". Proceedings of the Conference held in Tübingen, 27-29 November 2015.* DOI: 10.15496/publikation-10394.

– eds. (2018b). *Proof Theory. Proceedings of the Workshop held at Unilog'2018 in Vichy, 25 June 2018.* DOI: 10.15496/publikation-25225.

– eds. (2019a). *General Proof Theory.* Special issue of *Studia Logica* (Vol. 107 No. 1). DOI: 10.1007/s11225-018-9818-4.

– eds. (2019d). *Proof-Theoretic Semantics: Assessment and Future Perspectives. Proceedings of the Third Tübingen Conference on Proof-Theoretic Semantics, 27–30 March 2019.* DOI: 10.15496/publikation-35319.

Binder, D., T. Piecha, and P. Schroeder-Heister, eds. (2022b). *The Logical Writings of Karl Popper.* Berlin: Springer. DOI: 10.1007/978-3-030-94926-6.

(D) Dissertations supervised, selected dissertations examined

Keronen, S. R. (1991). *Computational Natural Deduction*. Australian National University, Canberra: Doctoral Dissertation [Supervised by Robin Stanton]. DOI: 10.25911/5d78dc1842b02.

Mortari, C. A. (1991). *Issues in Epistemic and Modal Logics and their Applications*. University of Tübingen: Doctoral Dissertation [Supervised by Walter Hoering]. URL: https://philpapers.org/archive/MORIIE.pdf.

Soldati, G. (1991). *Bedeutung und psychischer Gehalt: Eine Untersuchung zur sprachanalytischen Kritik von Husserls früher Phänomenologie*. University of Tübingen: Doctoral Dissertation [Supervised by Manfred Frank]. Published (rev.) Paderborn: Schöningh, 1994.

Eriksson, L.-H. (1993). *Finitary Partial Inductive Definitions and General Logic*. Royal Institute of Technology, Stockholm: Doctoral Dissertation [Supervised by Lars Hallnäs].

Oestermeier, U. (1993). *Bildliches und logisches Denken. Eine Kritik der Computertheorie des Geistes*. University of Tübingen: Doctoral Dissertation. Published Wiesbaden: Deutscher Universitätsverlag.

Zimmermann, E. (1995). *Spezielle Systeme der Modallogik, deren relationale Semantik und philosophische Deutung*. University of Tübingen: Doctoral Dissertation [Supervised by Walter Hoering].

Widebäck, F. (2001). *Identity of Proofs*. University of Stockholm: Doctoral Dissertation [Supervised by Dag Prawitz]. Published Stockholm: Almqvist & Wiksell.

Falkman, G. (2003). *Issues in Structured Knowledge Representation: A Definitional Approach with Application to Case-Based Reasoning and Medical Informatics*. Chalmers University of Technology and Göteborg University: Doctoral Dissertation [Supervised by Lars Hallnäs]. URL: https://research.chalmers.se/en/publication/?id=369.

Bean, J. M. L. (2005). *Ribbon Proofs: A Proof System for the Logic of Bunched Implications*. Queen Mary and Westfield College: Doctoral Dissertation [Supervised by David Pym and Edmund Robinson]. URL: http://www.eecs.qmul.ac.uk/tech_reports/RR-06-01.pdf.

Więckowski, B. (2006). *Modality without Reference: An Alternative Semantics for Substitutional Quantified Modal Logic and its Philosophical Significance*. University of Tübingen: Doctoral Dissertation. URL: https://hdl.handle.net/10900/46279.

Arndt, M. (2008). *Logical Tomography: Exposing the Structural Constituents of Logic*. University of Tübingen: Doctoral Dissertation. URL: https://hdl.handle.net/10900/49337.

Tranchini, L. (2010). *Proof and Truth: An Anti-Realist Perspective*. University of Siena and University of Tübingen: Doctoral Dissertation [Joint Siena-Tübingen degree. Siena supervisor: Gabriele Usberti]. URL: https://hdl.handle.net/10900/46853. Published (rev.) Pisa: Edizioni ETS, 2013.

Lüdecke, R. (2012). *Infinite-Valued Least Model and Game Semantics for Formula-Based and Normal Logic Programs*. University of Tübingen: Doctoral Dissertation. Published München: Verlag Dr. Hut.

Piecha, T. (2012). *Formal Dialogue Semantics for Definitional Reasoning and Implications as Rules*. University of Tübingen: Doctoral Dissertation. URL: https://hdl.handle.net/10900/49704.

Yim, A. (2012). *On Galois Correspondences in Formal Logic*. University of Oxford: Doctoral Dissertation [Supervised by Jochen Koenigsmann]. URL: https://ora.ouls.ox.ac.uk/objects/uuid:b47d1dda-8186-4c81-876c-359409f45b97.

Maurer, H. (2014). *Integrative (Synchronisations-)Mechanismen der (Neuro-)Kognition vor dem Hintergrund des (Neo-)Konnektionismus, der Theorie der nichtlinearen dynamischen Systeme, der Informationstheorie und des Selbstorganisationsparadigmas*. University of Tübingen: Doctoral Dissertation. Published Norderstedt: Books on Demand.

Spahn, S. A. (2018). *Axiomatizations of Compositional Inductive-Recursive Definitions*. University of Tübingen: Doctoral Dissertation [Supervised by Neil Ghani]. DOI: 10.15496/publikation-26071.

Oliveira, H. (2019). *Inference Rules and the Meaning of the Logical Constants: An Elaboration of Dummett's Notions of Proof-Theoretic Validity*. University of Tübingen: Doctoral Dissertation. DOI: 10.15496/publikation-27634.

Piccolomini d'Aragona, A. (2019). *Dag Prawitz's Theory of Grounds*. Aix-Marseille University and Sapienza University of Rome: Doctoral Dissertation [Joint Aix-Marseille-Rome degree. Supervisors: Gabriella Crocco and Cesare Cozzo]. URL: https://hal.archives-ouvertes.fr/tel-02482320.

Rezende de Castro Alves, T. (2019). *Synonymy and Identity of Proofs: A Philosophical Essay*. University of Tübingen: Doctoral Dissertation. DOI: 10.15496/publikation-32950.

Gazzari, R. (2020). *Formal Theories of Occurrences and Substitutions*. University of Tübingen: Doctoral Dissertation. DOI: 10.15496/publikation-47553.

Lolić, A. (2020). *Automated Proof Analysis by CERES*. Technical University Vienna: Doctoral Dissertation [Supervised by Alexander Leitsch and Matthias Baaz]. DOI: 10.34726/hss.2020.47184.

Gheorghiu, A. (2023). *Investigations into Semantics in Reductive Logic*. University College London: Doctoral Dissertation [Supervised by David Pym].

(R) Reviews

Schroeder, P. (1981). Review of: R. H. Wettstein, Eine Gegenstandstheorie der Wahrheit. Argumentativ-rekonstruierender Aktualisierungs- und Erweiterungsversuch von Kants kritischer Theorie (Königstein/Ts.: Forum Academicum 1980). *Dialectica* 35, 361–362. DOI: 10.1111/j.1746-8361.1981.tb00789.x.

Schroeder-Heister, P. (1983b). Review of: M. D. Resnik, Frege and the Philosophy of Mathematics (Ithaca/London: Cornell University Press 1980). *History and Philosophy of Logic* 4, 99–102. DOI: 10.1080/01445348308837049.
– (1984e). Review of: E.-H. W. Kluge, The Metaphysics of Gottlob Frege: An Essay in Ontological Reconstruction (The Hague/Boston/London: Martinus Nijhoff 1980). *History and Philosophy of Logic* 5, 241–245. DOI: 10.1080/01445348408837073.
– (1984f). Review of: G. Currie, Frege: An Introduction to His Philosophy (Brighton: Harvester Press; Totowa, NJ: Barnes & Noble 1982). *History and Philosophy of Logic* 5, 239–241. DOI: 10.1080/01445348408837073.
Friedrichsdorf, U. and P. Schroeder-Heister (1985). Review of: G. E. Hughes & M. J. Cresswell, A Companion to Modal Logic (London: Methuen 1984). *Journal of Semantics* 4, 389–390. DOI: 10.1093/jos/4.4.389.
Schroeder-Heister, P. (1986). Essay review of: G. Frege, Collected Papers on Mathematics, Logic, and Philosophy. Ed. by B. McGuinness. Trans. by M. Black, V. H. Dudman, P. Geach, H. Kaal, E.-H. W. Kluge, B. McGuinness and R. H. Stoothoff (Oxford/New York: Blackwell1984). *History and Philosophy of Logic* 7, 187–193. DOI: 10.1080/01445348608837101.
– (1988–1994). Reviews no. MR0869008, MR0875532, MR0877309, MR0902364, MR0920132, MR0934575, MR0990559, MR1029453, MR1172519. In: *Mathematical Reviews*. American Mathematical Society. URL: https://mathscinet.ams.org/mathscinet.
– (1989b). Review of: W. Schüler, Grundlegungen der Mathematik in transzendentaler Kritik. Frege und Hilbert (Hamburg: Meiner 1983). *Journal of Symbolic Logic* 54, 622. DOI: 10.2307/2274880.

(E) Encyclopedia articles

Schroeder-Heister, P. (1980–2018). Ca. 220 entries. In: *Enzyklopädie Philosophie und Wissenschaftstheorie. Vol. I–IV (1980–1996). 2nd Edition: Vol. I–VIII (2005–2018).* Ed. by J. Mittelstraß. Stuttgart: Metzler.
– (1992c). Schließen, natürliches. In: *Historisches Wörterbuch der Philosophie, Vol. 8*. Ed. by J. Ritter, K. Gründer, and G. Gabriel. Basel, 1300–1303. DOI: 10.24894/HWPh.3706.
– (1995a). Popper, Karl Raimund. In: *Enzyklopädie Philosophie und Wissenschaftstheorie. Vol. 3*. Ed. by J. Mittelstraß. Stuttgart: Metzler, 289–296. *2nd Edition: Vol. 6*, 2016, 369–377.
– (1995b). Selbstreferenz (I. Logik). In: *Historisches Wörterbuch der Philosophie, Vol. 9*. Ed. by J. Ritter, K. Gründer, and G. Gabriel. Basel, 515–516. DOI: 10.24894/HWPh.5435.
Herre, H. and P. Schroeder-Heister (1998). Formal languages and systems. In: *Routledge Encyclopedia of Philosophy*. London: Routledge. DOI: 10.4324/9780415249126-Y011-1.

Schroeder-Heister, P. (2001a). Popper, Sir Karl Raimund. In: *Neue Deutsche Biographie, Vol. 20*. Berlin: Duncker & Humblot, 625–628. URL: https://www.deutsche-biographie.de/pnd118595830.html#ndbcontent.

– (2005b). Dummett, Michael Anthony Eardley. In: *Enzyklopädie Philosophie und Wissenschaftstheorie. 2nd Edition, Vol. 2*. Ed. by J. Mittelstraß. Stuttgart: Metzler, 258–262.

– (2012c). Proof-Theoretic Semantics. In: *Stanford Encyclopedia of Philosophy*. Ed. by E. N. Zalta. Metaphysics Research Lab, Stanford University. URL: https://plato.stanford.edu/entries/proof-theoretic-semantics/. Revised 2018 and 2023.

– (2015b). Popper, Karl Raimund (1902–94). In: *International Encyclopedia of the Social & Behavioral Sciences*. Ed. by J. D. Wright. Second Edition. Vol. 18. Oxford: Elsevier, 522–527. DOI: 10.1016/B978-0-08-097086-8.61109-1. First Edition 2001.

(M) Miscellaneous including selected unpublished manuscripts

Hallnäs, L. and P. Schroeder-Heister (1992a). Girard's "fixpoint theorem". In: *Linear Logic Mailing List (Included in TYPES Forum)*. DOI: 10.15496/publikation-70816. URL: https://www.seas.upenn.edu/~sweirich/types/archive/1992/msg00043.html.

Schroeder-Heister, P. (1998a). Ein Logik-Lotse geht von Bord. Zur Verabschiedung des Naturwissenschaftlers und Philosophen Walter Hoering. In: *Schwäbisches Tagblatt, 4.12.1998*. Tübingen, 27. DOI: 10.15496/publikation-68456.

– (2001b). Prof. Dr. Walter Felscher (Obituary). *Tübinger Universitätsnachrichten, Vol. 21, No. 100, 9 April 2001*. DOI: 10.15496/publikation-68455.

– (2003b). *Proof-Theoretic Semantics: Some Basic Ideas (Manuscript)*. Published online together with the manuscript "Proof-Theoretic and Constructive Consequence" (2003). University of Tübingen. DOI: 10.15496/publikation-72548.

– (2006c). Prof. Dr. Gisbert Hasenjaeger (1.6.1919–2.9.2006). In: *Chronik und Bericht der Rheinischen Friedrich-Wilhelms-Universität Bonn über das Akademische Jahr 2005/2006*. University of Bonn, 57–59. URL: https://nbn-resolving.de/urn:nbn:de:hbz:5:2-251940.

– (2008c). The categorical and the hypothetical: an inferentialist critique of the transmission view of consequence. Draft submitted for a special issue of *Erkenntnis* on inferentialism that did not materialize. DOI: 10.15496/publikation-70814.

– (2011c). *Proof-Theoretic Semantics (Manuscript)*. University of Tübingen. DOI: 10.15496/publikation-72550.

Hallnäs, L. and P. Schroeder-Heister (2016). General proof theory and definitional reflection. Notes dedicated to Dag Prawitz on his 80th birthday. DOI: 10.15496/publikation-70815.

Heidelberger, M. and P. Schroeder-Heister (2019). *Walter Hoering (1933–2019). Grabrede und Nachruf*. Universität Tübingen. DOI: 10.15496/publikation-69428.

Schroeder-Heister, P. (2019a). Die Logik in allen ihren Aspekten. Zum Tode von Professor Dr. Dr. Walter Hoering (Obituary). *Newsletter Uni Tübingen aktuell No. 4/2019*. DOI: 10.15496/publikation-69428.

– (2019b). Kosta Došen (1954–2017). *Studia Logica* 107. Special issue *General Proof Theory*. Ed. by T. Piecha and P. Schroeder-Heister, 7–9. DOI: 10.1007/s11225-018-9814-8.

Popper, K. R. and P. Schroeder-Heister (2022). Popper's correspondence with Peter Schroeder-Heister. In: *The Logical Writings of Karl Popper*. Ed. by D. Binder, T. Piecha, and P. Schroeder-Heister. Berlin: Springer, 509–522. DOI: 10.1007/978-3-030-94926-6_32.

Schroeder-Heister, P. (2022c). *Kosta Došen 1954–2017. Obituaries*. Universität Tübingen. DOI: 10.15496/publikation-72328.

– (2022d). Pandemic ponderings: reflections of a pencil and paper academic. In: *Pandemic Ponderings: Fellow Reflections beyond Corona*. Ed. by C. Garsten and S. M. Rekanovic. Uppsala: Swedish Collegium for Advanced Study (SCAS), 25–31.

Grundlagen der Arithmetik, §17: Part 1. Frege's Anticipation of the Deduction Theorem*

Göran Sundholm

Abstract A running commentary is offered on the first half of Frege's *Grundlagen der Arithmetik*, §17, and suggests that Frege anticipated the method of demonstration used by Paul Bernays for the Deduction Theorem.

The Natural Deduction community owes Peter Schroeder-Heister a large debt of gratitude, not only for his seminal extension of the Natural Deduction techniques to higher-level derivations, and as the champion advocate of Proof-Theoretic Semantics.[1] He has also been a tireless and successful organizer of successful workshops and conferences at Tübingen that, over the past quarter of a century, have served to put Natural Deduction firmly on the map as a viable alternative for proof-theoretic bookkeeping. It is a pleasure to contribute to a Profile volume devoted to his work, and in his honour. However, by the side of his Natural Deduction expertise, Peter Schroeder-Heister has also followed other lines of research; thus he has also dealt

Göran Sundholm
Institute for Philosophy, Leiden University, The Netherlands, e-mail: goran.sundholm@gmail.com

* I have spoken on this material at the workshop *Mathematics and its Philosophy between the 18th and the 19th Century*, Amsterdam, July 13–14, 2018, and am grateful to Jamie Tappenden for insightful comments. I am also indebted to Dr. Joan Bertran-San Millan, of the Czech Academy of Science in Prague for helpful conversations on Frege's *Begriffsschrift* during the *2nd Prague Workshop on Frege's Logic*, July 1–2, 2019 that he organised, and on other occasions, as well as for his generous help with derivations in the *Begriffsschrift* system. I am also indebted to Matthias Wille for a clarifying account of how Frege passed from a Leibnizian epistemological framework in BS to a Kantian one in GLA. In recent years my Leiden colleague Maria van der Schaar has been a constant sparring partner on all things Fregean. The two referee reports that were offered to me by the Editors were most helpful and I am obliged to the referees for their careful commentary.

[1] Schroeder-Heister's elegant extension of Natural Deduction was presented in (1984d), whereas the reference to (2018d) is self-explanatory.

T. Piecha and K. F. Wehmeier (eds.), *Peter Schroeder-Heister on Proof-Theoretic Semantics*, Outstanding Contributions to Logic 29,
https://doi.org/10.1007/978-3-031-50981-0_2

seriously with the technical, proof-theoretical aspects of Frege's logic, in which work he has found few colleagues and fewer peers.

The Editors of the present volume extended to me an invitation for a contribution on Proof-Theoretic Semantics, and one may reasonably expect that many contributions to the present volume will deal with topics and issues chosen from within that field. Accordingly a paper on a proof-theoretic theme in Frege might also be welcome, touching as it does on a longstanding topic of interest for Peter Schroeder-Heister.

1 How to read the "Frege conditional"?

Today, after customary predicate-logic standardization, the "Frege conditional"

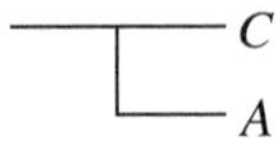

is read as an implication $\supset$ between propositions A and C, whereas an iterated Frege conditional $(*)$

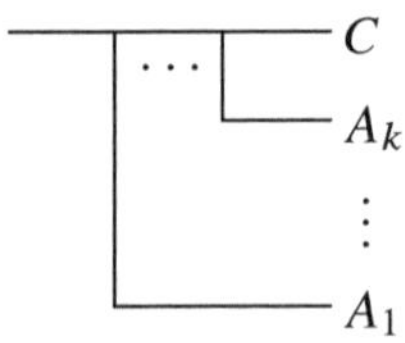

is accordingly read as the iterated implication

(i) $(A_1 \supset (A_2 \supset (\ldots \supset (A_k \supset C)\ldots)),$

or, after the fashion of Schütte (1951), as the tautologically equivalent formula

(ii) $\neg A_1 \vee \neg A_2 \vee \ldots \vee \neg A_k \vee C.$

Such Schütte disjunctions correspond exactly to Gentzen sequents,

(iii) $A_1, A_2, \ldots, A_k \Rightarrow C,$

where, in Schütte's (ii), the negated A's match the antecedent formulae of Gentzen's sequent and C, of course, is the succedent of the sequent (where more succedent side formulae are possible, as well). That the iterated Frege implication (*) could also alternatively be read as a Gentzen sequent was made implicitly clear already in Schütte (1951), where reformulations of Gentzen's sequent calculi, and their syntactic cut-elimination theorems, are given using the representation (i) for the *intuitionistic* calculus, and (ii) for the classical version. This representation of classical sequents was further refined by Tait (1968), where instead of treating sequents as disjunctive propositions, they are now cast as disjunctively read sets of propositions.

The sequent rendering of iterated Frege implications was, to the best of my knowledge, explicitly formulated first by the late Pavel Tichý (1988a, pp. 248–252), but also Franz von Kutschera (1996a) and Peter Schroeder-Heister (1997c; 1999b;

2002c; 2014b) have stressed the fact. Tichý made the connection especially clear also graphically by rotating the Frege conditional (∗) 90° clockwise, while altering the notation slightly, thereby producing a familiar result, namely the sequent (iii)

$$A_1, A_2, \ldots, A_k \Rightarrow C.$$

Tichý, and especially Peter Schroeder-Heister, further noted that the parallel between iterated Frege implications and Gentzen sequents works well, also down to quite a fine-grained level, regarding rules, axioms, and notations.[2] Schroeder-Heister has also studied Frege's permutation argument from *Grundgesetze*, §10, jointly with Kai Wehmeier. His writings on Frege are listed in the references.

2 Fact and condition

I take as my text the first half of *Grundlagen der Arithmetik*, §17:

> Statt eine Schlussreihe unmittelbar an eine Thatsache anzuknüpfen, kann man, diese dahingestellt sein lassend, ihren Inhalt als Bedingung mitführen. Indem man so alle Thatsachen in einer Gedankenreihe durch Bedingungen ersetzt, wird man das Ergebnis in der Form erhalten, dass von einer Reihe von Bedingungen ein Erfolg abhängig gemacht ist. Diese Wahrheit wäre durch Denken allein, oder, um mit Mill zu reden, durch kunstfertiges Handhaben der Sprache begründet. Es ist nicht unmöglich, dass die Zahlgesetze von dieser Art sind. Sie wären dann analytische Urtheile, obwohl sie nicht durch Denken allein gefunden zu sein brauchten; denn nicht die Weise des Findens kommt hier in Betracht, sondern die Art der Beweisgründe . . .[3]

Here we must consider the proper interpretation of *Thatsache* and of *Bedingung*. The importance of *Thatsache* — "fact" — in this context goes back to Frege's

[2] Per Martin-Löf (in conversation) was convinced that Gentzen knew Frege's writings. He pointed to the matching rules, the naming of rules and axioms, and the notations, especially Gentzen's use of capital Greek letters. In the absence of a direct citation, he considered the available indications about as strong evidence as one could expect to have for claiming that Gentzen knew Frege's *Grundgesetze*.

[3] Instead of linking our chain of deductions directly to any matter of fact, we can leave the fact on one side, while adopting its content in the form of a condition. By substituting in this way conditions for facts throughout the whole of a train of reasoning, we shall finally reduce it to a form in which a certain result is made dependent on a certain series of conditions. This truth would be established by thought alone or, to use Mill's expression, by an artful manipulation of language. It is not impossible that the laws of number are of this type. This would make them analytic judgements, despite the fact that they would not normally be discovered by thought alone; for we are concerned here not with the way in which they are discovered but with the kind of ground on which their proofs rest . . .

English translations of Frege's texts, as a rule, are of *very* uneven quality, though the recent scholarly translation of *Grundgesetze* makes for a pleasant exception here. The horror story of the many diverging renderings of *beurteilbarer Inhalt* in *The Collected Papers* is as discouraging an example as can be. Thus, against this background of inferior translations, I prefer to quote Frege in German. Those who wish to read Frege's *Grundlagen* in English can readily do so in the omnipresent Austin translation, and may benefit from Schirn's (2010) examination of the more recent effort by the late Dale Jacquette (2007).

characterization of analytic and synthetic judgements in GLA, §3 that we shall have occasion to discuss in some detail. A *Thatsache* is not general, or "lawlike", but is particularized, with regard to a specific individual or object: it is an *Einzelthatsache*. This distinction between fact and (general) laws is common to the tradition of demonstrative knowledge, from Aristotle's *Analytica Posterioria*, Bk I, Chapter 4, onwards, where we do not have demonstrative, or "scientific", knowledge of individual facts concerning particulars. Frege, no doubt, was familiar with this tradition, from the *Posterior Analytics*, or from Trendelenburg's *Erläuterungen*, a work that most likely was put into his hands by Rudolf Eucken, his mentor at Jena, and later also neighbour across the Forstweg. Furthermore, a *Thatsache* is an item of knowledge and it will be asserted, rather than merely assumed.

Concerning *Bedingung* we take note of a little observed fact regarding the *implication* $(A \supset B)$ between propositions A and B. It is customary to render this implication into the vernacular as the *conditional*(ization) "If A, then B" or "B, on condition that A". Today, however, the customary vernacular rendering of propositions turn them *by definition* into (referents of) *that clauses*: that snow is white, that grass is green, etc. ...[4] The *implication* accepts that-clauses *salva significatione*. When A and B are taken to be that-clauses, putting them on either side of *implies* works. For instance, "that snow is white *implies* that grass is green" is grammatically well formed. *Conditionalization* does not work, however, with that-clauses: "If that snow is white, then that grass is green" is nonsensical and contains too many main verbs. Here we need complete sentences: "if snow is white, *then* grass is green", or "snow is white, on *condition* that grass is green", both work perfectly well. If we wish to deploy the same propositions as above here, the conditional will have to be: "*If* A is true, *then* B is true", or "B is true, on *condition* that A is true", for instance, "that snow is white is true, on *condition* that grass is green is true".

Accordingly, *implication* and *conditional* are not the same. We have the proposition $A \supset B$ and the conditional judgement "If A is true, then B is true", or equivalently, "B is true, on condition that A is true". We may also follow Gentzen, and allow also relations of *consequence* among propositions, in the form of "sequents", German *Sequenzen*, such as $A \Rightarrow B$, to be read: "B follows from A", or, equivalently: "B is a consequence of A".

Propositions, and thus also the implication $(A \supset B)$, may be held to be true. Then "$(A \supset B)$ is true" is a judgement, as is "$[A \Rightarrow B]$ holds": consequence holds from A to B. Similarly, the conditional "If A is true, then B is true" is not a proposition, but a conditional judgement in which truth under the condition that proposition A is true, is ascribed to proposition B: "B is true, on condition that A is true".

The three judgements, of respectively, truth of the implication proposition, conditional — "hypothetical" — truth of a proposition, and holding of consequence,

[4] Indeed, both Frege and Bolzano avail themselves of an *independent* notion of proposition, known as a (Fregean) *Gedanke* or (Bolzanian) *Satz an sich*. Frege's and Bolzano's independent "Platonic" notions of proposition are available in their respective ontologies and are made to serve as referents of *that clauses*, but they are not defined by that meaning-theoretical role. The customary current notion of proposition, on the other hand, is not language-independent, like those of Frege and Bolzano, but is *defined* as that for which a *that clause* stands.

(i) $(A \supset B)$ is true;
(ii) If A is true, then B is true, or, in synonymous formulations,
B is true, on condition (under the assumption) that A is true,
B is an *open consequence* of A;
(iii) $[A \Rightarrow B]$ holds,
the *closed consequence* from A to B holds;

are different.[5] They have different meanings, but are *equiassertible*: when the assertion condition for one of these judgements is fulfilled by means of a certain demonstration, so that one is entitled to make it, then so may the other two readily be fulfilled as well. We should note that the judgement (iii) is a higher-level judgement of the kind that Peter Schroeder-Heister has introduced: it expresses the rule that we may pass to the judgement "B is true" from the judgement "A is true".[6] Frege's terminological choices in the *Begriffsschrift*, to wit *Bedingung* and *Bedingungsstrich*, make it clear that also the conditional is a viable alternative reading of what today is known as the "Frege implication".

3 A problem of GLA, §17

After these prefatory observations, I now return to the quoted text from GLA, §17. Instead of linking a chain of inferences directly to an (individual) fact, Frege claims, one may instead consider the implication $\vdash E \supset C$, where $\vdash C$ is the conclusion of an inference drawn to $\vdash C$, from the fact $\vdash E$ in question, because we may let $\vdash E$, that is the assertion that the fact E holds, remain standing and use its content as a condition (*Bedingung*) in the implication $E \supset C$. The conclusion $\vdash C$, after all, can be readily obtained by modus ponens from the premises $\vdash E$ and $\vdash E \supset C$. If we carry out this transformation repeatedly, inserting antecedent conditions with respect to all the facts in a deduction, we obtain an implication

$$(*) \qquad (E_1 \supset (E_2 \supset (\ldots \supset (E_k \supset C)\ldots)).$$

The demonstration of C that draws upon the facts $E_i, i = 1, \ldots, k$, can then be cast in a "normal" form:

$$\dfrac{\dfrac{\mathcal{D}}{\vdash (E_1 \supset (E_2 \supset (\ldots \supset (E_k \supset C)\ldots))} \qquad \vdash E_1, \vdash E_2, \ldots, \vdash E_k}{\dfrac{k \text{ applications of modus ponens, each time dropping one fact } E_i}{\vdash C}}$$

where $\mathcal{D}$ is a demonstration of the iterated implication with the individual fact-contents as antecedents, and the other premises are assertions of the facts $E_1, E_2, \ldots, E_k$. Of

[5] The notions open and closed consequences, and the difference between them, are explained in my (1997, §7, pp. 205–209).

[6] In his dissertation (1981c) and the JSL article (1984d).

the implication (∗), which is obtained through the repeated additions of (singular) fact-contents as *Bedingungen* (conditions), that is, as antecedents of implications, since $\vdash C$, of course, we get $\vdash E \supset C$ as a theorem by an application of modus ponens from the *Begriffsschrift* axiom $\vdash C \supset (E \supset C)$, and by repetition, also the implication (∗). We see that (∗) is trivially demonstrable, but Frege claims that it is *analytic*, that is, analytic in *his* own sense that was introduced in GLA §3, rather than that of Kant. Demonstrability is plain to view, since $\vdash C$ is demonstrable, but analyticity, under whatever characterization of that notion, is not. That the implication in question is analytic, however, is not plain to view, under whatever characterization that we might use for analytic.

Frege's *Grundlagen*, §17, has met with astonishingly little commentary. I am aware of only three discussions. In his *Grundlagen* book Michael Dummett (1991c, p. 70) mentions "*the problem of the value of analytic judgements . . .*":

> Frege . . . contents himself, with pointing out that it coincides with that of the fruitfulness of deductive reasoning, since we can always transform a piece of reasoning into an analytic truth by framing the conditional whose antecedent is the conjunction of the premises and whose consequent is the conclusion.

The problem of the fruitfulness of deductive reasoning is also known as the "*Paradox of Inference*" that results from the tension between the usefulness and the validity of an inference. The validity of an inference appears to presuppose that the conclusion is *already* "epistemically contained" in the premises, whereas its fruitfulness — its utility, if you wish — seems to demand that the conclusion *does go beyond* what is contained in the premises. Dummett dealt *in extenso* with this vexing issue in his 1973 British Academy Lecture on *The Justification of Deduction*. In the quotation above, he does not solve it, though. In particular, Dummett offers no reason as to *why* the transformation works, no explanation as to why its final result should be *analytic*, nor does he explain why the problem of the value of analytic judgements coincides with the Paradox of Inference.

The components of a *Fregean* "piece of reasoning", that is, I take it, an *inference*, are judgements that carry assertoric force: every inferential component in sight carries an assertion sign that is composed out of a content stroke and a judgement stroke. They are not propositions, whence they *cannot* be conjoined into a conjunction the way Dummett wishes: the conjuncts of a conjunction(-proposition) have to be *propositions* and not (asserted) premise-judgements. Similarly, the consequent of an implicational "horseshoe" (⊃) proposition has to be a proposition, but not the asserted conclusion-judgement. The idea of drawing conclusions, from unasserted premises, that is, from "*open assumptions*" in the sense of Natural Deduction, to an unasserted conclusion, was anathema to Frege. Michael Dummett routinely held to the maxim: *Frege was (always) right*. There are, I think, only two major points where, according to Dummett, Frege was not absolutely right, and the first, which need not detain us here, concerns truth-values in their role as objectual *Bedeutungen* of sentences.

The second point where Frege, according to Dummett, did not get it right was his construal of inference, and it took the genius of Gentzen to find the proper way to deal with this:

> Frege's account of inference allows no place for a[n] . . . act of supposition. Gentzen later had the highly successful idea of formalizing inference so as to leave a place for the introduction of hypotheses . . .

Indeed,

> [i]t can be said of Gentzen that it was he who showed how proof theory should be done.[7]

In this vein, Dummett uses the format of Natural Deduction from Gentzen's 1932 doctoral dissertation to criticise Frege. However, Gentzen's system, when read strictly according to the letter of the Hilbertian proof-theoretical school, is an uninterpreted one, and, for a fair comparison with Frege, it should be supplied both with meaning explanations for the primitive vocabulary, say, the arithmetical one, and signs that serve as force indicators, such as $\vdash$, that clarify the pragmatic dimension of his system.[8] Already in my inaugural lecture (1988) at Leiden University I claimed that Bolzano (1837), the Tractarian Wittgenstein (1921), Tarski (1936), and Quine (1950), adopt what I (later came to) call the "*second Bolzano reduction*" of the epistemic notion *validity of inference* (from judgement to judgement) to the ontological — "alethic" — notion of (logical) consequence (between antecedent propositions and a consequent proposition). Subsequently, in (2006b) (that was presented at the 1999 Tübingen workshop on Proof-Theoretic Semantics, and published in its Proceedings) as well as in the survey article (2009), I came to hold that Frege was substantially right in his views on inference: the criticisms of his views on inference arise from taking him, via the Bolzano reduction, to be concerned with (logical) consequence among propositions, rather than with inference from judgement to judgement.[9] *A* fair comparison between Frege and Gentzen will have to make explicit also the pragmatic level and incorporate it into the Gentzen system, and it will force us, I argued in (2006b), to deploy Natural Deduction in its *sequential form* from that was introduced by Gentzen (1936).

Apart from Dummett's I have been able to find only two further comments on *Grundlagen*, §17. The first was made by Ian Proops (2002, p. 293) in his article "The *Tractatus* on Inference and Entailment":

> So when we have a derivation of $p \supset q$ from basic logical laws, we have a case when the chain of deductions beginning with p and ending with q has been "reduced to the form in which" q is dependent on p. But since Frege is here in effect (tacitly) assuming one direction of the deduction theorem: (viz. if $p \vdash q$, then $\vdash p \supset q$), this means that for him q is dependent on p whenever q is derivable from p. Moreover, if we regard Frege here as envisaging the process of "substituting conditions for facts and so on to be reversible (and so as in effect tacitly committed to both directions of the deduction theorem), we shall have a

---

[7] Dummett (1981b, p. 309, and p. 435, respectively), where also several textual references to Frege for his rejection of unasserted, merely assumed premises of inferences can be found.

[8] In (2006b), that is, my contribution to the first workshop on Proof-Theoretic Semantics at Tübingen, 1999, I spell out this perspective on Gentzen's (1933a) framework in considerable detail and offer a comparison with his re-casting of it as a sequential version of Natural Deduction in his "First consistency" paper (1936).

[9] The articles (2009) and (2006b) were drafted in 1998, and presented at meetings in Helsinki and Tübingen shortly thereafter. Unfortunately both had to wait an inordinately long time to appear in print.

case of the notion of dependence being cashed in terms of derivability: q is dependent on p iff $\vdash p \supset q$ iff $p \vdash q$.[10]

The *Deduction Theorem* that Proops refers to is commonly credited to Herbrand (1930), and to Tarski (1930); *sub specie aeterni*, those attributions may be justified, but the well-known demonstration of the Deduction Theorem by means of the rearrangement of a concretely given derivation was first given by Paul Bernays in Hilbert and Bernays (1934a, pp. 157–158).[11] Be that as it may, an appeal to the Deduction Theorem in this context cannot be right. From a Fregean perspective Proops' formula "$p \vdash q$" does not make sense.[12] In Frege's work, as amply borne out by, e.g. BS, §6, inferences all have asserted (judgements as) premises, and such

[10] Proops' passage is part of a discussion of Frege's notion of dependence from the second series of articles on the *Grundlagen der Geometrie*. I have dealt with that Frege material at some length in connection with Bolzano in my (2000).

[11] Kleene (1952a, Ch. V, §§21–23, pp. 90–102) gives a meticulous exposition of Bernays' demonstration of the Deduction Theorem, as does Alonzo Church (1956b, Ch. III, §36, pp. 196–201). Kneebone (1963, Ch. 3, §5, pp. 78–86) follows Bernays' style of exposition closely, and unhurriedly explains the hurdles and pitfalls very well. The story behind the Deduction Theorem and the genesis of Natural Deduction is quite complex. Already in (1916) Lesniewski uses what amounts to the Deduction Theorem, or "*implication introduction*", in *informal* reasoning. Subsequently, in his Warsaw lectures, he used that mode of proceeding in formal derivations as well, while his student Tarski (1923; 1924a; 1924b) does so also in print. Tarski claims, in footnote † on p. 32 of the English translation of (1930) in Tarski (1956/1983), that in 1921 he established the Deduction Theorem for the formalism of *Principia Mathematica*, and he refers for a published proof to a location in a paper from 1933, translated at Tarski (1956/1983, p. 286). It should be noted that the footnote † occurs only in the English translation of (1930), but not in the original. The (1956/1983) edition(s) ought not to be seen as just a translation(s), since a number of its papers have undergone considerable editorial retouches. For serious historical work it is not reliable, and one is well advised always go back and check the text against the originals.

However, there are more Polish twists to our tale. Jaśkowski (1934c), in which article Natural Deduction appeared in print for the first time, simultaneously with Gentzen's paper, opens as follows:

> In 1926 Prof. J. Łukasiewicz called attention to the fact that mathematicians in their proofs do not appeal to the theses of the theory of deduction, but make use of other methods of reasoning. The chief means employed in their method is that of an arbitrary supposition.

Jaśkowski also reports that he had spoken on this method of deduction in Łukasiewicz's seminar, also in 1926, and elsewhere. Łukasiewicz's connection to Natural Deduction is quite close and goes back a long way. In 1910, as *Privatdozent*, he spent time at Graz when Meinong was preparing his much-enlarged second edition of *Über Annahmen*. In that work Meinong considers *Annahmeschlüsse* ("assumption inferences"). Is this the source that in 1926 prompted Łukasiewicz's suggestion, and through it, the Polish branch of Natural Deduction?

[12] Indeed, here in the quotation from Proops, the turnstile is used as a predicate for the holding of (syntactic) *consequence* between formulae p and q that here should be taken as meta-variables for wff's rather than as "sentence letters" in the propositional calculus. Frege's assertion sign ("content stroke plus judgement stroke") is applied to judgeable contents ("propositions") and will yield a judgement; Frege's *symbol*, but not its intended use, was retained by Barkley Rosser who turned it into a theorem predicate applicable to formal wff's. Kleene then extended this use to have also side-formulae to the left, when it has been turned into a predicate expressing syntactic consequence among formal wff's. Kleene (1952a, p. 88, p. 526) gives details and references.

a passage, from an unasserted formula p to an unasserted formula q, simply is not possible, and nor is it contemplated by Frege.

4 Ambiguity in logical notation

The confusion is made worse owing to the threefold ambiguity involved in our current logical notation within Natural Deduction. A formula A may serve in at least three different roles:

1. As a proposition, e.g. when it is a component of the complex proposition $A \supset B$.
2. As the asserted conclusion of a closed proof-tree, when it is an asserted theorem that a proposition is true, rather than a proposition tout court.
3. When in leaf-position in a Natural Deduction derivation-tree, the formula A stands for an (unasserted) assumption that a proposition is true.

Against this background we can understand that confusions may arise when one attempts to slide a contemporary modern Natural Deduction grid over the works of Frege. However, owing to the peculiar absence of meaning in the formalized object languages of Gentzen, the resulting ambiguities will not be acutely felt there. His languages are not designed for use, but are there to be spoken about. Thus, when it comes to rendering a Natural Deduction derivation into the vernacular, opening a deduction, one customarily says: "Assume A!" (with the intention of deriving B). This really introduces yet a fourth reading of the calculus letters, and one that is mainly concerned with manipulating ("metamathematical expressions", that is, a certain kind of) objects. But how, in the absence of meaning, can one "assume A"? A well-formed formula is a *thing*, a "metamathematical expression" that does not itself express, but gets expressed using an expression. Owing to the absence of meaning it is not something that can be assumed (except in this derived, "symbol manipulating" sense). It is an object, an entity, that may be discussed, but it does not say anything. One does not assume, nor assert, objects such as Paris, 14, or BBC.

The first of the three renderings above can also be ruled out: one does not assume a *proposition A*, but one may, as in (3), assume that *the proposition A is true*. However, one may also assume *that one knows, or is granted, that proposition A is true*. Assumptions of the kind (3) are not used in Frege's works, whereas the rules of inference all use *epistemic assumptions* that *beurtheilbare Inhalte* are known (while prefixed with assertion signs). The modern notion of a predicate-logic proposition is the inheritor of Frege's *beurtheilbarer Inhalt*, and in Fregean inferences they are prefixed by the *Inhaltsstrich-cum-Urtheilsstrich*. The premisses of Fregean inferences are presupposed to be known. In order to validate an inference, rather than establish that consequence, be it logical or not, holds from antecedent proposition(s) to consequent proposition, that is, the Bolzano method, one assumes that one is granted assertions of the premise judgements, and will then have to take responsibility for asserting the conclusion judgment:

When I say "*Therefore*" I give others my permission to assert the conclusion on my authority, provided that they grant me their authority for asserting the premisses.

This view on the validation of inference steps, rather than on the holding of consequence, renders Frege's insistence on inferring from (possibly hypothetically) *known* premisses understandable and right. Criticism of Frege's view of inference commonly takes him to be speaking about the modern notion of *logical consequence* (among propositions), that is, holding in all variations, after the fashion of Bolzano, Wittgenstein's *Tractatus*, and Tarski. That, however, is a notion that, just as logical truth, plays no role in Frege's writings.

Finally, Wolfgang Kienzler (2009, §4.6.3) gave a rich epistemological analysis of the background to Frege's *Grundlagen* (including its §17) in the German translation of Mill's *Logic*, but does not enter into the technicalities of Frege's analyticity claim. Kienzler deserves praise for thus drawing attention to the Schiel translation of Mill. It constitutes a source that has not yet been exhausted by current Frege research. For instance, Mill's discussion of facts (*Thatsachen*) is likely to provide some insight into Frege's use of that term in GLA.

5 "Frege analyticity"

In order to understand and evaluate GLA, §17 we need to consider what Frege means by *analytic*. In Frege (and in Kant), "analytic" is applied to judgements, and not to propositions, that is, Fregean *beurtheilbare Inhalte*. The *Grundlagen der Arithmetik* is not the only work where Frege makes use of analyticity. Both the earlier *Begriffsschrift*, as well as the later essay *Über Sinn und Bedeutung* from 1892, make explicit and acknowledged use of the *Kantian* notion, rather than his own notion of analyticity from *Grundlagen*.[13] The *Jäsche Logik*, to which Frege refers in the *Grundlagen*, is a likely source for his account of the Kantian notion of analyticity. However, only in *Grundlagen* does Frege make public use in print of his own notion of analyticity.

The modern (post-Quinean) discussion of analyticity takes it to be a classification of propositions, that is, of contents. An analytic proposition is there seen as one that is *logically true*, come what may, independently of what is the case, in or under, all "variations", or reduces to such a logically true proposition through definitions. Bolzano, Wittgenstein, Quine, and Tarski are well-known adherents to (versions of) this view. That, however, is not how Frege views matters, and Quine, who brought the *logical truth under substitutions of synonyms* concept to the fore, and, even more so, Paul Boghossian, who, as far as I know, very deliberately coined the unfortunate misnomer "*Frege analyticity*", have to carry the responsibility.

Frege, *Grundlagen der Arithmetik*, §3:

1. Jene Untersuchungen von apriori und aposteriori, synthetisch und analytisch betreffen . . . nicht den Inhalt des Urtheils, sondern die Berechtigung zur Urtheils-

[13] *Begriffsschrift* (1879, §8, pp. 14–15) and *Über Sinn und Bedeutung* (1892, p. 25, p. 50), but also in the letter to Stumpf (published as being to Marty) from 29 VIII 1882.

fällung. . . . Wenn man einen Satz in meinem Sinne a posteriori oder analytisch nennt so urtheilt man . . . darüber worauf im tiefsten Grunde die Berechtigung der Führwahrhaltens beruht.

(These investigations of apriori and aposteriori, synthetic and analytic, are not concerned with the content of the judgement (i.e. the Thought, the proposition G. S.), but with the right to judge . . . When one calls a judgement made aposteriori or analytic in my sense, one judges . . . about whereupon the deepest ground for the justification of its affirmation rests.)

Frege deploys his versions of the Kantian dichotomies, analytic/synthetic and a priori/a posteriori at the level of the *justifications* for the judgements made: they "are not concerned with the content of the judgement". *So much for Boghossian's "Frege analyticity"!*

In the case of mathematics that means we are now considering the demonstrations that serve to establish their conclusions, that is, the demonstrated theorems. Whether a theorem is analytic or not depends on the kind(s) of demonstration that it has, in particular on the starting points of the reasoning, that is, its *Urwahrheiten.*

2. Dadurch wird die Frage dem Gebiete . . . der Mathematik zugewiesen, wenn es sich um eine mathematische Wahrheit handelt. Es kommt nun darauf an den Beweis zu finden und ihn bis auf Urwahrheiten zu zurückzuverfolgen.

   (Thereby, when one is concerned with a mathematical truth, . . . the question is assigned to the domain of mathematics. The task now is to find the demonstration and trace it back to basic truths.)

The question as to whether a certain "mathematical truth" is analytic is accordingly a *mathematical* one: it is incumbent upon us to find its demonstration. *Urwahrheit* — "basic truth" — is Frege's favoured term for axioms. Frege here appears to presuppose that a "mathematical truth" has a demonstration, and that it can be found. Hence, perhaps somewhat surprisingly, and in contradistinction to latter-day realists, he is committed to the principle that mathematical truths can be known. In familiar current terminology, Fregean "mathematical truths" are — demonstrated or demonstrable — *theorems*.

3. Stösst man auf diesem Wege nur auf die allgemeinen logischen Gesetze und auf Definitionen, so hat man eine analytische Wahrheit, wobei vorausgesetzt wird, dass auch die Sätze mit in Betracht gezogen werden, auf denen etwa die Zulässigkeit einer Definition beruht.

   (When along this way one encounters only the general logical laws and definitions, one has an analytical truth, whereby it is presupposed that also those theorems are taken into consideration upon which the admissibility of a definition depends.)

The crucial step in Frege's treatment arrives when we have reached the "leaves" positions in the fully developed justificatory, backwards-search tree. It is strongly reminiscent of Aristotle's treatment at the outset of the *Posterior Analytics*, Book I, Chapter 4: the ultimate principles should be *general, topic neutral*, and in Latin *per se* that is, knowable *in and from themselves*: "self-evident". All three are directly at issue here in Frege's formulations. Just as Aristotle demands that the Principles of Demonstrative knowledge should be *general*, so does Frege. The leaf positions in his trees of justification — "grounding trees" — must be occupied by assertions with general propositions as contents. A single particular judgment in leaf position is enough to rule out Fregean analyticity.

6 General logical laws: self-evident and topic-neutral

In the Quotation (3) above Frege draws upon "the general logical laws". What are they? In 1884, Frege's short answer would be: the rules and axioms of my *Begriffsschrift*. In the light of this ready answer, Frege's "*Theory of Consequence*", that is, his treatment of the topics in the *Prior Analytics*, does not match Aristotle's in its answers to questions about consequence. Frege's form of judgement is different from Aristotle's *P belongs to S*, so their respective answers to the question of consequence: *What follows from what?* are different. His *Theory of Demonstration*, though, is taken straight out of the *Posterior Analytics*, and his answer to the question: *What is a demonstration* ("proof")? is very Aristotelian. The generality constraint on the *Urwahrheiten* of a mathematical demonstration, as well as their topic neutrality and applicability across the board, irrespective of subject matter, are taken straight out of Aristotle's account in the *Posterior Analytics*.

Also the third Aristotelian criterion, namely, *perseity*, we may find in Frege. Aristotle has four kinds of perseity and the two first are of interest to us.[14] Perseity of the first kind concerns a "belonging" of the form *P* belongs to *S* and holds when the predicate *P* is part of the "*formula*" (*logos*) of the subject *S*. The knowledgeable reader will here have recognized a version of Kantian analyticity: the predicate *P* is contained in the notion of the subject *S*. Aristotle applies perseity to "belongings", and in the Scholastic version of his account, say in the *Summa Theologiae* of St Thomas Aquinas, Q2, art. 2, where the "belonging" *P* belongs to *S* becomes the predication *S* is *P*, there is now talk of a *propositio per se nota*, that is, a judgment known in, or from, itself. Thomas elucidates this by saying that a *propositio per se nota* of this first kind, where "the predicate lies contained in the subject", is also *self-evident*, in the sense of *quae statim, cognitis terminis, cognoscuntur*, that, is *known as soon as their terms are known.*[15]

The second of the four Aristotelian kinds of *per se* connection between terms is exemplified by "even" and "number": number must occur in the *logos* of a term to which one may apply "even". Today we would say that "*a* is a number" is a *presupposition* for the *meaningfulness* of "*a* is even": only if "*a* is a number" does "*a* is even" make sense. Aristotle also gives *straight* and *oblique* as a *per se* connection of the second kind to *line*: the meaningfulness of "*a* is oblique" presupposes that "*a* is a line". This second kind of *perseity*, I suggest, is the source of Frege's rider on

[14] An. Post, I;4, 73a35 ff.

[15] Aquinas ST Q2:art2, SCG Ch. 10.

definitions in Quotation 3 above: "it is presupposed that also those theorems are taken into consideration upon which the admissibility of a definition depends". Without it being known that x is a number, the definition of "x is even" would be inadmissible. The *propositio per se nota* of Thomas Aquinas is matched by Frege with certain items of that are known from meaning alone, be it a self-evident judgment or knowledge of presuppositions.

In §5 Frege uses a well-known characterization of axioms: *unbeweisbar und unmittelbar klar*, that is axioms should be *indemonstrable and immediately clear*. The clarity in question is familiar from Descartes' characterization of the notion of *evidence*.[16] Evident knowledge has to be "clear and distinct". The immediacy here is not temporal — an axiom, that is, a self-evident judgement, need not be obvious, or at all easy to grasp. It is immediate in that it has no predecessors in the epistemic order. The immediacy terminology stems from Aristotelian syllogistic. An "*S* is *P*" judgement is immediate when there is no middle term to be found that could mediate between the major and minor terms in order to syllogize.

We also find several occurrences of another terminological item from traditional epistemology in GLA, §5, namely *einleuchten*, which is notoriously difficult to translate into English.

> Und ist es dann *unmittelbar einleuchtend*, dass 135664 + 37863 = 173527 ist?
>
> wie sollen sie anders eingesehen werden als durch einen Beweis, da sie *unmittelbar nicht einleuchten*?
>
> so müsste die Richtigkeit unserer Gleichung *sofort einleuchten*
>
> durch die Anschauung *unmittelbar einleuchten*
>
> (*my emphasis* G. S.)

and in §90

> weil der Mathematiker zufrieden ist, wenn jeder Übergang zu einem neuen Urtheile als richtig einleuchtet, ohne nach der Natur dieses Einleuchtens zu fragen, ob es logisch oder anschaulich sei.

The novel Ebert-Rossberg translation of GGA uses "obvious" here. I would prefer *evident*, but not *self-evident*, since *einleuchten* is coupled with *unmittelbar* and with *sofort*. Obviousness is always immediate: the obviousness of what is obvious should also be obvious, and so "immediately obvious" has a pleonastic ring to it, whereas "*unmittelbar* self-evident" *is* pleonastic. The evidence (of what is evident), on the other hand, can be immediate, but also *mediate*.

> Es handelt sich um mein Grundgesetz (V). Ich habe mir nie verhehlt, dass es nicht so einleuchtend ist, wie die andern, und wie es eigentlich von einem logischen Gesetze verlangt werden muss.[17]

---

[16] *Evidence* — Dutch *klaarblijkelijkheid* — in the sense *evidence of* (what is evident) but not *evidence for* (a claim). English is the only language that uses *evidence for*, particularly in legal contexts; nevertheless, also in the OED, *evidence (of)* is given as the first meaning.

[17] GGA, Vol II, p. 253.

To demand *obviousness* of an axiom is asking too much. It need not be at all easy to grasp the terms out which an axiom is built, or the way they contribute to render self-evident the result of joining them together in that particular way. It, as an axiom, must be *self*-evident, but that evidence need not be obvious, or at all easy to fathom: it can demand considerable experience in working with the concepts out of which an axiom is built before the penny drops and one realizes its self- evidence. As remarked already the immediacy (*Unmittelbarkeit*) in question is conceptual, but not temporal. *Temporal* immediacy, on the other hand, Frege marks by the use of *sofort*.

> Wenn es nicht möglich ist, den Beweise zu führen ohne Wahrheiten zu benutzen, welche nicht allgemein logischer Natur sind, sondern sich auf ein besonderes Wissensgebiet beziehen, so ist der Satz ein synthetischer.
>
> (When it is not possible to conduct the demonstration without using truths that are not of a general logical nature, but pertain to a particular domain of knowledge, the theorem is a synthetic one.)

Here Frege's Aristotelian background comes to the fore. His *Theory of Consequence* — "what follows from what?" — is not Aristotelian since the form of judgement in the Fregean ideography — *A ist eine Thatsache* — is different from the Aristotelian "*P* belongs to *S*". The difference between analytic and synthetic judgments is based on what laws are used in leaf-positions in the trees of demonstration. An analytic judgement may draw *only* upon general principles that are, in Gilbert Ryle's happy phrase, "topic neutral", that is, applicable, within all fields of inquiry, irrespective of subject matter. When the demonstration makes use of a general law of restricted applicability that holds only within a specific area of scientific inquiry, say geometry, or biology, the demonstrated theorem is synthetic. Frege's formulations, when explaining his dichotomy analytic and synthetic, really pertain to demonstrations, rather than to their theorems. Hence, it cannot be ruled out that one may find also an analytic demonstration for a judgement that has been established through a synthetic demonstration. The limitations imposed on synthetic judging stem from the Aristotelian need to avoid *metábasis eis állo génos*, that is, topic transferring principles from one domain of scientific discourse to another, where their applicability is not uncontested, for example, trying to use biological principles for establishing geometrical theorems.

> Damit einen Wahrheit aposteriori sei, wird verlangt, dass ihr Beweis nicht ohne Berufung auf Thatsachen auskomme; dass heisst, auf unbeweisbare Wahrheiten ohne Allgemeinheit die Aussagen von bestimmten Gegenstände enthalten.
>
> (For a truth to be a posteriori, it is required that its demonstration cannot be conducted without recourse to facts, that is, indemonstrable truths without generality that contain statements about certain objects.)

A judgement is *a posteriori* if it cannot be established without recourse to individual, specific facts. Both analytic and synthetic judgements have to be demonstrated by logical means alone from *general* laws, and in virtue of this they are *a priori*.

7 Axiom *eines Beweises weder fähig noch bedürftig*

> Ist es dagegen möglich, den Beweis ganz aus allgemeinen Gesetzen zu führen, die selber eines Beweises weder fähig noch bedürftig sind, so ist die Wahrheit apriori.
>
> (If, to the contrary, it is possible to conduct the demonstration entirely from general laws that neither admit nor need demonstration, the truth is a priori.)

We should note here that the arresting phrase "*eines Beweises weder fähig noch bedürftig*" is not original with Frege, but places him squarely in the rationalist epistemological tradition. Scholarly opinion varies as to where Frege took it from. Thus Gottfried Gabriel holds that the "formulation is an acknowledged quotation from Lotze":

> Sätze ... deren Gültigkeit für uns unmittelbar feststeht, die daher eines Beweises weder bedürftig noch fähig sind.[18]

I have not been able to find such an acknowledgement in Frege.

However, by the side of Lotze's *Logik*, there are several more places known to Frege, from where he could have taken the formulation. Thus, according to Tyler Burge, Leibniz in the *Noveaux essais*, Book IV, Ch. 9, §3 is a likely source:

> The assumption that axioms are basic laws or basic truths can be explicated in terms of Frege's characterization of primitive general laws as being "*neither capable nor in need of proof*" (FA §3). This phrase comes directly from Leibniz, from whom Frege probably got it (Leibniz [Nov. Es.] IV,ix,2).[19]
>
> (*my emphasis*, here and in the other quotations G. S.)

Leibniz's French reads very much the same as its English translation:

> une évidence entière qui *n'est point capable provée et n'en a point besoin.*
>
> (Erdmann, 373)

This is a comment on Locke's *An Essay Concerning Human Understanding*, Book IV, chapter xvii, §14, where we read:

> And this, therefore, as has been said, I call *Intuitive Knowledge*; which is certain, beyond all Doubt, and *needs no Probation, nor can have any*; this being the highest of all Humane certainty.

However, Frege's main philosophical source for his *Grundlagen*, namely Baumann's *Die Lehren von Raum, Zeit und Mathematik in der neueren Philosophie*, contains the above passage from Locke in German translation:

> So nimmt der Geist wahr, dass ein Bogen des Kreises kleiner ist als der ganze Kreis, ebenso klar, wie er die Idee des Kreises wahrnimmt; und dies demnach nenne ich anschauende Erkenntnis, welche gewiss ist, über alle Zweifel erhaben, *keines Beweises bedarf und keinen haben kann*; dies ist die höchste aller menschlichen Gewissheit.[20]

---

[18] Gabriel (2002, p. 47), refers to Lotze (1874, §200). I have not been able to find any direct quotation of Lotze's §200 in GLA. Gabriel and Schlotter (2017, Ch. 4, §3) offer more details as to Frege and Lotze on the status of logical laws.

[19] Burge (1998, p. 313). My *emphasis*, here and in the other quotations.

[20] Baumann (1868, Band I, p. 362).

Apart from the one quoted by Burge, other Leibniz passages (from the same source) are given by Baumann:

> Die unmittelbare Apperception unseres Daseins und unserer Gedanken liefert uns die ersten Wahrheiten a posteriori oder thatsächlicher Art, d. h. die ersten Erfahrungen. Sie sind *unfähig bewiesen zu werden* und können unmittelbar genannt werden, — weil Unmittelbarkeit zwischen Verstand und Object bei ihnen stattfindet.[21]

and Baumann himself also uses the formulation freely:

> Dies Prinzip, das ich soll erschlichen haben, ist das vom Bedürfnis eines zureichenden Grundes, damit eine Sache existiert, ein Ereignis eintritt, eine Wahrheit statt hat. Ist das *ein Prinzip, das Beweise bedarf*?
>
> (Erdmann p. 778, 125.)

and

> Der Gegensatz ist einleuchtend: Leibniz stellt das Prinzip [i.e. Satz vom Grunde] in der ihm eigenthümlichen Fassung auf, rundweg auf *als keines Beweises bedürftig* . . .[22]

Against the background of such plenitude, Baumann seems to me a more likely source than either Lotze or Leibniz.

At time when Frege was writing the *Begriffsschrift*, and his *Grundlagen*, Brentano lectured on *Logik* in Vienna:

> Man kann das evidente Urteil als solches bezeichnen, welches in sich als richtig charakterisiert ist. Solche Urteile sind *keines Beweises fähig und keines Beweises bedürftig*.[23]

This Frege will not have known, but it shows how widespread formulations in terms of *fähig* and *bedürftig* have already become by 1880.

To my mind, none of the above, however, was Frege's source for his formulation:

> Alle Gewissheit ist entweder eine unvermittelte oder eine vermittelte, d. h. sie bedarf eines Beweises, oder ist *keines Beweises fähig oder bedürftig*.
>
> Apodicticity is either immediate or mediate, that is, it stands in need of a demonstration, or it is neither capable nor in need of demonstration.

In this quotation we have all three components *Beweis*, *fähig* and *bedürftig* in close juxtaposition, which are not found in the quotations from Baumann.

The author is Kant, and the place is the *Jäsche Logik*, Ch. IX *Logische Vollkommenheit der Erkenntnis*, A108-109, a work we know that Frege read and quoted at the time of writing of *Grundlagen*. The English translators (1988, p. 79) give:

> "All certainty is either *mediated* or not *mediated*, that is, it either requires proof or is neither susceptible nor in need of any proof."

---

[21] Baumann (1868, Band II, p. 225).

[22] Baumann (1868, Band II, p. 292, and 293, respectively).

[23] Brentano (1956, §29, art 85, g, p. 111). The fragment in question stems from a set of notes for logic lectures delivered in Vienna in the second half of the 1880's.

I definitely prefer *immediate or mediate* over their choice, as well as the term *demonstration* (that is cognate to *Beweis*) rather than "proof", which is related to *tests* (German *prüfen*, *Prüfung*) rather than to *Beweise*, and is cognate with *probe* and *approve*. *Gewissheit* is notoriously difficult to translate into English. The fine verb *to wit* (that is cognate to *wissen*) has been jettisoned, which leaves us a term short for convenient translation of *scire* (French *savoir*, German *wissen*), since *know* obviously has to translate *cognoscere* (French *connaitre*, German *kennen*). Upon reflection, the best I can do is translating *gewiss* with *apodictic*, which yields the apposite *apodicticity* for *Gewissheit*, and is in harmony with Frege's adoption of Kantian terminology, at BS, §4, concerning *apodictic* and *assertoric* judgements.

Kant returns to the formulation in §33 as well:

> Demonstrable Sätze sind die, welche *eines Beweises fähig* sind; die keines Beweises fähig sind, werden indemonstrable genannt.
>
> Demonstrable statements are those which are capable of demonstration; those not capable of demonstration are called indemonstrable.

That starting points of demonstrations are not capable of proof is clear; they are indemonstrable. But on the other hand, nor are they in need of demonstration, since they are self-evident; as soon as their terms are known, according to the Thomistic formula quoted above, they can be known as such.

8 The *Begriffsschrift*

With the benefit of hindsight one may find the genesis of Frege's approach to analyticity in the *Preface* to the *Begriffsschrift*. Already here Frege runs the Aristotelian regress of asking for justifications, and the grounding of his judgements eventually bring him back to first principles, in much the same way that we came across previously in GLA, §3, above. Frege is concerned to place demonstrative "scientific" truths — truths of *reason* — after a Leibnizian fashion, at one side of a *dichotomy*, with truths of *fact* as the other side. When the Leibnizian resolution, of replacing expressions by their definitions, comes to an end it yields an a priori demonstration of its *S* is *P* conclusion; and the leaf positions at which the resolution stops are "*judgements of identity*": An *A* is *A*, or an *AB* is *A*.[24] In this case the truth is a priori and necessary. In the case of an empirical, contingent "*S* is *P*" judgment the *resolution*—Latin for Greek *analysis*—will not come to an end, because if it did we would read the resolution tree as a demonstration, in the opposite direction, from top to bottom, whence the judgement would be rendered necessary. The terms of a contingent judgement are fully available only to God (by direct acts of grasping). God does not come to know discursively through analysis, but we humans will have to relay of empirical research, such as acts of perception, in order to come to know the judgement in question. The analysis does not work here, since owing to infinite complexity, it

[24] The fragment *Primae veritates*, Eng. tr. "First Truths", in: Leibniz (1973, pp. 87–92), gives a clear exposition of Leibnizian resolution.

will be non-terminating, and does not reach primitive terms and the concomitant judgements of identity. However, even though Frege does draw attention to analyticity, in §8, and in §24, there it is not his own later notion that is brought into play but the Kantian one.

Frege, Preface to *Begriffsschrift*:

> Das Erkennen einer Wissenschaftliche Wahrheit durchläuft in der Regel mehrere Stufen der Sicherheit.
>
> (The *recognition* of a *scientific* truth, as a rule, runs through several levels of *certainty*.)

Note that Frege uses *Sicherheit*, and not *Gewissheit*. Here it is the degree of conviction that is at issue, and not absolute apodicticity; it may be different, more or less strong, at different stages in scientific inquiry. The desired apodicticity is reached only after a demonstration has been given:

> Das apodiktische Urtheil unterscheidet sich von assertorischen dadurch, dass das Bestehen allgemeiner Urtheile angedeutet wird, aus denen der Satz geschlossen werden kann, während bei der assertorischen eine solche Andeutung fehlt.[25]
>
> (The apodictic judgement differs from the assertoric judgement in that the existence of general judgments is indicated, from which the sentence may be inferred, while for the assertoric there is no such indication.)

This is very close to what becomes a priori judgments in the *Grundlagen*. The tree of justifications ends in general laws, but they are not required to be topic-neutrally applicable across the board, as will be the case in GLA, for analytic judgments.

> Die festeste Beweisführung ist offenbar die rein logische, welche, von der besonderen Beschaffenheit der Dinge absehend, sich allein auf Gesetze gründet, auf denen alle Erkenntnis beruht.
>
> (The firmest mode of *demonstration* is patently the purely logical one, which, disregarding the particular nature of objects, grounds itself only upon those *laws* on which all *knowledge* depends.)

How, we may wonder, does Frege know that there is a *firmest* mode? Perhaps all modes of demonstration are equally firm, or given a mode of demonstration there is always a firmer to be found? A formulation along the lines: *There is no firmer mode* . . . might seem more prudent. Furthermore, if firmness be related to degree of conviction, Frege's claim is highly doubtful, at least to me. A trivial computation, as taught in first grade elementary school, establishes $25 + 42 = 67$ and it will produce considerably more conviction than a huge derivation in predicate logic, involving, at least, 67 variables of quantification; with numbers above 1000 the computations are still perfectly feasible and can be kept under full control, whereas the corresponding derivations will sorely tax the patience and meticulousness of even the most diligent computist. Crucially important here is Frege's insistence on the topic-neutral applicability of logical laws.

Already at the very outset of his *Begriffsschrift* Frege isolated the special character of knowledge that is obtained from purely logical, general and topic-neutral laws.

---

[25] BS, §4.

> Wir theilen danach alle Wahrheiten, die einer Begründung bedürfen, in zwei Arten, indem der Beweis bei den einen rein logisch vorgehen kann, bei den anderen sich auf Erfahrungsthatsachen stützen muss.
>
> (We then divide all truths that demand a justification into two kinds such that for one the demonstration can proceed purely logically, while for the other it has to rest upon facts of experience.)

This is the Leibnizian dichotomy of (necessary) *truths of reason* versus (contingent) *truths of fact*. We should, however, note that not every truth demands a justification. Some truths, it would appear, are *per se* nota and can be known of, or from, themselves. For truths that demand justification, on the other hand, the dichotomy between *purely logical* versus *relying on experiential facts* applies, but in every case it is possible to (come to) know the truth as such. *Fregean truths are knowable!*

However, even though Frege does draw attention to the analytic/synthetic distinction also in *Begriffsschrift*, §8, there it is not his own later notion that is brought into play, but the Kantian one. To a name there is associated a *Bestimmungsweise* — "mode of determination" — of a content.[26] Two different names a and b may be associated with different modes of determination that yield one and the same content. Then $\vdash a \equiv b$, that is, the judgement of *Inhaltsgleichheit* between a and b, is a *synthetic* one in the sense of Kant. To this way of reading the *Inhaltsgleichheit* sign "≡" as expressing that the associated *Bestimmungsweisen* determine identical contents. (In later modern terminology that is inspired by SuB, the names a and b are *co-referential*.) To this co-determinability of the associated *Bestimmungsweisen*, in §24, Frege adds definitional equality as a second reading of the sign "≡".

Frege returns to his example at the opening of SuB:

> $a = a$ und $a = b$ sind offenbar Sätze von verschiedenem Erkenntniswert: $a = a$ gilt *a priori* und ist nach Kant analytisch zu nennen, während Sätze von der Form $a = b$ oft sehr wertvolle Erweiterungen unserer Erkenntnis enthalten und *a priori* nicht immer zu begründen sind.
>
> ($a = a$ und $a = b$ are patently sentences of different worth for the cognition: $a = a$ holds *a priori* and according to Kant is to be labelled analytic, while sentences of the form $a = b$ often contain very valuable amplifications of our knowledge and are not always to be justified *a priori*.)[27]

The Leibnizian organization of the material in *Begriffsschrift* gives way to the

---

[26] Subsequently, in FuB, and SuB, the mode of determination/content dichotomy, with respect to names, is replaced by the distinction between the *Sinn* and the *Bedeutung* of a name. The abstract discussion here in BS, §8, is there sharpened by consideration of the trivial $2^4 = 4^2$ arithmetical theorem and the famous example concerning the planet Venus as Morning Star and Evening Star. The *Bestimmungsweise* (of a content) in BS is turned into the *Art des Gegebenseins* (of a *Bedeutung*) in SuB.

[27] Apart from the notorious difficulty of how best to render *Bedeutung* into English, the opening passage of SuB into English, poses tricky challenges for the translator. I prefer *worth* over *value* in the translation of *Erkenntniswert*: the "cognitive value" that is preferred in the usual translations, to my mind, has a philosophical contemporary technical ring to it that is absent in the German. Furthermore, the choice to use *extension* for Frege's *Erweiterung* obscures the obvious Kantian context of the example. Here the proper English translation of *Erweiterung* is not *extension*, but *amplification*, being cognate with *ampliative*, as used in the standard translation of Kant's *Erweiterungsurteil*.

Kantian framework of *Grundlagen*. In Frege's letter to Marty(?) of August 29, 1882, the transition seems to have been effected.

> Ich habe jetzt ein Buch nahezu vollendet, in welchem ich den Begriff der Anzahl behandele und nachweise, dass die ersten Sätze über das Zählen der Zahl , die man bisher als unbeweisbare Axiome anzusehen geneigt war, sich nur mittels der logischen Gesetze aus Definitionen bewiesen lassen, sodass sie im Kantischen Sinnen wohl als analytisch Urteile zu betrachten sind.
>
> (I have now nearly completed a book in which I treat the concept of Number and demonstrate that the first theses on computation that one has hitherto tended to regard as indemonstrable axioms can now, so that they should be regarded as Kantian analytic judgements.)

Frege's gloss on analyticity — "be demonstrated from definitions by means of the logical laws alone" — reads as a summary of his GLA, §3, account, and his choice of a Kantian stance is made a few sentences further down in the same letter:

> Denn, während Leibniz [die Kraft des diskursiven Denkens] wohl überschätzt hat, indem er alles aus Begriffen bewiesen möchte, scheint mir Kant umgekehrt die Bedeutung der analytischen Urteile zu gering zu achten, indem er sich an zu einfachen Bespiele hält. Ich sehe es als ein grosses Verdienst Kants darin dass er die Sätze der Geometrie als synthetische Urteile erkannt hat, aber ich kann ihn für die Arithmetik nicht folgen.
>
> (For while Leibniz may well have overestimated [the power of discursive thought] when he wished to demonstrate everything from concepts, Kant on the contrary seems to me to place too low an estimate on the significance of analytic judgements because he sticks to too simple examples. I consider it as greatly to Kant's credit that he recognized the theorems of geometry as synthetic judgements, but I cannot follow him regarding arithmetic.)

Frege's assertion that the laws of Number as he describes them are analytic *in the sense of Kant* is hard to understand. What Frege described is much closer to his *own* conception of analyticity. Furthermore, a Kantian analytic judgement is of "*S* is *P*" form, where the predicate *P* is "contained in" the notion of the subject, whereas Frege's arithmetical laws are not readily cast in such a form. On the other hand, in GLA, §3, p. 3, footnote *, Frege claims that he does not wish to introduce a novel sense here, but only wants to hit what earlier writers, in particular Kant, have meant. Be that as it may: Frege's §3 account of analyticity appears be substantially different from Kant's.

There remains also to be answered the vexing question as to why Frege, after having gone to some length in order to set up his own novel version of analyticity never uses it again after *Grundlagen*.[28] Today it is a commonplace that Frege's "Logicism" began with the *Begriffsschrift*. The traditional current formulation of Logicism was codified by C. G. Hempel in (1945, §10):

> [T]he *thesis of logicism concerning the nature of mathematics*:
>
> Mathematics is a branch of logic. It can be derived from logic in the following sense:
>
> a. All the concepts of mathematics, i.e. of arithmetic, algebra and analysis, can be defined in terms of four concepts of pure logic.

[28] In spite of Frege's disingenuous declaration, in the footnote GLA, §3, that he did not wish to endow the Kantian terms with novel sense, surely he must have been aware that that is precisely what he has done.

b. All the theorems of mathematics can be deduced from those definitions by means of the principles of logic (including the axioms of infinity and choice).

Mathematico-philosophical staple diet though it may be, and as admirably lucid it is, it does not match Fregean reality. Hempel's folklore picture is painted with the benefit of hindsight, after Frege's informal, non-technical presentation was offered in GLA, and draws extensively on the elaboration provided by Russell in his (post *Principia Mathematica*) *Introduction to Mathematical Philosophy*.

In *Begriffsschrift* Frege's Logicism had not yet been formulated. In the *Grundlagen* it takes the form of showing that the arithmetical theorems are analytic in the sense he has so painstakingly spelled out in §3. This primarily is an epistemological investigation, since analyticity is, after all, an epistemic notion, but the attempted execution of that programme also had ontological consequences.[29] The aim of the Begriffsschrift was epistemological. Frege wanted to have a means of ensuring that chains of inference were *lückenlos*, that is, "gap-free", and he hoped to find one by using a formal language, with the allowable forms of inference clearly delineated and circumscribed.[30] Codifying logic in such a fashion one could be certain that no extraneous assumptions, or modes of inference, were illegitimately deployed in the course of demonstrations. Perhaps the BS system does not comprise all valid inferences, nor all self-evident axioms, but Frege, from his point of view, could claim

---

[29] The, admittedly very neat, customary modern arrangement of the concepts *analytic*, *a priori*, and *necessary* as being, respectively, *semantic*, *epistemic*, and *metaphysical* in character, is post WW II. The first two labelling's gained prominence with Quine (1953), and its final codification is due to Saul Kripke in *Naming and Necessity* from 1970.

[30] The term *lückenlos* has recently had a good innings, as "gap-free", in modern Philosophy of Mathematics, but then applied to *demonstrations*. Frege used if for *chains of inferences*, and his use was proper, since a chain of inferences remains a chain of inferences, even though it may contain a gap. The modern use is incoherent: *ein lückenhafter Beweis*, that is, a *gappy demonstration*, is a contradiction in terms. *Lückenhaft* applied to demonstrations is not a qualifying term, but a *modifying* one: a demonstration *with gaps* is no demonstration. Similarly *invalid* is such a modifying term with respect to demonstrations. Such modifying terms come in pairs comprising a modifying term and a restorative one, for instance *false* and *true*, say, when applied to friends. "His behaviour made me think he was a false friend, but then I learned the full background and understood that he was a true friend after all." Similarly *invalid* (*incorrect*), and *valid* (*correct*), when applied to demonstrations. The dichotomy correct/incorrect is not a proper partition of demonstrations. We give a demonstration and that is enough; we do not then have to demonstrate as a further theorem that the demonstration we have given is correct (valid). If and when a challenge is made towards the demonstration in question, we may legitimately use the term *incorrect* (or *invalid*) when formulating our retraction, or as the case may be, use the term *valid* when stating that the criticism of the demonstration was illfounded, because it is a valid demonstration after all. *Validity applied to inferences* and *validity applied to demonstrations* are different notions. When a challenge is made towards a demonstration, the presupposition for using a modifying term has been met and we may use the term *incorrect* (or *invalid*) without incoherence, when formulating our retraction, or, as the case may be, use the term *valid* when stating that the criticism of the demonstration in question was ill-founded, because it is a *valid* demonstration after all. The term *lückenlos* is the *restitutive* term that matches the modifying term *lückenhaft*.

A more elaborated exposition of these matters relating to different meanings of *validity* can be found in my (2019).

this much: if something has been derived by (mechanically) following the rules of the *Begriffsschrift* system, the theorem thus derived is correct.[31]

With respect to the question as to why Frege did not use his own notion of analyticity, it is easy to answer for BS. In 1879 Frege had not yet formulated his own version of analyticity, so it obviously was not available for him to use in the *Begriffsschrift.* We saw it *in nuce* at work in the letter to Marty(?) from 1882. Accordingly, between 1879 and 1882, Frege formulated his own version of analyticity. His use of the Kantian notion in BS is not without its own difficulties. Frege's form of judgement rules out the use of the Kantian "*S* is *P*" containment explanation, and the other characterizations of being elucidatory only, but not ampliative, and resting solely on the Law of (Non-)Contradiction are not explicitly spelled out by Frege in *Begriffsschrift.* The further characterization used by Locke, Leibniz and St Thomas Aquinas as *trifling* judgments (Locke), *propositions frivoles* (Leibniz) and *nugatoriae* (Thomas) are not discussed by Frege, and with good reason. Frege most definitely does not want those characteristics since his analytic judgments are meant to be ampliative. During the writing of SuB that was published in 1892, Frege nevertheless uses the Kantian version of analyticity, and not his own. That is because of a major difference between him and Kant regarding analyticity. Thus, in Frege's extensive list of contents, GLA §99 carries the description *Kants Unterschätzung der analytischen Urtheile*, that is, "Kant's disparagement of analytic judgements". According to Frege analytic judgements are not void of epistemic worth; they are not *folgeleer.* True analytic identities, when the predicate is contained *impliciter* in the subject, may well be ampliative, contrary to the Kantian characteristics of analytic judgements.[32] In this context we should also take note of the III:rd of Frege's Habilitation Theses from 16.5.1874:

> III. Zahl ist nicht ein ursprünglich Gegebenes, sondern läßt sich definieren.[33]
> (Number is not a primitive given, but can be defined.)

9 The general laws of logic in *Begriffsschrift* and the transformation in *Grundlagen*, §17

The BS laws of logic that according to Frege are meant to yield the desired apodicticity with respect to the theorems derived are the universal closures, with respect to quantification of what today is called first and second order, regarding the following Urwahrheiten.[34]

[31] "Begriffsschrift" is an overburdened term in Frege research. I use *Begriffsschrift* for Frege's booklet. Following Jonathan Barnes (2022) I use ideography for its language, and BS system for the deductive apparatus in which the ideography is deployed.

[32] The *Jäsche Logik*, §37 *Tautologische Sätze* reads like a preliminary study for the opening paragraph of Frege's article SuB.

[33] Kreiser (2001, p. 123).

[34] In BS Frege has only one kind of quantification that is deployed indiscriminately with respect to all kinds of domains.

L1: $A \supset (B \supset A)$
L2: $(C \supset (B \supset A)) \supset ((C \supset B) \supset (C \supset A))$
L3: $(D \supset (B \supset A)) \supset (B \supset (D \supset A))$
L4: $(B \supset A) \supset (\neg A \supset \neg B)$
L5: $\neg\neg A \supset A$
L6: $A \supset \neg\neg A$
L7: $(c \equiv d) \supset (F(c) \supset F(d))$
L8: $(c \equiv c)$
L9: $\forall a F(a) \supset F(c)$, for any c

The permitted rules of inference are

(MP) Modus Ponens,
(Gen) Generalization.

Frege, in *Begriffsschrift*, also uses a second version (Gen*) of Generalization that allows for the inference

$$\vdash C \supset A(x)$$
$$\vdash C \supset \forall x A(x)$$

where the free variable does not occur in C.

This rule, however, is readily derivable by adding an additional axiom, to wit,

($) $$\vdash \forall x[C \supset A(x)] \supset [C \supset \forall x A(x)],$$

by using (Gen) on the premiss of (Gen*) to get

$$\vdash \forall x[C \supset A(x)]$$

and from this and ($), MP yields

$$\vdash C \supset \forall x A(x).$$

A demonstration in the system will now start off with, in leaf positions, suitable closures of some of the nine axioms — ten if we opt to include the axiom ($) — and then Modus Ponens and axiom L9 will gives us desired substitution instances of the axioms.[35]

So we are presented with an assertion of an empirical fact E, to which will be attached a chain of logical inferences. For Frege, as we saw in the previous section, the general logical inferences are those that can be carried out for his ideography, while using the formal apparatus displayed in the *Begriffsschrift*, without drawing either on facts or on scientific general laws that pertain to a special science such as arithmetic, biology, or astronomy.

So we are given a "demonstration tree" T for our theorem $\vdash A$ where in top "leaf" position we may find:

[35] Frege, of course, uses a rule of substitution in BS without formulating it explicitly. It is a highly non-trivial task of giving exact substitution rules and today formulations in terms of axioms plus substitution rule have given way to formulations in terms of axiom-schemata. Church (1956b, p. 289–270) carefully reports the vicissitudes of the substitution rule.

1. the *Einzelthatsache* E;
2. general, topic neutral laws that are applicable across the board, irrespective of subject matter.

We then continue downwards from these "leaves" using

(MP) Modus Ponens,
(Gen) Generalization.

We now eliminate *Einzelthatsachen* from the demonstration-tree T by replacing each formula C that occurs in T by the E-transformed formula $E \supset C$.

Thus, in general, the E-transformed tree "$E \supset$" T, will no longer be a BS derivation. Our task, accordingly, is now to restore BS deducibility at each place.

We begin with the leaf formulae in the tree T. There are two possibilities: we may have the empirical fact E, or we may have an instance of a general, topic-neutral logical law, that is, a BS *axiom*.

Case i: The formula E has been replaced by $E \supset E$, which can be derived as follows:

1. $\vdash E \supset ((E \supset E) \supset E)$	by L1 and $\forall E$
2. $\vdash (E \supset ((E \supset E) \supset E) \supset ((E \supset (E \supset E)) \supset (E \supset E))$	L2
3. $\vdash (E \supset (E \supset E)) \supset (E \supset E)$	MP on 1 and 2
4. $\vdash (E \supset (E \supset E))$	L1
5. $\vdash E \supset E$	MP on 3 and 4

Accordingly $\vdash E \supset E$ has been demonstrated from topic-neutral logical laws and is thus an analytic theorem ("truth" in Frege's terminology).[36]

Case ii: When M is one of the a logical laws L1–L9, or ($), its E-transform $\vdash E \supset M$ is BS demonstrable as follows:

1. $\vdash M$	is a BS axiom
2. $\vdash M \supset (E \supset M)$	L1
3. $\vdash E \supset M$	by MP on 1 and 2

Thus the two leaf-positions in the demonstration are general, topic-neutral logical laws, and the judgement $\vdash E \supset M$ is analytic in Frege's sense.

In order to cope with inferences further down the "demonstration tree" T we consider an instance of *modus ponens*:

$$\frac{\vdash (A \supset B) \qquad \vdash A}{\vdash B}$$

and assume that we have demonstrations for the two transformed premises:

$$\vdash E \supset A$$

[36] Frege's demonstration of this is found as Theorem 27 on p. 43 of the *Begriffsschrift*.

and

$$\vdash E \supset (A \supset B).$$

From these we must give a demonstration of $\vdash E \supset B$ using only topic neutral laws and inferences.

However,

$$\vdash (E \supset (A \supset B)) \supset ((E \supset A) \supset (E \supset B))$$

is an instance of BS axiom L2, whence by Modus Ponens,

$$\vdash (E \supset A) \supset (E \supset B),$$

and one more application of Modus Ponens yields the desired theorem

$$\vdash E \supset B.$$

Finally we address the inference rule of Generalization. We are given a demonstration of a theorem

$$\vdash E \supset A(x), \text{ where } x \text{ does not occur in } E.$$

By Generalization

$$\vdash \forall x(E \supset A(x))$$

but

$$\vdash \forall x(E \supset A(x)) \supset (E \supset \forall x A(x))$$

is an instance of axiom ($) and by a further Modus Ponens

$$\vdash E \supset \forall x A(x)$$

we are done.

My discerning readers will, no doubt, already have recognized the inspiration provided by Paul Bernays' elegant (1934a) demonstration of the Deduction Theorem: Frege's hypothetically asserted fact $\vdash E$ here takes the place of the *assumption* formula A in Bernays' demonstration.

A sequence of applications of this little result will yield the required implicational formula that will be analytic in Frege's sense if the original train of inferences ends in only everywhere applicable, topic-neutral laws.

An alternative treatment is needed for the case where one does not wish to avail oneself of the additional axiom ($), but prefers to remain with Frege's third rule of inference (Gen*):

$$\frac{\vdash C \supset A(x)}{\vdash C \supset \forall x A(x)}$$

where the free variable does not occur in C. In order to complete the demonstration we accordingly assume that the judgement

$$(*) \qquad \vdash E \supset (C \supset A(x))$$

is demonstrated and have to obtain a demonstration of

$$\vdash E \supset (C \supset \forall x A(x)).$$

The trick needed here is well known from the treatments in Bernays (1934) and in Kleene (1951); we note that

$$(\#) \qquad \vdash P \supset (C \supset R) \leftrightarrow (P \,\&\, C) \supset R$$

and so one reasons, from $(*)$ to

$$(**) \qquad \vdash E \,\&\, C \supset A(x)$$

and from here by (Gen*) to

$$(***) \qquad \vdash E \,\&\, C \supset \forall x A(x).$$

An application of (#) in the opposite direction yields the required conclusion

$$\vdash E \supset (C \supset \forall x A(x)).$$

Alas, this rather pleasing slight-of-hand, which also I used already in my treatment of the Deduction Theorem in (1983b), is not open to us here, because the conjunction & is not a primitive sign in the ideography of Frege's *Begriffsschrift* that uses only $\neg$ and $\supset$. But

$$E \,\&\, C \leftrightarrow \neg(\neg E \vee \neg C) \leftrightarrow \neg(E \supset \neg C),$$

The *Begriffsschrift* is semantically complete in its propositional part, whence the treatment can be carried out in the ideographical system of the *Begriffsschrift*. To do so, however, was not an enticing prospect. Going from $(*)$ to $(***)$ using in $(**)$ not $E \,\&\, C$, but an ideographical rendering of $\neg(E \supset \neg C)$, as middle formula, using only Frege's axioms and rules, would to the modern reader rather hide and complicate what is going on rather than illuminate it. Accordingly, above I instead opted for the easy way out, adding the ($) axiom, thereby obviating awkward derivational work inside the *Begriffsschrift* ideography.[37] With this my little exercise in Fregean proof theory and on the philosophy behind it has come to an end. The projected Part 2 of the running commentary, with the title

An "objective order of demonstrations"?

and comprising discussion of Frege, Bolzano, Port-Royal, Thomas Aquinas, and Aristotle, shall have to wait for another occasion.

[37] Here I am very much indebted to Dr. Joan Bertran-San Millán, of the Czech Academy of Science in Prague, who, with his intimate knowledge of the *Begriffsschrift* and its details, drew my attention to the fact that Frege does point out precisely these derivational steps in §11, page 22. I, of course, now suggest that he did this with the express purpose of justifying (Gen*) by reducing it to (Gen 1), along the lines that also I used above. He generously shared his meticulous derivations in the *Begriffsscrhift*-like $\{\neg, \supset\}$ fragment of propositional calculus with me, and they are now reproduced in the appendix.

Appendix

This appendix is devoted to a demonstration that the propositions

$$(C \supset (B \supset A)) \quad \text{and} \quad \neg(C \supset \neg B) \supset A,$$

are equivalent in the *Begriffsschrift*. For ease of exposition, and of typing, I use the $\{\neg, \supset\}$ fragment of propositional calculus that corresponds to the *Begriffsschrift* system.

The proposition $\neg(C \supset \neg B) \supset A$ is, of course, $C \,\&\, B \supset A$ in *Begriffsschrift* disguise.

These derivations are needed for a full treatment of the second of the two rules for the universal quantifier and were designed by Dr. Joan Bertran-San Millán, of the Czech Academy of Science, Prague. I am indebted to him for pointing out Frege's use of the equivalence in BS, §11, and for sharing with me his meticulous Begriffsschrift-like derivations in the $\{\neg, \supset\}$ fragment.

Theorem (i) $\vdash (D \supset (C \supset (B \supset A))) \supset (D \supset (C \supset (\neg A \supset \neg B)))$

Demonstration:

1. $\vdash (B \supset A) \supset ((C \supset B) \supset (C \supset A))$ BS Thm 5
2. $\vdash [(C \supset (B \supset A)) \supset (C \supset (\neg A \supset \neg B))] \supset [(D \supset (C \supset (B \supset A)) \supset (D \supset (C \supset (\neg A \supset \neg B)))]$ Subst. 1.: $C \supset (B \supset A)/B, C \supset (\neg A \supset \neg B)/A, D/C$
3. $\vdash (C \supset (B \supset A)) \supset (C \supset (\neg A \supset \neg B))$ BS Thm 29
4. $\vdash (D \supset (C \supset (B \supset A))) \supset (D \supset (C \supset (\neg A \supset \neg B)))$ MP on 2 and 3

Theorem (ii) $\vdash (\neg B \supset \neg A) \supset (A \supset B)$

Demonstration:

1. $\vdash (C \supset B) \supset ((B \supset A) \supset (C \supset A))$ BS Thm 9
2. $\vdash ((\neg B \supset \neg A) \supset (\neg\neg A \supset B)) \supset (((\neg\neg A \supset B) \supset (A \supset B)) \supset ((\neg B \supset \neg A) \supset (A \supset B)))$ Subst. 1: $(\neg B \supset \neg A)/C$, $(\neg\neg A \supset B)/B$, $(A \supset B)/C$
3. $\vdash ((\neg B \supset A) \supset (\neg A \supset \neg\neg B)) \supset ((\neg B \supset A) \supset (\neg A \supset B))$ BS Thm 32
4. $\vdash ((\neg B \supset \neg A) \supset (\neg\neg A \supset \neg\neg B)) \supset ((\neg B \supset \neg A) \supset (\neg\neg A \supset B))$ Subst. in 3: $\neg A/A$
5. $\vdash (B \supset A) \supset (\neg A \supset \neg B)$ BS Axiom L4
6. $\vdash (\neg B \supset \neg A) \supset (\neg\neg A \supset \neg\neg B)$ Subst. in 5: $\neg B/B$, $\neg A/A$
7. $\vdash ((\neg B \supset \neg A) \supset (\neg\neg A \supset B))$ MP on 4 and 6
8. $\vdash ((\neg\neg A \supset B) \supset (A \supset B)) \supset ((\neg B \supset \neg A) \supset (A \supset B))$ MP on 2 and 7
9. $\vdash (C \supset B) \supset (B \supset A) \supset (C \supset A))$ BS Thm 9
10. $\vdash (A \supset \neg\neg A) \supset ((\neg\neg A \supset B) \supset (A \supset B))$ Subst. 9: A/C, $\neg\neg A/B$, B/A
11. $\vdash A \supset \neg\neg A$ BS Axiom L6
12. $\vdash ((\neg\neg A \supset B) \supset (A \supset B))$ MP on 10 and 11
13. $\vdash (\neg B \supset \neg A) \supset (A \supset B)$ MP on 8 and 12

Theorem (iii) $\vdash (D \supset (C \supset (\neg B \supset \neg A))) \supset (D \supset (C \supset (A \supset B)))$

Demonstration:

1. $\vdash (B \supset A) \supset ((D \supset (C \supset B)) \supset (D \supset (C \supset A)))$ BS Thm 7
2. $\vdash ((\neg B \supset \neg A) \supset (A \supset B)) \supset ((D \supset (C \supset (\neg B \supset \neg A)) \supset (D \supset (C \supset (A \supset B)))$ Subst 1: $(\neg B \supset \neg A)/B$, $(A \supset B)/A$
3. $\vdash (\neg B \supset \neg A) \supset (A \supset B)$ BS Thm 22
4. $\vdash (D \supset (C \supset (\neg B \supset \neg A))) \supset (D \supset (C \supset (A \supset B)))$ MP on 2 and 3

Theorem (iv) $\vdash (C \supset (B \supset A)) \supset (\neg(C \supset \neg B) \supset A))$

Demonstration:

1. $\vdash A \supset A$ BS Thm 27
2. $\vdash (C \supset (B \supset A)) \supset (C \supset (B \supset A))$ Subst. 1: $(C \supset (B \supset A))/A$
3. $\vdash (D \supset (C \supset (B \supset A))) \supset (D \supset (C \supset ((\neg A \supset \neg B))))$ Theorem (ii)
4. $\vdash ((C \supset (B \supset A)) \supset (C \supset (B \supset A))) \supset ((C \supset (B \supset A)) \supset (C \supset ((\neg A \supset \neg B))))$ Subst. 3: $C \supset (B \supset A)/D$
5. $\vdash (C \supset (B \supset A)) \supset (C \supset ((\neg A \supset \neg B))$ MP on 2 and 4
6. $\vdash (D \supset (C \supset (B \supset A))) \supset (D \supset (B \supset (C \supset A)))$ BS Thm 12
7. $\vdash ((C \supset (B \supset A)) \supset (C \supset (\neg A \supset \neg B))) \supset ((C \supset (B \supset A)) \supset (\neg A \supset (C \supset \neg B)))$ Subst. 6: $C \supset (B \supset A)/D$, $\neg A/B$, $\neg B/A$
8. $\vdash (C \supset (B \supset A)) \supset (\neg A \supset (C \supset \neg B))$ MP on 5 and 7
9. $\vdash (C \supset (\neg B \supset A)) \supset (C \supset (\neg A \supset B))$ BS Thm 34
10. $\vdash (C \supset (B \supset A)) \supset (\neg A \supset (C \supset \neg B)) \supset ((C \supset (B \supset A)) \supset (\neg(C \supset \neg B) \supset A))$ Subst. 9: $C \supset (B \supset A)/C$, A/B, $C \supset \neg B/A$
11. $\vdash (C \supset (B \supset A)) \supset (\neg(C \supset \neg B) \supset A))$ MP 9 and 10

Theorem (v) $\vdash (\neg(C \supset \neg B) \supset A)) \supset (C \supset (B \supset A))$

Demonstration:

1. $\vdash (D \supset (C \supset (\neg B \supset \neg A))) \supset (D \supset (C \supset (A \supset B)))$ Theorem (iii)
2. $\vdash ((\neg(C \supset \neg B) \supset A)) \supset (C \supset ((\neg A \supset \neg B))) \supset ((\neg(C \supset \neg B) \supset A)) \supset (C \supset (B \supset A)))$ Subst. in 1: $(\neg(C \supset \neg B) \supset A)/D$, A/B, B/A
3. $\vdash (D \supset (C \supset (B \supset A))) \supset (D \supset (B \supset (C \supset A)))$ BS Thm 12
4. $\vdash ((\neg(C \supset \neg B) \supset A) \supset (\neg A \supset (C \supset \neg B))) \supset ((\neg(C \supset \neg B) \supset A) \supset (C \supset (\neg A \supset \neg B)))$ Subst. in 3: $(\neg(C \supset \neg B) \supset A)/D$, $\neg A/C$, C/B, $\neg B/A$
5. $\vdash (\neg B \supset A) \supset (\neg A \supset B)$ BS Thm 33
6. $\vdash (\neg(C \supset \neg B) \supset A) \supset (\neg A \supset (C \supset \neg B))$ Subst. in 5: $C \supset \neg B/B$
7. $\vdash (\neg(C \supset \neg B) \supset A) \supset (C \supset (\neg A \supset \neg B))$ MP on 4 and 6
8. $\vdash (\neg(C \supset \neg B) \supset A)) \supset (C \supset (B \supset A))$ MP on 2 and 7

References

Aquinas, T. (n.d.[a]). *Summa Contra Gentiles*. URL: https://aquinas.cc/la/en/~SCG1.

– (n.d.[b]). *Summa Theologiae*. URL: https://aquinas.cc/la/en/~ST.I.

Baumann, J. J. (1868). *Die Lehren von Raum, Zeit und Matehmatik in der neueren Philosophie*. Berlin: Georg Reimer Verlag.

Boghossian, P. (1997). Analyticity. In: *The Blackwell Companion to Philosophy of Language*. Ed. by B. Hale and C. Wright. Oxford: Blackwell, 331–638.

Bolzano, B. (1837). *Wissenschaftslehre. Versuch einer ausführlichen und grösstentheils neuen Darstellung der Logik mit steter Rücksicht auf deren bisherige Bearbeiter, 4 volumes*. Sulzbach: J. E. v. Seidel.

Brentano, F. (1956). *Lehre vom richtigen Urteil*. Bern: Francke Verlag.

Burge, T. (1998). Frege on knowing the foundation. *Mind* 107, 307–347.

Church, A. (1956b). *Introduction to Mathematical Logic, Vol. I*. Princeton University Press.

Dummett, M. (1973b). *The Justification of Deduction, Annual Philosophical Lecture (Henriette Herz Trust), British Academy*. London: Oxford University Press.

– (1981b). *Frege. Philosophy of Language*. 2nd ed. London: Duckworth.

– (1991c). *Frege. Philosophy of Mathematics*. London: Duckworth.

Frege, G. (1879). *Begriffsschrift*. Cited as BS, after the facsimile edition in Matthias Wille (ed.), *Begriffsschrift*, Springer, Berlin, 2018. Halle: Lois Nebert.

– (1884b). *Grundlagen der Arithmetik*. Breslau: Verlag von Wilhelm Koebner. English translation by John Austin, (1952), The Foundations of Arithmetic, (second edition) Basil Blackwell, Oxford. Dutch translation by A. Phalet, (1981), De grondslagen der arithmetica, Het Wereldfenster, Baarn. Swedish translation by Fredrik Stjernberg, (2001), Aritmetikens grundvalar, Thales, 2002. English translation by Dale Jacquette, (2007), The Foundations of Arithmetic, Pearson, Longman, New York. Cited as GLA.

– (1892). Über Sinn und Bedeutung. *Zeitschrift für Philosophie und philosophische Kritik* 100, 25–50.

– (1976). Letter to Marty[?] In: *Gottlob Frege Wissenschaftlicher Briefwechsel*. Ed. by G. Gabriel et al. Hamburg: Felix Meiner, 163–165.

– (1984). *Collected Papers on Mathematics, Logic, and Philosophy*. Ed. by B. McGuinness. Oxford: Blackwell.

– (2007). *The Foundations of Arithmetic*. Translated by Dale Jacquette. New York: Pearson, Longman.

– (1893/1903). *Grundgesetze der Arithmetik, I, II*. Jena: Herman Pohle. English translation by P. Ebert and M. Rossberg, as Basic Laws of Arithmetic, Oxford University Press, 2013. Cited as GGA.

Gabriel, G. (2002). Frege, lotze and the continental roots of analytic philosophy. In: *From Frege to Wittgenstein. Perspectives on Early Analytic Philosophy*. Ed. by E. H. Reck. Oxford: Oxford University Press, 39–51.

Gabriel, G. and S. Schlotter (2017). *Frege und die kontinentalen Urspünge der analytischen Philosophie*. Münster: Mentis.

Gentzen, G. (1933a). Über die Existenz unabhängiger Axiomensysteme zu unendlichen Satzsystemen. *Mathematische Annalen* 107, 329–350.
– (1936). Die Widerspruchsfreiheit der reinen Zahlentheorie. *Mathematische Annalen* 112, 493–565.
Herbrand, J. (1930). Recherches sur la théorie de la démonstration. PhD thesis. Faculté des sciences de Paris.
Hilbert, D. and P. Bernays (1934a). *Die Grundlagen der Mathematik*. Berlin: Springer.
Jaśkowski, S. (1934c). On the Rules of Suppositions in Formal Logic. *Studia Logica* 1.
Kant, I. (1800/1904a). *Logik*. Ed. by G. B. Jäsche. English translation by R. S. Hartman and W. Schwartz, *Logic*, Dover, N.Y. 1988. Leipzig: Felix Meiner.
Kienzler, W. (2009). *Begriff und Gegenstand*. Frankfurt am Main: Vittorio Klostermann.
Kleene, S. C. (1952a). *Introduction to Metamathematics*. Amsterdam: North-Holland.
Kneebone, G. T. (1963). *Mathematical Logic and the Foundations of Mathematics*. Also Dover Publications, New York, 2001. London: Van Nostrand.
Kreiser, L. (2001). *Gottlob Frege. Leben – Werk – Zeit*. Hamburg: Felix Meiner.
Kutschera, F. von (1996a). Frege and Natural Deduction. In: *Frege: Importance and Legacy*. Ed. by M. Schirn. Berlin: Walter de Gruyter, 301–307.
Leibniz, G. W. (1973). *Philosophical Writings, edited and translated by G. H. R. Parkinson*. London: J. M. Dent and Sons.
– (1704/1962). *Nouveaux Essais sur lentendement humain*. Berlin: Akademieausgabe, VI.6.
Mill, J. S. (1843). *A System of Logic, Ratiocinative and Inductive, Being a Connected View of the Principles of Evidence and the Methods of Scientific Investigation, Vol. 1*. London: John Parker.
– (1868). *System der deductiven und inductiven Logik*. German translation by J. Schiel, of Mill (1843).
Millán, J. B.-S. (2018). Frege's Begriffsschrift, and Logicism. In: *LOGICA Yearbook 2017*. London: College Publications.
Millán, J. B.-S. and C. Badesa (2017). Function and Argument in Begriffsschrift. *History and Philosophy of Logic* 38 (4), 316–341.
– (2020). Begriffsschrift's Logic. *Notre Dame Journal of Formal Logic* 61, 409–440.
Proops, I. (2002). The Tractatus on Inference and Entailment. In: *From Frege to Wittgenstein. Perspectives on Early Analytic Philosophy*. Ed. by E. H. Reck. Oxford: Oxford University Press.
Quine, W. V. O. (1950). *Methods of Logic*. New York: Henry Holt and Co.
Schirn, M. (2010). On Translating Frege's *Die Grundlagen der Arithmetik. History and Philosophy of Logic* 31 (1), 47–72.
Schroeder-Heister, P. (1981c). Untersuchungen zur regellogischen Deutung von Aussagenverknüpfungen. PhD thesis. Universität Bonn.
– (1984d). A natural extension of natural deduction. *Journal of Symbolic Logic* 49, 1284–1300.
– (1984h). Freges Permutationsargument. Zu §10 der 'Grundgesetze der Arithmetik'. In: *Frege Conference 1984. Proceedings of the International Conference held at*

Schwerin (GDR), September 10–14, 1984. Ed. by G. Wechsung. Berlin: Akademie-Verlag, 182–188.
– (1987b). A model-theoretic reconstruction of Freges permutation argument. *Notre Dame Journal of Formal Logic* 28, 69–79.
– (1997c). Frege and the resolution calculus. *History and Philosophy of Logic* 18, 95–108.
– (1999b). Gentzen-Style Features in Frege. In: *Abstracts of the 11th International Congress of Logic, Methodology, and Philosophy of Science (Cracow, August 1999), Cracow, 1999*, 499.
– (2002c). Resolution and the origins of structural reasoning: Early proof-theoretic ideas of Hertz and Gentzen. *Bulletin of Symbolic Logic* 8, 246–265.
– (2014b). Frege's sequent calculus. In: *Trends in Logic XIII: Gentzen's and Jaśkowski's Heritage – 80 Years of Natural Deduction and Sequent Calculi*. Ed. by A. Indrzejczak, J. Kaczmarek, and M. Zawidzki. Łódż University Press, 233–245. DOI: 10.15496/publikation-72324.
– (2018d). Proof-Theoretic Semantics. In: URL: https://plato.stanford.edu/entries/proof-theoretic-semantics/.
Schroeder-Heister, P. and K. F. Wehmeier (2005). Frege's permutation argument revisited. *Synthese* 147, 43–61.
Schütte, K. (1950). Schlussweisen-Kalküle der Prädikatenlogik. *Mathematische Annalen* 122, 47–65.
Sundholm, G. (1983b). Systems of deduction. In: *Handbook of Philosophical Logic*. Ed. by D. Gabbay and F. Guenthner. Vol. I. Dordrecht: Reidel. Chap. I:2, 133–188.
– (1986a). Proof Theory and Meaning. In: *Handbook of Philosophical Logic*. Ed. by D. Gabbay and F. Guenthner. Vol. III. Dordrecht: Reidel. Chap. III:8, 471–506.
– (1988). *Oordeel en Gevolgtrekking. Bedreigde Species? (Judgement and Inference: Endangered Species?)* An inaugural lecture in the Chair of Logic, at Leiden University, delivered on September 9, 1988, and published in pamphlet form by that university.
– (1994). Proof-theoretical Semantics and Fregean Identity-criteria for Propositions. *The Monist* 77, 294–314.
– (1997). Implicit Epistemic Aspects of Constructive Logic. *Journal of Logic, Language, and Information* 6, 192–212.
– (1998a). Inference versus Consequence. In: *LOGICA Yearbook 1997*. Prague: Filosofia Publishers, Czech Academy of Sciences, 26–35.
– (2000). When, and why, did Frege read Bolzano? In: *Logica Yearbook 1999*. Prague: Filosofia Publishers, Czech Academy of Science, 164–174.
– (2006b). Semantic values of natural deduction derivations. *Synthese* 148 (3), 623–638.
– (2009). A Century of Judgment and Inference: 1837–1936. In: *The Development of Modern Logic*. Ed. by L. Haaparanta. Oxford University Press, 262–317.
– (2012). 'Inference versus consequence' revisited: inference, consequence, conditional, implication. *Synthese* 187, 943–956.
– (2019). Validity. In: *Proof-Theoretic Semantics: Assessment and Future Perspectives. Proceedings of the Third Tübingen Conference on Proof-Theoretic Semantics,*

27–30 March 2019. Ed. by T. Piecha and P. Schroeder-Heister. University of Tübingen, 237–252. URL: http://dx.doi.org/10.15496/publikation-35319.

Tarski, A. (1923). Sur le terme primitive de la Logistique. *Fundamenta Mathematicae* 4, 196–200.

– (1924a). Sur les truth-functions au sense de MM. Russell and Whitehead. *Fundamenta Mathematicae* 5, 59–74.

– (1924b). Über einige fundamentale Begriffe der Metamathematik. *Comptes Rendus des séances de la Socité des Sciences et des Lettres de Varsovie, XXIII, Classe III*, 22–29.

– (1930). Über einige fundamentale Begriffe der Metamathematik. In: *Comptes rendus des Séances de la Société des Sciences et des Lettes de Varsovie, Classe III*. Vol. 23, 22–29. English translation in Tarski (1956/1983).

– (1936). Über den Begriff der logischen Folgerung. In: *Actes du Congrès international de philosophie scientifique, Sorbonne, Paris 1935*. Fasc. VII, Logique. Paris: Hermann, 1–11. English translation in Tarski (1956/1983).

– (1956/1983). *Logic, Semantics, Metamathematics*. Indianpolis: Hackett Publishing Company.

Tichý, P. (1988a). *Freges Foundations of Logic*. Berlin: De Gruyter.

Tichý, P. and J. Tichý (1999). *On Inference*. Prague: Philosophical Institute, Czech Academy of Science, FILOSOFIA Publishers, 73–85.

Wittgenstein, L. (1921). Logisch-Philosophische Abhandlung. *Annalen der Naturphilosophische* XIV (3/4).

Frege's Class Theory and the Logic of Sets

Neil Tennant

Abstract We compare Fregean theorizing about sets with the theorizing of an ontologically non-committal, natural-deduction based, inferentialist. The latter uses free Core logic, and confers meanings on logico-mathematical expressions by means of rules for introducing them in conclusions and eliminating them from major premises. Those expressions (such as the set-abstraction operator) that form singular terms have their rules framed so as to deal with *canonical identity statements* as their conclusions or major premises. We extend this treatment to *pasigraphs* as well, in the case of set theory. These are defined expressions (such as 'subset of', or 'power set of') that are treated as basic in the *lingua franca* of informal set theory. Employing pasigraphs in accordance with their own natural-deduction rules enables one to 'atomicize' rigorous mathematical reasoning.

1 Introduction

Our honoree Peter has had abiding and deep interests both in Frege's work in logic, and in proof-theoretic semantics, a field in which he has played an important founding role. I thought it fitting, then, to combine a bit of both in this paper in his honor.

In his recent study Schroeder-Heister (2016), Peter's abstract reads as follows:

> I present three open problems the discussion and solution of which I consider relevant for the further development of proof-theoretic semantics: (1) The nature of hypotheses and the problem of the appropriate format of proofs, (2) the problem of a satisfactory notion of proof-theoretic harmony, and (3) the problem of extending methods of proof-theoretic semantics beyond logic.

This study will address (3), by venturing beyond logic to set theory. In seeking to provide a *natural and free* logic of sets, we shall also have some things to say about

Neil Tennant
Department of Philosophy, The Ohio State University, Columbus, OH, United States of America, e-mail: tennant.9@osu.edu

T. Piecha and K. F. Wehmeier (eds.), *Peter Schroeder-Heister on Proof-Theoretic Semantics*, Outstanding Contributions to Logic 29,
https://doi.org/10.1007/978-3-031-50981-0_3

(1) and (2). The first part of our journey will involve revisiting Frege, to examine why such a logic is called for, and how to set it up. We shall find that certain proof-theoretic constraints will make that 'it' — that logic — unique ... or so the line of exposition and development on offer here should lead one to believe.

The eventual goal, which will be reached by the end of this study, is to show how the logic of sets (which consists entirely of natural-deduction rules of inference) can be made *ontologically non-committal*. Its rules of inference will nevertheless be fully *meaning-conferring*. This observation applies not only to the central primitive notions — the variable-binding term-forming operator $\{x \mid \ldots x \ldots\}$ for set abstraction, and the binary relation $\in$ for membership — but also to all those ancillary expressions such as $\subseteq$ ('subset of'), $\bigcup$ ('union of'), $\mathscr{P}$ ('power set of'), etc. The latter notions — though of course logically definable in terms of the primitive notions — are so familiar and 'practically primitive' in the *lingua franca* of informally rigorous set theory that they call for a more focused rule-theoretic treatment. We shall call them *pasigraphs*, and furnish rules for them. Those rules will be meaning-conferring, but *still* incur no ontological commitments at all.

This means that we can furnish set theorists with a framework of logical rules for set-theoretic notions without committing them to an ontology. We can leave to them the job of specifying which sets exist outright, and which sets exist conditionally on the existence of which other sets.

Our foray in this study into the logic of sets is a protean, if rather ambitious, first step in a more general and unifying study of both *natural deduction and truthmaker semantics for pasigraphs*. The motivating idea is that every pasigraph will have introduction and elimination rules in a system of natural deduction governing one's deductive reasoning to and from sentences with the pasigraph in question suitably dominant.[1] In addition, every pasigraph will have (model-relative) verification- and falsification-rules for constructing logical truthmakers and falsitymakers of the kind described in Tennant (2018) and Tennant (2010). Such rules afford the pasigraphs what is essentially a proof-theoretic semantics. In the language of the pasigraphs, the notion of logical consequence will be defined in terms of how verifications for premises may be transformed into verifications for conclusions. The aim will then be to show how proofs in Classical Core Logic $\mathbb{C}^+$ afford 'quasi-'effective methods for carrying out such transformations of (model-relative) truthmakers. (The scare-quoted prefix will be able to be dropped in the constructive case where Core Logic $\mathbb{C}$ affords all the transformations required.)

---

[1] 'Suitable dominance' is plain dominance in the case of sentence-forming operators such as connectives and quantifiers. In the case of term-forming operators @, such as the set-abstraction operator $\{x \mid \Phi(x)\}$, the natural-deduction rules will govern inferences to and from 'canonical identity statements' of the form $t = @x\Phi(x)$. We shall expand on this below.

2 The natural and free logic of sets

There can be an analytically valid logic of sets, even if sets themselves are not *logical* objects. For the purposes of this study, the words 'set' and 'class' will be treated as synonyms. No von Neumann–Bernays–Gödel distinction will be countenanced, according to which sets are those classes small enough to be able to belong to yet other sets or classes, whereas (proper) classes are too big to do so, even though they exist.

Frege is the natural starting point for our study. His legacy of complete formalization, both of his logical resources and of the proofs he provided for his results, is invaluable when it comes to considering exactly what the *logic* of sets really is. Two other Fregean themes are of great importance here too.

1. Is this logic to be framed for a so-called 'logically perfect' language in Frege's sense (the most important feature of which, for the purposes of this study, is that every singular term denotes)? — or should one use instead a free logic that can handle non-denoting singular terms?
2. Is this logic to be formulated in such a way as to allow for the possibility that the universe of discourse also contains *Urelemente*?[2]—or should it concern only the more restricted mathematicians' universe of (hereditarily) *pure* sets?

Frege, as is well known, plumped for the first option in each of these cases. And as is also well known, his system suffered the disaster of Russell's paradox. That (in our view) was entirely owing to Frege's answer to question (1) — that every singular term denotes. His answer to question (2) — that we should allow for *Urelemente* — threatens no inconsistency at all, and is well worth implementing in any universally applicable logic of sets that recognizes that some things are *not* sets, and that some sets can have non-sets as members.

It will be argued here that the disaster of Russell's paradox stemmed solely from the misguided choice of a 'logically perfect' language for theorizing about sets, regardless of whether one speaks only of pure sets or allows for the 'impurities' of *Urelemente* in their membership pedigrees.

It will also be argued that the logic of sets that emerges for the revisionary Fregean who adopts a free logic is optimally formulated in terms of introduction and elimination rules (in natural deduction) for the set-abstraction operator

$$\{x \mid \ldots x \ldots\}.$$

Such rules will be stated in due course. This pulls one from the set-theoretic frontiers of Zermelo (1908), back to Fregean origins. The usual story about set theory is one

[2] *Urelemente* — if one's theory permits them — are individuals in the domain of discourse that are *not* sets (or classes). Simple examples would be ordinary physical objects, such as Hilbert's beer mugs, chairs and tables; or, in more sophisticated vein, the fundamental particles of subatomic physics. Not all *Urelemente*, however, have to be concrete individuals. They can be abstract, without being sets (or classes). One could, for example, treat the *natural numbers* as *sui generis* mathematical (or *logical*) objects, not to be identified with any 'set-theoretical surrogates' such as the finite von Neumann ordinals. One could then 'build sets' on top of them, as Weyl sought to do.

of the logicist being utterly vanquished, and the transition being made to a purely *mathematical* (synthetic *a priori*, at best) theory of abstract objects known as pure sets, characterized (as had been the natural, rational and real numbers) by an appropriate effectively decidable set of axioms and axiom schemata. The main implication of the investigation that will unfold here is that this 'mathematization' of set theory by Zermelo and his followers can be regarded as overly precipitous. It abandoned too early, and too pessimistically, the logicist's aim of characterizing at least the *logic* of our talk about sets. This logic embodies just the constraints governing or constituting the *concept* of set, rather than the existential or ontological commitments of any particular set theories.

At the very least, Zermelo's set theory makes it impossible to deal with *Urelemente* alongside sets. This is because its Axiom of Extensionality identifies any two things that have no members. The empty set (which Zermeloan set theory says exists) has no members; and no *Urelement* can have any members. Thus every *Urelement is* the empty set. But no *Urelement* is a set. So there are no *Urelemente*. Zermelo can be talking only about (hereditarily) *pure* sets. And it would remain a mystery how his set theory can find application in our talk about 'the real world' of physical objects, which are the paradigm examples of *Urelemente*. Another such example would be the natural numbers taken as objects *sui generis*, as they are in Reverse Mathematics. These too are really *Urelemente*, a subtlety often overlooked.

There is a line to be drawn between what is logico-analytically valid in our theorizing about sets in general, and which of them have to be specifically postulated, outright or conditionally, as existing. We shall learn that the natural-deduction theorist who is sympathetic to the pursuit of a logic of sets can make a distinctive contribution by taking a very careful look at what was going on in Frege's first systematic stab at the problem. The *Core Logicist* can sharpen the tools Frege left us in a way that is interestingly and significantly short of total mathematizing surrender to the disaster that was Russell's paradox.

The Core Logicist is the theorist who follows the methodological maxim that rules of inference serving to fix the meanings of primitive logico-mathematical expressions have their natural niche in the constructive and relevant deductive reasoning characterized by Core Logic. Conceptual interconnections articulated by definitional rules of inference are constructive and relevant. The aforementioned 'logic of sets' will be generated by using the rules of (free) Core Logic for the usual logical operators, along with well-chosen rules of natural deduction governing set-abstraction.

As explained in Tennant (2017), Core Logic, in its natural deduction formulation, has all its elimination rules in 'parallelized' form. Moreover, their major premises always *stand proud*, with no non-trivial proof-work above them. This ensures two important features: (i) all core natural deductions are in normal form; and (ii) they are also, in a naturally definable sense, isomorphic to the corresponding sequent-calculus proofs. In Core Logic, sequent proofs use Reflexivity as their only structural rule; and otherwise consist only of applications of Right rules and/or Left rules for the operators involved. So core sequent proofs are both cut-free and thinning-free. Right rules in sequent calculus correspond to introduction rules in natural deduction; while Left rules correspond to elimination rules.

A simple example of a natural deduction and its corresponding sequent proof will serve to fix these ideas. Note how the step of $\wedge$-Elimination labeled (2) (and with major premise $A \wedge B$) is in parallelized form, and discharges the conjuncts A and B at their assumption occurrences.

Natural Deduction

$$\dfrac{\dfrac{\overset{(i)}{\overline{A\wedge B}} \qquad \dfrac{\neg A\vee\neg B \qquad \dfrac{\overline{\neg A}^{\,(3)}\quad \overline{A}^{\,(2)}}{\bot} \qquad \dfrac{\overline{\neg B}^{\,(3)}\quad\overline{B}^{\,(2)}}{\bot}}{\bot}\,(3)}{\bot}\,(2)}{\dfrac{\bot}{\neg(A\wedge B)}\,(1)}$$

Sequent Proof

$$\dfrac{\dfrac{\dfrac{\dfrac{A:A}{\neg A, A:}\qquad\dfrac{B:B}{\neg B,B:}}{\neg A\vee\neg B, A, B:}}{\neg A\vee\neg B, A\wedge B:}}{\neg A\vee\neg B:\neg(A\wedge B)}$$

3 A look at some Fregean basics

Consider this formal sentence in Frege's now archaic notation:

$$\Delta \frown \grave{\varepsilon}\Phi(\varepsilon)$$

Today it would be written

$$\Delta \in \{x \mid \Phi(x)\}.$$

For Frege, Δ stood for an individual, and Φ for a first-level concept. Frege stipulated in his *Grundgesetze* that the sentence of his displayed form above was to be co-referential[3] with

$$\Phi(\Delta).$$

This would mean, for the modern inferentialist, that Frege would regard as logically or analytically valid the two inference rules[4]

$$\text{F1}\quad \dfrac{\Phi(t)}{t \in \{x \mid \Phi(x)\}} \qquad \text{and} \qquad \text{F2}\quad \dfrac{t \in \{x \mid \Phi(x)\}}{\Phi(t)}.$$

Here we use t instead of Frege's Δ as a placeholder for singular terms. We shall do this throughout, when couching things in natural-deduction terms.

Frege wanted also to have his first-order binary membership relation

$$\xi \frown \zeta$$

explained 'für alle möglichen Gegenstände als Argumente'. The explanatory definition he offered was as follows (here, for $a \frown u$):[5]

[3] The German term was 'gleichbedeutend' (Frege, 1893, §34, at p. 52). All English translations of material quoted from Frege are taken from Frege (2013).

[4] See, for example, the Appendix in Prawitz (1965).

[5] *Ibid.*, p. 53.

(Def.$\frown$) 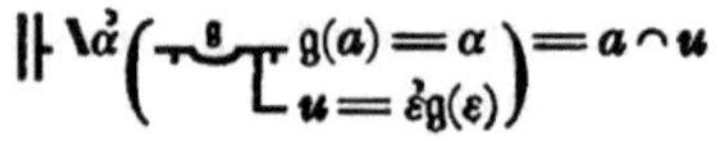

We seek to render (Def.$\frown$) in notation we use today. In preparing to do so, we need to remind ourselves that any singular term of the form

$$\backslash\dot{\varepsilon}\,\Phi(\varepsilon)$$

was Frege's version of a definite description ('the x such that $\Phi(x)$'), but with the strange twist — in fulfillment of Frege's strict self-imposed requirement that all well-formed singular terms should denote — that, should there *not* be exactly one Φ, the denotation of the displayed term is the class of all Φs. So, if Φ is an empty concept, then the denotation of the displayed term is the empty class; while if more than one object falls under the concept Φ, then the denotation is the class (or set) that they form.[6] In contemporary notation (using iota as the modern definite-description operator) we may render Frege's definition of the displayed term as follows:

$$\backslash\dot{\varepsilon}\,\Phi(\varepsilon) =_{df} \begin{cases} \iota x\Phi(x) & \text{if } \exists x\forall y(x = y \leftrightarrow \Phi(y)); \\ \{x \mid \Phi(x)\} & \text{otherwise.} \end{cases}$$

Let us now turn to the task of translating into modern notation Frege's definition

(Def.$\frown$) 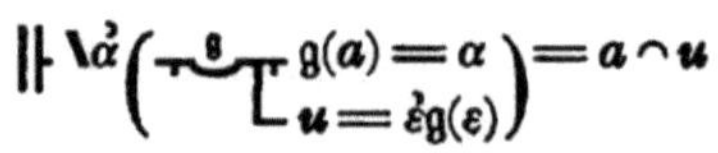

Remember that this definitional identity is an identity between the truth-value of the left-hand side:

$$\backslash\dot{\alpha}\,(\ldots\alpha\ldots)$$

and the truth-value of the right-hand side:

$$a \frown u.$$

The parenthetically enclosed material on the left-hand side is Frege's way of rendering the following second-order sentence in modern notation:

$$\neg\forall G(u = \{x \mid G(x)\} \rightarrow \neg G(a) = \alpha).$$

This is classically equivalent to

$$\exists G\neg(u = \{x \mid G(x)\} \rightarrow \neg G(a) = \alpha),$$

[6] See Frege (1893), §11. As Roy Cook puts it in his Appendix in Frege (2013) (at p. A-19),

> [...] '$\backslash\dot{\varepsilon}\,\Phi(\varepsilon)$' denotes the unique object that is mapped to the True by the concept named by '$\Phi(\xi)$', if there is such, and denotes the object named by '$\dot{\varepsilon}\,\Phi(\varepsilon)$' otherwise.

which in turn is classically equivalent to

$$\exists G(u=\{x \mid G(x)\} \wedge G(a)=\alpha).$$

So (Def. $\frown$) is asserting an identity between the truth-value of $a \frown u$ and the truth-value denoted by the Fregean definite description (here rendered in mixed notation)

$$\backslash\dot{\alpha}\,(\exists G(u=\{x \mid G(x)\} \wedge G(a)=\alpha)).$$

For this to be the True (*das Wahre*), the term u has to denote an object — the extension (*Werthverlauf*) of some appropriate concept G. Thus the innocent-looking first-order binary predication

$$a \frown u.$$

commits one to the existence of a denotation for the term u. This is of a piece with Frege's insistence that a 'logically perfect' language has all its singular terms denoting.

4 Moving on from Frege

Suppose we abandon that very imperfect conception of logical perfection, and work with a free logic. Let $\exists!t$ be the familiar abbreviation for $\exists x\, x=t$, for any singular term t. Free logic has the Rule of Atomic Denotation for atomic predicates A:[7]

$$\text{RAD} \quad \frac{A(\ldots t \ldots)}{\exists!t};$$

and expresses the Reflexivity of Identity by the rule

$$\text{Ref=} \quad \frac{\exists!t}{t=t}.$$

Note that these are, respectively, logical *weakenings* of the zero-premise rules

$$\overline{\exists!t} \qquad \text{and} \qquad \overline{t=t}$$

to which Frege, with his 'logically perfect' language, was already committed. So one could invite Frege to recognize the validity of these weakened rules (along with those about to be stated) even though one is now working in a logically 'imperfect' language. In the transition to free logic, the rule of Substitutivity of Identicals remains unchanged:

---

[7] See Tennant (1978), Ch. 7 §10 for a detailed treatment of free logic and the rules for set theory that we are presenting again here. The rule RAD captures the Russellian requirement that an atomic proposition is true only if all its singular terms denote. Of course it is required in addition that the denotations stand in the relation represented by the predicate of the atomic proposition concerned. The rule RAD captures just the existential presuppositions concerning the singular terms involved.

$$\text{Sub.=}\quad \frac{\varphi \qquad t = u}{\psi}$$

where ψ results from φ by at least one substitution of an occurrence of t for one of u or of an occurrence of u for one of t.

The natural-deduction rules about to be stated aim to characterize no more than the *logic* of talk about sets. To this end, one needs to clarify the interrelationships among sethood (i.e., existence of a set), set-abstraction, predication, and membership. This is a theoretical or foundational aim that the natural-deduction theorist can *share* with the Fregean. This study will investigate how the two theorists can pursue that aim, and whether one of them can claim to have achieved it in a more satisfactory fashion. Our answer, of course, will be that the natural-deduction theorist, with her free logic, is the winner in this comparison.

The rules that the two theorists can formulate are for a language with the set-term forming operator $\{x \mid \ldots x \ldots\}$ primitive. Also primitive is the two-place predicate $\in$ of membership. In due course another primitive (but one-place) predicate S will be added to the language (for 'is a set'). The working assumption will be that the same formal-linguistic resources are available to both theorists (Fregean and natural-deduction), so that the comparison of their approaches will be based on the primitive logical rules that they postulate for the same language. Bear in mind that an axiom is here construed as a zero-premise rule.

4.1 Natural-deduction rules for pure sets

Some notational preliminaries. Where t is a closed term (which of course could be a parameter) and Φ is a formula with just the variable x free, we denote by Φ^x_t the result of replacing every free occurrence of x in Φ with an occurrence of t. Where Φ is a sentence involving at least one occurrence of the parameter a, and with none of those occurrences within the scope of a variable-binding operator applied to the variable x, we denote by Φ^a_x the result of replacing every occurrence of a in Φ with an occurrence of x. Note that every such occurrence of x in Φ^a_x is free.

The rule of introduction (in free logic) for the variable-binding abstraction operator that forms set-terms from predicates is

$$\{\,\}\mathrm{I}\qquad \frac{\begin{array}{c}\underbrace{{}^{(i)}\overline{\exists!a}\;,\;\overline{\Phi^x_a}^{(i)}}\\ \vdots\\ a\in t\end{array}\qquad \exists!t \qquad \begin{array}{c}\overline{a\in t}^{(i)}\\ \vdots\\ \Phi^x_a\end{array}}{t=\{x\mid\Phi\}}{}^{(i)}\quad\text{, where } a \text{ is parametric.}^{8}$$

Note how the canonical conclusion

[8] Note that since $\in$ is an *atomic* binary predicate, the assumption $a \in t$ in the rightmost subordinate proof implies $\exists!a$ (by the rule RAD). So is it not necessary to have $\exists!a$ as a further dischargeable assumption in that subordinate proof.

$$t = \{x \mid \Phi\}$$

of { }I has t on its left-hand side, as a placeholder for *any singular term whatsoever*, including the *parameters* (conventionally $a, b, c, \ldots$) that can be used for reasoning involving existentials and universals. On the right-hand side of the identity is a set-abstraction term, formed by means of a *dominant* occurrence of the variable-binding abstraction operator $\{x \mid \ldots x \ldots\}$. This operator may be applied to a formula Φ to form the set-abstraction term $\{x \mid \Phi\}$ if, but only if, the variable x has a free occurrence in Φ.

The elimination rules corresponding to the introduction rule stated above for { } are the following three, each one employing the canonical identity statement

$$t = \{x \mid \Phi\}$$

as its major premise (to the left, immediately above the inference stroke). The minor premises (or subproofs) of the first and third rules correspond, respectively, to the first and third immediate subproofs of the introduction rule. This is a convincing sign that the elimination rules are in harmony with the introduction rule that begets them. Bear in mind that the major premise $t = \{x \mid \Phi\}$ stands proud in any application of an elimination rule for { }.

$$\{\,\}\mathrm{E}_1 \quad \frac{t = \{x \mid \Phi\} \qquad \exists! v \qquad \Phi^x_v}{v \in t}$$

$$\{\,\}\mathrm{E}_2 \quad \frac{t = \{x \mid \Phi\}}{\exists! t}$$

$$\{\,\}\mathrm{E}_3 \quad \frac{t = \{x \mid \Phi\} \qquad v \in t \qquad \begin{array}{c}\overline{\Phi^x_v}^{\,(i)} \\ \vdots \\ \theta\end{array}}{\theta}\,(i)$$

The rule { }E$_1$ has an atomic conclusion, so it is not necessary to parallelize it for the purposes of Core Logic. This is because no atomic conclusion can feature as the major premise of any elimination. The rule { }E$_2$ is a special case of the Rule of Atomic Denotation. The rule { }E$_3$ needs to be parallelized, in order to avoid having non-trivial proof-work above Φ^x_v should it happen to stand as the major premise of an elimination.[9]

The introduction rule and the elimination rules just stated for { } are, as just intimated, in harmony. Harmony requires that there at least be reduction procedures that will eliminate from proofs 'maximal sentence occurrences' — conclusions of introductions that are also major premises of eliminations. (Whether harmony requires *more* than this is a more complicated issue that we shall not broach here.) What follow now are the three reduction procedures that are required for harmony (one for each

[9] This risk would be incurred if { }E$_3$ were to be stated in the serial form $\dfrac{t = \{x \mid \Phi\} \qquad v \in t}{\Phi^x_v}$.

elimination rule). We state them in the notation of Tennant (2017)[10] rather than in our original format in Tennant (1978). We shall refrain from providing reduction procedures for any other Introduction-Elimination pairs, since (as a referee was kind enough to observe) they 'write themselves'.

$$\left[\dfrac{\begin{array}{c}\underbrace{\Delta_1 ,\ {}^{(i)}\overline{\exists!a} , \overline{\Phi^x_a}^{(i)}}\\ \Pi_1 \\ a\in t\end{array} \quad \begin{array}{c}\Delta_2\\ \Pi_2\\ \exists!t\end{array} \quad \begin{array}{c}\underbrace{\Delta_3 , \overline{a\in t}^{(i)}}\\ \Pi_3\\ \Phi^x_a\end{array}}{t=\{x\mid\Phi\}}{\scriptstyle (i)} \; , \quad \dfrac{t=\{x\mid\Phi\} \quad \begin{array}{c}\Gamma_1\\ \Sigma_1\\ \exists!v\end{array} \quad \begin{array}{c}\Gamma_2\\ \Sigma_2\\ \Phi^x_v\end{array}}{v\in t}{\scriptstyle \{\,\}\mathrm{E}_1}\right]$$

$$= \left[\begin{array}{c}\Gamma_1\\ \Sigma_1\\ \exists!v\end{array} , \left[\begin{array}{c}\Gamma_2\\ \Sigma_2\\ \Phi^x_v\end{array} , \begin{array}{c}\underbrace{\Delta_1 , \exists!v , \Phi^x_v}\\ \Pi_1{}^{a}_{v}\\ v\in t\end{array}\right]\right]$$

$$\left[\dfrac{\begin{array}{c}\underbrace{\Delta_1 ,\ {}^{(i)}\overline{\exists!a} , \overline{\Phi^x_a}^{(i)}}\\ \Pi_1 \\ a\in t\end{array} \quad \begin{array}{c}\Delta_2\\ \Pi_2\\ \exists!t\end{array} \quad \begin{array}{c}\underbrace{\Delta_3 , \overline{a\in t}^{(i)}}\\ \Pi_3\\ \Phi^x_a\end{array}}{t=\{x\mid\Phi\}}{\scriptstyle (i)} \; , \quad \dfrac{t=\{x\mid\Phi\}}{\exists!t}{\scriptstyle \{\,\}\mathrm{E}_2}\right] = \begin{array}{c}\Delta_2\\ \Pi_2\\ \exists!t\end{array}$$

$$\left[\dfrac{\begin{array}{c}\underbrace{\Delta_1 ,\ {}^{(i)}\overline{\exists!a} , \overline{\Phi^x_a}^{(i)}}\\ \Pi_1 \\ a\in t\end{array} \quad \begin{array}{c}\Delta_2\\ \Pi_2\\ \exists!t\end{array} \quad \begin{array}{c}\underbrace{\Delta_3 , \overline{a\in t}^{(i)}}\\ \Pi_3\\ \Phi^x_a\end{array}}{t=\{x\mid\Phi\}}{\scriptstyle (i)} \; , \quad \dfrac{t=\{x\mid\Phi\} \quad \begin{array}{c}\Gamma_1\\ \Sigma_1\\ v\in t\end{array} \quad \begin{array}{c}\underbrace{\Gamma_2 , \overline{\Phi^x_v}^{(i)}}\\ \Sigma_2\\ \theta\end{array}}{\theta}{\scriptstyle \{\,\}\mathrm{E}_3}\right]$$

$$= \left[\left[\begin{array}{c}\Gamma_1\\ \Sigma_1\\ v\in t\end{array} , \begin{array}{c}\underbrace{\Delta_3 , v\in t}\\ \Pi_3{}^{a}_{v}\\ \Phi^x_v\end{array}\right] , \begin{array}{c}\underbrace{\Gamma_2 , \Phi^x_v}\\ \Sigma_2\\ \theta\end{array}\right]$$

---

[10] Note that in Core Logic the reduction procedures are used only in proving the *admissibility* of 'Cut with potential epistemic gain'. All core proofs are in normal form. Reductions therefore do not eliminate maximal occurrences from core proofs, because there aren't any such occurrences in core proofs. Reductions come into the picture only when core proofs are 'strung together', with the conclusion of one core proof occurring as a premise of another. The applicable reductions, when carried out by the core logician, will then furnish a core proof of some subsequent of the 'target sequent' that the follower of Gentzen would be happy to prove by stringing proofs together and repeatedly applying his structural rule of CUT. For the core logician, CUT is not a rule *of* or *in* the system. Nor is THINNING.

A degenerate application of { }I ensures that everything is the set of its members:

Theorem 4.1 *If t exists, then t is the set of all things bearing ∈ to t.*

Proof

$$\dfrac{\overset{(1)}{\overline{a \in t}} \quad \exists! t \quad \overset{(1)}{\overline{a \in t}}}{t = \{x \mid x \in t\}}\ (1)\ \{\ \}\mathrm{I}$$

□

So one needs to bear in mind that, with { }I in its present form, the universe of discourse is presumed to consist only of sets and to have no *Urelemente*. This is why the title of this subsection indicates that our rules are for theorizing about (hereditarily) *pure* sets.[11]

4.2 Natural-deduction rules for impure sets

If *Urelemente are* to be countenanced — meaning that one has to allow for the possibility of (hereditarily) *impure* sets — then the second premise (currently $\exists!t$) of the rule { }I will have to take the form St, to be interpreted as 't is a set'. By the Rule of Atomic Denotation this will also secure $\exists!t$, since S is an atomic predicate.

The S-modified introduction rule for { } is as follows.

$$\{\ \}\mathrm{I} \qquad \dfrac{\begin{matrix} \underbrace{\overset{(i)}{\overline{\exists! a}}\ ,\ \overset{(i)}{\overline{\Phi^x_a}}} \\ \vdots \\ a \in t \end{matrix} \qquad St \qquad \begin{matrix} \overset{(i)}{\overline{a \in t}} \\ \vdots \\ \Phi^x_a \end{matrix}}{t = \{x \mid \Phi\}}\ (i)$$

, where a is parametric.

As a result of this modification, a corresponding change is needed only to { }E_2 among the elimination rules:

$$\{\ \}\mathrm{E}_2 \qquad \dfrac{t = \{x \mid \Phi\}}{St}.$$

Note that the S-modified { }E_2 is *not* an instance of the Rule of Atomic Denotation. With S-modification the rule { }E_2 has, as it were, 'come into its own' as an elimination rule making its own distinctive contribution.

The S-modified introduction rule for { } is in harmony with its corresponding elimination rules.

4.3 Single-barreled vs. double-barreled abstraction

The kind of introduction rule being considered here (with or without S-modification) will be called a *single-barreled* rule, because of the single occurrence of the set-

[11] The introduction and elimination rules just stated were first given in Tennant (1978).

abstraction operator in the rule's conclusion, dominant on the right-hand side. The intended contrast is with a *double-barreled* rule, such as Frege's ill-fated[12]

$$\text{(Va)}\quad \frac{\forall x(\Phi x \leftrightarrow \Psi x)}{\{x \mid \Phi x\} = \{x \mid \Psi x\}},$$

or even its 'free-logical' modification

$$\text{(fVa)}\quad \frac{\forall x(\Phi x \leftrightarrow \Psi x) \qquad \exists!\{x \mid \Phi x\}}{\{x \mid \Phi x\} = \{x \mid \Psi x\}}.$$

Such rules venture to 'introduce' (in the conclusion) *two* occurrences of the set-abstraction operator, one on each side of the identity sign. This represents a prima-facie limitation on the intended range of identifications afforded by the rule, since it does not involve the more general placeholder t that would be replaceable in licit applications by names or parameters in addition to set-abstraction terms.

If one is countenancing the possibility of *Urelemente*, and accordingly using the predicate S for '. . .is a set', then Theorem 4.1 will read 'If t is a set, then t is the set of all things bearing $\in$ to t.' And in the proof of this one will simply substitute St for $\exists!t$ in the proof of Theorem 4.1. This is worth stating as a separate theorem.

Theorem 4.2 (for the S-modified rule $\{\ \}$I) $\dfrac{St}{t = \{x \mid x \in t\}}$.

Proof

$$\frac{\overset{(1)}{\overline{a \in t}} \qquad St \qquad \overset{(1)}{\overline{a \in t}}}{t = \{x \mid x \in t\}}\ (1)\ \{\ \}\text{I, } S\text{-modified form.}$$

□

The introduction and elimination rules are *ontologically neutral* — they characterize only the *logic* of one's talk about set abstraction, membership, and predication, not one's theory about what sets actually exist. Depending on a mutually agreed decision to confine one's theorizing to pure sets, or, alternatively, to allow for impure sets, *Frege would have conceded the validity of the corresponding natural-deduction rules — especially* the validity of the *elimination* rules for the set-term forming operator.

[12] The rules stated here as Va (on this page) and Vb (on page 102) are respectively the inferential equivalents of Frege's own Va and Vb on p. 69 of the *Grundgesetze*, in §53. From our free-logical perspective it is Va that is disastrous, in permitting easy derivation of Russell's paradox. Ironically, Frege himself, in his *Nachwort*, presented a regimentation of Russell's reasoning in the formalism of the *Grundgesetze*, which ended up laying the blame for Russell's paradox on Vb. (I am grateful here to Peter, for drawing this to my attention.) I rather suspect, though, that if one were to regiment *in natural deduction* Frege's own *Nachwort* reconstruction, within his own formal system, of Russell's reasoning, we would find that Frege was blaming the 'wrong half' of Basic Law V. Vindicating this suspicion, however, is beyond the scope of the present paper.

5 Results provable by the Fregean or by the natural-deduction theorist

In what follows, the *Theorems* stated (at least, up to and including Theorem 8.3) are results to the effect that such-and-such rules of the natural-deduction theorist allow one to derive so-and-so principle of the Fregean; and the *Lemmas* (at least, up to and including Lemma 8.9) are to the effect that so-and-so principles of the Fregean allow one to derive such-and-such rule of the natural-deduction theorist. The convenient abbreviation

$$R_1, \ldots, R_n \Rightarrow R$$

will be used to state these results. The rules $R_1, \ldots, R_n$ will be primitive for one of the theorists, and the rule R will be primitive or derivable for the other. Any other rules not mentioned, but which are used in the derivation, will be ones that are primitive for both of them (such as, for example, the Rule of Atomic Denotation). The aim, in the first instance, is to see whether the two theoretical approaches (roughly: Fregean vs. natural-deduction) are essentially equivalent.

The reader should be aware that we shall apply { }E$_3$ in its serial form (see footnote 9) rather than its parallelized form whenever it is convenient to do so. A good example of this is at the final step of the formal proof that follows, of Theorem 5.1.

Theorem 5.1 { }E$_3$ $\Rightarrow$ F2.

Proof

$$\dfrac{\dfrac{\dfrac{t \in \{x \mid \Phi(x)\}}{\exists!\{x \mid \Phi(x)\}}\,\text{RAD}}{\{x \mid \Phi(x)\} = \{x \mid \Phi(x)\}}\,\text{Ref=} \qquad t \in \{x \mid \Phi(x)\}}{\Phi(t)}\,\{\ \}\text{E}_3$$

□

If Frege had been instructed on how to construct proofs using rules of inference in the manner employed here, he would have derived the rule

$$\{\ \}\text{E}_1 \quad \dfrac{t = \{x \mid \Phi\} \qquad \exists!v \qquad \Phi^x_v}{v \in t}$$

in the following even stronger form (by not availing himself of the premise $\exists!v$).

Lemma 5.2 F1 $\Rightarrow$ { }E$_1$.

Proof

$$\dfrac{\dfrac{\Phi^x_v}{v \in \{x \mid \Phi\}}\,\text{F1} \qquad t = \{x \mid \Phi\}}{v \in t}$$

□

Bear in mind: one is talking in this instance of the Frege who is committed to each singular term's enjoying a denotation. That is why he would have eschewed the free logician's needed extra premise $\exists!v$ in the rule { }E$_1$.

Frege would also have been able to derive the rule { }E$_3$:

$$\{\ \}\text{E}_3 \quad \dfrac{t = \{x \mid \Phi\} \qquad v \in t}{\Phi^x_v}$$

Lemma 5.3 F2 $\Rightarrow$ { }E_3.

Proof

$$\dfrac{\dfrac{v \in t \qquad t = \{x \mid \Phi\}}{v \in \{x \mid \Phi\}}}{\Phi^x_v}\,\text{F2}$$

□

Suppose, for the sake of some imaginary and counterfactual speculation, that Frege could have been induced to consider the possibility (in advance of Russell's paradox) that certain kinds of singular terms might not always be secured denotations. He could have been invited to consider the possibility that the extensions of certain concepts were impossible to comprehend as individual, completed entities. Using Kripke's metaphor: such an 'extension' (because of some peculiarity of the concept whose extension it would erroneously be supposed to be) might resist being captured within the limited embrace of any 'intellectual lasso' trying to draw together all its members.

This speculative Frege, one presumes, would have recognized the free-logical validity of the rule { }E_1, and would have remained content — with the free logician's concurrence — with the derivation which uses the rule F2 and which was supplied on his behalf in Lemma 5.3 for the rule { }E_3. But he would have modified his erstwhile rule F1 to become fF1 ('fF' here for 'free-logical Frege'), furnished with the two existential presuppositions that are needed in the free-logical context:

$$\text{fF1} \quad \dfrac{\Phi(t) \qquad \exists! t \qquad \exists!\{x \mid \Phi(x)\}}{t \in \{x \mid \Phi(x)\}}$$

Lemma 5.4 fF1 $\Rightarrow$ { }E_1.

Proof

$$\dfrac{\dfrac{\Phi(v) \qquad \exists! v \qquad \dfrac{t = \{x \mid \Phi\}}{\exists!\{x \mid \Phi\}}}{v \in \{x \mid \Phi\}}\,\text{fF1} \qquad t = \{x \mid \Phi\}}{v \in t}$$

□

The natural-deduction theorist can return the favor, with the following converse.

Theorem 5.5 { }$E_1 \Rightarrow$ fF1.

Proof

$$\dfrac{\begin{matrix}\exists!\{x \mid \Phi(x)\},\ \text{i.e.}\\ \exists y\, y = \{x \mid \Phi(x)\}\end{matrix} \qquad \dfrac{\dfrac{\dfrac{}{a = \{x \mid \Phi(x)\}}^{(1)} \qquad \exists! t \qquad \Phi(t)}{t \in a}\,\text{E}_1 \qquad \dfrac{}{a = \{x \mid \Phi(x)\}}^{(1)}}{t \in \{x \mid \Phi(x)\}}}{t \in \{x \mid \Phi(x)\}}\,(1)$$

□

6 Russell's paradox

Here we shall be scrupulous in using the 'official', parallelized form of the rule { }E_3. This is in order to ensure the correctness of the claim that the Core logician can show that the Russell set does not exist.

Let us use the abbreviation r for the set-term $\{x \mid \neg x \in x\}$. Consider the following proof

$$\begin{array}{c}\exists! r\\ \Sigma\\ \neg r \in r\end{array} \ : \quad \dfrac{\dfrac{\dfrac{\exists! r}{r = \{x \mid \neg x \in x\}}\,\text{Ref.=} \qquad \dfrac{}{r \in r}\,(1) \qquad \dfrac{(2)\,\dfrac{}{\neg r \in r} \qquad \dfrac{}{r \in r}\,(1)}{\bot}}{\bot}\,\{\ \}\mathrm{E}_3}{\neg r \in r}\,(1)$$

Now use Σ to construct the proof

$$\begin{array}{c}\exists! r\\ \Xi\\ r \in r\end{array} \ : \quad \dfrac{\dfrac{\exists! r}{r = \{x \mid \neg x \in x\}} \qquad \exists! r \qquad \begin{array}{c}\exists! r\\ \Sigma\\ \neg r \in r\end{array}}{r \in r}\,\{\ \}\mathrm{E}_1$$

Finally, embed Ξ twice as follows, to form the (dis)proof

$$\begin{array}{c}\exists! r\\ \Pi\\ \bot\end{array} \ : \quad \dfrac{\dfrac{\exists! r}{r = \{x \mid \neg x \in x\}}\,\text{Ref.=} \qquad \begin{array}{c}\exists! r\\ \Xi\\ r \in r\end{array} \qquad \dfrac{(1)\,\dfrac{}{\neg r \in r} \qquad \begin{array}{c}\exists! r\\ \Xi\\ r \in r\end{array}}{\bot}}{\bot}\,(1)\ \{\ \}\mathrm{E}_3$$

The disproof Π is in normal form. It avails itself of only the following rules:

1. Rule of Atomic Denotation,
2. Rule of Reflexivity of Identity,
3. $\neg$-Introduction,
4. $\neg$-Elimination,
5. $\{\ \}\mathrm{E}_1$,
6. $\{\ \}\mathrm{E}_3$,

and all of these would be acceptable to Frege. Conspicuously absent is any appeal to Basic Law V (or, more accurately, (Va)). So Frege was in error in concluding

> The error can only lie in our Law (Vb) which must therefore be false.
> (Der Fehler kann allein in unserm Gesetze (Vb) liegen, das also falsch sein muss.)

(See the *Nachwort* in Frege (1903), at p. 257.) To be sure, as will be seen by the end of this study, one can use (Va) to get into Russellian trouble; but Russell's paradox can (as just seen) be derived from much more basic logical materials, in a manner whose strict formalization makes no use at all of the beknighted (Va). Moreover, Russell's result, in this free-logical setting, is not a paradox at all. Rather, Π is a normal-form disproof of the claim $\exists! r$, that is, of $\exists y\, y = \{x \mid \neg x \in x\}$. By the rule $\neg$I it straightforwardly yields the negative existential theorem $\neg\exists y\, y = \{x \mid \neg x \in x\}$ in the logic of sets.

Constructivism and intuitionism in logic and mathematics, especially as formalisms, came well after Volume 2 of Frege's *Grundgesetze*. So Frege cannot be expected to have sought a constructive reductio, such as Π above, of the assumption that the Russell set exists. But it is worth pointing out that Π is indeed constructive. There is no application within Π of any strictly classical rule for negation. Frege was barking up a wrong tree when he informally invoked the Law of Excluded Middle in his informal presentation (in his *Nachwort* to Volume 2 of the *Grundgesetze*) of the Russellian reasoning. As he pondered the right revisionary response to Russell's paradox, he considered the possibility that one might have to abandon the Law of Excluded Middle. He seriously posed the question

> Should we assume the law of excluded middle fails for classes?
> (Sollen wir annehmen, das Gesetz vom ausgeschlossenem Dritten gelte von den Klassen nicht?)

(See the *Nachwort* in Frege (1903), at p. 254.) We can see now, however, that placing the blame on Excluded Middle, and abandoning it as a logical law, would have been futile. *For the Russell Paradox is a problem for the constructivist*, not just the classicist. Frege could have performed the formal reasoning in Π (even suppressing the middle premise of each application of $\{\ \}E_1$ therein), to reduce the assumption $\exists!\{x \mid \neg x \in x\}$ to absurdity, without any appeal to Excluded Middle.

Frege's ultimate mistake, on this analysis of the Russell Paradox, is his assumption that every grammatically well-formed singular term must denote something. It is ironic that he required this of any 'logically perfect' language. The real folly of Basic Law V is how it visits Fregean 'logical perfection' on set-abstraction terms, *even in the context of an explicitly free logic.* The folly can be seen at work in the direction given by (Va), from coextensiveness of concepts to identity of their extensions. Recall that (Va) can be expressed as a rule as follows:

$$\frac{\forall x(\Phi x \leftrightarrow \Psi x)}{\{x \mid \Phi x\} = \{x \mid \Psi x\}}$$

Take Φ for Ψ. Then *in free logic* we have the proof

$$\cfrac{\cfrac{\vdots\ \text{Logic} \atop \forall x(\Phi x \leftrightarrow \Phi x)}{\{x \mid \Phi x\} = \{x \mid \Phi x\}}\ \text{Va}}{\exists!\{x \mid \Phi x\}}\ \text{RAD}$$

So (Va) commits one to the existence of the set $\{x \mid \Phi x\}$ for *every* concept (or predicate) Φ.

It is frequently remarked that Frege's error was to believe in the *Axiom Schema of Naive Comprehension*:

$$\forall\Phi\exists X\forall y(y \in X \leftrightarrow \Phi y).$$

This is indeed derivable — even in free logic — by appeal to (Va). It is worth pointing out that the derivation makes use only of the *elimination* rules $\{\ \}E_1$ and $\{\ \}E_3$ for the set-abstraction operator. No use is made either of $\{\ \}E_2$ or of the introduction rule. This in turn means that what the following proof reveals is invariant across the 'pure' vs. 'impure' divide.[13] Nothing turns, that is, on whether the middle premise of the rule { }I takes the form $\exists! t$ or St. (We shall have occasion to remark once again on 'pure vs. impure invariance' in due course.)

$$\dfrac{\dfrac{\dfrac{\vdots\,\text{Logic}}{\forall x(Fx\leftrightarrow Fx)}}{\{x\,|\,Fx\}=\{x\,|\,Fx\}}\,\text{(Va)}}{\exists!\{x\,|\,Fx\}} \qquad \dfrac{\dfrac{\dfrac{\vdots\,\text{Logic}}{\forall x(Fx\leftrightarrow Fx)}}{\{x\,|\,Fx\}=\{x\,|\,Fx\}}\,\text{(Va)}\quad \dfrac{}{a\in\{x\,|\,Fx\}}\,(1)}{Fa}\,\{\,\}\text{E}_3 \qquad \dfrac{\dfrac{\dfrac{\vdots\,\text{Logic}}{\forall x(Fx\leftrightarrow Fx)}}{\{x\,|\,Fx\}=\{x\,|\,Fx\}}\,\text{(Va)}\quad \dfrac{}{\exists! a}\,(2)\quad \dfrac{}{Fa}\,(1)}{a\in\{x\,|\,Fx\}}\,\{\,\}\text{E}_1$$

$$\dfrac{\dfrac{\exists!\{x\,|\,Fx\} \qquad \dfrac{\dfrac{Fa \qquad a\in\{x\,|\,Fx\}}{a\in\{x\,|\,Fx\}\leftrightarrow Fa}\,(1)}{\forall y(y\in\{x\,|\,Fx\}\leftrightarrow Fy)}\,(2)}{\exists X\forall y(y\in X\leftrightarrow Fy)}}{\forall\Phi\exists X\forall y(y\in X\leftrightarrow \Phi y)}$$

7 Extensionality

One might wonder whether the introduction rule for the set-term forming operator plays any philosophically important role. The answer is that it makes a crucial contribution in proving the *extensionality* of sets — that two sets are identical if they have exactly the same members. The proof of this result (Theorem 7.1 below) invokes Theorem 4.1 — whose own proof involved a degenerate application of the rule { }I — and then makes use of a further, *non*-degenerate application of { }I. The natural-deduction theorist proves Theorem 7.1 using the rules for pure set theory (favoring $\exists! t$ over St as the second premise of { }I). The reader can be left the exercise of modifying the proof so that with the rules involving St the result can be established in a suitably 'S-restricted' form.[14]

Theorem 7.1 *Sets are identical if they have exactly the same members.*

Sentential version: $\forall x\forall y(\forall z(z\in x\leftrightarrow z\in y)\rightarrow x=y)$.

Inferential version:

$$\dfrac{\exists! t \qquad \exists! u \qquad \begin{matrix}\dfrac{}{a\in t}\,(i)\\ \vdots \\ a\in u\end{matrix} \qquad \begin{matrix}\dfrac{}{a\in u}\,(i)\\ \vdots \\ a\in t\end{matrix}}{t=u}\,(i)$$

[13] This remark is about only the sufficiency, and not necessarily the necessity, of eschewal of { }E_1 and { }E_3 for the invariance in question.

[14] The S-modified result to be proved in the sentential version in Theorem 7.1 is

$$\forall x(Sx\rightarrow\forall y(Sy\rightarrow\forall z(z\in x\leftrightarrow z\in y)\rightarrow x=y))).$$

Its inferential version is

$$\dfrac{St \qquad Su \qquad \begin{matrix}\dfrac{}{a\in t}\,(i)\\ \vdots \\ a\in u\end{matrix} \qquad \begin{matrix}\dfrac{}{a\in u}\,(i)\\ \vdots \\ a\in t\end{matrix}}{t=u}\,(i)\,.$$

Proof Sentential version:

$$
\dfrac{
\dfrac{
\dfrac{
\dfrac{\dfrac{\overline{a \in d}^{\,(1)}}{\exists! a} \quad \overline{\forall z(z \in c \leftrightarrow z \in d)}^{\,(2)}}{a \in c \leftrightarrow a \in d} \quad \overline{a \in d}^{\,(1)}
}{a \in c}
\quad \overline{\exists! c}^{\,(4)} \quad
\dfrac{
\dfrac{\dfrac{\overline{a \in c}^{\,(1)}}{\exists! a} \quad \overline{\forall z(z \in c \leftrightarrow z \in d)}^{\,(2)}}{a \in c \leftrightarrow a \in d} \quad \overline{a \in c}^{\,(1)}
}{a \in d}
}{c = \{x \mid x \in d\}}\,(1)\{\ \}\mathrm{I}
\qquad
\dfrac{\overline{\exists! d}^{\,(3)}}{d = \{x \mid x \in d\}}\,(\mathrm{Th.4.1})
}{
\dfrac{c = d}{\dfrac{\forall z(z \in c \leftrightarrow z \in d) \rightarrow c = d}{\dfrac{\forall y(\forall z(z \in c \leftrightarrow z \in y) \rightarrow c = y)}{\forall x \forall y(\forall z(z \in x \leftrightarrow z \in y) \rightarrow x = y)}\,(4)}\,(3)}\,(2)
}
$$

Inferential version:

$$
\dfrac{
\dfrac{
\begin{array}{c}\overline{a \in u}^{\,(i)} \\ \vdots \\ a \in t\end{array}
\quad \exists! t \quad
\begin{array}{c}\overline{a \in t}^{\,(i)} \\ \vdots \\ a \in u\end{array}
}{t = \{x \mid x \in u\}}\,(1)\{\ \}\mathrm{I}
\qquad
\dfrac{\exists! u}{u = \{x \mid x \in u\}}\,(\mathrm{Th.4.1})
}{t = u}
$$

□

Note that Extensionality as a *derived* result here — in either its sentential or its inferential version — does not itself contain any occurrences of the set-abstraction operator. In conventional (first-order) set theory, in the usual stripped-down language with $\in$ as a predicate but *without* the set-abstraction operator primitive, one would need of course to follow Zermelo in *postulating* Extensionality as an axiom. One of the virtues of the natural-deduction rules essayed here is that Extensionality is 'built in' to the resulting conception of set, whether one is theorizing about pure or about impure sets.

8 Natural-deduction for set abstraction vs. Fregean abstraction

The question now arises: just how radical a departure from the (incoherent) Fregean conception of class (or set) is represented by the conception captured by the natural deduction rules for the set-abstraction operator, in a free logic? What is the relationship between the latter rules, and Frege's Basic Law V (whose half called Vb is what Frege — mistakenly, in our free-logical view — blamed for Russell's paradox)?

The following completely formal derivations will reveal the answer.

The innocent half of Basic Law V (call it Vb):

$$
(\mathrm{Vb}) \qquad \dfrac{\{x \mid Fx\} = \{x \mid Gx\}}{\forall x(Fx \leftrightarrow Gx)}
$$

can be derived by the natural-deduction theorist (see Theorem 8.1) *and by the Fregean* (see Lemma 8.2). This latter result is interesting; it stems from just the rules F1 and F2 adopted by the Fregean who assumes a 'logically perfect' language. So the

Frege who would venture to adopt both F1 and F2 as logically correct and primitive inferences need not have bothered to state Vb as a basic law, or as a conjunctive part of any other basic law (such as the relevant half, in one direction, of the biconditional Basic Law V).

Theorem 8.1 { }E$_1$, { }E$_3$ ⇒ Vb.

Proof Note the symmetry in the proof (to be expected with '=' and '↔'), and the use of the two main elimination rules for { }, but not of the introduction rule. This is another instance of 'pure vs. impure invariance'.

$$\dfrac{\dfrac{\dfrac{\dfrac{\{x|Fx\}=\{x|Gx\}}{\exists!\{x|Fx\}}}{\{x|Fx\}=\{x|Fx\}} \quad \dfrac{\{x|Fx\}=\{x|Gx\} \quad \overline{\exists!a}^{\,(2)} \quad \overline{Ga}^{\,(1)}}{a\in\{x|Fx\}}\,\{\,\}\mathrm{E}_1}{Fa}\,\{\,\}\mathrm{E}_3 \qquad \dfrac{\dfrac{\dfrac{\{x|Fx\}=\{x|Gx\}}{\exists!\{x|Gx\}}}{\{x|Gx\}=\{x|Gx\}} \quad \dfrac{\{x|Fx\}=\{x|Gx\} \quad \overline{\exists!a}^{\,(2)} \quad \overline{Fa}^{\,(1)}}{a\in\{x|Gx\}}\,\{\,\}\mathrm{E}_1}{Ga}\,\{\,\}\mathrm{E}_3}{\dfrac{Fa\leftrightarrow Ga}{\forall x(Fx\leftrightarrow Gx)}\,(2)}\,(1)$$

□

Lemma 8.2 F1, F2 ⇒ Vb.

Proof

$$\dfrac{\dfrac{\dfrac{\dfrac{\overline{Ga}^{\,(1)}}{a\in\{x\mid Gx\}}\,\mathrm{F1} \quad \{x\mid Fx\}=\{x\mid Gx\}}{a\in\{x\mid Fx\}}}{Fa}\,\mathrm{F2} \qquad \dfrac{\dfrac{\dfrac{\overline{Fa}^{\,(1)}}{a\in\{x\mid Fx\}}\,\mathrm{F1} \quad \{x\mid Fx\}=\{x\mid Gx\}}{a\in\{x\mid Gx\}}}{Ga}\,\mathrm{F2}}{\dfrac{Fa\leftrightarrow Ga}{\forall x(Fx\leftrightarrow Gx)}\,(2)}\,(1)$$

□

Can Lemma 8.2 be strengthened by using the 'free logical' Fregean rule fF1 in place of F1? The answer is affirmative. See Lemma 8.9.

The natural-deduction theorist can prove not the converse of Vb — which would be the inconsistent Va — but a slight weakening of it, by adding an existential presupposition. The inference to be established is

$$\text{fVa}\quad \dfrac{\forall x(Fx\leftrightarrow Gx) \qquad \exists!\{x\mid Fx\}}{\{x\mid Fx\}=\{x\mid Gx\}}\,.$$

Theorem 8.3 { }E$_1$, { }E$_3$, { }I ⇒ fVa.

Proof

$$\dfrac{\dfrac{\dfrac{\dfrac{\exists!\{x|Fx\}}{\{x|Fx\}=\{x|Fx\}} \quad \overline{\exists!a}^{\,(1)} \quad \dfrac{\dfrac{\overline{\exists!a}^{\,(1)} \quad \forall x(Fx\leftrightarrow Gx)}{Fa\leftrightarrow Ga} \quad \overline{Ga}^{\,(1)}}{Fa}}{a\in\{x|Fx\}}\,\{\,\}\mathrm{E}_1 \qquad \exists!\{x|Fx\} \qquad \dfrac{\dfrac{\dfrac{\exists!\{x|Fx\}}{\{x|Fx\}=\{x|Fx\}} \quad \overline{a\in\{x|Fx\}}^{\,(1)}}{Fa}\,\{\,\}\mathrm{E}_3 \quad \dfrac{\dfrac{\overline{a\in\{x|Fx\}}^{\,(1)}}{\exists!a} \quad \forall x(Fx\leftrightarrow Gx)}{Fa\leftrightarrow Ga}}{Ga}}{\{x|Fx\}=\{x|Gx\}}\,(1)\,\{\,\}\mathrm{I}$$

□

Note that easy re-lettering will enable one to use the premise $\exists!\{x\mid Gx\}$ instead of $\exists!\{x\mid Fx\}$. The proof of Theorem 8.3 uses only the two main elimination rules for { }, along with its introduction rule.

The natural-deduction theorist's proof of Theorem 8.3 just given can be adapted so as to employ the rule { }I in its S-modified form. The adapted proof follows. Note that it appeals to Observation 8.4 in its middle immediate subproof. The natural-deduction theorist's proof of Observation 8.4 is presumed to be available here, and will be found on p. 106. It involves only two primitive steps.

$$\dfrac{\dfrac{\dfrac{\exists!\{x|Fx\}}{\{x|Fx\}=\{x|Fx\}}\quad \overline{\exists!a}^{\,(1)}\quad \dfrac{\dfrac{\overline{\exists!a}^{\,(1)}\quad \forall x(Fx\leftrightarrow Gx)}{Fa\leftrightarrow Ga}\quad \overline{Ga}^{\,(1)}}{Fa}}{a\in\{x|Fx\}}\,\{\,\}\mathrm{E}_1\qquad \dfrac{\exists!\{x|Fx\}}{S\{x|Fx\}}\,\text{Obs. 1}\qquad \dfrac{\dfrac{\dfrac{\exists!\{x|Fx\}}{\{x|Fx\}=\{x|Fx\}}\quad \overline{a\in\{x|Fx\}}^{\,(1)}}{Fa}\,\{\,\}\mathrm{E}_3\quad \dfrac{\dfrac{\overline{a\in\{x|Fx\}}^{\,(1)}}{\exists!a}\quad \forall x(Fx\leftrightarrow Gx)}{Fa\leftrightarrow Ga}}{Ga}}{\{x|Fx\}=\{x|Gx\}}\,(1)\ \{\,\}\text{I, } S\text{-modified form}$$

So we see that Basic Law V, suitably conditioned in its problematic direction with a much-needed existential premise, is derivable in our free logic for the set-term forming operator { }. It is fVa that is the (modified) Fregean way to express the fact that sets are extensional; and in deriving it, the natural-deduction theorist needs to use { }I.

Is the natural-deduction theorist's logic of sets tantamount to nothing more than Basic Law V thus modified? The answer would presumably be affirmative if, but only if, by using the inferences

$$\frac{\{x \mid Fx\}=\{x \mid Gx\}}{\forall x(Fx \leftrightarrow Gx)} \qquad\qquad \frac{\forall x(Fx \leftrightarrow Gx) \quad \exists!\{x \mid Fx\}}{\{x \mid Fx\}=\{x \mid Gx\}}$$

the free-logical Fregean could derive (in free logic) the introduction and elimination rules that have been stated for { }. The rules { }E_1 and { }E_3 have already been furnished with derivations of the requisite kind (see Theorems 5.2 and 5.3). The reader will recall that (in the case of pure set theory) { }E_2 is a special case of the rule RAD of free logic. And in the case of a set theory countenancing *Urelemente*, the S-modified rule

$$\{\,\}\mathrm{E}_2 \quad \frac{t = \{x \mid \Phi\}}{St}$$

will in due course be adopted by (us, on behalf of) the Fregean as a primitive rule, to be called F3 (see below). The remaining task, then, for the Fregean, is to derive { }I. Can this be done?

The answer is a cautious affirmative. The caution is occasioned by the residual need *on the part of the Fregean* to supply the inferential transition occurring twice in the following proof, as indicated by the descending dots:

$$\dfrac{\dfrac{\begin{array}{c}\Delta_2\\ \Pi_2\\ \exists!t\\ \vdots\\ t=\{x\mid x\in t\}\end{array}}{\exists!\{x\mid x\in t\}}\qquad \dfrac{\dfrac{\begin{array}{c}\underbrace{{}^{(1)}\overline{\varphi a},\ \overline{\exists!a}^{\,(2)},\ \Delta_1}\\ \Pi_1\\ a\in t\end{array}\qquad \begin{array}{c}\underbrace{{}^{(1)}\overline{a\in t},\ \Delta_3}\\ \Pi_3\\ \varphi a\end{array}}{a\in t\leftrightarrow\varphi a}\,(1)}{\forall x(x\in t\leftrightarrow\varphi x)}\,(2)}{\{x\mid x\in t\}=\{x\mid\varphi x\}}\,\text{fVa}\qquad \begin{array}{c}\Delta_2\\ \Pi_2\\ \exists!t\\ \vdots\\ t=\{x\mid x\in t\}\end{array}$$
$$\overline{\qquad\qquad\qquad\qquad\qquad\qquad\qquad\qquad}$$
$$t=\{x\mid\varphi x\}$$

That inferential transition is of course guaranteed by Theorem 4.1, to which the free-logical, natural-deduction theorist about sets is entitled. But would either the original Frege, or the speculatively 'free-logicized' Frege, be thus entitled?

Let us explore some conceptual and logical possibilities on Frege's behalf. We know that he conceived of his *Begriffsschrift* as a language for the pursuit of truths not just about the abstract realm, but also about concrete reality. So he would not have been satisfied with being confined to talk only of sets. 'Pure-set' theorizing would have been too restrictive, expressively, for Frege. So it is reasonable to infer that he would have been prepared to adopt a primitive predicate like S, so as to be able to express the informal idea, concerning any supplied argument (*Gegenstand*), that it is a set — that is, (for Frege) the completed extension of some concept. For any *Urelement*, of course, what is thus expressed is false.

Countenancing the possibility of *Urelemente*, Frege would have refused — correctly — to adopt the inferential principle

$$\frac{\exists!t}{St}.$$

But he would have been happy — rightly — with its converse:

$$\frac{St}{\exists!t}.$$

This rule says that every set exists. And, since St is an atomic predication, the rule is a special case of the Rule of Atomic Denotation.

The distinction was drawn earlier between *pure set* theorizing and *impure set* theorizing. Frege's theorizing was of the latter kind, since he regarded the universe of discourse as truly universal, and therefore containing all concrete objects (*Urelemente*, such as Julius Caesar) along with sets formed from them (such as the singleton {Julius Caesar}), *and* along with sets that happened to be hereditarily pure (such as $\{y \mid y=\{x \mid \neg x=x\}\}$). Suppose, however, that Frege had been asked to theorize in a more focused way about the (sub-)universe (for him) of hereditarily pure sets. It would have been both obvious and natural for him to give expression to this expressive focus by adopting the following rule π (for 'purity'):

$$\pi \quad \frac{\exists!t}{St}$$

where the intended reading of 'St' is (as always here) 't is a set'. It will follow from this that t is *hereditarily pure* as a set, since every member of t exists, hence (by rule π) is a set. Since no *Urelement* is a set:

$$\frac{Ut \qquad St}{\bot}$$

the membership pedigree of t does not contain any *Urelemente*, and t is accordingly pure. A little bit of sethood, given mere existence, goes a long way. The foregoing rule of contrariety is primitive for both the Fregean and the natural-deduction theorist when they allow for *Urelemente*.

What would it take, then, on the part of some *Gegenstand* t, to earn the honorific sortal S from Frege? Surely it would be enough that t really *be* the extension of some concept Φ. That is, the following inferential principle would be logico-analytically valid for the Fregean, and self-evident:

$$\text{F3} \quad \frac{t = \{x \mid \Phi(x)\}}{St}$$

This rules says that if t is the set of all Φs, then t is a set. It is primitive for the ND-theorist (for it is the set-abstraction elimination rule $\{\ \}E_2$), and the Fregean has every right to adopt it as a primitive rule too.

Observation 8.4 $\dfrac{\exists!\{x \mid \Phi(x)\}}{S\{x \mid \Phi(x)\}}$.

Proof (by the Fregean)

$$\dfrac{\exists!\{x \mid \Phi(x)\} \qquad \dfrac{\dfrac{\overline{a = \{x \mid \Phi(x)\}}^{\,(1)}}{Sa}\,\text{F3} \qquad \overline{a = \{x \mid \Phi(x)\}}^{\,(1)}}{S\{x \mid \Phi(x)\}}}{S\{x \mid \Phi(x)\}}\,(1)$$

□

Proof (by the ND-theorist)

$$\dfrac{\dfrac{\exists!\{x \mid \Phi(x)\}}{\{x \mid \Phi(x)\} = \{x \mid \Phi(x)\}}\,\text{Ref.=}}{S\{x \mid \Phi(x)\}}\,\{\ \}\text{E}_2$$

□

Observation 8.4 tells us that even though, as pointed out above, the inference

$$\frac{\exists! t}{St}$$

does not hold in general, it *does* hold when, more specifically, we have a set-abstraction term in place of t.

We have been considering a Frege who distinguishes between sets and *Urelemente*. So for fair comparison of his system with that of the free-logical natural-deduction theorist, the latter must give up commitment to theorizing only about sets. This means that the rule $\{\ \}I$ must have its second premise in the form St; and, correspondingly, the elimination rule $\{\ \}E_2$ will be

$$\frac{t = \{x \mid \Phi\}}{St}$$

The sought derivation of the S-modified rule $\{\ \}I$ using Fregean principles will eventually be found. See Lemma 8.8 below.

Now consider the prospect of having some sort of converse of principle F3. Suppose one is given just the premise St. Then one knows that t is the extension of *some* concept or other. And what might be the most general — indeed canonical — concept that would fit this bill? Why,

'... is a member of t',

of course. So the following inferential principle would be logico-analytically valid for the Fregean:

$$\text{F4}\quad \frac{St}{t = \{x \mid x \in t\}}$$

Note that the primitive rule F4 here being accepted by (us, on behalf of) the Fregean is Theorem 4.2 of the natural-deduction theorist. Both theorists have it *as part of their logic of sets*. The only difference is that for the Fregean it is primitive — because it has to be — whereas for the natural-deduction theorist it is derived.

Why do we say that F4 *has to be* primitive for the Fregean? The answer is that inspection of all of the Fregean's other primitive rules — RAD, Sub=, Ref=, fF1, F2, F3, fVa, and π — reveals that one cannot use them to derive the conclusion $t = \{x \mid x \in t\}$ from the premise St.

The 'pure-set' Fregean, by adopting the rules π and F4, is able to mimic the natural-deduction theorist's Theorem 4.1 as follows.

Lemma 8.5 $\dfrac{\exists!t}{t=\{x \mid x \in t\}}$.

Proof

$$\dfrac{\dfrac{\exists!t}{St}\,\pi}{t=\{x \mid x \in t\}}\,\text{F4}$$

□

The question arises: can the Fregeans prove Theorem 7.1 using their own rules thus far, so as to parallel and emulate what the natural-deduction theorist did? The answer is affirmative. We shall deal with Extensionality in its inferential form.

Lemma 8.6 F4, π, fVa $\Rightarrow$

$$\dfrac{\exists!t \qquad \exists!u \qquad \begin{array}{c}\overline{a \in t}^{\,(i)}\\ \vdots \\ a \in u\end{array} \qquad \begin{array}{c}\overline{a \in u}^{\,(i)}\\ \vdots \\ a \in t\end{array}}{t=u}\,(i)$$

.

Proof

$$\dfrac{\dfrac{\dfrac{\dfrac{\dfrac{\begin{array}{c}\overline{a \in t}^{\,(1)}\\ \vdots \\ a \in u\end{array} \qquad \begin{array}{c}\overline{a \in u}^{\,(1)}\\ \vdots \\ a \in t\end{array}}{a \in t \leftrightarrow a \in u}\,(1)}{\forall x(x \in t \leftrightarrow x \in u)} \qquad \exists!\{x \mid x \in t\}}{\{x \mid x \in t\}=\{x \mid x \in u\}}\,\text{fVa} \qquad \dfrac{\exists!t}{t=\{x \mid x \in t\}}\,\text{L8.5}}{t=\{x \mid x \in u\}} \qquad \dfrac{\exists!u}{u=\{x \mid x \in u\}}\,\text{L8.5}}{t=u}$$

□

Fregean rule F4 is potent in another important regard. Teaming up with fVa, it yields the natural-deduction theorist's introduction rule { }I — both in its original version (ensuring purity of sets) and in its S-modified version (allowing for *Urelemente*) as shown, respectively, by Lemma 8.7 and Lemma 8.8.

Lemma 8.7 F4, fVa, $\pi \Rightarrow \{\ \}$I (pure-set version).

Proof Recall that the proof of Lemma 8.5 used F4.

$$\dfrac{\dfrac{\dfrac{\dfrac{\begin{matrix}\Delta_2\\ \Pi_2\\ \exists! t\end{matrix}}{t=\{x \mid x\in t\}}\,\text{L8.5}}{\exists!\{x \mid x\in t\}} \qquad \dfrac{\dfrac{\begin{matrix}\underbrace{\overset{(1)}{\overline{\varphi a}}\,,\,\overline{\exists! a}^{\,(2)}\,,\,\Delta_1}\\ \Pi_1\\ a\in t\end{matrix} \qquad \begin{matrix}\underbrace{\overset{(1)}{\overline{a\in t}}\,,\,\Delta_3}\\ \Pi_3\\ \varphi a\end{matrix}}{a\in t\leftrightarrow\varphi a}\,(1)}{\forall x(x\in t\leftrightarrow\varphi x)}\,(2)}{\{x \mid x\in t\}=\{x \mid \varphi x\}}\,\text{fVa} \qquad \dfrac{\begin{matrix}\Delta_2\\ \Pi_2\\ \exists! t\end{matrix}}{t=\{x \mid x\in t\}}\,\text{L8.5}}{t=\{x \mid \varphi x\}}$$

□

Lemma 8.8 F4, fVa $\Rightarrow \{\ \}$I (S-modified version).

Proof

$$\dfrac{\dfrac{\dfrac{\dfrac{\begin{matrix}\Delta_2\\ \Pi_2\\ St\end{matrix}}{t=\{x \mid x\in t\}}\,\text{F4}}{\exists!\{x \mid x\in t\}} \qquad \dfrac{\dfrac{\begin{matrix}\underbrace{\overset{(1)}{\overline{\varphi a}}\,,\,\overline{\exists! a}^{\,(2)}\,,\,\Delta_1}\\ \Pi_1\\ a\in t\end{matrix} \qquad \begin{matrix}\underbrace{\overset{(1)}{\overline{a\in t}}\,,\,\Delta_3}\\ \Pi_3\\ \varphi a\end{matrix}}{a\in t\leftrightarrow\varphi a}\,(1)}{\forall x(x\in t\leftrightarrow\varphi x)}\,(2)}{\{x \mid x\in t\}=\{x \mid \varphi x\}}\,\text{fVa} \qquad \dfrac{\begin{matrix}\Delta_2\\ \Pi_2\\ St\end{matrix}}{t=\{x \mid x\in t\}}\,\text{F4}}{t=\{x \mid \varphi x\}}$$

□

By Lemma 8.7, the Fregean rules F4, fVa, and π suffice for proof of $\{\ \}$I in its original form (ensuring purity); and by Lemma 8.8, F4 and fVa suffice for proof of $\{\ \}$I in its S-modified form (allowing for *Urelemente*). No matter which form of it is used, $\{\ \}$I in turn suffices for proof of extensionality in the relevant form (with unrestricted or S-restricted quantifiers, as seen from Theorem 7.1 and the comment thereon in footnote 14).

Recall Theorem 8.1:

$$\{\ \}\text{E}_1,\ \{\ \}\text{E}_3 \Rightarrow \text{Vb},$$

Lemma 5.4:

$$\text{fF1} \Rightarrow \{\ \}\text{E}_1,$$

and Lemma 5.3:

$$\text{F2} \Rightarrow \{\ \}\text{E}_3.$$

It follows by 'rule transitivity' that

$$\text{fF1},\ \text{F2} \Rightarrow \text{Vb}.$$

Here is a more direct proof of this last result. It can be obtained by accumulating the proofs of Theorem 8.1, Lemma 5.4, and Lemma 5.3, and applying to that accumulation the two 'shrinking reductions' that the reader will find are obviously called for.

Lemma 8.9 fF1, F2 ⇒ Vb.

Proof

$$\dfrac{\dfrac{\dfrac{\dfrac{\dfrac{\overline{Ga}^{(1)}\quad \overline{\exists!a}^{(2)}\quad \dfrac{\{x|Fx\}=\{x|Gx\}}{\exists!\{x|Gx\}}}{a\in\{x|Gx\}}\,\text{fF1}\qquad \{x|Fx\}=\{x|Gx\}}{a\in\{x|Fx\}}}{Fa}\,\text{F2}\qquad \dfrac{\dfrac{\dfrac{\overline{Fa}^{(1)}\quad \overline{\exists!a}^{(2)}\quad \dfrac{\{x|Fx\}=\{x|Gx\}}{\exists!\{x|Fx\}}}{a\in\{x|Fx\}}\,\text{fF1}\qquad \{x|Fx\}=\{x|Gx\}}{a\in\{x|Gx\}}}{Ga}\,\text{F2}}{Fa\leftrightarrow Ga}\,(1)}{\forall x(Fx\leftrightarrow Gx)}\,(2)$$

□

So Vb is redundant for the (modified) Fregean. Note how no use is made of Vb in the (modified) Fregean's derivations of the ND-rules.

9 Taking stock

Let us take stock of the progress made thus far, in our comparison of the 'free-logical' Fregean with the (likewise 'free-logical') natural-deduction set theorist. Let us henceforth call each of them simply *free*, rather than 'free-logical'.

Bear in mind the following three crucial points of methodological agreement between them.

1. Both of these theorists are countenancing the possibility of *Urelemente*. So they both employ the atomic predicate S, for '. . . is a set'.
2. Both of them have the same conception of the truth conditions of atomic statements (including, of course, statements of identity). An atomic statement is false if any of its immediate constituent terms fails to denote.
3. Both of them use the usual introduction and elimination rules for free first-order logic with identity, based on ¬, ∧, ∨, →, ∀, ∃, and =. Included among these rules are the Rule of Atomic Denotation, and the modified Rule of Reflexivity of Identity. The quantifier rules are also embellished with existential presuppositions in the usual well-understood way. (See Tennant, 1978, Ch. 7.)

Here are the basic inferential principles *for sets* espoused by these two theorists.

The free Fregean's basic inferential principles for sets:

$$\text{fF1}\quad \dfrac{\Phi(t)\qquad \exists!\{x\mid\Phi(x)\}\qquad \exists!t}{t\in\{x\mid\Phi(x)\}}\qquad\qquad \text{F4}\quad \dfrac{St}{t=\{x\mid x\in t\}}$$

$$\text{F2}\quad \frac{t \in \{x \mid \Phi(x)\}}{\Phi(t)} \qquad\qquad \text{fVa}\quad \frac{\forall x(\Phi x \leftrightarrow \Psi x) \qquad \exists!\{x \mid \Phi x\}}{\{x \mid \Phi x\} = \{x \mid \Psi x\}}$$

$$\text{F3}\quad \frac{t = \{x \mid \Phi(x)\}}{St} \qquad\qquad \pi\quad \frac{\exists!t}{St}$$

The free natural-deduction theorist's basic inferential principles for sets:

$$\{\ \}\text{I}\quad \frac{\begin{array}{c} \underbrace{{}^{(i)}\overline{\exists!a}\ ,\ \overline{\Phi^x_a}^{(i)}} \\ \vdots \\ a \in t \end{array} \qquad \exists!t \qquad \begin{array}{c} \overline{a \in t}^{(i)} \\ \vdots \\ \Phi^x_a \end{array}}{t = \{x \mid \Phi\}}\,(i)$$

, where a is parametric.

$$\{\ \}\text{E}_1\quad \frac{t = \{x \mid \Phi\} \qquad \exists!v \qquad \Phi^x_v}{v \in t} \qquad\qquad \{\ \}\text{E}_2\quad \frac{t = \{x \mid \Phi\}}{\exists!t}$$

$\{\ \}\text{E}_3$ serial form:

$$\frac{t = \{x \mid \Phi\} \qquad v \in t}{\Phi^x_v}$$

parallelized form:

$$\frac{t = \{x \mid \Phi\} \qquad v \in t \qquad \begin{array}{c} \overline{\Phi^x_v}^{(i)} \\ \vdots \\ \theta \end{array}}{\theta}\,(i)$$

A remarkable contrast strikes the eye. The (free) Fregean states rules that make no provision for discharge of assumptions, whether in sentential form or in rule form. The natural-deduction theorist, however, states { }I so as to allow discharge of certain assumptions. It is also a single-barreled rule. These are crucial reasons why the latter's rules, overall, provide a more succinct and unified account of the interrelations among the concepts involved. This is the case even when both theorists are employing free logic and are achieving pure vs. impure invariance in their theorizing about sets. We have learned the lesson that the Gentzenian approach, allowing for rules that effect discharge of assumptions, is an essential advance over the Fregean one, and frees the set-logician to do more with less.

Summary of the Fregean's results: Lemma 5.2: F1 $\Rightarrow$ $\{\ \}\text{E}_1$.
Lemma 5.3: F2 $\Rightarrow$ $\{\ \}\text{E}_3$.
Lemma 5.4: fF1 $\Rightarrow$ $\{\ \}\text{E}_1$.
F3 is, hence $\Rightarrow$ $\{\ \}\text{E}_2$.
Lemma 8.2: F1, F2 $\Rightarrow$ Vb.
Lemma 8.5: $\dfrac{\exists!t}{t=\{x \mid x \in t\}}$.

Lemma 8.6: F4, π, fVa $\Rightarrow$

$$\frac{\exists!t \qquad \exists!u \qquad \begin{array}{c} \overline{a \in t}^{(i)} \\ \vdots \\ a \in u \end{array} \qquad \begin{array}{c} \overline{a \in u}^{(i)} \\ \vdots \\ a \in t \end{array}}{t=u}\,(i)\ .$$

Lemma 8.7: F4, fVa, $\pi \Rightarrow$ { }I in its original form (ensuring purity).
Lemma 8.8: F4, fVa $\Rightarrow$ { }I in its S-modified form (allowing *Urelemente*).
Lemma 8.9: fF1, F2 $\Rightarrow$ Vb.

Summary of the ND-theorist's results: Theorem 4.1: $\dfrac{\exists! t}{t=\{x \mid x \in t\}}$.

Theorem 4.2: { }I $\Rightarrow$ F4.
Theorem 5.1: { }E_3 $\Rightarrow$ F2.
Theorem 5.5: { }E_1 $\Rightarrow$ fF1.
{ }E_2 is, hence $\Rightarrow$ F3.

Theorem 7.1: { }I $\Rightarrow$

$$\dfrac{\exists! t \qquad \exists! u \qquad \begin{array}{c}\overline{a \in t}^{\,(i)} \\ \vdots \\ a \in u\end{array} \qquad \begin{array}{c}\overline{a \in u}^{\,(i)} \\ \vdots \\ a \in t\end{array}}{t=u}\,(i) .$$

Theorem 8.1: { }E_1, { }E_3 $\Rightarrow$ Vb.
Theorem 8.3: { }E_1, { }E_3, { }I $\Rightarrow$ fVa.

10 Summary of our comparison of the free Fregean approach with the free ND-approach

Clearly, the free Fregean approach is equivalent to the free ND-approach. Each primitive rule of the one theorist is either primitive for, or derivable by, the other theorist. The free Fregean, however, adopts as primitive the rule F4, which the ND-theorist easily but non-trivially derives. Furthermore, comparison of their respective proofs of the non-trivially derivable result of Extensionality reveals that the ND-proof is more succinct than the Fregean one. They tie, however, in proving Vb — taking ten primitive steps each.

There is a satisfying unity to the ND-approach that is lacking in the Fregean approach. Having harmoniously balanced introduction and elimination rules for set-abstraction is a definite plus. These rules require only minor tweaks to toggle between the pure and the impure conceptions of sets. The Fregean, by contrast, resorts to adopting two new primitive rules—F4 and π—to ensure restriction to pure sets.

The extension of Gentzenian methods from the usual logical operators so as to include also the operator for set-abstraction appears to yield a methodological advantage. If the methods of natural deduction had been available to Frege, the tradition could arguably have delivered an ontologically non-committal *logical foundation* for abstraction, membership, sethood and predication. On that foundation Zermelo and his successors could then have built further, by supplying axioms and axiom schemata for outright existence (e.g., the empty set) and conditional existence (e.g., the pair set of any two things).

The original sin revealed by Russell's paradox can be viewed, through this new lens, as Frege's insistence that every well-formed singular term denotes. The fateful Va codified that insistence as it concerned set-abstraction terms in particular. If Frege's erroneous conception of 'logical perfection' could have been eliminated earlier, the route would have been cleared to his acceptance of all the principles of modern free logic. He could have had a logic of sets, but without any sets as logical objects. That the set-term $\{x \mid \neg x \in x\}$ does not denote would then have been no more disastrous a discovery than that the definite descriptive term $\iota x(\Phi x \wedge \neg\Phi x)$ does not either.

Foundationalists know, as practicing mathematicians themselves, that all our mathematical reasoning is (intuitively) *relevant*, and therefore should be able to be regimented in a formal logic devoid of the paradoxes of irrelevance. There is foundational dispute over whether mathematical reasoning should be *constructive*; and we know now, even from a constructive standpoint, that the classical extensions of constructive theories are consistent if the latter are. The lesson that emerges here is that adopting a *free* logic for the foundations of mathematics appears to be crucial for the *constructivist* to ensure consistency.

This brings us to the end of our comparison of the Fregean and the natural-deduction theorist's approaches to the primitive rules for the *logic* of sets.

We turn now to explain how the natural-deduction theorist can pursue the project of framing introduction and elimination rules even further, so as to capture all the familiar 'compiled' concepts of set theory. The main philosophical lesson of the remaining part of our investigation is that one can capture the meanings of set-theoretic predicates and operators without incurring any ontological commitments. The latter commitments result only from subsequent existential postulations, either outright or conditional. But the concepts embedded in such postulates are *already understood*, thanks to the ontologically non-committal rules of introduction and elimination that govern them in a *free* logic. So the main 'takeaway' is that the postulates of set theory do *not* serve 'implicitly to define' the concepts involved. Rather, those concepts are already available to be grasped *before* any existential postulation is undertaken. Such is the power of harmoniously balanced introduction and elimination rules.

11 An inferentialist treatment of set-theoretic pasigraphs

At present our official list of primitive expressions contains only the logical operators $\neg$, $\wedge$, $\vee$, $\rightarrow$, $\exists$, and $\forall$, the identity predicate $=$, the membership predicate $\in$, and the (singular) term-forming variable-binding set-abstraction operator $\{x \mid \ldots x \ldots\}$.

Set theory, however, is not laid out in such austerely primitive vocabulary. Set theorists and ordinary mathematicians making use of set-theoretic ideas employ a host of already familiar-looking *defined* expressions (such as '$\subseteq$' for 'is a subset of', and '$\mathscr{P}$' for 'the power set of'). These defined expressions are indispensable for communicating in a conveniently condensed fashion what would otherwise be extremely cumbrously expressed set-theoretical thoughts. The inferentialist seeks to frame rules of introduction and elimination for these defined expressions, so that they can be understood as being employed as 'local primitives' in mathematical discourse of the normal explicit texture.

When these defined function-operators and predicates are supplied with rules of inference, we call those operators and predicates *pasigraphs*, because they are so readily recognizable as pieces of notation in use, despite being absent from the official list of primitive expressions. In some cases we shall invent new pasigraphs, in order to express in yet more succinct symbolic form what practicing set theorists often render only in 'logician's English'. This enables one to be uniform and thorough in rigorously regimenting informal set-theoretical proofs as formal, logical proofs.

Every defined notion in set theory (as in any other branch of mathematics) stands at the apex of its own 'pyramid of preceding definitions'. The official primitive expressions form its base. There are of course only finitely many preceding definitions in any such pyramid; the process of constructing new concepts is always well-founded. Those definitions 'lower down' in the pyramid will be of pasigraphs that can be employed in the statement of introduction and elimination rules for the new notion at the apex. Not all of them need be thus employed; but they are eligible to be. Typically, the new notion at the apex will have its rules framed by using earlier notions just a layer or two down. But this is not necessarily the case in general.

We recall the special rule of free logic called the *Rule of Atomic Denotation*:

$$\frac{A(\ldots,t,\ldots)}{\exists!t}$$

We saw also that in free logic for languages containing function signs as primitives, there is the *Rule of Functional Denotation*:

$$\frac{\exists!f(\ldots,t,\ldots)}{\exists!t}$$

We shall adopt the following constraint on the formulation of any new concepts represented by our pasigraphs:

1. predicate-pasigraphs are to qualify as instances of A in the Rule of Atomic Denotation;
2. operator-pasigraphs are to qualify as instances of f in the Rule of Functional Denotation.

11.1 Pasigraphs for restricted quantifications

We shall start with a pasigraph that is neither an operator- nor a predicate-pasigraph. We are very familiar with the following form of generalization:

$$\forall x \in t\, \varphi(x).$$

But strictly speaking the universal quantifier $\forall$, as a logical primitive, is a *unary* quantifier; so the foregoing form is that of a binary-quantifier *pasigraph*, which needs either to be defined explicitly:

$$\forall x(x \in t \rightarrow \varphi(x))$$

or to have its meaning specified by rules that can govern it as an apparent primitive:

$$\forall\text{I}\quad \dfrac{\begin{array}{c}\overline{a\in t}^{\,(i)}\\ \vdots\\ \varphi^x_a\end{array}}{\forall x\in t\,\varphi}\,(i)\qquad\qquad \forall\text{E}\quad \dfrac{\forall x\in t\,\varphi\qquad u\in t}{\varphi^x_u}$$

We are very familiar also with the following form of generalization:

$$\exists x\in t\,\varphi(x).$$

But strictly speaking the existential quantifier $\exists$, as a logical primitive, is a *unary* quantifier; so the foregoing form is that of a binary-quantifier *pasigraph*, which needs either to be defined explicitly:

$$\exists x(x\in t\wedge\varphi(x))$$

or to have its meaning specified by rules that can govern it as an apparent primitive:

$$\exists\text{I}\quad \dfrac{u\in t\qquad \varphi^x_u}{\exists x\in t\,\varphi}\qquad\qquad \exists\text{E}\quad \dfrac{\exists x\in t\,\varphi\qquad \begin{array}{c}\underbrace{{}^{(i)}\overline{a\in t}\;,\;\overline{\varphi^x_a}^{\,(i)}}\\ \vdots\\ \psi\end{array}}{\psi}\,(i)$$

11.2 Pasigraph ∅ for empty set

In primitive set-theoretic vocabulary the Axiom of Empty Set is as follows:

$$\exists!\{x\mid\neg x=x\}.$$

We shall now introduce the *constant pasigraph* $\emptyset$ for the empty set, governed by the following (existentially non-committal) rules.

$$\emptyset\text{I}\quad \dfrac{\exists!t\qquad \begin{array}{c}\overline{a\in t}^{\,(i)}\\ \vdots\\ \bot\end{array}}{t=\emptyset}\,(i)\qquad\qquad \emptyset\text{E}\quad \dfrac{t=\emptyset}{\exists!t}\qquad \dfrac{t=\emptyset\qquad u\in t}{\bot}$$

The Axiom of Empty Set can now take the form $\exists!\emptyset$. This is an outright existence postulate. Its inferential form is the zero-premise rule

$$\overline{\exists!\emptyset}\,.$$

11.3 Pasigraph for separated sets

Suppose φ has x free. Then

$$\{x \in t \mid \varphi\} =_{df} \{x \mid x \in t \wedge \varphi\}$$

The Axiom Scheme of Separation can now take the rule form

$$\frac{\exists ! t}{\exists ! \{x \in t \mid \varphi\}}$$

Instead of stipulating that $\{x \in t \mid \varphi\}$ is an abbreviation of $\{x \mid x \in t \wedge \varphi\}$, we could adopt as a grammatical primitive the term-formation operation on a term t and a formula φ with x free, that produces $\{x \in t \mid \varphi\}$ as a genuine term of the language, from the two constituents mentioned. We could then furnish the operation with its own introduction and elimination rules as follows.

$$\frac{\begin{array}{c} \underbrace{{}^{(i)}\overline{\varphi^x_a}\ ,\ \overline{a \in v}^{(i)}} \\ \vdots \\ a \in t \end{array} \qquad \exists ! t \qquad \begin{array}{c} \overline{a \in t}^{(i)} \\ \vdots \\ a \in v \end{array} \qquad \begin{array}{c} \overline{a \in t}^{(i)} \\ \vdots \\ \varphi^x_a \end{array}}{t = \{x \in v \mid \varphi\}}\,(i)$$

$$\frac{t = \{x \in v \mid \varphi\} \qquad \varphi^x_u \qquad u \in v}{u \in t} \qquad\qquad \frac{t = \{x \in v \mid \varphi\}}{\exists ! t}$$

$$\frac{t = \{x \in v \mid \varphi\} \qquad u \in t}{u \in v} \qquad\qquad \frac{t = \{x \in v \mid \varphi\} \qquad u \in t}{\varphi^x_u}$$

By virtue of the foregoing Introduction and Elimination Rules for the Separation Pasigraph, along with the Introduction and Elimination Rules for the Set-Abstraction Operator, we have

$$t = \{x \in v \mid \varphi\} \vdash t = \{x \mid x \in v \wedge \varphi\}$$

and its converse

$$t = \{x \mid x \in v \wedge \varphi\} \vdash t = \{x \in v \mid \varphi\}.$$

11.4 Pasigraph for pair-sets

In primitive set-theoretic vocabulary the Axiom of Pairing is as follows:

$$\forall x \forall y \exists ! \{z \mid z = x \vee z = y\}$$

We shall now introduce a (binary) *operator pasigraph.*

We shall write $\mathbb{P}(t, u)$ for the pair set $\{z \mid z = t \vee z = u\}$. Note that the 'pair' set $\mathbb{P}(t, u)$ is a *singleton* if $t = u$ (whence both t and u exist).

The Axiom of Pairing can now also be expressed as follows:

$$\forall x \forall y \exists ! \mathbb{P}(x, y).$$

The inferentialist working in free logic requires the interdeducibility

$$t = \mathbb{P}(u, v) \dashv\vdash t = \{z \mid z{=}u \vee z{=}v\}$$

rather than the provability of the identity

$$\mathbb{P}(u, v) = \{z \mid z{=}u \vee z{=}v\},$$

The latter identity commits one to the existence of the pair set of u and v:

$$\exists ! \mathbb{P}(u, v).$$

The interdeducibility, however, does not carry such existential commitment. It allows one to pin down the meaning of $\mathbb{P}$ as an operator on sets without committing one to its being everywhere (or indeed: *any*where) defined. One can grasp what $\mathbb{P}$ means without yet adopting the Axiom of Pair Sets. And when we *do* adopt that axiom, it does not serve implicitly to define the meaning of $\mathbb{P}$. For that meaning will already have been defined by the Introduction and Elimination rules that we frame for $\mathbb{P}$.

We propose the following introduction and elimination rules for the pairing operator $\mathbb{P}$.

$$\mathbb{P}\text{-I} \qquad \frac{u \in t \qquad v \in t \qquad \begin{matrix} \overline{a \in t}^{\,(i)} \\ \vdots \\ a{=}u \vee a{=}v \end{matrix}}{t{=}\mathbb{P}(u, v)}\,(i)$$

$$\mathbb{P}\text{-E}_1 \ \frac{t{=}\mathbb{P}(u, v)}{u \in t} \qquad \mathbb{P}\text{-E}_2 \ \frac{t{=}\mathbb{P}(u, v)}{v \in t} \qquad \mathbb{P}\text{-E}_3 \ \frac{t{=}\mathbb{P}(u, v) \qquad w \in t}{w{=}u \vee w{=}v}$$

Lemma 11.1 *The operator pasigraph $\mathbb{P}$ obeys the Rule of Functional Denotation; that is, the following are provable:*

$$\frac{\exists ! \mathbb{P}(u, v)}{\exists ! u} \qquad \frac{\exists ! \mathbb{P}(u, v)}{\exists ! v}.$$

Proof

$$\frac{\begin{matrix}\exists ! \mathbb{P}(u, v), \text{ i.e.,} \\ \exists x\, x{=}\mathbb{P}(u, v)\end{matrix} \qquad \dfrac{\dfrac{\overline{a{=}\mathbb{P}(u, v)}^{\,(1)}}{u \in a}\,\mathbb{P}\text{-E}_1}{\exists ! u}\,\text{RAD}}{\exists ! u}\,(1) \qquad \frac{\begin{matrix}\exists ! \mathbb{P}(u, v), \text{ i.e.,} \\ \exists x\, x{=}\mathbb{P}(u, v)\end{matrix} \qquad \dfrac{\dfrac{\overline{a{=}\mathbb{P}(u, v)}^{\,(1)}}{v \in a}\,\mathbb{P}\text{-E}_1}{\exists ! u}\,\text{RAD}}{\exists ! u}\,(1)$$

□

Lemma 11.2 $t = \mathbb{P}(u, v) \vdash t = \{z \mid z{=}u \vee z{=}v\}$.

Proof

$$\dfrac{\dfrac{\dfrac{}{a=u\lor a=v}\,(1)\qquad \dfrac{\dfrac{}{a=u}\,(2)\quad \dfrac{t=\mathbb{P}(u,v)}{u\in t}\,\text{P-E}_1}{a\in t}\qquad \dfrac{\dfrac{}{a=v}\,(2)\quad \dfrac{t=\mathbb{P}(u,v)}{v\in t}\,\text{P-E}_2}{a\in t}}{a\in t}\,(2)\qquad \dfrac{t=\mathbb{P}(u,v)}{\exists!t}\quad \dfrac{t=\mathbb{P}(u,v)\quad \dfrac{}{a\in t}\,(1)}{a=u\lor a=v}\,\text{P-E}_3}{t=\{z\mid z=u\lor z=v\}}\,(1)\,\{\,\}\text{-I}$$

□

Lemma 11.3 $t=\{z\mid z{=}u\lor z{=}v\},\ \exists!u,\ \exists!v\vdash t=\mathbb{P}(u,v)$.

Proof Abbreviate $z{=}u\lor z{=}v$ as Φz, where convenient, to reduce sideways spread.

$$\dfrac{\dfrac{t{=}\{z\mid\Phi z\}\quad \dfrac{\dfrac{\exists!u}{u{=}u}}{\Phi u}}{u\in t}\,\{\,\}\text{-E}_{\text{PM}}\qquad \dfrac{t{=}\{z\mid\Phi z\}\quad \dfrac{\dfrac{\exists!v}{v{=}v}}{\Phi v}}{v\in t}\,\{\,\}\text{-E}_{\text{PM}}\qquad \dfrac{t{=}\{z\mid\Phi z\}\quad \dfrac{}{a\in t}\,(1)}{a{=}u\lor a{=}v}\,\{\,\}\text{-E}_{\text{MP}}}{t=\mathbb{P}(u,v)}\,(1)\,\text{P-I}$$

□

11.5 Pasigraph for singletons

A special case of pairs $\mathbb{P}(u,v)$ arises when $u=v$. Here, the talk is of *singletons*. $\mathbb{P}(u,u)$ is often abbreviated as $\{u\}$. We, however, shall introduce a special single operator σ, and write σu for $\{u\}$.

The introduction and elimination rules for σ arise from the obvious simple modifications of the rules for $\mathbb{P}$.

$$\sigma\text{-I}\quad \dfrac{u\in t\qquad \begin{matrix}\dfrac{}{a\in t}\,(i)\\ \vdots\\ a{=}u\end{matrix}}{t{=}\sigma u}\,(i)\qquad\qquad \sigma\text{-E}_1\quad \dfrac{t{=}\sigma u}{u\in t}\qquad\qquad \sigma\text{-E}_2\quad \dfrac{t{=}\sigma u\qquad w\in t}{w{=}u}$$

The next two Lemmas answer a query from Ethan Brauer: "Do the rules enable one to prove that $\mathbb{P}(u,u)$ is σu, given the existence of either one of them?".

Lemma 11.4 $\dfrac{\exists!\mathbb{P}(u,u)}{\mathbb{P}(u,u)=\sigma u}$.

Proof

$$\dfrac{\dfrac{\dfrac{\exists!\mathbb{P}(u,u)}{\mathbb{P}(u,u)=\mathbb{P}(u,u)}\,\text{Ref=}}{u\in\mathbb{P}(u,u)}\,\text{PE}_1\qquad \dfrac{\dfrac{\dfrac{\exists!\mathbb{P}(u,u)}{\mathbb{P}(u,u)=\mathbb{P}(u,u)}\,\text{Ref=}\quad \dfrac{}{a\in\mathbb{P}(u,u)}\,(2)}{a=u\lor a=u}\,\text{PE}_3\quad \dfrac{}{a=u}\,(1)\quad \dfrac{}{a=u}\,(1)}{a=u}\,(1)}{\mathbb{P}(u,u)=\sigma u}\,(2)\,\sigma\text{I}$$

□

Lemma 11.5 $\dfrac{\exists!\sigma u}{\sigma u=\mathbb{P}(u,u)}$.

Proof

$$\dfrac{\dfrac{\dfrac{\exists!\sigma u}{\sigma u = \sigma u}\,\text{Ref=}}{u \in \sigma u}\,\sigma\text{E}_1 \qquad \dfrac{\dfrac{\exists!\sigma u}{\sigma u = \sigma u}\,\text{Ref=}}{u \in \sigma u}\,\sigma\text{E}_1 \qquad \dfrac{\dfrac{\dfrac{\exists!\sigma u}{\sigma u = \sigma u}\,\text{Ref=} \qquad \dfrac{}{a \in \sigma u}\,(1)}{a = u}\,\sigma\text{E}_2}{a = u \vee a = u}}{\sigma u = \mathbb{P}(u,u)}\,(1)\ \mathbb{P}\text{I}$$

□

11.6 The binary predicate pasigraph $\in^2$ for 'is a member of a member of'

Now consider the simple notion that *t is a member of a member of u*. Let us use for this the (binary) *predicate* pasigraph

$$t \in^2 u.$$

Here are the introduction and elimination rules for this new pasigraph:

$$\in^2\text{-I}\quad \dfrac{t \in v \qquad v \in u}{t \in^2 u} \qquad\qquad \in^2\text{-E}\quad \dfrac{t \in^2 u \qquad \begin{matrix} \underbrace{\overline{t \in a}^{(i)}\ ,\ \overline{a \in u}^{(i)}} \\ \vdots \\ \psi \end{matrix}}{\psi}\,(i)$$

Lemma 11.6 $\dfrac{t \in^2 u}{\exists!t}, \quad \dfrac{t \in^2 u}{\exists!u}.$

Proof

$$\dfrac{t \in^2 u \qquad \dfrac{\dfrac{}{t \in a}\,(1)}{\exists!t}\,\text{RAD}}{\exists!t}\,(1)\ \in^2\text{-E} \qquad\qquad \dfrac{a \in u \qquad \dfrac{\dfrac{}{t \in a}\,(1)}{\exists!u}\,\text{RAD}}{\exists!u}\,(1)\ \in^2\text{-E}$$

□

Lemma 11.7 $t \in^2 u \dashv\vdash \exists x(t \in x \wedge x \in u)$.

Proof

$$\dfrac{t \in^2 u \qquad \dfrac{\dfrac{\overline{a \in u}^{(1)}}{\exists!a} \qquad \dfrac{\overline{t \in a}^{(1)} \qquad \overline{a \in u}^{(1)}}{t \in a \wedge a \in u}}{\exists x(t \in x \wedge x \in u)}}{\exists x(t \in x \wedge x \in u)}\,(1)$$

$$\dfrac{\exists x(t \in x \wedge x \in u) \qquad \dfrac{\overline{t \in a \wedge a \in u}^{(2)} \qquad \dfrac{\overline{t \in a}^{(1)} \qquad \overline{a \in u}^{(1)}}{t \in^2 u}}{t \in^2 u}}{t \in^2 u}\,(2)$$

□

11.7 Pasigraph for unions

We shall next choose the Axiom of Unions to illustrate further the inferentialist's method of treating *operator* pasigraphs. The pasigraph $\bigcup$, for the Union of a given set, is *unary* (unlike the pasigraph $\mathbb{P}$ for Pairing, which, as we have seen, is binary). Set-theorists handle the pasigraph $\bigcup$ with ease, as though it were a familiar primitive expression of their language.

The genuinely primitive form of the Axiom of Unions:

$$\forall x \exists ! \{y \mid \exists z(y \in z \wedge z \in x)\}$$

can be re-written

$$\forall x \exists ! \bigcup x$$

provided only that $\bigcup$ is furnished with Introduction and Elimination rules so that

$$t = \bigcup v \dashv\vdash t = \{y \mid \exists z(y \in z \wedge z \in v)\}.$$

We choose to require the interdeducibility

$$t = \bigcup v \dashv\vdash t = \{y \mid \exists z(y \in z \wedge z \in v)\}$$

rather than require the provability of the identity

$$\bigcup v = \{y \mid \exists z(y \in z \wedge z \in v)\}$$

because we are working in a *free* logic. The latter identity commits one to the existence of the union of v:

$$\exists ! \bigcup v.$$

The interdeducibility, however, does not. It allows one to pin down the meaning of $\bigcup$ as an operator on sets without committing one to its being everywhere (or indeed: *any*where) defined. One can grasp what $\bigcup$ means without yet adopting the Axiom of Unions. And when we *do* adopt that axiom, it does not serve implicitly to define the meaning of $\bigcup$. For that meaning will already have been defined by the following Introduction and Elimination rules for $\bigcup$.

$$\bigcup\text{-I} \quad \dfrac{\exists ! t \qquad \exists ! v \qquad \begin{array}{c} \scriptstyle (i)\,\overline{\quad} \\ a \in^2 v \\ \vdots \\ a \in t \end{array} \qquad \begin{array}{c} \scriptstyle (i)\,\overline{\quad} \\ a \in t \\ \vdots \\ a \in^2 v \end{array}}{t = \bigcup v}\,(i)$$

$$\bigcup\text{-E}_1 \quad \frac{t = \bigcup v \qquad w \in^2 v}{w \in t} \qquad\qquad \bigcup\text{-E}_2 \quad \frac{t = \bigcup v \qquad w \in t}{w \in^2 v}$$

$$\bigcup\text{-E}_3 \quad \frac{t = \bigcup v}{\exists ! t} \qquad\qquad \bigcup\text{-E}_4 \quad \frac{t = \bigcup v}{\exists ! v}$$

Lemma 11.8 *The operator pasigraph* $\bigcup$ *obeys the Rule of Functional Denotation; that is, the following is provable:*

$$\frac{\exists!\bigcup u}{\exists!u}.$$

Proof

$$\dfrac{\exists!\bigcup u,\ \text{i.e.,}\ \exists x\, x=\bigcup u \qquad \dfrac{\overline{a=\bigcup u}^{\,(1)}}{\exists!u}\,{\textstyle\bigcup}\text{-E}_1}{\exists!u}\,(1)$$

□

Lemma 11.9 $t=\bigcup v \vdash t=\{y \mid \exists z(y\in z\wedge z\in v)\}$.

Proof

$$\dfrac{\dfrac{t=\bigcup v \qquad \dfrac{\overset{(1)}{\overline{a\in b\wedge b\in v}} \qquad \dfrac{\overset{(2)}{\overline{a\in b}} \quad \overline{b\in v}^{\,(2)}}{a\in^2 v}\,\in^2\text{-I}}{a\in^2 v}\,(1)}{a\in t}\,{\textstyle\bigcup}\text{-E}_1 \qquad \dfrac{t=\bigcup v}{\exists!t} \qquad \dfrac{\dfrac{t=\bigcup v \quad \overline{a\in t}^{\,(4)}}{a\in^2 v} \qquad \dfrac{\dfrac{\overset{(3)}{\overline{a\in c}}}{\exists!c} \qquad \dfrac{\overline{a\in c}^{\,(3)} \quad \overline{c\in v}^{\,(3)}}{a\in c\wedge c\in v}}{\exists z(a\in z\wedge z\in v)}}{\exists z(a\in z\wedge z\in v)}\,(3)\ \in^2\text{-E}}{t=\{y \mid \exists z(y\in z\wedge z\in v)\}}\,(4)\ \{\}\text{-I}$$

□

Lemma 11.10 $t=\{y \mid \exists z(y\in z\wedge z\in v)\},\ \exists!v \vdash t=\bigcup v$.

Proof Abbreviate $\exists z(y\in z\wedge z\in v)$ as Φy, where convenient, to reduce sideways spread. Let Π_1 be the following fragment of the final proof that we shall construct. Note that the two discharge strokes labeled (2) are being put in place in advance of the eventual step labeled (2) (in the final proof) that will effect that discharge.

$$\dfrac{t=\{y \mid \Phi y\} \qquad \dfrac{\overset{(2)}{\overline{a\in^2 v}} \qquad \dfrac{\dfrac{\overset{(4)}{\overline{a\in c}}}{\exists!c} \qquad \dfrac{\overline{a\in c}^{\,(4)} \quad \overline{c\in v}^{\,(4)}}{a\in c\wedge c\in v}}{\exists z(a\in z\wedge z\in v)}}{\exists z(a\in z\wedge z\in v)}\,(4)\ \in^2\text{-E} \qquad \dfrac{\overline{a\in^2 v}^{\,(2)}}{\exists!a}\,\text{L11.6}}{a\in t}\,\{\}\text{-E}_{\text{PM}}$$

Now let Π_2 be the following fragment of the final proof that we shall construct. Note once again that the discharge stroke labeled (2) has already been put in place, in advance of the eventual step labeled (2) (in the final proof) that will effect that discharge.

$$\dfrac{\dfrac{t=\{y \mid \Phi y\} \qquad \overline{a\in t}^{\,(2)}}{\Phi a}\,\{\}\text{-E}_{\text{MP}} \qquad \dfrac{\overset{(3)}{\overline{a\in b\wedge b\in v}} \qquad \dfrac{\overset{(1)}{\overline{a\in b}} \quad \overline{b\in v}^{\,(1)}}{a\in^2 v}\,\in^2\text{-I}}{a\in^2 v}\,(1)}{a\in^2 v}\,(3)\ \exists\text{-E}$$

Now we can form the final proof

$$\cfrac{\cfrac{t=\{y \mid \Phi y\}}{\exists! t} \quad \exists! v \quad \begin{array}{c}\Pi_1 \\ a \in t\end{array} \quad \begin{array}{c}\Pi_2 \\ a \in^2 v\end{array}}{t=\bigcup v} \ (2)\ \bigcup\text{-I}$$

□

11.8 Pasigraph for inclusion, or subset

The notion ⊆ of inclusion is one of the most familiar and frequently used binary, but *ancillary*, or *defined*, relations in set theory. The usual reading is 't is a subset of u'. The usual definition in primitive vocabulary would be

$$t \subseteq u \equiv_{df} \forall x(x \in t \rightarrow x \in u).$$

The inferentialist, however, working in free logic, lays down instead the following introduction and elimination rules for this pasigraph:

$$\subseteq\text{I} \quad \cfrac{\exists! t \quad \exists! u \quad \begin{array}{c}\overline{a \in t}^{\,(i)} \\ \vdots \\ a \in u\end{array}}{t \subseteq u}\,(i)$$

$$\subseteq\text{E}_1 \quad \frac{t \subseteq u}{\exists! t} \qquad \subseteq\text{E}_2 \quad \frac{t \subseteq u}{\exists! u} \qquad \subseteq\text{E}_3 \quad \frac{t \subseteq u \quad v \in t}{v \in u}$$

That the predicate pasigraph ⊆ obeys the Rule of Atomic Denotation is obvious from $\subseteq\text{E}_1$ and $\subseteq\text{E}_2$.

11.9 The unary predicate pasigraph 'trans'

A *transitive* set is one that contains as members all members of its members. That is, every member of a transitive set is a subset of it. Thus we have the following introduction rule:

$$\text{trans-I} \quad \cfrac{\exists! t \quad \begin{array}{c}\overline{a \in t}^{\,(i)} \\ \vdots \\ a \subseteq t\end{array}}{\mathsf{trans}(t)}\,(i)$$

matched by these elimination rules:

$$\text{trans-E} \quad \frac{\mathsf{trans}(t)}{\exists! t} \qquad \frac{\mathsf{trans}(t) \quad u \in t}{u \subseteq t}$$

11.10 Pasigraph for power sets

We can now turn our attention to the Axiom of Power Sets to illustrate further the inferentialist's method for set theory.

We shall apply our earlier method to the unary operator-pasigraph $\mathscr{P}$ which set-theorists handle with ease, as though it were a familiar primitive expression of their language. The genuinely primitive form of the Axiom of Power Sets:

$$\forall x \exists ! \{y \mid \forall z(z \in y \rightarrow z \in x)\}$$

i.e.,

$$\forall x \exists ! \{y \mid y \subseteq x\}$$

can be re-written

$$\forall x \exists ! \mathscr{P} x$$

provided only that $\mathscr{P}$ is furnished with Introduction and Elimination rules so that

$$t = \mathscr{P} v \dashv\vdash t = \{y \mid \forall z(z \in y \rightarrow z \in x)\}.$$

As with Unions, we choose with Power Sets to require the 'general term t'-involving interdeducibility

$$t = \mathscr{P} v \dashv\vdash t = \{y \mid \forall z(z \in y \rightarrow z \in v)\}$$

rather than require the provability of the identity

$$\mathscr{P} v = \{y \mid \forall z(z \in y \rightarrow z \in v)\},$$

because we are working in a *free* logic. The latter identity commits one to the existence of the power set of v:

$$\exists ! \mathscr{P} v.$$

The interdeducibility, however, does not. It allows one to pin down the meaning of $\mathscr{P}$ as an operator on sets without committing one to its being everywhere (or indeed: *any*where) defined. One can grasp what $\mathscr{P}$ means without yet adopting the Axiom of Power Sets. And when we *do* adopt that axiom, it does not serve implicitly to define the meaning of $\mathscr{P}$. For that meaning will already have been defined by the following Introduction and Elimination rules for $\mathscr{P}$.

$$\mathscr{P}\text{-I} \quad \dfrac{\exists ! t \qquad \exists ! v \qquad \begin{array}{c} {}^{(i)}\overline{a \subseteq v} \\ \vdots \\ a \in t \end{array} \qquad \begin{array}{c} {}^{(i)}\overline{a \in t} \\ \vdots \\ a \subseteq v \end{array}}{t = \mathscr{P} v}\,{\scriptstyle (i)}$$

$$\mathscr{P}\text{-E}_1 \quad \dfrac{t = \mathscr{P} v}{\exists ! t} \qquad\qquad \mathscr{P}\text{-E}_2 \quad \dfrac{t = \mathscr{P} v}{\exists ! v}$$

$$\mathscr{P}\text{-E}_3 \quad \frac{t = \mathscr{P}v \qquad u \subseteq v}{u \in t} \qquad\qquad \mathscr{P}\text{-E}_4 \quad \frac{t = \mathscr{P}v \qquad u \in t}{u \subseteq v}$$

Lemma 11.11 $\dfrac{u \in \mathscr{P}v}{\exists! v}$.

Proof

$$\dfrac{\exists!\mathscr{P}v,\ \text{i.e.,}\ \exists x\, x = \mathscr{P}v \qquad \dfrac{\overline{a = \mathscr{P}v}^{\,(1)}}{\exists! v}\,\mathscr{P}\text{-E}_2}{\exists! v}\,(1)$$

□

Lemma 11.12 $\dfrac{\exists!\mathscr{P}v \qquad u \subseteq v}{u \in \mathscr{P}v}$.

Proof

$$\dfrac{\exists!\mathscr{P}v,\ \text{i.e.,}\ \exists x\, x = \mathscr{P}v \qquad \dfrac{\overline{a = \mathscr{P}v}^{\,(1)} \qquad u \subseteq v}{u \in \mathscr{P}v}\,\mathscr{P}\text{-E}_3}{u \in \mathscr{P}v}\,(1)$$

□

Lemma 11.13 $\dfrac{u \in \mathscr{P}v}{u \subseteq v}$.

Proof

$$\dfrac{\dfrac{u \in \mathscr{P}v}{\exists!\mathscr{P}v,\ \text{i.e.,}\ \exists x\, x = \mathscr{P}v} \qquad \dfrac{\overline{a = \mathscr{P}v}^{\,(1)} \qquad u \in \mathscr{P}v}{u \subseteq v}\,\mathscr{P}\text{-E}_4}{u \subseteq v}\,(1)$$

□

Lemma 11.14 $t = \mathscr{P}v \vdash t = \{y \mid \forall z(z \in y \rightarrow z \in v)\}$.

Proof

$$\dfrac{\dfrac{t=\mathscr{P}v \qquad \dfrac{\overline{\exists! a}^{\,(3)} \quad \dfrac{t=\mathscr{P}v}{\exists! v} \quad \dfrac{\overline{c \in a}^{\,(1)} \quad \dfrac{\overline{\forall z(z\in a \rightarrow z \in v)}^{\,(3)} \quad \dfrac{\overline{c\in a}^{\,(1)}}{\exists! c}}{c \in a \rightarrow c \in v}}{c \in v}\,(1)\subseteq\text{I}}{a \subseteq v}}{a \in t} \qquad \dfrac{t=\mathscr{P}v}{\exists! t} \qquad \dfrac{\dfrac{\dfrac{\dfrac{t=\mathscr{P}v \quad \overline{a\in t}^{\,(3)}}{a \subseteq v}\,\mathscr{P}\text{E}_4 \quad \overline{b \in a}^{\,(2)}}{b \in v}\,\subseteq\text{E}_3}{b\in a \rightarrow b \in v}\,(2)}{\forall z(z\in a\rightarrow z \in v)}}{t=\{y \mid \forall z(z \in y \rightarrow z \in v)\}}\,(3)\,\{\,\}\text{I}$$

□

Lemma 11.15 $\exists! v,\ t = \{y \mid \forall z(z \in y \rightarrow z \in v)\} \vdash t = \mathscr{P}v$.

Proof Abbreviate $\forall z(z \in y \rightarrow z \in v)$ by Φy and $\forall z(z \in a \rightarrow z \in v)$ by Φa, where convenient, to reduce sideways spread. Moreover, let Π be

$$\dfrac{\dfrac{\overline{b \in a}^{\,(3)} \qquad \dfrac{\dfrac{\overline{a \in t}^{\,(1)} \quad t=\{y \mid \Phi y\}}{\Phi a} \qquad \dfrac{\overline{b \in a}^{\,(3)}}{\exists! b}}{b \in a \rightarrow b \in v}}{b \in v}}{a \subseteq v}\,(3)\subseteq\text{I}$$

Then form the proof

$$
\dfrac{\dfrac{t=\{y\,|\,\Phi y\}}{\exists!t} \qquad \exists!v \qquad \dfrac{t=\{y\,|\,\Phi y\} \qquad \dfrac{\dfrac{\dfrac{\overline{a\subseteq v}^{\,(1)} \quad \overline{b\in a}^{\,(1)}}{b\in v}\,{\subseteq}\mathrm{E}_3}{b\in a\rightarrow b\in v}\,(2)}{\Phi a} \qquad \dfrac{\overline{a\subseteq v}^{\,(1)}}{\exists!a}}{a\in t}\,\{\,\}\mathrm{E}_{\mathrm{PM}} \qquad \Pi}{t=\mathscr{P}v}\,(1)
$$

□

Lemma 11.16 $\exists!\mathscr{P}(t),\ \mathsf{trans}(t) \vdash \mathsf{trans}(\mathscr{P}(t))$.

Proof

$$
\dfrac{\dfrac{\dfrac{\exists!\mathscr{P}(t) \qquad \dfrac{\mathsf{trans}(t) \qquad \dfrac{(1)\,\overline{c\in d} \qquad \dfrac{\overline{d\in\mathscr{P}(t)}^{\,(2)}}{d\subseteq t}\,\mathrm{L}11.13}{c\in t}\,{\subseteq}\mathrm{E}_3}{c\subseteq t}\,\mathsf{trans\text{-}E}}{c\in\mathscr{P}(t)}\,\mathrm{L}11.12}{d\subseteq\mathscr{P}(t)}\,(1)\,{\subseteq}\mathrm{I}}{\mathsf{trans}(\mathscr{P}(t))}\,(2)\,\mathsf{trans\text{-}I}
$$

□

11.11 The binary predicate pasigraph ⅂ for 'is disjoint from'

We now introduce a new binary-relation pasigraph: $t \,\text{⅂}\, u$ is to mean that t is disjoint from u — that is, they have no member in common. This pasigraph will be useful in the formulation of the Axiom (or Rule) of Regularity. The introduction and elimination rules for ⅂ are as follows.

⅂-I

$$
\dfrac{\exists!t \qquad \exists!u \qquad \begin{array}{c} \underbrace{(i)\,\overline{a\in t}\ ,\ \overline{a\in u}\,(i)} \\ \vdots \\ \bot \end{array}}{t\,\text{⅂}\,u}\,(i)
$$

$$
\text{⅂}\mathrm{E} \quad \dfrac{t\,\text{⅂}\,u}{\exists!t} \qquad \dfrac{t\,\text{⅂}\,u}{\exists!u} \qquad \dfrac{t\,\text{⅂}\,u \qquad v\in t \qquad v\in u}{\bot}
$$

Lemma 11.17 $t\,\text{⅂}\,u \dashv\vdash t\cap u=\emptyset$.

11.12 Pasigraph for ranges

Suppose φxy has the variables x, y free, and is functional from x to y, at least for x in t. The Replacement Pasigraph $\varphi_{xy}[t]$ can then be explicitly defined as follows.

$$\varphi_{xy}[t] =_{df} \{y \mid \exists x(x \in t \wedge \varphi)\}.$$

The subscripting with x and y registers the fact that this pasigraph binds those two variables.

[The Axiom Scheme of] Replacement, due to Fraenkel, can be formulated as the following 'conditional existence' rule (where '$\exists_1$' is the uniqueness quantifier):

$$\dfrac{\begin{array}{c}\overline{a \in t}^{\,(i)}\\ \vdots\\ \exists_1 y \varphi a y\end{array}}{\exists! \varphi_{xy}[t]}\,(i)$$

The pasigraph could also be taken as a grammatical primitive, furnished with the following rules of introduction and elimination.

$$\ldots[\ldots]\text{-Intro}\qquad \dfrac{\begin{array}{c}{}^{(i)}\overline{a \in v}\\ \vdots\\ w \in t\end{array}\quad \begin{array}{c}{}^{(i)}\overline{a \in v}\\ \vdots\\ \varphi w a\end{array}\quad \begin{array}{c}\underbrace{{}^{(i)}\overline{c \in t}\ ,\ \overline{\varphi c b}^{\,(i)}}\\ \vdots\\ b \in v\end{array}\quad \begin{array}{c}\underbrace{{}^{(i)}\overline{c \in t}\,,\overline{\varphi c b}^{\,(i)}\,,\overline{\varphi c d}^{\,(i)}}\\ \vdots\\ b = d\end{array}}{v = \varphi_{xy}[t]}\,(i)$$

Note that the first two subproofs call for some same term w in their conclusions. This means that the elimination rule will have a part that corresponds to these two subproofs taken together. The final subproof ensures the functionality of φxy on t as its domain.

$$\ldots[\ldots]\text{-Elim}\qquad \dfrac{v = \varphi_{xy}[t]\qquad u \in v\qquad \begin{array}{c}\underbrace{{}^{(i)}\overline{a \in t}\ ,\ \overline{\varphi a u}^{\,(i)}}\\ \vdots\\ \theta\end{array}}{\theta}\,(i)$$

$$\dfrac{v = \varphi_{xy}[t]\qquad u \in t\qquad \varphi u w}{w \in v}\qquad\qquad \dfrac{v = \varphi_{xy}[t]\qquad u \in t\qquad \varphi u v\qquad \varphi u w}{v = w}$$

In set theory, the successor of a set u is defined as $u \cup \{u\}$ — or, in the notation we have thus far introduced, as $\bigcup \mathbb{P}(u, \sigma u)$. We shall now introduce a unary operator pasigraph s to represent successor, and furnish it with introduction and elimination rules that secure for it the same meaning.

$$\mathsf{sI}\qquad \dfrac{u \in t\qquad \begin{array}{c}\overline{a \in u}^{\,(i)}\\ \vdots\\ a \in t\end{array}\qquad \begin{array}{c}\overline{b \in t}^{\,(i)}\\ \vdots\\ b \in u \vee b = u\end{array}}{t = \mathsf{s}u}\,(i)$$

$$\mathsf{sE}\qquad \dfrac{t = \mathsf{s}u}{u \in t}\qquad \dfrac{t = \mathsf{s}u\qquad v \in u}{v \in t}\qquad \dfrac{t = \mathsf{s}u\qquad v \in t}{v \in u \vee v = u}$$

Lemma 11.18 $t = u \cup \{u\} \dashv\vdash t = \bigcup \mathbb{P}(u, \sigma u)$.

11.13 The unary predicate pasigraph 'comp'

$\mathsf{comp}(t, u)$ is to mean that t and u are comparable in terms of the membership relation:

$$\mathsf{comp}(t, u) \equiv_{df} t \in u \vee t{=}u \vee u \in t.$$

Introduction and elimination rules that secure this meaning directly are as follows.

$$\mathsf{comp\text{-}I} \quad \frac{t \in u}{\mathsf{comp}(t, u)} \qquad \frac{t = u}{\mathsf{comp}(t, u)} \qquad \frac{u \in t}{\mathsf{comp}(t, u)}$$

$$\mathsf{comp\text{-}E} \quad \frac{\mathsf{comp}(t, u) \quad \begin{array}{c} \overline{t \in u}^{\,(i)} \\ \vdots \\ \theta/\bot \end{array} \quad \begin{array}{c} \overline{t = u}^{\,(i)} \\ \vdots \\ \theta/\bot \end{array} \quad \begin{array}{c} \overline{u \in t}^{\,(i)} \\ \vdots \\ \theta/\bot \end{array}}{\theta/\bot}\,(i)$$

11.14 The unary predicate pasigraph 'conn'

$\mathsf{conn}(t)$ is to mean that t is connected by the membership relation:

$$\mathsf{conn}(t) \equiv_{df} \forall x \forall y((x \in t \wedge y \in t) \rightarrow (x \in y \vee x{=}y \vee y \in x)).$$

Introduction and elimination rules that secure this meaning directly are as follows.

$$\mathsf{conn\text{-}I} \quad \frac{\exists! t \qquad \begin{array}{c} \underbrace{{}^{(i)}\overline{a \in t}\ ,\ \overline{b \in t}^{\,(i)}} \\ \vdots \\ \mathsf{comp}(a, b) \end{array}}{\mathsf{conn}(t)}\,(i)$$

$$\mathsf{conn\text{-}E} \quad \frac{\mathsf{conn}(t)}{\exists! t} \qquad \frac{\mathsf{conn}(t) \quad u \in t \quad v \in t}{\mathsf{comp}(u, v)}$$

11.15 The unary predicate pasigraph 'O' (for 'is an ordinal')

$\mathsf{O}(t)$ is to mean that t is an ordinal. The introduction and elimination rules are as follows.

$$\text{O-I}\ \frac{\mathsf{trans}(t) \quad \mathsf{conn}(t)}{\mathsf{O}(t)} \qquad \text{O-E}\ \frac{\mathsf{O}(t)}{\mathsf{trans}(t)} \quad \frac{\mathsf{O}(t)}{\mathsf{conn}(t)}$$

11.16 The unary predicate pasigraph 'IoS' (for 'is an initial or successor ordinal')

$\mathsf{IoS}(t)$ is to mean that t is an *initial or successor* ordinal. The introduction and elimination rules are as follows.

$$\text{IoS-I}\ \frac{t=\emptyset}{\mathsf{IoS}(t)} \quad \frac{\mathsf{O}(u) \quad t=\mathsf{s}u}{\mathsf{IoS}(t)} \qquad \text{IoS-E}\ \frac{\mathsf{IoS}(t) \quad \begin{array}{c}\overline{t=\emptyset}^{\,(i)} \\ \vdots \\ \theta/\bot\end{array} \quad \begin{array}{c}\underbrace{\overline{\mathsf{O}(a)}^{\,(i)}\ ,\ \overline{t=\mathsf{s}a}^{\,(i)}} \\ \vdots \\ \theta/\bot\end{array}}{\theta/\bot}\,(i)$$

11.17 The unary predicate pasigraph 'fO' (for 'is a finite ordinal')

$\mathsf{fO}(t)$ is to mean that t is a *finite* ordinal. The introduction and elimination rules are as follows.

$$\text{fO-I}\ \frac{\mathsf{O}(t) \quad \mathsf{IoS}(t) \quad \begin{array}{c}\overline{a\in t}^{\,(i)} \\ \vdots \\ \mathsf{IoS}(a)\end{array}}{\mathsf{fO}(t)}\,(i) \qquad \text{fO-E}\ \frac{\mathsf{fO}(t)}{\mathsf{O}(t)} \quad \frac{\mathsf{fO}(t)}{\mathsf{IoS}(t)} \quad \frac{\mathsf{fO}(t) \quad u\in t}{\mathsf{IoS}(u)}$$

11.18 The constant pasigraph ω

The set ω of finite ordinals:

$$\omega =_{df} \{x \mid \mathsf{fO}(x)\}$$

is the canonical choice among set theorists of a *countably infinite* set. The Axiom of Infinity is usually formulated as the statement that ω exists:

$$\exists!\omega.$$

Theorem 11.19 $\exists!t \ \vdash\ t = \bigcup \mathscr{P}t$.

Proof The following proof uses the new rules for $\bigcup$, $\subseteq$, $\mathscr{P}$, and σ. It also appeals to the existence of singletons, and the existence of power sets.

$$\dfrac{\exists!t \quad \dfrac{\exists!t}{\exists!\mathscr{P}t} \quad \dfrac{\dfrac{\dfrac{\dfrac{\exists!t}{\exists!\mathscr{P}t}}{\mathscr{P}t=\mathscr{P}t} \quad \dfrac{}{b\in\mathscr{P}t}{}^{(1)}}{b\subseteq t}\,\mathscr{P}\mathrm{E}_4 \quad \dfrac{}{a\in b}{}^{(1)}}{a\in t}\,\subseteq\mathrm{E}_3 \quad \dfrac{\dfrac{\dfrac{\exists!t}{\exists!\mathscr{P}t}}{\mathscr{P}t=\mathscr{P}t} \quad \dfrac{\dfrac{\dfrac{\dfrac{\dfrac{\dfrac{\dfrac{}{c\in t}{}^{(1)}}{\exists!c}}{\exists!\sigma c}}{\sigma c=\sigma c} \quad \dfrac{}{d\in\sigma c}{}^{(2)}}{d=c}\,\sigma\mathrm{E}_2 \quad \dfrac{}{c\in t}{}^{(1)}}{d\in t}}{\sigma c\subseteq t}\,(2)\subseteq\mathrm{I}}{\sigma c\in\mathscr{P}t}\,\mathscr{P}\mathrm{E}_3}{t=\bigcup\mathscr{P}t}\,(1)\bigcup\mathrm{I}$$

□

We do *not* have the operator-commutation of Theorem 11.19, namely $\exists!t \vdash t = \mathscr{P}\bigcup t$. Here is a counterexample: *Take* $t = \{\{a\}\}$, *where* $\exists!a$, *whence* $\exists!t$. *Then* $\bigcup t = \{a\}$. *So* $\mathscr{P}\bigcup t = \{\emptyset, \{a\}\} \neq \{\{a\}\} = t$.

12 New rules for identity and existence in free logic

In our deployment of free first-order logic thus far, we have used the abbreviation $\exists!t$ to express the thought that t exists, or is defined. Moreover, $\exists!t$ has been taken as a mere abbreviation of the longer, 'official' sentence $\exists x\, x=t$.

We shall now provide new introduction and elmination rules for the *separate* expressions !, =, $\exists$ and $\forall$, which do a better job of capturing their logical roles.

The formal sentence $!t$ will now, officially, be a well-formed sentence produced by our logical grammar. $!t$ will *take over* the role formerly played by $\exists!t$. Since ! will be treated as a *primitive* expression, $!t$ will *not* be a definitional abbreviation.

$!t$ means "t exists", or, as mathematicians often put it, "t is defined".

12.1 Introduction and elimination rules for !

! has two parts to its introduction rule:

$$\frac{A(\ldots t\ldots)}{!t}, \text{ where } A \text{ is an atomic predicate;} \qquad \frac{!f(\ldots t\ldots)}{!t}.$$

So: $!t$ is a consequence of any atomic fact involving (the denotation of) t; and is a consequence also of the existence of any function's yielding a value on arguments among which is (the denotation of) t.

When a proposition can be inferred from each of such a wide range of propositions, it must be extremely weak; and its own consequences will be at least as weak.

So when the question arises: *What might legitimately be inferred from* $!t$*?*, given its own two-part introduction rule, the answer must be: an atomic proposition, involving t as its only constituent term, that is bound to be true no matter what 'positive' atomic facts might obtain (involving the denotation of t), and no matter what mappings might be effected by what functions involving the denotation of t as an input. An excellent candidate for such an atomic proposition would be $t=t$. The Elimination Rule for ! is, accordingly:

$$!\mathrm{E}\quad \frac{!t}{t=t}.$$

12.2 Introduction and elimination rules for =

$$=\mathrm{I}\quad \frac{\begin{matrix}\overline{F(t)}^{\,(i)}\\ \vdots\\ F(u)\end{matrix}\qquad !t\qquad !u}{t=u}\,{\scriptstyle (i)}\ \text{, where } F \text{ is parametric;}\qquad =\mathrm{E}\quad \frac{t=u\qquad \psi(t)}{\psi(u)}\quad \frac{t=u}{!t}\quad \frac{t=u}{!u}.$$

Note how the last two parts of =E are already covered by the rule !I — since identity statements are atomic.

Theorem 12.1 $!t \vdash t=t$.

Proof $$\frac{\overline{Ft}^{\,(i)}\qquad !t\qquad !t}{t=t}\,{\scriptstyle =\mathrm{I}}$$ □

It is interesting that this derived result using =I simply *is* the rule !E.

$$\dfrac{\dfrac{\begin{matrix}\Delta\\ \Pi\\ A(t)\end{matrix}}{!t}\,{\scriptstyle !\mathrm{I}}}{t=t}\,{\scriptstyle !\mathrm{E}}\ \rightsquigarrow\ \dfrac{{\scriptstyle !1}\ \overline{Ft}\qquad \dfrac{\begin{matrix}\Delta\\ \Pi\\ A(t)\end{matrix}}{!t}\,{\scriptstyle !\mathrm{I}}\qquad \dfrac{\begin{matrix}\Delta\\ \Pi\\ A(t)\end{matrix}}{!t}\,{\scriptstyle !\mathrm{I}}}{t=t}\,{\scriptstyle (1)=\mathrm{I}}\qquad \dfrac{\dfrac{\begin{matrix}\Delta\\ \Pi\\ !f(t)\end{matrix}}{!t}\,{\scriptstyle !\mathrm{I}}}{t=t}\,{\scriptstyle !\mathrm{E}}\ \rightsquigarrow\ \dfrac{{\scriptstyle !1}\ \overline{Ft}\qquad \dfrac{\begin{matrix}\Delta\\ \Pi\\ !f(t)\end{matrix}}{!t}\,{\scriptstyle !\mathrm{I}}\qquad \dfrac{\begin{matrix}\Delta\\ \Pi\\ !f(t)\end{matrix}}{!t}\,{\scriptstyle !\mathrm{I}}}{t=t}\,{\scriptstyle (1)=\mathrm{I}}$$

We see, then, that the two inferences

$$\frac{A(t)}{t=t}\quad\text{and}\quad\frac{!f(t)}{t=t}$$

have normal proofs. We shall henceforth adopt them as primitive inferences, while mindful that they are actually derived rules.

With $!t$ as our preferred way of expressing the existence or definedness of t, our earlier proof of the sequent $\exists !t : t=\bigcup\mathscr{P}t$ can be rewritten as follows:

$$\dfrac{!t\qquad \dfrac{!t}{!\mathscr{P}t}\qquad \dfrac{\dfrac{\dfrac{\dfrac{!t}{!\mathscr{P}t}}{\mathscr{P}t=\mathscr{P}t}\qquad \overline{b\in\mathscr{P}t}^{\,(1)}}{b\subseteq t}\,{\scriptstyle \mathscr{P}\mathrm{E}_4}\qquad \overline{a\in b}^{\,(1)}}{a\in t}\,{\scriptstyle \subseteq\mathrm{E}_3}\qquad \dfrac{\dfrac{\dfrac{!t}{!\mathscr{P}t}}{\mathscr{P}t=\mathscr{P}t}\qquad \dfrac{\dfrac{\dfrac{\dfrac{\dfrac{\dfrac{\overline{c\in t}^{\,(1)}}{!c}}{!\sigma c}}{\sigma c=\sigma c}\qquad \overline{d\in\sigma c}^{\,(2)}}{d=c}\,{\scriptstyle \sigma\mathrm{E}_2}\qquad \overline{c\in t}^{\,(1)}}{d\in t}}{\sigma c\subseteq t}\,{\scriptstyle (2)\subseteq\mathrm{I}}}{\sigma c\in\mathscr{P}t}\,{\scriptstyle \mathscr{P}\mathrm{E}_3}}{t=\bigcup\mathscr{P}t}\,{\scriptstyle (1)\,\bigcup\mathrm{I}}$$

12.3 Some results for !, =, ∃, and ∀

Theorem 12.2 $!t \vdash \exists x\, x{=}t$.

Proof
$$\dfrac{!t \qquad \dfrac{!t}{t{=}t}\,{!}\mathrm{E}}{\exists x\, x{=}t}\,\exists\mathrm{I}$$
□

Theorem 12.3 $\exists x\, x{=}t \vdash\, !t$.

Proof
$$\dfrac{\exists x\, x{=}t \qquad \dfrac{\overline{a{=}t}^{\,(1)}}{!t}\,{!}\mathrm{I}}{!t}\,(1)\,\exists\mathrm{E}$$
□

Theorem 12.4 $\vdash \forall x\, !x$.

Proof
$$\dfrac{\overline{!a}^{\,(1)}}{\forall x\, !x}\,(1)\,\forall\mathrm{I}$$
□

Theorem 12.5 $\vdash \forall x\, x{=}x$.

Proof
$$\dfrac{\dfrac{\overline{!a}^{\,(1)}}{a{=}a}\,{!}\mathrm{E}}{\forall x\, x{=}x}\,(1)\,\forall\mathrm{I}$$
□

12.4 Using the new pasigraph ! to reformulate the introduction and elimination rules for set-theoretic pasigraphs

The rules for set-theoretical pasigraphs, rewritten: ∅.

$$\emptyset\mathrm{I}\quad \dfrac{!t \qquad \begin{array}{c}\overline{a \in t}^{\,(i)}\\ \vdots \\ \bot\end{array}}{t{=}\emptyset}\,(i) \qquad\qquad \emptyset\mathrm{E}\quad \dfrac{t{=}\emptyset}{!t} \qquad \dfrac{t{=}\emptyset \qquad u \in t}{\bot}$$

The rules for set-theoretical pasigraphs, rewritten: ⊆.

$$\subseteq\mathrm{I}\quad \dfrac{!t \qquad !u \qquad \begin{array}{c}\overline{a \in t}^{\,(i)}\\ \vdots \\ a \in u\end{array}}{t \subseteq u}\,(i)$$

$$\subseteq\mathrm{E}_1\quad \dfrac{t \subseteq u}{!t} \qquad \subseteq\mathrm{E}_2\quad \dfrac{t \subseteq u}{!u} \qquad \subseteq\mathrm{E}_3\quad \dfrac{t \subseteq u \qquad v \in t}{v \in u}$$

The rules for set-theoretical pasigraphs, rewritten: ℙ. These rules are unchanged.

$$\mathbb{P}\text{-}\mathrm{I}\quad \dfrac{u \in t \qquad v \in t \qquad \begin{array}{c}\overline{a \in t}^{\,(i)}\\ \vdots \\ a{=}u \vee a{=}v\end{array}}{t{=}\mathbb{P}(u,v)}\,(i)$$

$$\mathbb{P}\text{-E}_1\ \frac{t=\mathbb{P}(u,v)}{u\in t}\qquad \mathbb{P}\text{-E}_2\ \frac{t=\mathbb{P}(u,v)}{v\in t}\qquad \mathbb{P}\text{-E}_3\ \frac{t=\mathbb{P}(u,v)\quad w\in t}{w=u\vee w=v}$$

The rules for set-theoretical pasigraphs, rewritten: σ. These rules are unchanged.

$$\sigma\text{-I}\ \frac{u\in t\qquad \begin{array}{c}\overline{a\in t}^{\,(i)}\\ \vdots\\ a\in u\end{array}}{t=\sigma u}\,(i)\qquad\qquad \sigma\text{-E}_1\ \frac{t=\sigma u}{u\in t}\qquad \sigma\text{-E}_2\ \frac{t=\sigma u\quad w\in t}{w=u}$$

The rules for set-theoretical pasigraphs, rewritten: $\subseteq$.

$$\bigcup\text{-I}\ \frac{!t\quad !v\qquad \begin{array}{c}\underbrace{{}^{(i)}\overline{a\in b}\ ,\ \overline{b\in v}^{\,(i)}}\\ \vdots\\ a\in t\end{array}\qquad \begin{array}{c}{}^{(i)}\overline{c\in t}\\ \vdots\\ c\in w\end{array}\qquad \begin{array}{c}{}^{(i)}\overline{c\in t}\\ \vdots\\ w\in v\end{array}}{t=\bigcup v}\,(i)\quad ,\ a,b,c\ \text{parametric}$$

$$\bigcup\text{-E}_1\ \frac{t=\bigcup v\qquad w\in u}{u\in v}\qquad\qquad \bigcup\text{-E}_3\ \frac{t=\bigcup v}{!t}$$

$$\bigcup\text{-E}_2\ \frac{t=\bigcup v\quad u\in t\qquad \begin{array}{c}\underbrace{{}^{(i)}\overline{u\in a}\ ,\ \overline{a\in v}^{\,(i)}}\\ \vdots\\ \theta\end{array}}{\theta}\,(i)\quad ,\ a\ \text{parametric;}\qquad \bigcup\text{-E}_4\ \frac{t=\bigcup v}{!v}$$

The rules for set-theoretical pasigraphs, rewritten: Separation.

$$\text{Intro}\quad \frac{\begin{array}{c}\underbrace{{}^{(i)}\overline{\varphi^x_a}\ ,\ \overline{a\in v}^{\,(i)}}\\ \vdots\\ a\in t\end{array}\qquad !t\qquad \begin{array}{c}{}^{(i)}\overline{a\in t}\\ \vdots\\ a\in v\end{array}\qquad \begin{array}{c}{}^{(i)}\overline{a\in t}\\ \vdots\\ \varphi^x_a\end{array}}{t=\{x\in v\mid\varphi\}}\,(i)$$

$$\text{Elim}\quad \frac{t=\{x\in v\mid\varphi\}\qquad \varphi^x_u\qquad u\in v}{u\in t}\qquad\qquad \frac{t=\{x\in v\mid\varphi\}}{!t}$$

$$\frac{t=\{x\in v\mid\varphi\}\qquad u\in t}{u\in v}\qquad\qquad \frac{t=\{x\in v\mid\varphi\}\qquad u\in t}{\varphi^x_u}$$

The rules for set-theoretical pasigraphs, rewritten: $\mathscr{P}$.

$$\mathscr{P}\text{-I}\quad \frac{!t\quad !v\qquad \begin{array}{c}{}^{(i)}\overline{a\subseteq v}\\ \vdots\\ a\in t\end{array}\qquad \begin{array}{c}{}^{(i)}\overline{a\in t}\\ \vdots\\ a\subseteq v\end{array}}{t=\mathscr{P}v}\,(i)$$

$$\mathscr{P}\text{-E}_1 \quad \frac{t = \mathscr{P}v}{!t} \qquad\qquad \mathscr{P}\text{-E}_2 \quad \frac{t = \mathscr{P}v}{!v}$$

$$\mathscr{P}\text{-E}_3 \quad \frac{t = \mathscr{P}v \qquad u \subseteq v}{u \in t} \qquad\qquad \mathscr{P}\text{-E}_4 \quad \frac{t = \mathscr{P}v \qquad u \in t}{u \subseteq v}$$

The rules for ontologially committal — and *classical* — set-theory, rewritten.

Extensionality

This is now derivable, given the I- and E-rules for $\{x \mid \ldots x \ldots\}$.

Regularity, or Foundation

$$\frac{u \in t \qquad \begin{matrix} \underbrace{{}^{(i)}\overline{a \in t}\ ,\ \overline{t \rceil a}^{(i)}} \\ \vdots \\ \theta \end{matrix}}{\theta}\ (i)$$

Existence of Empty Set

$$\frac{}{!\emptyset}$$

Separation Schema

$$\frac{!t}{!\{x \in t \mid \varphi\}}$$

Replacement Schema

$$\frac{\begin{matrix} \overline{a \in t}^{(i)} \\ \vdots \\ \exists_1 y \varphi a y \end{matrix}}{!\varphi[t]}\ (i)$$

Pairing

$$\frac{!t \qquad !u}{!\mathbb{P}(t,u)}$$

Infinity

$$\frac{}{!\omega}$$

Union

$$\frac{!t}{!\bigcup t}$$

Choice

$$\frac{(\forall x \in t)(\exists y \in u)R(x,y)}{(\exists \phi : t \mapsto u)(\forall x \in t)R(x, \phi x)}$$

Power Set

$$\frac{!t}{!\mathscr{P}t}$$

13 Concluding remarks

It is important to remind the reader of the methodological underpinnings of this study. We have sought to illuminate the meanings of set-theoretical expressions in such a way as to secure agreement on those meanings from classicists, intuitionists, and constructivists alike. This we have done by laying down ontologically non-committal rules for set-theoretical expressions, no matter whether they are, conventionally, either primitive or defined. This captures the 'analytical core' of set-theoretical talk — what Quine once called 'virtual set theory'. Interestingly, all the proofs thus far involved in delivering this analytical core are proofs in Core Logic.

It is then a *further* question what sets actually exist — either outright or conditionally. Two simple examples, respectively, will illustrate this. That the empty set exists is an *outright* existential assertion. That, given any two sets, their pair set exists, is a *conditional* existential assertion. On such simple, finitistic assertions it is no surprise that no theorist from any of the competing camps — classical, intuitionistic, or constructivist — demurs. Disagreements arise only when one begins to deal with such matters as *completed infinities*; sets being specified by means of *effectively undecidable* formulae; sets being specified by means of *impredicative* formulae; and/or whether there can be sets answering to formulae whose extensions would be *too extensive*. Constructive set theorists also have to be vigilant about their choices of constructively distinguishable (i.e., non-equivalent) formulations of axioms or axiom-schemes that the classicist is able to regard as equivalent (possibly, *modulo* other, 'more basic', or 'secure' axioms already laid down). Such is the case with various possible forms of the Axiom of Choice; of the Axiom of Regularity, or Foundation; and (so this author contends) with the Axiom Scheme of Separation.[15]

In the conduct of the further investigations to which these latter considerations give rise, the present author offers the *common parlance* of the pasigraphs treated above. They provide the *lingua franca* within which classicists, intuitionists, and constructivists can subsequently disagree, or agree to differ, given their respective doctrinal grounds concerning the nature of mathematical existence, the bivalence of mathematical truth, whether such truth is epistemically constrained, etc.

References

Frege, G. (1893). *Grundgesetze der Arithmetik. I. Band.* reprinted 1962. Hildesheim: Georg Olms Verlagsbuchhandlung.

– (1903). *Grundgesetze der Arithmetik. II. Band.* reprinted 1962. Hildesheim: Georg Olms Verlagsbuchhandlung.

– (2013). *Basic Laws of Arithmetic. Derived using concept-script. Volumes I & II.* Oxford: Oxford University Press.

Prawitz, D. (1965). *Natural Deduction: A Proof-Theoretical Study.* Stockholm: Almqvist & Wiksell.

Schroeder-Heister, P. (2016). Open Problems in Proof-Theoretic Semantics. In: *Advances in Proof-Theoretic Semantics.* Ed. by T. Piecha and P. Schroeder-Heister. Berlin: Springer, 253–83.

Tennant, N. (1978). *Natural Logic.* Edinburgh University Press.

– (2010). Inferential Semantics. In: *The Force of Argument: Essays in Honor of Timothy Smiley.* Ed. by J. Lear and A. Oliver. Routledge: London, 223–57.

– (2017). *Core Logic.* Oxford: Oxford University Press.

– (2018). A logical theory of truthmakers and falsitymakers. In: *Handbook on Truth.* Ed. by M. Glanzberg. Oxford: Oxford University Press, 355–393.

– (2020). Does Choice Really Imply Excluded Middle? Part I: Regimentation of the Goodman–Myhill result, and its immediate reception. *Philosophia Mathematica* 28, 139–171.

[15] See Tennant (2020; 2021).

Tennant, N. (2021). Does Choice Really Imply Excluded Middle? Part II: Historical, philosophical and foundational reflections on the Goodman–Myhill result. *Philosophia Mathematica* 29.

Zermelo, E. (1908). Untersuchungen über die Grundlagen der Mengenlehre I. *Mathematische Annalen* 65, 261–81.

The Validity of Inference and Argument*

Dag Prawitz

It has been common in contemporary logic and philosophy of logic to identify the validity of an inference with its conclusion being a (logical) consequence of its premisses. This identification pays attention to at most a necessary condition for an inference being acceptable in a deductive argument or proof. An inference is not acceptable unless the conclusion becomes evident because of being supported by the premisses. Can we define this condition in a stringent way so that we get a concept of valid inference allowing us to characterize a proof or valid deductive argument as a chain of valid inferences? This is the main question that I shall be concerned with in this essay. By the validity of an inference I understand henceforth a concept of that kind, which I shall strive to explicate here.

1 The concept of inference

Before entering the main discussion of what should be required of a valid inference, we should pay attention to the concept of inference. It is reasonable to think that the validity of an inference should be connected with the activity of making inferences

Dag Prawitz
University of Stockholm, Sweden, e-mail: dag.prawitz@philosophy.su.se

* It is a great pleasure to me to submit this essay to a volume devoted to Peter Schroeder-Heister in the series of "Outstanding Contributions to Logic". I have had the privilege to follow part of Peter's career from his doctoral dissertation. His work in proof-theory has influenced my own in substantial ways. In the last years we have had intensive discussions on the topic of this essay. As indicated in the essay, I am now following a path partly different from the one he argues for. The concept of valid inference is however a difficult one to explicate, and as can be seen from this essay, I have tried several blind alleys in the past. I hope for a continued discussion and co-operation with Peter on these issues and a deepened understanding of validity.

T. Piecha and K. F. Wehmeier (eds.), *Peter Schroeder-Heister on Proof-Theoretic Semantics*, Outstanding Contributions to Logic 29,
https://doi.org/10.1007/978-3-031-50981-0_4

and in particular with what we expect to achieve when we perform this activity and infer a conclusion from some premisses.[1]

Regarded as a mental act, an inference comprises a number of judgments and consists in a transition to one of them, the conclusion, from the other ones, the premisses. In this transition, the premisses are held to support the conclusion, which thereby is taken to be justified. In a *deductive* inference the conclusion is held to get a *conclusive* support or, as one also says, to be provided with a *binding* ground; these attributes will usually be left out since we are concerned here with deductive inference exclusively.

When an inference is verbalized, it becomes a compound speech act comprising a number of assertions. They can be seen as manifestations of judgements, performed by uttering sentences with assertive force. That the conclusion is taken to be supported by the premisses is then typically indicated by inserting a prefix, like "therefore" or "hence", in front of the conclusion, or when the conclusion is stated first, by beginning the premisses with a word like "since" or "because". Inferences will here be seen primarily as speech acts of that kind.

An inference is thus not just a succession of assertions. Its crucial feature is that one of the assertions, the conclusion, is held to be supported by the other assertions, the premisses.[2] To support the assertion that appears as conclusion and to justify it thereby is also the very aim of the inference. An individual inference may of course be driven by all kinds of different individual aims. But the *characteristic aim* (to use a term from speech act theory) of inference seen as an act-type is to obtain a support for the conclusion. When one holds the conclusion to be supported by the premisses, one thus understands the inference act as having been successful in attaining its aim. What precisely it amounts to for an inference to provide its conclusion with a conclusive support is a question that we have to come back to when trying to explicate the concept of valid inference; the person who makes an inference need not have an answer to this question but may nevertheless be right in holding the premisses to support the conclusion.

This view of inferences as transitions from categorical assertions already justified to conclusions that become justified agrees essentially with how Frege saw inferences. Established deductive practice knows however plenty of inferences that do not conform to Frege's picture. Reductio ad absurdum, frequently used already at the time of the Greek antiquity, is an example. It presupposes reasoning from assumptions not considered by Frege (although his two-dimensional way of writing formulas could be said to be a way to represent assertions made under assumptions[3]). Such reasoning got an explicit and regimented form only later with Gentzen (1935) and Jaśkowski (1934).

A full account of deductive inferences should pay attention also to such reasoning. We shall therefore allow not only categorical assertions but also hypothetical ones, even called assertions made under assumptions. Furthermore, we shall allow assertions

---

[1] That inferences are primarily acts has been emphasized in contemporary logic by Martin-Löf and Sundholm in particular; see for instance Martin-Löf (1985) and Sundholm (1998).

[2] This feature of inferences is also stressed by Boghossian (2014).

[3] See Tichý (1988), von Kutchera (1996), and Schroeder-Heister (2014).

and assumptions that are unsaturated or open, expressed by sentences containing free variables.

This makes inferences more complicated as compared to what was said above, since in addition to being transitions from premisses to conclusions they may also discharge or, as I shall say, *bind* assumptions that the premisses depend on. They may also bind variables that occur free in asserted sentences. The terminology is to hint to how we understand reasoning with assumptions and variables that are free in the argument, not bound by any inference, namely as a kind of schematic reasoning intended to remain correct if free variables are replaced by closed terms and free assumptions are replaced by valid arguments for them; we shall return to this later to make it clear.

2 Arguments

To describe how inferences bind variables and assumptions we have to consider *arguments*, by which we shall understand reasoning that proceeds by making a number of inferences chained to each other so that the conclusion of one inference also becomes a premiss of another.

The validity of an inference or argument should of course not depend on who makes the inference or in what situation or at what time the inference is made, provided no indexicals are involved,[4] which I presuppose here. We may therefore restrict ourselves here to *generic acts* where we have abstracted from such features of individual inferences or arguments.

An argument, that is, a generic argument act, is determined by its inferences, their ordering, and information concerning initial premisses about whether they are asserted outright, categorically, or occur as assumptions, a category of speech acts of its own. For convenience I shall sometimes count initial premisses that are asserted outright as inferred by inferences from zero premisses.

The inferences of the argument are in turn determined by their premisses and conclusions and by the variables and assumption occurrences that they bind; they may be seen as determined by yet other factors, but here I restrict myself to the mentioned ones.

When we make an argument its inferences become of course ordered linearly by time, but for the purpose of logic it is sufficient and in fact more to the point to require that the ordering is a strict partial order such that for each inference, except for one last inference, its conclusion is also the premiss of the immediately succeeding inference. This ordering gives rise to a strict partial ordering of the assertion occurrences of the argument, too. There is thus one last conclusion of an argument, called its *final conclusion*. Each occurrence of an assertion determines a *subargument*, namely the initial part of the argument that has the occurrence as its final conclusion. By

[4] I owe this proviso to Cesare Cozzo.

the *immediate subarguments* of a given argument are understood the subarguments determined by the premisses of the last inference of the argument.

In ordinary deductive practice, an inference that binds an assumption binds all its occurrences. But when studying inferences on a meta-level, considering among other things operations on them, it is important to allow an inference to bind only some occurrences of an assumption.

The dependency on assumption occurrences is defined inductively: An assumption occurrence *depends* on itself; the conclusion of an inference *depends on* every assumption occurrence that a premiss depends on and is not being bound by the inference. An occurrence of an assertion in an argument depending on the set of assumptions Γ is said to be an *assertion under the (set of) assumptions* Γ. An argument Π whose final conclusion is an occurrence of $\mathcal{A}$ depending on the set of assumptions Γ is said to be an *argument for* the assertion $\mathcal{A}$ *from (the set of assumptions)* Γ; when the final conclusion does not depend on any assumption, Π is said to be an *argument* for $\mathcal{A}$.

When an inference binds a variable, it binds all occurrences of the variable that are free in the assertions of the subargument determined by a premiss and are not bound by an inference in the subargument. Such binding is not to occur if the variable has free occurrences in an assumption that the conclusion depends on; a restriction imposed in order that the replacement described at the end of the previous section is to give the desired result.

An argument is said to be *closed* when all its assumption occurrences as well as all the variables that occur free in an assertion of the argument are bound. It is said to be *open* otherwise. Note that an occurrence of a variable in an assertion of an argument may be free *in the assertion but bound in the argument.*

If Π is an argument for the assertion $\mathcal{A}$ and Σ is an argument in which $\mathcal{A}$ occurs as a free assumption, we understand by a *composition of* Π *and* Σ a result of putting the two arguments together by letting one or several free occurrences of the assumption $\mathcal{A}$ in Σ come after an occurrence of Π; these assumption occurrences in Σ are in this way replaced by arguments for them.

3 Representations of inferences and arguments

We have to distinguish between generic inferences considered in isolation and inferences occurring in an argument. For inferences that do not bind assumptions or variables — let us call them *simple inferences* — there is no difference: A simple inference is determined by its premisses and conclusion. It can be represented by a figure of the form

$$\frac{A_1 \quad A_2 \quad \dots \quad A_n}{B}$$

where B is the sentence asserted by the conclusion and $A_1, A_2, \dots, A_n$ are the sentences asserted by the premisses of the inference. (This common way of representing generic inferences introduces an order between the premisses that is insignificant but will actually be used sometimes as a way of reference.)

In the general case, a generic inference is determined also by the variables that it binds and the assumptions that it may bind. It can be represented by a figure of the form

$$\frac{\begin{matrix}[\Gamma_1] & [\Gamma_2] & \ldots & [\Gamma_n] \\ \Big|\, v_1 & \Big|\, v_2 & \ldots & \Big|\, v_n \\ A_1 & A_2 & \ldots & A_n\end{matrix}}{B}$$

where $A_1, A_2, \ldots, A_n$, and B are again sentences, v_i is a set of variables, and Γ_i is a set of sentences. When applied within an argument, all free occurrences of the variables of v_i in assertions of the argument for A_i not already bound by other inferences become bound, and furthermore occurrences of assumptions that have the shape of a sentence in Γ_i and have A_i in its scope may become bound ($i \leq n$); not until the generic inference is applied in a specific argument is it determined which assumption occurrences become bound — the generic inference determines only which assumptions *may* become bound. For convenience, I presume that the variables in v_i do not occur in B.

Figures of the form exhibited above that represent generic inferences can be called *inference figures*. An *inference rule* or *schema* is like an inference figure but instead of containing sentences, predicates, individual variables, and individual terms it may contain schematic letters for them. An *instance* of an inference rule is obtained by replacing the schematic letters by specimens of the kind that they stand for, and is thus a generic inference (figure).

A generic argument act can be represented conveniently by a tree of assertions, which in turn may be represented by writing Frege's assertion sign in front of the sentences asserted — or, dropping the assertions sign, by just the sentences asserted. At the top of the tree are put sentences representing the initial premisses with information about whether they represent categorical assertions or assumptions; in the latter case the sentence represents both the assumption made and the assertion of the sentence under that assumption. Going down in the tree, we put successively sentences that represent the assertions inferred. The binding of variables and assumptions are to be marked at the inference where it occurs (e.g., one may attach the same numeral to an assumption and to the inference that binds it).

Note that when an argument is represented in this way by a tree of sentences, an assertion of a sentence A under the assumptions Γ is represented by just the sentence A; thus, it is only A that appears as a premiss or conclusion of an inference — the assumptions Γ that A depends on are easily read off from the tree. This is Gentzen's original way of arranging his natural deductions; arguments differ from them only in the respect that its inferences need not be instances of predetermined rules but can be of any kind.[5,6]

When we want to distinguish the representations of arguments from the arguments that they represent, we may call them *argument figures*. They may be seen as protocols of argument acts in which all the features of the acts that matter logically are noted down.

The inferences of an argument can be seen as *applications* of generic inferences. We shall allow that a substitution σ of terms for free variables is made at such applications; as usual it is here taken for granted that the terms do not contain free variables that become bound in the result A^σ of carrying out σ on A. Let $\mathcal{G}$ be a generic inference represented by the inference figure exhibited above. An inference of an argument is an *application of* $\mathcal{G}$, if and only if, for some (possibly empty) substitution σ of terms for variables different from the ones of v_i and occurring free in B, A_i or sentences of Γ_i it holds: 1) the premisses and conclusion of the inference are $A_1^\sigma, A_2^\sigma, \ldots, A_n^\sigma$, and B^σ, respectively, 2) the inference binds all occurrences of the variables of v_i that stand free in assertions of the argument for the premiss A_i^σ and 3) the inference binds at most some occurrences of assumptions of the form of a sentence of Γ_i^σ in the argument for A_i^σ that A_i^σ depends on ($i \leq n$). Note that there can be several different generic inferences of varying generality that an inference of an argument is the application of; replacing an individual term of a sentence of a generic inference with a variable, we get a new more general generic inference that has as applications all the applications of the first more specific generic inference. By a *result of applying* $\mathcal{G}$ to a sequence or a set of arguments $\{\Pi_1, \Pi_2, \ldots, \Pi_n\}$ we shall understand an argument whose last inference is an application of $\mathcal{G}$ and whose immediate subarguments are Π_1, Π_2, . . ., and Π_n; their final conclusions have to be $A_1^\sigma, A_2^\sigma, \ldots, A_n^\sigma$ for some substitution σ.

Since all particular features of generic acts of inferences or arguments that are logically significant are present also in the figures that represent them, we may as well make the syntactical representations instead of the acts themselves the object of

[5] An alternative is to represent the assertions of the argument by sequents $\Gamma \Longrightarrow A$ where Γ is a sequence of sentences representing the assumptions that A is asserted under. Premisses and conclusions of inferences will then be represented by sequents instead of sentences. In some later publications Gentzen adopted this way of writing natural deductions. At this level it is only a question of alternative representations of one and the same argument act. However, if we follow Sundholm (2006) and understand a sequent $A_1, A_2, \ldots, A_n \Longrightarrow B$ as representing an assertion saying "if A_1 is true, A_2 is true, . . ., and A_n is true, then B is true", the tree of sequents will primarily represent not reasoning from assumptions but reasoning starting from axioms of the form "If A is true, then A is true". The inferences too take different forms; cf.

$$\frac{Pc \quad \text{(assumption)}}{\exists x Px} \qquad\qquad \frac{Pc \Longrightarrow Pc}{Pc \Longrightarrow \exists x Px}.$$

[6] The Curry-Howard isomorphism suggests that there is also an alternative representation of arguments by terms in an extended lambda calculus containing parameters for functions corresponding to different arbitrary inferences of an argument. However, to establish really that this is a possible way to represent arguments, we have to pin down what it is to argue from assumptions, which is what is attempted here, partly by using a representation that is closer at hands.

study. The figures may simply be called inferences and arguments, respectively, as is customary in logic, the representation of an argument for the assertion of a sentence A may for simplicity be called an argument for A, and so on. It remains however that when discussing their validity, one should recall that the syntactical objects are representations of acts with aims; this general feature of the generic acts is of course lost when they are represented by figures.

4 Soundness and validity of inferences. A first approximation of validity

As remarked in the introduction, the validity of an inference has commonly been identified with the holding of the relation of (logical) consequence. The inferences or rather inference figures that one has in mind here are the simple ones at which no assumptions or variables are bound. Such an inference figure

$$\frac{A_1 \quad A_2 \quad \dots \quad A_n}{B}$$

where A_i and B are closed sentences, is valid, one has said, when the inference is necessarily truth preserving, spelled out by saying either that it is impossible that all the premisses A_i $(i \leq n)$ are true while the conclusion B is false, or that necessarily if all A_i $(i \leq n)$ are true, then so is B. These two conditions are of course equivalent classically and have also been used alternatively in the traditional definition of the relation of entailment or consequence.

With Bolzano and Tarski the modal notion of necessity or impossibility is replaced with a variation of the meaning of the non-logical terms of the sentences involved and of the individual domain. We then get the well-known definition saying that a sentence A is a logical consequence of a set of sentences Γ, when A is true under each variation of assignments to the non-logical terms and of the domain of the individual variables under which all the sentences of Γ are true. This has become the dominant definition also of the validity of an inference from Γ to A in contemporary philosophy and logic.

The concept of valid inference that we are concerned with in this essay must obviously be very different from that notion of valid inference. Even if one considers only simple inferences, that notion demands both too much and too little from the perspective of this essay. Although clearly a valid inference cannot have true premisses and a false conclusion, inferences by for instance mathematical induction come out as non-valid when it is required that a valid inference preserve truth under all variations of the meaning of the non-logical terms and of the domain of the variables. More importantly, the prevalent notion demands too little. To establish that something is a logical consequence of something else we usually need a proof, often a long proof with many inferences. What is to be required of these inferences cannot then be that the sentence asserted in conclusion is a logical consequence of the sentences asserted in the premisses; if that was sufficient, there would never be a need of proofs containing more than one inference step.

The property of inference that has commonly been called validity is nevertheless a significant one, and I propose that it is called *soundness*. This would be in agreement with established terminology in connection with deductive systems, which are called sound when their inference rules preserve truth.

Already at the beginning of logic one was interested in distinguishing a kind of inferences that satisfied stronger demands than soundness. Aristotle distinguished between *syllogisms* in general and *perfect syllogisms* saying:

> A syllogism is a form of speech in which, certain things being laid down, something follows of necessity from them.
>
> A perfect syllogism is one that needs nothing other than the premisses to make the conclusion evident.[7]

Aristotle's general notion of syllogism (not restricted to the particular inferences that he studied in detail) has been a common point of departure for discussions and different proposals about what later became called valid inference. In contrast, there has been little interest in trying to develop his narrower notion of perfect syllogism; the attention it is has received seems mostly to have been of an exegetical kind about what Aristotle intended with that notion. Having an epistemic ingredient, it seems to be in the same direction as the concept that is focused on in this essay.

The term "evident" used in the above translation of Aristotle's definition of perfect syllogism may seem to be natural to use here in view of its etymology: when an inference is valid, it should be "seen" that the conclusion is right given that the premisses are.[8] However, the term is not to be understood here as referring to the actual state of mind of a person when something is obvious to her. We do not want to say that an inference is valid for a person, nor is it likely that Aristotle meant that a syllogism can be perfect for one person but not for another. The term must therefore be understood here not primarily as a psychological term but as referring to an objective property: A valid inference or perfect syllogism gives evidence to the assertion made in the conclusion in the sense that it gives a ground for the assertion, which thereby becomes justified. When understood in this way "evidence of an assertion" may be used interchangeably with "ground for an assertion". It is another thing that the existence of a ground for an assertion can in principle become known and therefore makes the assertion potentially evident to a person.

The notion of ground has of course a broader use. We are here interested in epistemic grounds. A speaker is normally expected to have some kind of epistemic ground for what he or she asserts. The nature of what is counted as such grounds for assertions varies with different kinds of assertions. For assertions with empirical content a ground may be obtained by observations under suitable circumstances. A ground for the assertion of an arithmetical identity may be got from a computation. How good the ground is required to be varies with the context. In some contexts, for

---

[7] Ross (1949, p. 287).

[8] The term is used by Martin-Löf and Sundholm too in their writings on the validity of inference, see, e.g., Martin-Löf (1985) and Sundholm (2004).

instance in mathematics, the ground is expected to be binding, and this is the case that concerns us now.

When we make an inference, it is understood that the ground for the conclusion comes from the premisses. But how? The premisses of an inference are sometimes called grounds for the conclusion, but clearly the premisses themselves do not constitute grounds for the premisses. The ground for the conclusion must rather come somehow from their grounds, which we take implicitly to exist since they are asserted.

For the inference to be valid, there must be some immediacy in how the ground for the conclusion comes from the grounds for the premisses. No further inferences should be needed to obtain the ground. This is a point that Aristotle perhaps wants to make when he says that the perfect syllogism "needs nothing other than the premisses".

The validity of an inference thus requires that a ground for the conclusion appears directly given any grounds for the premisses; "given" in the sense of being at least assumed to exist. One should expect furthermore that the meanings of the sentences involved could be crucial for the validity of the inference.

The concept of validity is to be tied primarily to generic inferences. What has been said so far may be put together as a first rough approximation of their validity:

> For a generic inference to be valid it must be required that in virtue of the meanings of the involved sentences, it appears directly, without any further inferences, that given any grounds for the premises, there is a ground for the conclusion.

If the stated requirement is not satisfied, the assertion that occurs in the conclusion is made without a ground, or at least not with a ground coming from the premisses, and the inference cannot then be valid. The requirement is also sufficient for the validity of an inference: when it is satisfied a person who makes the inference is being provided with a ground for the conclusion, or can at least easily provide herself with such a ground, given that she knows the meaning of the involved sentences and has grounds for the premisses; the assertion appearing as the conclusion of the inference thereby becomes justified and what was aimed at when making the inference is thus achieved. An argument consisting of inferences that all satisfy the proposed requirement gives a ground for its final conclusion, and if closed it can then be called a proof. To call such inferences valid is thus in accordance with my introductory declaration.

One may object to the requirement of directness and remark that a challenge of an inference is typically met by inserting a number of other inferences between the premisses and the conclusion. If these inferences are accepted as valid, one normally accepts as valid also the challenged inference. But this objection is built on another concept of validity than the one we are now concerned with. We want to clarify what may be called *immediate validity* where the point is that the inference should satisfy certain requirements as it stands (without adjuncts). When that has been clarified, one can easily define what it is for a simple inference to be *mediately valid*: there is an argument for the conclusion of the inference from its premisses that uses only immediately valid inferences. The problem is to explicate immediate validity, and I shall continue to refer to it simply as validity, using the term mediate validity for the property that can then be defined in terms of it.

To get on with this task we must especially inquire what constitute grounds for assertions in the present context. In mathematics we expect since the time of Greek antiquity that grounds for categorical assertions come in the form of deductive proofs, valid closed arguments. At least in the case of categorical assertions of logically compound sentences, we know no other way to obtain conclusive justifications. In case the assertions are not categorical but hypothetical or open, the grounds take the form of valid open arguments.

However, if we explain proofs as valid closed arguments and valid arguments as arguments consisting of valid inferences, as I have suggested above, and then explain the validity of inferences in terms of grounds explicated as valid arguments, we are of course moving in a circle.[9] This may seem disastrous for the proposed explanation of valid inference and valid argument that I have just begun.

5 Other concepts of proof

Is there a way to avoid this circularity problem? To turn to other ways of understanding the concept of proof may be thought to be a possibility. In the philosophy of intuitionistic mathematics, a proof has been seen not as a chain of valid inferences but as a mathematical construction. A sentence is taken to express the intention of a construction and to prove the sentence is to realize this intention.[10] More precisely, a proof is the construction process that results in the intended construction expressed by the sentence.[11]

In the so-called BHK-interpretation[12] as usually understood, a shift occurs so that a proof becomes rather the intended construction itself, not the realization process that establishes the existence of the intended construction. The proofs are there defined by recursion over the build-up of the proved sentences; for instance, "a proof of $A \to B$ is a construction that permits us to transform any proof of A into a proof of B".[13] However, the construction intended by a sentence cannot in itself in general constitute a ground for asserting the sentence. To have defined a construction that in fact transforms any proof of A into a proof of B does not justify the assertion of

[9] This circularity problem was noted in several lectures by Martin-Löf in the last decade. The problem is noted in one of his earlier papers too (Martin-Löf, 1985), where he saw it as a mistake to take the concept of (valid) immediate inference as conceptually prior to the concept of proof and concluded: "inference and proof are the same".

[10] Heyting (1934).

[11] As Heyting (1958) puts it: "The steps of the proof are the same as the steps of the mathematical construction." See also Sundholm (1983) concerning the ambiguity of the term construction.

[12] Stated by Troelstra and Dalen (1988).

[13] Troelstra and Dalen (1988, p. 9).

the implication $A \rightarrow B$ unless we have some ground for holding that the defined construction does effect the transformation in question.[14]

When Heyting's general idea of proofs as the realizations of intended constructions is further developed as is done in Martin-Löf's type theory, we are lead to a construction process in which at each step it is also demonstrated that the construction obtained is of the right, intended kind.[15] The process will thus contain a chain of inferences. We are then back to a notion of proof that presupposes the notion of valid inference.

We thus find that a proof in intuitionistic mathematics is either an intended construction of a proposition or a demonstration establishing that a given construction is the intended construction of a certain proposition. In the first case it does not constitute in itself a ground for asserting the propositions, and in the second case it is a chain of valid inferences. In neither case does it offer a solution of our problem.[16]

6 The conceptual order between valid inference and valid argument

Sticking to the idea of proofs as chains of valid inferences, one may contemplate as a way out of the circularity problem the possibility of first defining the concept of valid argument without referring to the validity of inference and then the concept of valid inference in terms of it, thus turning up-side down what has seemed to be the natural conceptual order.[17] A proposal for how to take the last step from valid argument to valid inference is to define an inference as valid when any application of it to valid arguments is a valid argument.

The proposed defining condition should in fact be a necessary condition for the validity of an inference in accordance with the basic idea concerning how open

[14] For this reason, in the BHK-interpretation first presented by Troelstra (1977, p. 977), the proof of an implication $A \rightarrow B$ did not consist of just a construction c that in fact transforms any poof of A into a proof of B but contained also "the insight that c has the property: d proves $A \implies cd$ proves B". To include such an "insight" into the proofs is of course difficult to make compatible with the intuitionistic idea that proofs are mathematical objects, but by dropping this element from the proofs as was done in the later and more well-known BHK-interpretation presented by Troelstra and Dalen (1988), the BHK-proofs lost the general epistemic power to justify the assertion of sentences that they are proof of. See also fn. 18.

[15] See for instance Martin-Löf (1984).

[16] In several papers (see for instance Prawitz 2015b; 2019b), I have discussed in a positive vein the possibility of identifying the grounds for asserting sentences with the intended construction that the sentences are taken to express when understood in an intuitionistic sense, although noting that the way they are defined must be restricted. I do not anymore consider this approach as promising since I know no way of making the right restriction. It is true that when the constructions are restricted to be what can be defined in certain extended lambda calculi it can be seen directly from well-formed terms that they denote the intended constructions of certain sentences. But the constructions intended by sentences of sufficient complexity cannot be exhausted by what can be defined in formal systems.

[17] I proposed a definition of such a notion of valid argument at a fairly early stage (Prawitz, 1973). It was later taken up and somewhat modified by Dummett (1991) and Schroeder-Heister (2006) and was modified more recently and more radically by myself (Prawitz, 2019a).

valid arguments are to be understood as schematic reasoning, presented at the end of Section 1. That idea can be stated more precisely in the form of a principle as formulated below. To get a short formulation let us say that a *regular instance* of an argument is obtained by first replacing a number of variables that are free in the argument by terms and then, in the resulting argument, replacing a number of free assumptions by valid arguments for them; the order is important in case one wants to obtain a closed regular instance.

Principle concerning open arguments and their instances
All regular instances of valid arguments are valid.

Since an inference can be seen as a one-step open argument whose premisses are assumptions and the result of applying it to valid arguments is a regular instance of the argument, we find in particular that it is a necessary condition for the validity of an inference that applications of the inference to valid arguments result in valid arguments.

However, the condition is not sufficient. If there is no valid argument for the assertion of A, the inference from the assertion of A to any other assertion satisfies the condition vacuously.[18] For instance, since, as we know from the proof of Fermat's Theorem, there is no closed valid argument for the premiss of the following inference

$$\frac{\exists x \exists y \exists z\, x^3 + y^3 = z^3}{\bot}$$

(the variables being supposed to range over positive integers), the inference comes out as valid according to the proposed definition of validity. Of course, the inference should come out as mediately valid, but certainly not as (immediately) valid; contrary to what Fermat thought, we seem to need a quite long and complicated argument to refute the assumption $\exists x \exists y \exists z\, x^3 + y^3 = z^3$.

The proposed definition of valid inference in terms of valid argument thus fails, and I see no way of attaining such a definition. The validity of arguments seems best explained by saying that a valid argument is one whose inferences are all valid. If the validity of inferences in turn is explained in terms of grounds as proposed above and grounds for an assertion consist in valid arguments for them, we must conclude that the concepts of valid inference and valid argument depend on each other and cannot be defined in isolation. If so we have to be satisfied with stating principles about how they are related to each other and to some other concepts.[19]

[18] The converse of the principle, in particular the idea that an open argument is valid if all its closed, regular instances are, therefore fails too. It was a substantial part of the definition of valid argument mentioned in fn. 17. The notion of hypothetical proof proposed by Martin-Löf (1985) suffers from the same problem: any argument from the assumption of a false sentence satisfies vacuously his defining condition for being such a proof. Similarly every false sentence A comes out as having a BHK-proof: There is a construction that permits us (vacuously) to transform any proof of a false sentence A into a proof of $\bot$, namely the (empty) function that is defined for all proofs of A and, for each such proof, assumes as value a proof of $\bot$.

[19] In the sequel I shall take this mutual dependency between valid inference and valid argument as a working hypothesis. I want nevertheless to keep open that the validity of inferences could be

7 Principles and heuristic ideas on the validity of inferences and arguments

A basic principle, which expresses intuitions already referred to, is:

Principle 1. The relation between validity of inferences and validity of arguments
An argument is valid, if and only if, all its inferences are valid.

This principle establishes a relation between the validity of an argument and the validity of the inferences that the argument consists of. We have to relate the latter validity to the validity of generic inferences that the concept of validity is primarily tied to, as suggested above. An inference of an argument can always be seen as an application of a generic inference, but as noted above (Section 3), it may be the application of several different generic inferences of varying generality. To be counted as valid it should be sufficient that it is an application of one valid generic inference (which is the same as saying that the least general generic inference that it is an application of is valid). We thus define:

Definition 1.
An inference of an argument is valid, if and only if, it is an application of a valid generic inference.

Principle 1 does not amount to a definition of the concept of valid argument as long as we lack a definition of the concept of validity of generic inference not depending on the concept of valid argument, but it is still informative about the involved concepts and has several immediate corollaries. Some of them are noted below for later use:

Corollary 1.
All results of applying a valid generic inference to a set of valid arguments are valid arguments.

Proof. Consider a valid generic inference $\mathcal{G}$ and let Π be the result of an application of it to a set S of valid arguments. By Principle 1 all inferences of the arguments of S are valid, and by the definition above so is the application of $\mathcal{G}$. Since all the inferences of Π are hence valid, Π is valid by Principle 1, now used in the other direction.

Corollary 2.
A subargument of a valid argument is valid.

Proof. Let Π be a valid argument and let Π' be a subargument of Π. By Principle 1 all the inferences of Π are valid and hence so are all inferences of Π'. The validity

explained in other ways without reference to grounds for the involved assertions. For example, in discussions about the validity of inferences that Peter Schroeder-Heister and I have had, he has suggested that one should demand more of a valid inference than I have done here. It should not only give a ground for the conclusion in the form of a valid argument for it when valid arguments for the premisses are given, but should more generally guarantee an argument for the conclusion, good or bad, of the same quality as the given arguments for the premisses. This stronger requirement should be possible to express without referring to valid arguments, he suggests.

of the Π' follows by using Principle 1 in the other direction. By the same kind of reasoning we get:

Corollary 3.
A composition of two valid arguments is valid.

The part of the principle stated in the previous section that concerns the result of replacing free assumptions of a valid argument by valid arguments for them follows as a corollary of Principle 1 since they can be seen as the effect of an iterated composition — the result is thus valid by Corollary 3. The other part is an independent principle, which we note down here:

Principle 2. The relation between an argument and its substitution instances
The result of substituting terms for variables that are free in a valid argument is valid.

The principle of the previous section is thus obtained as a corollary of Principles 1 and 2:

Corollary 4.
All regular instances of valid arguments are valid.

I shall presuppose that the languages in which inferences are verbalized have closed terms for all individuals in the domain that the variables are intended to range over; as usual the domain is supposed to be non-empty. Consider under this presupposition the following somewhat strengthened converse of Principle 2: An open argument $\mathcal{A}(v)$ with free variables v is valid, if all results $\mathcal{A}(t)$ of substituting closed terms t for v are valid. Is it a reasonable principle?

The answer must clearly be no since that would be contrary to the idea of valid inference discussed here: the fact that all the arguments $\mathcal{A}(t)$ are valid cannot be sufficient for the validity of the open argument $\mathcal{A}(v)$, unless this fact appears directly from made inferences and the meanings of the involved sentences. We shall return to this issue when now returning to what I called a first approximation of the validity of inferences (Section 4).

This first approximation now amounts to another basic idea concerning the relation between validity of inference and validity of argument when having acknowledged that grounds for assertions consist of valid arguments for them. It now reads as follows when put in the form of an equivalence and restricted to simple inferences (which do not bind anything) where the premisses and conclusions are closed, valid arguments for them therefore amounting to proofs:

A simple generic inference whose premisses and conclusion are closed is valid, if and only if, in virtue of the meanings of the involved sentences, it appears directly, without any further inferences, that given any proofs of the premisses, there is a proof of the conclusion.

It follows from Principle 1 that a necessary condition for the validity of a generic inference is the existence of a valid argument for the conclusion given valid arguments for the premisses; indeed, the result of applying the inference to the valid arguments for the premisses is such a valid argument for the conclusion according to Corollary 1. The equivalence above states a necessary condition for the validity of a simple generic inference (whose premisses and conclusion are closed) that is in some respect weaker and in some respect stronger than the condition of Corollary 1. It is weaker since it requires only the existence of *some* valid argument for the conclusion. It is stronger since it requires that this existence appear directly in virtue of the meaning of the involved sentences.

Furthermore, the equivalence provides a sufficient condition for the validity of an inference. On the meaning theory presented in the next section there are cases, namely so-called introduction inferences, where the very application of a generic inference to valid arguments results in an argument that is valid in virtue of the meaning of the conclusion, and where the inference is thus valid according to the equivalence above. But in other cases we shall have to find another valid argument for the conclusion in order to establish the validity of an inference via the sufficient condition stated by the equivalence.

It is to be recalled that we are now concerned with the validity of generic inferences, in terms of which the validity of the inferences of an argument is defined. A valid argument for the conclusion of a particular inference in a given argument may appear directly from the arguments for the premisses, given that they are valid, but this is not sufficient for the validity of the inference; otherwise the inference from $A \vee B$ to A would come out as valid when $A \vee B$ has been inferred after having obtained a proof of A. The above condition of validity requires that given *any* proofs of the premisses, a proof of the conclusion appear.

When an application of the generic inference binds variables and may bind occurrences of assumptions at a premiss, the ground for that premiss assumed to exist in the condition for the inference to be valid takes the form of a valid argument for the premiss from the set of assumptions whose occurrences it may bind. However, when the premiss is an open assertion, the condition for validity must require more. Consider the inference represented by the figure

$$\frac{A(x)}{B(x)}$$

where $A(x)$ and $B(x)$ are sentences that contain one free variable x. For it to be valid, it must appear directly that given any proof of the assertion of $A(t)$, there is a proof of the assertion of $B(t)$, where t is any closed individual term. This condition is also sufficient for the validity of the inference under the presupposition made above that each individual in the intended domain is denoted by some term. That a valid argument for asserting $B(x)$ appears directly given a valid argument for asserting $A(x)$ is a weaker condition that would not guarantee that there is a proof of $B(t)$ given a proof o $A(t)$.

This means that the general condition for the validity of a generic inference is most conveniently formulated in terms of applications of the inference, as that notion was defined in Section 3: for any application of the inference to a set S of valid arguments resulting in a closed argument Π, there is to appear a proof of the final conclusion of

Π. We then arrive at the following reformulation of the first approximate explication of the validity of inference:

Heuristic idea on the validity of inference in terms of valid arguments
A generic inference is valid, if and only if, in virtue of the meanings of the involved sentences, it appears directly without any further inferences that, for any application of the inference to a set of valid arguments resulting in a closed argument Π, there is a proof of the final conclusion of Π.

I call it a heuristic idea because of the vagueness of the expression "it appears directly". A condition of that kind is needed for at least two reasons. One is again the need to avoid the problem of vacuity discussed in the previous section: We do not want inferences to come out valid vacuously just because there are no valid arguments for their premisses. Such an outcome is meant to be blocked when it does not appear directly from the meaning of the premisses that there are no valid arguments for them. To illustrate again with Fermat's theorem: Although it is right that when a generic inference has $\exists x \exists y \exists z\, x^3 + y^3 = z^3$ as premiss any result of applying it to valid arguments satisfies vacuously whatever condition we choose (there being no valid argument for the premiss), this fact does not appear directly in virtue of the meaning of $\exists x \exists y \exists z\, x^3 + y^3 = z^3$. In contrast, the generic inference

$$\frac{\perp}{A}$$

comes out as valid according to the heuristic idea above as it should if we have explained the meaning of $\perp$ by saying that there is no proof of $\perp$; it can then be said rightly that it appears directly in virtue of the meanings of the involved sentences that, for any application of the inference to a valid argument for $\perp$, whatever condition we choose is satisfied.

The requirement of directness is also meant to block that the mere existence of a closed valid argument for the conclusion is sufficient for the validity of a generic inference. For an illustration, consider the generic inference

$$\frac{\exists x \forall y A(x, y)}{\forall y \exists x A(x, y)}$$

Given any valid argument for the assertion of a sentence $\exists x \forall y A(x, y)$ (not depending on assumptions), an argument for the assertion of $\forall y \exists x A(x, y)$ can easily be constructed by using inferences that should certainly come out as valid, but since the construction uses additional inferences, it does not appear directly.

A heavy burden is thus put on the meaning of the vague term directness. In spite of these shortcomings, the stated equivalence will serve as a heuristic guide when searching for additional, more precise principles about how valid inferences and valid arguments are related to each other.

8 Meaning of assertions and validity of inferences

There are inferences whose validity is independent of the assertions involved. An example is the generic inference, represented by the inference figure

$$\dfrac{A \qquad \begin{matrix}[A]\\ \vert \\ B\end{matrix}}{B}$$

(sometimes called the rule of explicit substitution when taken as an inference rule). For any application of this inference to valid arguments that results in a closed argument Π, there is a proof of its final conclusion, because the composition of the two immediate subarguments of Π is such a proof when each free occurrence of the assumption A in the second subargument is replaced by the first subargument; a composition of two valid arguments being valid by Corollary 3 of Section 7. Provided the appearance of this proof by forming a composition of the two valid arguments to which the inference is applied is counted as direct, the inference is thus valid according to the heuristic idea formulated above regardless of what sentences A and B are.

It is surely more common that the validity of an inference depends on what the assertions involved mean. For instance, whether the two generic inferences represented by the figures

$$\frac{A}{A \vee B} \qquad\qquad \frac{B}{A \vee B}$$

are valid cannot but depend on the meaning of sentences of the form $A \vee B$.

As argued by Michael Dummett, the meaning theory for a language should account for all features of the use of the language that depend on the meanings of its sentences, including the acceptance of inferences like the ones above as valid. For it to fulfil this task it is essential how the meanings of sentences are given. A truth-conditional meaning theory of the usual kind states the condition for a sentence to be true in such a way that it may not be possible to derive from that on what kind of grounds an assertion is accepted as justified. In contrast, what Dummett calls a verificationist or justificationist meaning theory explains the meaning of a sentence directly in terms of what counts as a ground or valid argument for asserting the sentence.

Gentzen had an idea about the meanings of logical constants which is a forerunner to Dummett's idea of justificationism. A special feature of Gentzen's system of natural deduction is that for every logical constant c there are a number of inference rules called introduction rules for c, or simply *c-introductions*, where the conclusions are sentences whose outermost sign is c. After having set up his system, Gentzen remarked that the meaning of a logical constant c could be seen as being determined by the c-introductions.[20] It is not obvious how this suggestion is to be understood. One

[20] "The introductions present, so to say, the 'definitions' of the symbols concerned." (Gentzen, 1935, p. 189).

element is of course that, in virtue of the meanings of the logical constants, instances of the introduction rules are to be seen as yielding valid arguments when applied to valid arguments for its premisses. But c-introductions do not constitute the only valid ways of inferring sentences that have c as their outermost sign. So what is special about the introduction rules that gives cause for taking them as meaning constitutive?

One way to answer this question is to say that the c-introductions present the *direct* or *canonical* way of inferring sentences that have the constant c as the outermost sign. This is to be understood as implying, not only that applications of instances of introductions to valid arguments yield valid arguments as results, but also that if the assertion of a sentence is provable at all, then in principle its proof can be put in such a form that the last inference is an instance of an introduction. An argument whose last inference is an instance of an instance of an introduction is said to be in *canonical form*[21] (whether open or closed and irrespective of question of validity).

To exemplify: that the meaning of the disjunction sign is determined by the $\vee$-introductions whose instances are generic inferences of the form exhibited above, is to be understood as saying that the meaning of disjunction is such that 1) the results of applying to valid arguments generic inferences of the kind exhibited above are valid, and 2) if there is a proof of a sentence $A \vee B$, then there is such a proof in canonical form. Hence, if $A \vee B$ is provable, there is a proof of either A or B.[22]

As seen, this fits well with how $\vee$ is understood intuitionistically, but not with how it is understood classically since a disjunction may be provable classically while neither of the disjuncts is provable. Gentzen's $\vee$-introductions can thus be seen as determining the meaning of intuitionistic disjunction, but not of classical disjunction.

Adopting this idea to all the logical constants of the intuitionistic language of first order predicate logic, the meanings of their compound sentences are explained by telling how arguments for the assertions of them have to look to be in canonical form What must be told for different cases of compound sentences is what the immediate subargument or subarguments of an argument Π in canonical form are to consist of. In the case of:

$A \wedge B$, they are to consist of an argument for A and an argument for B;

$A \vee B$, it is to consist of an argument for A or an argument for B;

$A \to B$, it is to consist of an argument for B from A, where free occurrences of the assumption A may be bound by the last inference of Π;

$\forall x\, A(x)$, it is to consist of an argument for $A(x)$, where x is being bound by the last inference of Π;

$\exists x\, A(x)$, it is to consist of an argument for $A(t)$ for some term t.

This has to be completed by telling what constitute arguments in canonical form for assertions of atomic sentences. In the case of the atomic sentence $\perp$, the explanation

[21] A term earlier used by Brouwer in a different way. Its use in the above sense was proposed by Prawitz (1974) and Dummett (1975).

[22] Although Gentzen never developed his ideas about meaning more precisely, it is clear that he was thinking in this way when remarking: the assertion of "$A \to B$ attests (German: dokumentiert) the existence of a derivation of B from A". (Gentzen, 1935, p. 189)

is that there is no argument in canonical form for the assertion of $\bot$; there are no $\bot$-introductions. I shall assume that the meanings of all atomic sentences have been explained by telling what the arguments in canonical form for the assertions of them are.[23] One may want to vary what is to count as canonical arguments for atomic sentences of a language. Let us say that a base for a language L specifies the canonical arguments for the atomic sentences and the set of closed individual terms of L. Validity of inferences can then be relativized to such bases.

To extend Gentzen's idea to other languages, one must thus be able to specify the meaning of each different sentence form by giving introduction rules for that sentence form or, in other words, by stating what constitute arguments in canonical form for the assertions of such sentences.[24] It is an open question to what extent this is possible,[25] but there is no problem to give introduction rules that are adequate for the logical constants understood classically.[26]

To summarize the meaning theoretical view adopted here, we can say more generally: To know the meanings of the sentences of a language is to know

1) specifically for each sentence form how arguments in canonical form for the assertions of sentences of that form look, and
2) generally 2a) that an argument in canonical form is valid if its immediate subarguments are and 2b) that all proofs can be put in canonical form.

Knowledge according to clause 1 also determines introduction rules for all forms of sentences. The validity of all their instances is implied trivially by the heuristic idea stated in the previous section. Thus, we get:

Principle 3. Validity of introduction rules
All instances of introduction rules are valid. In particular, all instances of Gentzen's introduction rules for intuitionistic predicate logic are valid.

[23] An early extension of Gentzen's idea to atomic sentences is due to Martin-Löf (1971). He took Peano's first and second axiom for natural numbers as two introduction rules for sentences of the form Nt, one allowing the inference of $N0$ from no premisses, and the other allowing the inference of Nst from the premiss Nt (s standing for the successor operation).

[24] I have left open here general requirements that should be put on meaning explanations to guarantee for instance that they are not circular. They correspond to requirements that introduction rules are to satisfy discussed by Dummett (1991).

[25] If the language contains sentences with empirical content, we may have to broaden the concept of inference and think of introduction inferences as transitions not only from assertions to other assertions but also from other acts such as observations; they deliver what is commonly called direct evidence and may be seen as meaning constitutive.

[26] For instance, a possible introduction rule for classical disjunction is displayed here:

$$\frac{\begin{array}{c}[\neg\alpha,\ \neg\beta]\\ |\\ \bot\end{array}}{\alpha \vee \beta}$$

α and β being schematic letters for sentences. We could have a language that contains both classical and intuitionistic logical constants, kept distinct by, e.g., attaching different subscripts to them, and formulate introduction rules for all of them; see further Prawitz (2015a).

9 A precise sufficient condition for an inference to be valid

A more challenging problem is to give precise principles for how inferences can be valid in virtue of the meaning of the involved sentences without being meaning constitutive in the way the introduction rules are according to the previous section. The question is whether and to what extent we can state precise principles for such validity guided by the heuristic idea of Section 7.

Gentzen meant that the elimination rules (E-rules) of his system of natural deduction were valid because of their relation to the meaning constitutive introduction rules (I-rules). He suggested, "It should be possible to display the E-inferences as unique functions of the corresponding I-inferences, on the basis of certain requirements".[27] As a step in that direction and to explain why the elimination rules are valid, I have described them as the inverses of the corresponding introduction rules in the sense of satisfying the following:

Inversion principle
If the last inference of an argument Π for A from Γ is an E-inference whose major premiss is the conclusion of an I-inference, the argument for the major premiss thus being in canonical form, the immediate subarguments of Π already "contain" an argument for A from $\Delta \subseteq \Gamma$.[28]

The expression "contain", which was left undefined when I first stated the principle, can be defined as follows: Let us say that the argument Π is immediately extracted from the set S of arguments when either

(i) Π is an argument of S or is a subargument of an argument of S or
(ii) Π is the result got by substituting terms for free variables in an argument that satisfies (i), or
(iii) Π is the composition of two arguments that satisfy (i) or (ii).

Definition 2. Containment
The argument Π is *contained in* a set S of arguments if and only if there is a sequence $\Sigma_1, \Sigma_2, \ldots, \Sigma_n$, of arguments such that $\Pi = \Sigma_n$ and for each Σ_i $(i \leq n)$, Σ_i is immediately extracted from $S \cup \{\Sigma_j \mid j < i\}$.[29]

We note the following corollary:

Corollary 5.
An argument immediately extracted from a set of valid arguments is valid. A fortiori, an argument contained in a set of valid arguments is valid.

The corollary simply combines principles and corollaries stated in Section 7: For the extraction used in clause (i) see Corollary 2, for the one used in clause (ii) see in addition Principle 2, and for the one used in clause (iii) see in addition Corollary 3.

---

[27] Gentzen (1935, p. 189).

[28] Prawitz (1965); the formulation there refers to natural deductions in Gentzen's system instead of arguments as above, otherwise its content is the same.

[29] Due to Peter Schroeder-Heister and me; see Prawitz (2019a).

All instances of Gentzen's elimination rules satisfy the inversion principle when containment is defined as above. We can strengthen the principle in two ways, namely by requiring in clause (i) of the definition of containment that Π is an argument of S or is an *immediate* subargument of an argument of S and by requiring that the argument for the conclusion is not only contained in the set of arguments for the premisses but can in fact be obtained from the set by one immediate extraction. Let us call the result the *strong inversion principle*. It too holds for all instances of Gentzen's elimination rules:

Fact about E-rules
Gentzen's elimination rules satisfy the strong inversion principle.

To illustrate by an example, consider the case where an $\exists$-introduction is immediately followed by an $\exists$-elimination. It has the form

$$\dfrac{\dfrac{\begin{array}{c}\Pi\\ A(t)\end{array}}{\exists x A(x)} \qquad \begin{array}{c}{\scriptstyle (1)}\\ [A(x)]\\ \Sigma(x)\\ B\end{array}}{B}\,{\scriptstyle (1)}$$

Another argument for B from no more assumptions is obtained by two operations: first substitute t for x in $\Sigma(x)$ and then form the composition of Π and $\Sigma(t)$ where Π replaces all free assumptions $A(t)$ in $\Sigma(t)$ (the corresponding assumptions $A(x)$ in $\Sigma(x)$ were bound by the considered application of the instance of $\exists$-elimination). The result of carrying out these two operations is an immediate extraction in the strengthened sense from the set of the immediate subarguments of the argument displayed above.

The elimination rule for $\perp$ is a somewhat special case. Any instance of that rule satisfies the (strong) inversion principle vacuously since according to the meaning of $\perp$ it has no canonical argument.

Invoking the heuristic idea of Section 7 and applying Corollary 5 and the above fact, we can now state

Principle 4 (initial part). Validity of elimination rules
All instances of Gentzen's elimination rules for intuitionistic logic are valid.

Is this principle really implied by the heuristic idea on the validity of inference? Consider an arbitrary instance of an E-rule. Call this generic inference $\mathcal{G}$. To show that $\mathcal{G}$ is valid in accordance with the heuristic idea, we have to show that in virtue of the meanings of the involved sentences it appears directly that, for any application of $\mathcal{G}$ to a set S of valid arguments resulting in a closed argument Σ, there is a proof of the final conclusion of that argument.

To this end we may say the following. Let B be the final conclusion of Σ, let A be the premiss of the last inference of Σ that corresponds to the major premiss of $\mathcal{G}$ (A is thus the result of carrying out on the major premiss of $\mathcal{G}$ the substitution (if any) that yields the application in question), and let Π be the subargument of Σ determined by A. Since the inference does not bind anything in the argument for the major premiss, Π is a closed argument, and is thus a proof since it is assumed to be valid. In virtue of the meaning of A, there is a proof Π^* of A in canonical form. Replacing Π by Π^* in Σ we have an argument Σ^* for B on which the strengthened Inversion Principle has bearing. Hence by a suitable extraction in the strengthened sense from the set S^* of arguments resulting from S by replacing Π by Π^* another argument for B appears. By Corollary 5, this argument for B is valid and hence it is a proof of B.

There are three steps in this little piece of reasoning. As already said, the first step, which gives the argument Π^* for A in canonical form, is immediate in virtue of the meaning of A. In the third and last step we use Corollary 5, an immediate consequence of Principles 1 and 2, which only make explicit basic intuitions about inferences and arguments. What can be questioned is that the appearance in the intermediate step of the argument for B by an immediate extraction is sufficiently direct. Admittedly a person may know the meanings of the involved sentences and have the intuitions made explicit in Principles 1 and 2 without realizing that there is this operation of extraction yielding an argument for B.[30] Nevertheless only knowledge of the meaning of the major premiss A of the inference and a few reflexions are needed to recognize that there is a canonical proof Π^* of A and a proof of the final conclusion B of Σ that uses no other inferences than those already occurring in Π^* or the arguments for the other premisses (if any).

The elimination rule for $\perp$ is again a somewhat special case. In virtue of the meaning of $\perp$, there is no application of an instance of that rule to a valid argument for $\perp$ resulting in a closed argument, since there is no argument for $\perp$ in canonical form and hence no closed valid argument for $\perp$. What has to be shown for all such applications according to the heuristic idea is thus vacuously satisfied directly.

To generalize the Inversion Principle so that it concerns inferences in general, not only elimination rules, and holds more generally for languages given with other introduction rules than Gentzen's, we have to take into account that the inferences whose validity we want to establish may not have one major premiss that can be referred to in the way we did in the statement of the Inversion Principle. Instead of referring to one premiss as the major one, some of the premisses that will play a similar role as the major premiss will be distinguished and will be identified by their ordinal numbers; we shall thus be speaking of the i:th premiss of an inference.[31]

A particular feature of a generic inference $\mathcal{G}$ that satisfies the Inversion Principle is that the following holds for any application of $\mathcal{G}$ to arguments among which the one for the major premiss is in canonical form: another argument for the final conclusion of the result of the application can be obtained whose inferences occur already in the arguments to which $\mathcal{G}$ is applied or are substitution instances of such inferences. Instead of applying $\mathcal{G}$, one can therefore argue for the conclusion by applying those

---

[30] This is a point stressed by Cozzo (2021), who draws the conclusion that the validity of an elimination inference is synthetic and non-meaning-involved.

[31] This means that I am using the ordering of the premisses of an inference, which otherwise is without significance with respect to their identity (see parenthetical remark in Section 3). In another generalization of the inversion principle proposed by Schroeder-Heister (1983) the idea to distinguish some of the premisses of an inference is crucial in a similar way; instead of referring to them by ordinals he marks them by asterisks.

inferences, which have already been used essentially and have thus been accepted implicitly as valid. What can be argued for by such applications of $\mathcal{G}$ can thus be argued for by using already available inferences. It seems therefore fitting to call inferences that share this feature with those that satisfy the Inversion Principle *non-creative*.

To begin with I shall restrict this notion to inferences whose applications to a set S of arguments are seen to enjoy this feature of non-creativity by extractions from S in the same way as for inferences satisfying the Inversion Principle. It is defined as follows:

Definition 3. Non-creative inferences
A generic inference $\mathcal{G}$ is *non-creative* when it has a number of distinguished premisses at which no binding occurs and it holds for any closed argument Π resulting from an application of $\mathcal{G}$ to a sequence S of arguments such that the ones with the same ordinal numbers as the distinguished premisses are in canonical form that an argument for the final conclusion of Π is contained in S.

With the major premiss as the distinguished one, Gentzen's elimination rules are thus non-creative. So is the generic inference represented by the inference figure

$$\frac{A \quad A \to B \quad B \to C}{C}$$

with $A \to B$ and $B \to C$ as the distinguished premisses. Instances of the rule of explicit substitution considered at the beginning of Section 8 are examples of non-creative inferences with an empty set of distinguished premisses.

We can now state Principle 4 in a more general form:

Principle 4 (completed). Validity of non-creative inferences
All non-creative inferences are valid.

The reasoning to see that the principle is implied by the heuristic principle is essentially the same as that for the first part of the principle stated above, except that we are stretching the requirement of immediate appearance since the argument for the final conclusion of the result of applying the inference may now appear after a series of extractions instead of just one.

It is to be noted that a generic inference $\mathcal{G}$ may be vacuously non-creative because there is one or more distinguished premisses for which it holds that there is no application of $\mathcal{G}$ to a sequence of arguments such that the ones with the same ordinal number as the distinguished ones are in canonical form. But this can only happen in case no argument in canonical form has been specified in the explanation of the meaning of the distinguished premisses, as is the case for $\perp$. In such a case $\mathcal{G}$ is also valid according to the heuristic idea since in virtue of the meaning of the involved sentences the condition stated by the heuristic idea is vacuously satisfied.

One may contemplate extending the term non-creative to inferences that enjoy the feature of non-creativity without this necessarily being seen by making extractions; in other words, conclusions of applications of these inferences to certain sets S of arguments have again arguments using only inferences occurring in S or substitution instances of them but they can be freely combined and do not need to occur in combinations obtained by extractions from S.

10 Concluding remarks

On the basis of an initial discussion of the nature and aim of inferences and inspired by Gentzen's idea about how the meanings of logical constants are determined, we have arrived at some precise principles about the validity of inferences. According to them all instances of the inference rules of natural deduction for intuitionistic logic are valid. The result can be seen as a proposal for how Gentzen's idea is to be understood in order to yield such a result.

A question that naturally arises is whether all non-creative inferences expressible in the language of intuitionistic predicate logic, LIPL, are derivable in that logic, IPL. Whether an inference is non-creative may depend on the base (Section 8); as will be exemplified below, induction inferences are non-creative given a certain condition on the base, although they are of course not derivable in IPL. We should therefore relativize non-creativity to a base and ask whether inferences expressible in IPL that are non-creative relative to all bases is derivable in IPL. This is a precise logical question about a kind of completeness of intuitionistic predicate logic that may be possible to answer.

A more philosophical and vaguer question is whether the heuristic idea about the validity of inferences implies the validity of inferences in the language of intuitionistic predicate logic above what follows from Principles 3 and 4. Here the validity should also be relativized to a base. The question is thus whether inferences expressible in LIPL that are valid relative to all bases according to the heuristic idea are so in force of Principles 3 and 4. A positive answer to this question would be a reason for identifying the logically valid inferences in LIPL with inferences that are either non-creative or are instances of introduction rules.

Going outside of logic, it is of interest to consider the rule of mathematical induction. If the intended individual domain is the set of natural numbers, it may be given the form

$$\frac{A(0) \qquad \begin{array}{c}[A(x)] \\ \vdots\, x \\ A(sx)\end{array}}{A(t)}$$

where t is a schematic letter for individual terms and s is the successor operation (since t may be replaced by a free variable y in instances of the rule, the rule has the whole strength of mathematical induction, allowing us to infer $\forall y\, A(y)$). Provided that the closed individual terms of the language in question consist of numerals, all instances of this inference rule are easily seen to be non-creative, and they are hence valid according to Principle 4. However, they cease to be so when there are closed individual terms other than numerals.

If the intended individual domain contains other elements than natural numbers, the rule of induction has to be qualified by adding Nt (N for the predicate to be a natural number) as a third premiss. In this form the rule has instances that are not non-creative. Since they should of course be counted as valid, going outside of logic one needs to find an extension of Principle 4 if one is to cover by precise principles inferences that we consider valid.

Acknowledgements I am grateful to Cesare Cozzo, Valentin Gorenko, Per Martin-Löf, Luiz Carlos Pereira, Antonio Piccolomini d'Aragona, and Peter Schroeder-Heister for their commenting on an earlier version of this essay, which has helped me to improve several passages. I have also benefitted from remarks made in the discussion after my presenting part of this essay at a Nordic Online Logic Seminar in March 2021.

References

Boghossian, P. (2014). What is inference? *Philosophical Studies* 169 (1), 1–18.

Cozzo, C. (2021). Fallibility and fruitfulness of deductions. *Erkenntnis*. DOI: 10.1007/s10670-021-00487-6.

Dummett, M. (1975). The philosophical basis of intuitionistic logic. In: *Logic Colloquium '73*. Ed. by H. E. Rose et al. Amsterdam: North Holland, 5–40.

– (1991). *The Logical Basis of Metaphysics*. London: Duckworth.

Gentzen, G. (1935). Untersuchungen über das logische Schließen, I. *Mathematische Zeitschrift* 39 (1), 176–210.

Heyting, A. (1934). *Mathematische Grundlagenforschung, Intuitionismus, Beweistheorie*. Berlin: Springer.

– (1958). Intuitionism in mathematics. In: *Philosophy in the Mid-Century*. Ed. by R. Klibansky. Florence: La Nuova Italia, 101–115.

Jaśkowski, S. (1934). On the rules of suppositions in formal logic. *Studia Logica* 1, 5–32.

Martin-Löf, P. (1971). Hauptsatz for the intuitionistic theory of iterated inductive definitions. In: *Proceedings of the Second Scandinavian Logic Symposium*. Ed. by J. E. Fenstad. Amsterdam: North-Holland, 179–216.

– (1984). *Intuitionistic type theory*. Napoli: Bibliopolis.

– (1985). On the meanings of the logical constants and the justifications of the logical laws. In: *Atti degli Incontri di Logica Matematica, Vol. 2*. Ed. by C. Bernardi and P. Pagli. Siena: Scuola di Specializzazione in Logica Matematica, Dipartimento di Matematica, Università degli Studi di Siena, 203–281. Republished in *Nordic Journal of Philosophical Logic 1*, 11–60 (1996).

Prawitz, D. (1965). *Natural Deduction: A Proof-Theoretical Study*. Stockholm: Almqvist & Wiksell. Reprinted Mineola NY: Dover Publications (2006).

– (1973). Towards a foundation of general proof theory. In: *Logic, Methodology and Philosophy of Science IV*. Ed. by P. Suppes et al. Amsterdam: North-Holland, 225–250.

– (1974). On the idea of a general proof theory. *Synthese* 27, 63–77.

– (2015a). Classical versus intuitionistic logic. In: *Why is this a Proof? Festschrift for Luiz Carlos Pereira*. Ed. by E. H. Haeusler et al. London, 15–32.

Prawitz, D. (2015b). Explaining deductive inference. In: *Dag Prawitz on Proofs and Meaning*. Ed. by H. Wansing. Cham: Springer, 65–100.
– (2019a). The fundamental problem of general proof theory. *Studia Logica* 107. *Special Issue: General Proof Theory*. Ed. by T. Piecha and P. Schroeder-Heister. Cham: Springer, 11–29.
– (2019b). The seeming interdependence between the concepts of valid inference and proof. *Topoi* 38, 493–503.
Ross, W. D. (1949). *Aristotle's Prior and Posterior Analytics*. Oxford: Oxford University Press.
Schroeder-Heister, P. (1983). The completeness of intuitionistic logic with respect to a validity concept based on an inversion principle. *Journal of Philosophical Logic* 12, 359–377.
– (2006). Validity concepts in proof-theoretic semantics. *Synthese* 148, 525–571.
– (2014). Frege's sequent calculus. In: *Trends in Logic XIII: Gentzen's and Jaśkowski's Heritage – 80 Years of Natural Deduction and Sequent Calculi*. Ed. by A. Indrzejczak, J. Kaczmarek, and M. Zawidzki. Łódż University Press, 233–245. DOI: 10.15496/publikation-72324.
Sundholm, G. (1983). Constructions, proofs and the meaning of logical constants. *Journal of Philosophical Logic* 12 (2), 151–172.
– (1998). Inferences versus consequence. In: *The LOGICA Yearbook 1998*. Prague: Czech Academy of Sciences, 26–36.
– (2004). Anti-realism and the notion of truth. In: *Handbook of Epistemology*. Ed. by I. Niiniluoto et al. Kluwer Academic Publishers, 437–466.
– (2006). Semantic values for natural deduction derivations. *Synthese* 146, 623–638.
Tichý, P. (1988). *The Foundations of Frege's Logic*. Berlin: De Gruyter.
Troelstra, A. S. (1977). Aspects of Constructive Mathematics. In: *Handbook of Mathematical Logic*. Ed. by J. Barwise. Amsterdam: North-Holland, 973–1052.
Troelstra, A. S. and D. van Dalen (1988). *Constructivism in Mathematics, 2 vol.* Amsterdam: North-Holland.
von Kutchera, F. (1996). Frege and Natural Deduction. In: *Frege: Importance and Language*. Ed. by M. Schirn. Berlin: De Gruyter, 301–304.

Kolmogorov and the General Theory of Problems

Wagner de Campos Sanz

Abstract This essay is our modest contribution to a volume in honor of our dear friend and fellow logician Peter Schroeder-Heister. The objective of the article is to reexamine Kolmogorov's problem interpretation for intuitionistic logic and the basics of a general theory of problems. The task is developed by first examining the interpretation and presenting a new elucidation of it through Reduction Semantics. Next, in view of Kolmogorov's intentions concerning his problem interpretation, Reduction Semantics is employed in an brief epistemological analysis of Euclidean Geometry and its construction problems. Finally, on the basis of the previous steps, some theses are raised concerning intuitionistic logical constants and concerning proofs and hypotheses in Euclid's *Elements*.

1 Introduction

The epistemology of mathematics has gone through an important change in the last centuries: from focus in problems to focus in theorems. It is time for the pendulum to move back. A landmark was *The Foundations of Geometry* (Hilbert, 1899) which transformed *Propositiones* of Euclid's *Elements* into theorem assertions, while a good amount of them where the statement of a problem. This is the case of the first three *Propositiones* in book I.

The paper investigates *Kolmogorov's problem interpretation*. The objective is to unfold the basics of a general theory of problems, answering some very simple questions like: what are problems, why they are not assertions and how problems in geometry were structured.

Kolmogorov (1932, p. 151) states his purposes as follows:

> Along with the development of theoretical logic, which systematizes the schemes of proofs of

Wagner de Campos Sanz
Faculdade de Filosofia, Universidade Federal de Goiás, Goiânia, Brazil,
e-mail: wsanz@ufg.br

T. Piecha and K. F. Wehmeier (eds.), *Peter Schroeder-Heister on Proof-Theoretic Semantics*, Outstanding Contributions to Logic 29,
https://doi.org/10.1007/978-3-031-50981-0_5

> theoretical truths, it is also possible to systematize the schemes of solution of problems, for example, geometrical construction problems [. . .] The second section in which the general intuitionistic presuppositions are accepted, presents a critical analysis of intuitionistic logic. It is shown that this logic should be replaced by the calculus of problems, since the objects under consideration are in fact problems, rather than theoretical propositions.

Years later, in a comment about his paper, Kolmogorov evaluated it as follows (Tikhomirov, 1991, p. 452):

> *On the interpretation of intuitionistic logic* (IIL) was written with the hope that the logic of solutions of problems would later become a regular part of courses on logic. It was intended to construct a unified logical apparatus dealing with objects of two types — propositions and problems.

Kolmogorov (1932) belongs to a context where the nature of intuitionistic logic was being discussed. A crisis concerning some basic mathematical concepts had opposed Hilbert's school with his foundational program for mathematics and Brouwer's intuitionistic school. Intuitionists rejected certain means of proof in traditional mathematics like the proof of existential sentences over infinite collections by deducing and absurd from a negated universal hypotheses (*reductio*), as also the use of third middle excluded principle. Since the criticisms were directed to logical principles, it became important to elucidate what would be an acceptable intuitionistic logic. Heyting (1930) is now assumed as the standard formulation of intuitionistic logic. Kolmogorov (1932) is a later publication where the issue of how to interpret logical constants was still being debated. It is considered as a part of the so called BHK[1] interpretation of intuitionistic logical constants.

In the quotes above Kolmogorov remarks that logic has been historically occupied with schemes of theoretical truths. Next, he states some surprising claims. First, there is the claim about logic being occupied with schemes of solutions of problems, with geometrical constructions as a relevant example. Second, the claim that intuitionistic logic should be replaced by a calculus of problems. Third the statement about the unified logical apparatus for dealing with objects of two distinct types: propositions and problems. As declared, he had hopes that the logic of solutions of problems could become a regular part of courses on logic. Kolmogorov's problem conceptualization has usually been explained through the paradigm of propositions-as-types (Coquand, 2007). Here, a different elucidation will be developed.

The investigation of Kolmogorov's problem semantics explores each of these claims. His problem semantics is going to be reformulated as Reduction Semantics. The adequacy of this elucidation is dealt with in two separate steps. First, the formulation of the semantics and the issues of soundness and of completeness for intuitionistic logic are examined in the Section 2. Second, a partial anatomy of Euclid's Geometry by means of Reduction Semantics trying to unfold the schemes of solutions of problems in geometrical constructions is investigated in Section 3. To these two sections is added a third section focusing two specific subjects. In one

[1] Kolmogorov's interpretation is the third element in what has been conventionally called BHK interpretation of intuitionistic logical constants. The abbreviation corresponds to the initials of Brouwer, Heyting and Kolmogorov.

hand, the questions of what is a logical constant and if the usual set of intuitionistic propositional logical constants should be extended or not. On the other hand, a partial discussion concerning the role of hypotheses inside geometry.

Reduction Semantics is the further development of an initial investigation of problem interpretation in de Campos Sanz (2012). Reduction semantics turns to be similar to Hypo Semantics (de Campos Sanz, 2019) with a new dressing, since the basic objects are now problems.

2 A semantics of problems

2.1 Kolmogorov on problems

Kolmogorov's article is schematic and it does not contain an explanation of what is a problem, only a few carefully chosen examples (Kolmogorov, 1932, pp. 151–152):

1. Find four integers x, y, z and n such that $x^n + y^n = z^n, n > 2$.
2. Prove that Fermat's theorem is false.
3. Construct a circle passing through three given points (x, y, z).
4. Given one root of the equation $ax^2 + bx + c = 0$, find the other root.
5. Assuming that the number π has a rational expression, $\pi = m/n$, find a similar expression for the number e.

According to Coquand (2007, § 2.5), the perspective of interpreting intuitionistic logic as a calculus of problems is an important antecedent of a later distinction: that between formulae-as-types and λ-terms which would then correspond to a separation of problems and solutions, respectively. Although this can be a productive way of interpreting Kolmogorov about the semantics of problems, it is not clear that this author would subscribe such a sharp separation.

Kolmogorov (1932, p. 151) characterized his semantics of problems for logical molecular operators as follows:

> If a and b are problems, then $a \wedge b$ denotes the problem "solve both problems a and b", while $a \vee b$ stands for "solve at least one of the problems a and b". Further, $a \supset b$ is the problem [(FIRST)] "given a solution to problem a, solve problem b" or, which is the same, [(SECOND)] "reduce the solution of problem b to the solution of problem a. [...] $\neg a$ denotes the problem "assuming that there is a solution to problem a, derive a contradiction".[2]

In the five problems quoted above, the main verb is imperative. They are to be considered as the command of an action. The semantic explanations, in their turn, notoriously contain the command: "solve". Implication is the only exception. It has two explanations: (FIRST) and (SECOND). And the author regards them as equivalent. The expression "solution" is used in both explanations for naming, and the expression "reduce" in the (SECOND) explanation is used for commanding.

[2] Square brackets were added by us.

Some assumed that the problem interpretation required two different representational structures for problems and solutions. But there are reason for adopting another point of view, given that the inner nature of problems and solutions seems to be the same: they are about actions.

Consider Kolmogorov's description of what is a conjunction. Given any two problems a and b, $a \wedge b$ is the problem "solve both problems a and b". The action "solve" being commanded is iterated in order to form the description of a problem. When there is plain knowledge of how to proceed in order to bring about an action being commanded, then solving the problem corresponds to do the required action. Hence, in case a and b are problems that one knows how to solve immediately, and problem $a \wedge b$ represents the problem of solving a and solving b, then to solve $a \wedge b$ consists in doing what solves a and doing what solves b.

An example might help to clarify the point. Consider the following simple problem: construct a circle *BCD* of center A with radius AB and construct a circle *ACE* of center B with radius BA. This is a conjunctive problem. Employing Kolmogorov's description, it denotes the problem of solving both problems, that is, solve the problem of constructing *BCD* and solve the problem of constructing *ACE* as required. Both problems are immediately solvable. The solution consists in doing the construction of the circle *BCD* and doing the construction of the circle *ACE*.

Concerning implication now, it is obscure how the semantics could be formulated by using the command "solve" in the antecedent. By saying "first solve a, then solve b" another logical connective distinct of the usual implication seems to be employed once "solve" is an action and the expression "first solve ..., then solve ..." indicates a temporal ordering for the actions. Actually, in order to establish an intuitionistic implication, if we start by solving b, $a \supset b$ can be considered solved, but this violates the meaning explanation using the word "solve" which is truly a before-after conjunction. We will come back to this question later.

In the reverse direction, an attempt to employ the expression "solution" for explaining conjunction and disjunction would require the use of another command verb in place of "solve", as in: "produce the solution of ..." or just "do ...". This seems to be no real progress. Thus, a strange heterogeneity appears in the above semantical explanations. In one side we have one kind of explanation for disjunction and conjunction employing the verb "solve", in the other side another kind for implication employing the concept "solution" and/or the verb "reduce".

Actually, the (SECOND) explanation of implication above is partially misleading since the expression "reduce *the* solution of b ..." seems to make reference to a specific solution. But a problem could possess distinct solutions. Observe that the expression "reduce a solution of b ..." is also unclear, since it could be understood as meaning: "for some solution of b reduce it ...". The statement would be better rendered if it were formulated as "reduce *solvability* of b ...", which in turn might bring it closer to the (FIRST) explanation. Similar issues concerning the article "the" also affect the complement part of the (SECOND) explanation. It is important to stress that the usual intuitionistic interpretation of logical constants requires, for any construction of a, a corresponding construction of b, in order to have a true implication.

The (FIRST) explanation of implication can be seen as a characterization where solvability/demonstrability is transmitted, similar to what happens in the case of admissible rules. Under this interpretation the expression "suppose" is to be assumed as a hidden preamble like: "[suppose] given a solution to problem a ...".

The simplest and best way for rendering the (SECOND) explanation, in our opinion, would be to explain it as "reduce problem b to problem a", characterizing a reducibility junction between problems. But this is acceptable only under the proviso that a solution for b must be obtainable whenever a solution for a is provided. This seems the best explanation for implication from the point of view of problem semantics. It does not suffer from the issues pointed in (SECOND) and it can also be extended to other constants, as we are going to argue bellow. But it is distinct from the (FIRST).

In order to exemplify the reducibility junction, let's consider again the above example concerning circles. A new problem of finding a point C that is distant of points A and B as much as A and B are among themselves is a problem that is reducible to the problem of constructing circle BCD and constructing circle ACE as described above. That is, constructing both circles as described is a way of obtaining a solution for the problem of finding the required point C.

Concerning the notions of negation and contradiction we find in Kolmogorov's paper the following passages. The first is the text of a footnote accompanying the word "contradiction" in the last quotation above, (*ibid.*, p. 151):

> [...] $\neg a$ should not be read as the problem "prove the unsolvability of problem a". In the general case, if the "unsolvability of problem a" is considered as a completely defined notion, we only obtain that $\neg a$ implies the unsolvability of a, and not the converse assertion. If, for example, it were proved that a realization of the well-ordering of the continuum is beyond our possibilities, it would not be possible to assert that the existence of such a well-ordering implies a contradiction.

p. 156:

> It should also be mentioned that if $\vdash p$ is false in classical propositional logic, then the corresponding problem $\vdash p$ cannot be solved. Indeed, in view of the earlier accepted formulas and rules of the calculus of problems, this formula $\vdash p$ readily implies the contradictory formula $\vdash a \wedge \neg a$.

p. 157:

> [...] Brouwer suggests a new definition of negation, namely "a is false" should be understood as "a leads to a contradiction". Thus, the negation of a proposition a is transformed into the *existential sentence* "there exists a chain of logical inferences leading to a contradiction if a is assumed to be true".

Kolmogorov does not give further explanations of what he takes negation and contradiction to mean in the context of problems. So, what we have is the conflation of the concept of contradiction with the usual contradictory formula. But this formula already uses a negation as can be observed in the second quotation of the three above. We believe that in a good measure the concept was assumed as unproblematic by the author. In a similar attitude, when explaining the intuitionistic logical constants, Heyting (1956, p. 102) explicitly says that he takes it to be a primitive notion and complements by pointing that it is easy to recognize a contradiction when we are in front of one.

2.2 Reduction semantics

We claim that the explanations of Kolmogorov can be made homogeneous by adopting a new organization of the background concepts by taking the concept of reduction as the basic relation in the semantical explanation of problem constants.

The expression $\Gamma \Vdash b$ will mean from now on "problem b has been reduced to the multiset of problems Γ" but only if it is guaranteed that a solution of b is obtained in case of having solutions whatever for each problem in Γ. This proviso concerning how to define reduction is stated in Veloso (1984, p. 26) as a requirement for the acceptability of problem transformations.[3]

Based on the reduction relation, every logical constant that Kolmogorov was considering can be defined by semantical clauses, where the logical constant "$\perp$" represents a basic problem–the *impossible* problem.[4] Negation is explicitly defined as: $\neg a \equiv a \supset \perp$. The clauses are:

Clauses for using a logical constant in the repertoire:

$(\wedge^l)$: $\Gamma, c \wedge d \Vdash e \cong \Gamma, c, d \Vdash e$;
$(\vee^l)$: $\Gamma, c \vee d \Vdash e \cong \Gamma, c \Vdash e$ and $\Gamma, d \Vdash e$;
$(\supset^l)$: $\Gamma, c \supset d \Vdash e \cong$ given any problem $a : (a, c \Vdash d \Rightarrow \Gamma, a \Vdash e)$;
$(\perp^l)$: $\Gamma, \perp \Vdash a \cong$ always (where a is a basic problem).

Clauses for using a logical constant in the focus:

$(\wedge^r)$: $\Gamma \Vdash c \wedge d \cong \Gamma \Vdash c$ and $\Gamma \Vdash e$;
$(\vee^r)$: $\Gamma \Vdash c \vee d \cong$ given any problem a $((\Gamma, c \Vdash a$ and $\Gamma, d \Vdash a) \Rightarrow \Gamma \Vdash a)$;
$(\supset^r)$: $\Gamma \Vdash c \supset d \cong \Gamma, c \Vdash d$;
$(\perp^r)$: $\Gamma \Vdash \perp \cong$ given any basic problem $a : \Gamma \Vdash a$.

The *repertoire* of problems is in left side of the semantical symbol, and the *focal* problem is in the right side of the semantical symbol. Clauses are of two kinds: (i) those explaining the use of logical constants in repertoires, that is, finite multisets[5]; (ii) those explaining the use of logical constants as a focal problem. The symbol "$\cong$"

[3] He suggests that we should not consider arbitrary links between problems, but only those guaranteeing that solvability of b is obtained from solvability of a.

[4] Kolmogorov did not use it in his paper, we here follow a late tradition in intuitionistic logic.

[5] Finite sets admitting multiple copies of an element.

expresses a necessary and sufficient condition[6], that is, an explanation of meaning which has *explicandum* in the left side and *explicans* in the right side.[7]

As an exemplification, the reading of the clauses for implication are presented next. Consider clause $(\supset^r)$. As a semantical rule, it is read as follows: in order to reduce the reducibility problem $c \supset d$ to the multiset of problems Γ it is necessary and sufficient to show that d reduces to the multiset $\Gamma \cup \{c\}$. The reading of clause $(\supset^l)$ is a little bit more complex. As a semantical rule, it has to be read as follows: in order to reduce problem e to the multiset of problems $\Gamma \cup \{c \supset d\}$ it is necessary and sufficient to show that, for any given problem a, on the supposition that d reduces to the multiset of problems $\{a, c\}$ it follows that d reduces to the multiset of problems $\Gamma \cup \{a\}$.

Some expressions with logical content are unavoidable in the metalanguage, and they must be interpreted in their constructive sense. The other clauses are read in a similar way to those two just considered, with *explicans* quantified or not.

The clauses can be seen in action in the following proof.

Theorem 2.1 $\Vdash \neg(a \wedge \neg a)$.

Proof Suppose that $b, a \Vdash \bot$. Thus, $b, a \Vdash \bot \Rightarrow a, b \Vdash \bot$. Hence, for any given problem b, $(b, a \Vdash \bot \Rightarrow a, b \Vdash \bot)$. Now, by clause $(\supset^l)$, $a, \neg a \Vdash \bot$ according to the definition of negation. By clause $(\wedge^l)$, $a \wedge \neg a \Vdash \bot$. By the definition of negation and by clause $(\supset^r)$ we have finally $\Vdash \neg(a \wedge \neg a)$. □

Given any problem, we can fairly say that either it has been solved or it has not. If it was solved, then either it has been *positively* solved — when a correct solution is given to the problem — or it has been *negatively* solved — when a positive solution has shown to be impossible. The expression "to be solvable" is ambiguous and it can be employed in a strict sense or in large sense. The distinction is relevant for the discussion coming next.

Although the command "solve" could mean in a large sense either positively solve or negatively solve[8], the intended reading of the expression in the case of Kolmogorov is probably the strict sense, that of positively solve. Since the notion of reduction is at hand, the command "solve" can be dismissed. That is, categorically and positively solve $\neg(a \wedge \neg a)$ is the same as to show that it has been reduced to the empty repertoire of problems. There is an infinite set of problems that can also be reduced to the empty repertoire.[9] Negatively solve a problem means to show that the impossible problem has been reduced to it, like in $a \wedge \neg a \Vdash \bot$. An implication $a \supset b$ is positively solved categorically when b has been reduced to a according to clause $(\supset^r)$. The expression "has been reduced" means that an action took place.

[6] Harmony is an inherent property of explanations stated in terms of necessary and sufficient conditions.

[7] Thus, it is not an explicit definition once the constants being explained occur also in the metalanguage used for stating the clauses.

[8] Which then means that one of two possible conclusions is expected.

[9] This must be taken into account when reading the clauses containing a quantification, i.e., $(\supset^l)$ and $(\vee^r)$

A true simplification seems to be achieved in Reduction Semantics, since both verbs "solve" and "do" mentioned before can be substituted by the verb "reduce".

The fictitious impossible ($\perp$) problem is considered as negatively solved. Thus, since negation is defined via implication, $\neg a$ is the problem of reducing the impossible to problem a. Hence, at the same time that a is negatively solved, $\neg a$ is also positively solved by ($\supset^r$). The fictitious impossible problem is assumed as a basic problem itself and it is semantically characterized as a problem to which all basic problems reduce. That is, its solution is a *panacea*. Hence, the concept of contradiction originally used in Kolmogorov's paper can also be dismissed. As such, when the author says that a conditional problem is meaningless and then consider it as solved, this is equivalent to say that the antecedent problem is impossible and the whole conditional problem is reduced to the empty repertoire, since every basic problem reduces to the impossible problem by definition.

The above clauses must be complemented by the following *structural* semantical principles making explicit the properties of the semantical relation of reduction:

(basic problem Identity)	for basic a	$a \Vdash a$;
(Load problem)		$\Gamma \Vdash d \Rightarrow \Gamma, c \Vdash d$;
(Drop basic problem)	for basic a	$(\Gamma, a \Vdash d \;\&\; \Gamma \Vdash a) \Rightarrow \Gamma \Vdash d$.

The above semantics of problems, that is, the structural principles plus the clauses is called *Reduction Semantics*. When the variables a, b, c etc. are interpreted as sentences we call it *Hypo[thesis] Semantics* (de Campos Sanz, 2019).

Theorem 2.2 *(i) The full (Identity) principle holds, i.e., for any c,* $c \Vdash c$. *(ii) The full (Drop) principle holds, i.e., for any c,* $(\Gamma, c \Vdash d \;\&\; \Gamma \Vdash c) \Rightarrow \Gamma \Vdash d$.

Proof (i) and (ii) by induction in the logical degree of problem c, by using both clauses for the repertoire and clauses for the focus.[10] □

Semantical variations or extensions of the above set of problem constants given by the clauses are envisageable. In some cases it might well occur that for a new constant there are restrictions for the application of full drop. We think this might well be the case of some modal operators, and this is a reason for preferring the weaker and not the full formulation of the structural principles.

Going back to Kolmogorov's paper in order to determine if the above picture fits in, notice that the author intends with his calculus "... to systematize the schemes of solution of problems", as those in geometrical constructions. Then next, two issues are going to be considered. The first is the adequacy of Kolmogorov's calculus of problems with respect to Reduction Semantics. The second is the adequacy of Kolmogorov's notion of problem for describing constructions and proofs in ancient geometry. A case study already envisaged by him.

[10] In case the full principles were assumed from the beginning, it would be enough to possess one clause for each logical constant, once the other clause could be obtained by the use of the full principles. But, of course, in that case, we would miss the fact that the weaker structural principles are enough to formulate the semantics.

2.3 Adequacy of Reduction Semantics with respect to Kolmogorov's interpretation

The second part of Kolmogorov's paper presents a calculus of problems and describes the validation of its axioms and theorems. Here are the axioms:

[group] A

2.1 $\vdash a \supset a \wedge a;$
2.11 $\vdash a \wedge b \supset b \wedge a;$
2.12 $\vdash (a \supset b) \supset (a \wedge c \supset b \wedge c);$
2.13 $\vdash (a \supset b) \wedge (b \supset c) \supset (a \supset c);$
2.14 $\vdash (b \supset (a \supset b);$
2.15 $\vdash a \wedge (a \supset b) \supset b;$
3.1 $\vdash a \supset (a \vee b);$
3.11 $\vdash a \vee b \supset b \vee a;$
3.12 $\vdash (a \supset c) \wedge (b \supset c) \supset (a \vee b \supset c);$
4.1 $\vdash \neg a \supset (a \supset b);$
4.11 $\vdash (a \supset b) \wedge (a \supset \neg b) \supset \neg a.$

Kolmogorov gives as an example an argument containing a general method for solving the problem 2.12, valid for any a, b, c. That is, he is basically showing an intuitive semantical justification based on the problem interpretation for the validity of axiom 2.12 above:

> For example, in problem 2.12, assuming that the solution of b has already been reduced to the solution of a, one should reduce the solution of $b \wedge c$ to that of $a \wedge c$. Let a solution of $a \wedge c$ be given. This means that we are given both a solution of a and a solution of c. By the hypothesis, we can derive a solution of b from that of a, and, since a solution of c is known, we obtain solutions of both problems b and c and hence a solution of problem $b \wedge c$.

The expression "the solution of", we remember, has been dismissed as being inappropriate. The above argumentation can be reformulated as a validity proof of the above axioms with respect to Reduction Semantics.

Lemma 2.3 *All axioms in the calculus of problems are valid with respect to Reduction Semantics.*

Proof The axiom 2.12 with all parentheses becomes: $\vdash (a \supset b) \supset ((a \wedge c) \supset (b \wedge c))$. It is validated as follows. (1) $a \wedge c \Vdash a \wedge c$ by (Id); (2) $a \supset b, a \wedge c \Vdash a \wedge c$ from 1 by (Load); (3) $a \supset b, a \wedge c \Vdash a$ from (2) by the necessary condition of $(\wedge^r)$; (4) $a \supset b, a \wedge c \Vdash c$ from (2) by the necessary condition $(\wedge^r)$; (5) $a \supset b \Vdash a \supset b$ by (Id); (6) $a, a \supset b \Vdash b$ from (5) by the necessary condition of $(\supset^r)$; (7) $a, a \supset b, a \wedge c \Vdash b$ by (Load) from (6); (8) $a \supset b, a \wedge c \Vdash b$ from (3) and (7) by (Drop); (9) $a \supset b, a \wedge c \Vdash b \wedge c$ from (4) and (8) by the sufficient condition of $(\wedge^r)$; (10) $a \supset b \Vdash (a \wedge c) \supset (b \wedge c)$ from (9) by the sufficient condition of $(\supset^r)$; and finally (11) $\Vdash (a \supset b) \supset ((a \wedge c) \supset (b \wedge c))$ by the sufficient condition of $(\supset^r)$. That is, axiom 2.12 is semantically valid. The other axioms are validated in a similar fashion. □

As remarked before, each expression of form $\Gamma \Vdash e$ is to be read as "problem e reduces to the multiset of problems in Γ". This reading does not mention the word "solution" at all. Hence, the above proof can be read without using the word "solution".

Next, Kolmogorov presents three rules for extending the set of solved problems in the calculus. The equivalence between this calculus and the intuitionistic propositional calculus is now explicitly pointed in the passage:

> [Group B] We can now formulate the rules of our calculus of problems.
> 1. First, we include the problems of group (A) in the list of solved problems.
> 2. If the list includes $\vdash p \wedge q$, then we are allowed to replace it by $\vdash p$.
> 3. If both formulas $\vdash p$ and $\vdash p \supset q$ are in the list, then we can replace them by $\vdash q$.
> 4. If $\vdash p(a, b, c, \ldots)$ is in the list and $q, r, s, \ldots$ are arbitrary problem functions, then we are allowed to replace it by $\vdash p(q, r, s, \ldots)$ in the list.[11]
>
> Based on the above Postulates, it is easily seen that the formal calculus does in fact guarantee the solution of the corresponding problems.
>
> We are not going to develop this calculus further here, since all formal rules and a priori formulas above coincide with the computational rules and axioms suggested by [Heyting (1930)].[12] Hence, we can interpret all formulas of this paper as problems and assume that all problems are solved.

Validity of 2, 3 and 4 in Reduction Semantics requires that the following principles be proved:

Lemma 2.4 *The following principles hold in Reduction Semantics:*
2. $\Vdash p \wedge q \Rightarrow \Vdash p$;
3. $(\Vdash p$ *and* $\Vdash p \supset q) \Rightarrow \Vdash q$;
4. $\Vdash p \Rightarrow \Vdash p[^{a,b,c,\ldots}_{q,r,s,\ldots}]$, *where* $p[^{a,b,c,\ldots}_{q,r,s,\ldots}]$ *is the result of substituting* $a, b, c, \ldots$ *by* $q, r, s, \ldots$ *inside* p*, respectively.*

Proof Principle 2 is obtained by the necessary condition of clause $(\wedge^r) \Vdash p \wedge q \Rightarrow (\Vdash p$ and $\Vdash q)$ and, *a fortiori*, $\Vdash p \wedge q \Rightarrow \Vdash p$. Principle 3 is obtained by using clauses full (Id) and $(\supset^r)$. Principle 4 holds in virtue of the clauses being closed under homogeneous substitution of problems. □

Theorem 2.5 *Kolmogorov's calculus of problems is sound with respect to Reduction Semantics, that is:* $\vdash c \Rightarrow \Vdash c$.

Proof By Lemmas 2.3 and 2.4. □

There are reasons to believe that the usual set of propositional intuitionistic logical constants is incomplete for dealing with problems. It probably should be extended with other constants, among them the before-after conjunction. If this is correct,

---

[11] We suspect that the word "replace" employed in 2, 3 and 4 might be a poor translation of the original, since "add" would be a better way to formulate it. A question that might be raised is: does the passage suggest that "problem" and "function" are related concepts? We do not think so. We interpret the passage as just stating that $q, r, s, \ldots$ are variables for problems which can be substituted by problems.

[12] The square brackets substitutes the bibliographic note [1] in the original paper.

then completeness is not a so meaningful property. The before-after conjunction will be used and discussed ahead when problems in geometry are discussed. Finally, we notice that Kolmogorov did not give a completness proof, probably because his semantics was formulated in an intuitive way.

Completeness can be shown if the set of logical constants is restricted to $\wedge, \vee, \supset, \bot$. To say that Kolmogorov's calculus of problems is complete with respect to Reduction Semantics means that $\Vdash c \Rightarrow \vdash c$. The recipe for proving it is the following. First, define a sequent calculus with rules for the semantic principles of reduction — full (Id), (Load) and full (Drop) — and double inferences, that is, left to right and right to left inferences corresponding to the following clauses: $(\vee^l)$, $(\wedge^r)$, $(\supset^r)$ and $(\bot^r)$. Second, prove that each of the semantical clauses and principles of Reduction Semantics is a valid metatheoretical property of this calculus, that is, substitute the symbol "$\vdash$" in place of the semantical symbol "$\Vdash$" and prove each Reduction Semantics principle, be it a clause or a structural principle, as a metalanguage property of the calculus just defined.[13] This is almost immediate. Third, prove that the calculus just described is theoremhood-equivalent to Kolmogorov's calculus of problems.

The meaning of problem logical constants was explained in Reduction Semantics through two distinct uses: in the repertoire and in the focus. The fact that the full (Drop) principle as the full (Identity) principle holds for the set of logical constants given in the clauses of the semantics above does not mean that it will hold for any extension of the set of logical constants.

Reduction Semantics is our way to elucidate, clarify, the problem interpretation of intuitionistic logic as proposed by Kolmogorov.

2.4 Why problem semantics and intuitionistic logic?

Problem semantics was the conceptual way in which Kolmogorov interpreted intuitionistic logic. We think that there are two reasons why Kolmogorov adopted such approach.

The first is the fact that *tertium non datur* ($\vdash a \vee \neg a$) is not a valid principle in the calculus of problems and in intuitionistic logic. Kolmogorov (1932, p. 156) interprets the sign "$\vdash$", differently from Heyting, as meaning generality:

> For a function $p(a, b, c, \ldots)$ of undefined problems $a, b, c, \ldots$ we simply write $\vdash p(a, b, c, \ldots)$ instead of $(a)(b)(c) \ldots\ p(a, b, c, \ldots)$. Hence, $p(a, b, c, \ldots)$ denotes the problem *"find a general method for solving the problem $p(a, b, c, \ldots)$ for each individual choice of the problems $a, b, c, \ldots$"*.[14]

His interpretation is constructive in the measure it considers the problem of proving logical formulas as requiring a general method of solution. In this sense intuitionistic logic is a part of a theory of problems. In particular, $\vdash a \vee \neg a$ would be valid if

[13] Observe that $(\supset^r)$ is the semantical version of deduction theorem.

[14] Our emphasis.

(a) $(a \vee \neg a)$ were valid, which means that we should have a general method for solving problems of form $a \vee \neg a$ for each individual choice of a.[15]

The observation concerning the requirement of a method of solution brings us to the second reason. It concerns the interpretation of existential propositions, when the proof of existence does not exhibit the object. The quotation is curious since the problem calculus presented is considering only propositional constants (Kolmogorov, 1932, p. 157):

> Brouwer does not, however, intend to exclude existential propositions from mathematics completely. He only explains that an existential proposition should not be stated without presenting the corresponding construction. At the same time, according to Brouwer, an existential proposition is not a mere indication of the fact that we have already found the desired element of K. In this case the existential proposition would be false prior to the invention of the construction and true after that. Thus, propositions of a completely new type arise, which, although their content does not change in time, can nevertheless be stated only under certain conditions.
>
> The natural question which can arise is whether this specific type of proposition is a mere fiction. Indeed, the problem "find an element of a set K possessing a property A" is posed. This problem actually has a certain sense independent of the state of our knowledge. If this problem has been solved, that is, if the corresponding element x is found, we obtain the empirical proposition "our problem is now solved". Thus, Brouwer's existential proposition is partitioned into two elements: an objective component (problem) and a subjective component (its solution).

Intuitionists do not accept [classical] negation of an universal to be a basis for the inference of an existential judgement involving an infinite collection. According to the conceptual background set by Kolmogorov, the meaning of an existential problem cannot depend on the possession of a construction exhibiting the required element because a previous understanding of the problem is needed in order to even look for the solution. Thus problems have an objectivity that solutions might lack. Problems have to be understood in order to be solved, they must be meaningful. Solutions require ingenuity in order to be obtained, and meaning cannot be made to depend on them.

3 Problems and solutions in geometry

After offering an analysis of Kolmogorov's problem interpretation in terms of Reduction Semantics, it is time to consider more closely the concepts of problem and solution taking as a reference for the analysis that piece of mathematical knowledge where problems seem to play a central role: Euclidean Geometry.

---

[15] Kolmogorov (1932, p. 156): "formula [$\vdash a \vee \neg a$] reads as follows: find a general method which for any problem a allows one either to find its solution or to derive a contradiction from the existence of such a solution." Since a formalized language has to be previously presented before defining its semantics, all basic problems in a calculus of problems would also be exhaustively enumerated beforehand. The only hope of validating *tertium non datur* principle occurs when language is such that we possess a general method for solving all basic problems enumerated on it. For a non-specific language such a general method is obviously impossible.

3.1 Problems and practical principles

The first three Postulates of Euclidean Geometry have been regarded historically as practical principles. There is a modern tradition culminating in Kant going in this sense (Critique of Pure Reason, A234/B287, our emphasis):[16]

> Now in mathematics *a postulate is the practical proposition* that contains nothing except the synthesis through which we first give ourselves an object and generate its concept, e.g., to describe a circle with a given line from a given point on a plane; and *a proposition of this sort cannot be proved*, since *the procedure that it demands is precisely that through which we first generate the concept of such a figure*

On Kant's interpretation, the same procedure that is going to generate a circle under the Postulate I.3 is also the procedure behind the definition of what is a circle. And it is fair to suppose that he holds a similar opinion about straight-lines. Each of the first three Postulates involves a practical proposition establishing some of the most basic elements of geometry. Although his description of what is a postulate can be understood for the first three, it is not clear that the same holds for the fourth and the fifth Postulates.

Actually, *problemata* were seen by Kant as practical propositions too (*Logik Hechsel*, 1992, p. 88):

> [...], and problemata, practical propositions which require a solution.

Moreover (*Jäsche Logik*, 1800/1904, § 38, our emphasis):

> A postulate is a practical, immediately certain proposition, or a principle that determines a possible action, in the case of which it is presupposed that the way of executing it is immediately certain. Problems (*problemata*) are demonstrable propositions that require a directive, or ones that express an action, the manner of whose execution is not immediately certain. [...] Note 2: A problem involves (1.) the *question*, which contains what is to be accomplished, (2.) the *resolution*, which contains the way in which what is to be accomplished can be executed, and (3.) the *demonstration* that when I have proceeded thus, what is required will occur.

Beyond any doubt, the treatment of problems as practical questions had already been considered in the history of philosophy. And the canonical examples of problems Kant is refereeing are those in geometry. He underlines three elements involved in problems in the footnote accompanying the quotation above. This is a good starting point for considering the nature of problems. They involve: (1) the question, or what is asked; (2) the resolution seen as something that is going to be executed, (3) a proof that the resolution accomplishes what was asked. The first two elements are common to all problems. The third is characteristic of mathematical problems. These three elements are present in many of the geometric *Propositiones* in Euclid's *Elements*, mainly in those asking a construction like I.1, for example.

---

[16] See Lassalle-Casanave (2019).

3.2 What is a problem?

As remarked before, Kolmogorov did not answer the question of what is a problem. We can try to organize some ideas departing from Kant's considerations.

Many of the Euclid's *Propositiones* are construction problems. One example is problem III.1: *to find the center of a given circle.*[17] According to Definition I.15, a circle is a plane figure contained by a single line, the circumference, such that all of the straight-lines radiating towards the circumference from one point amongst those lying inside the figure are equal to one another. Thus, historically at least, problem III.1 is not the problem of showing the existence of the center of the circle, since this is already guaranteed to exist according to Euclid's definition of a circle. Also, strictly speaking, there is no construction to be effected, once the circle is already given. The problem for which a solution is then asked is that of finding or determining the center when this is not clear from the figure given, supposing it to be a circle.

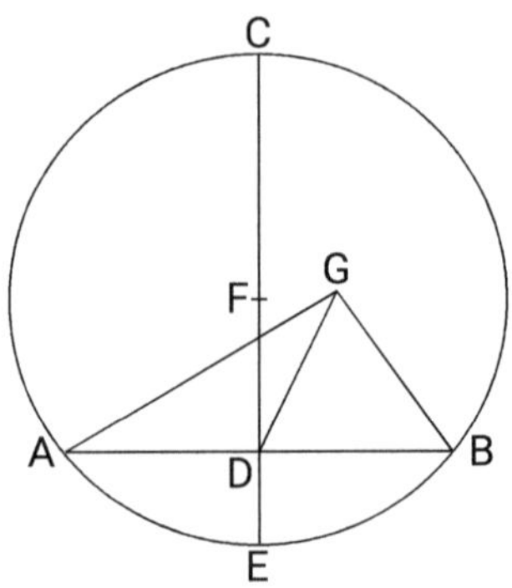

Fig. 1 Prop. III.1

By a series of actions the center of a circle can be found. Why do we say actions? Because the problem is to be solved in the end by employing the first three Postulates which, as Kant pointed, are practical principles. Indeed, the starting step in providing a solution to problem III.1 consists in the production of a chord AB of the circle, that is, the production of a straight-line according to Postulate I.1. And anyone of the endless chords would serve for the purpose.

Actually, the construction given as a solution to problem III.1 is only a description of which actions and in which order they should be taken for determining the center of the circle. It constitutes then a recipe of construction, or resolution in kantian terms.

What is problem III.1 about? It is about "finding the center" and the verb "to find" is a verb used to describe an action. How the problem can be put to someone, a student let's say? It can be put by uttering a command or asking an action: Find the center of circle ABC! Only actions can be commanded. Clearly, it does not make

[17] All quotations of Euclidean Geometry are taken from Fitzpatrick's (2008) translation of Heiberg (1885).

sense to command a usual proposition or a sentence. We assert sentences. The two speech acts are of different nature.

The same terminology that Kant used for describing the content of a command or invitation involving an action is going to be employed here: *practical proposition*. A *problem* is stated by asking or commanding an action whose result and/or execution is unknown or assumed to be unknown. We can separate two aspects of a problem: one corresponding to the speech act involved in uttering the problem as such, asking or commanding; and another concerning the content or practical proposition formulating the action being asked. This distinction seems to be suggested in Kant's passages quoted above. Kolmogorov is consistent in using the concept of problem in connection with actions, but he does not make explicit the two aspects just pointed.

To formulate a problem is the same as to ask or to command an action: to find something, to draw something and — we have reasons to think also that — to prove something. A *solution* to a problem is in consequence a certain organized sequence of actions accomplishing the action commanded or asked. In particular, Postulates can be interpreted as problems of a very specific nature: they are such that the action being asked is assumed to be immediately feasible or supposed to be immediately solvable. This explains why the first three Postulates are not proved. They are the simplest problems to which others are reduced.

But Euclid's *Elements* contain other kinds of *Propositiones* that do not ask a construction. Most of them are what we could call proving problems. *Propositio* I.47 — Pythagora's theorem — is an example. This is a curious case, since I.47 could also have been stated as a construction problem asking to obtain a square equivalent to the sum of other two given squares.

Geometry, in the hands of Hilbert (1899), has received a definite assertional turn in which construction problems as those of Euclid's *Elements* were left out. Nonetheless, the opposite movement of traveling back home might be considered, giving way to a *problemational turn*. That is, a turn in the focus of epistemology making it more sensitive to the original formulation of Euclidean Geometry.

Solutions can be divided in two kinds, which implies that problems are themselves divided in two kinds: token-problems and type-problems. For example, a problem like (1) in Kolmogorov's quotation enumerating problems, the one asking to find four numbers, is a token-problem. It requires a specific action. Its solution, when there is one, involves an act of exhibition of a result or a state-of-affairs. A problem like (3) is a type-problem, since the solution expected is a recipe/algorithm showing how to fulfill what is being commanded or asked for given x, y and z, as parameters.

Actually, all solutions can be characterized as recipes. Any token-problems can be assimilated for simplification to a type-problem requiring a recipe containing a final act of exhibition. In other terms, here the recipe is to be thought as being recoverable from the succession of actions that produced the result being exhibited.

From a philosophical point of view, the investigation of problems belongs at the same time to theoretical philosophy and to practical philosophy. A solution is a recipe describing actions which when realized will solve a problem, while a problem is equivalent to ask or command an action. This seems to be the borning star of mathematics.

3.3 The nature of solutions and problems in Kolmogorov's perspective

Reduction Semantics above has been proposed as an elucidation of the problem interpretation by Kolmogorov. In Reduction Semantics, problems and solutions are treated in an unified fashion and this fact is going to be used for explaining bellow the relative order among Postulates and *Propositiones* in ancient geometry.

Nonetheless, there is a question that requires our attention before proceeding with the analysis of problems. The question concerns the two kinds of objects that Kolmogorov's calculus of problems should deal with: propositions and problems.

Problem (5) in the list of examples quoted above is a conditional problem asking to find a way to express number e as a rational expression. It has in the condition position a proposition: "that the number π has a rational expression". Problems are uttered with imperative mood while sentences and propositions are uttered with the indicative mood. Example (5) has a condition in the indicative mood and the conditioned in the imperative mood. No doubt, (5) states a problem and its main verb is in the imperative mood. How propositions should be dealt with in the context of problems?

As mentioned before, typed λ-calculus disposes of two different structures for representing problems and solutions in one same formalism. Formulas are types for λ-terms. Solutions are λ-terms. Problems are regarded as formulas, that is, as types. Propositions or sentences are normally considered a subset of the formulas. Did Kolmogorov intented this classification? We have reasons to doubt it, among other things because different moods are used when stating problems and propositions.

Propositions or sentences can be dealt with in the context of problem interpretation inside some specific kinds of problems. Actually, Kolmogorov stated, although only in an implicit way, what looks like a unifying perspective. It appears when he is discussing the principle of excluded middle. Recall that for him the expression $\vdash a \vee \neg a$ should be read as asking to (*ibid.*, p. 156):

> [...] find a general method which for any problem a allows one either to find its solution or to derive a contradiction from the existence of such a solution! In particular, *if the problem a consists of proving a proposition*[18], one must have a general method which allows one either to prove each proposition or to reduce it to a contradiction.

Proving propositions then seems to be one important kind of problems and the answer to our question.

Assuming that proving propositions is a legitimate distinct kind of problems, then the concept of a problem becomes more general than the concept of theorem. Now, all *Propositiones* in Euclidean Geometry can be regarded as problems, some are mainly construction problems and others are mainly proving problems. In this sense, a calculus of problems would embrace all *Propositiones* without requiring any substantial change in the way they were formulated, instead of transforming construction *Propositiones* into existence statements or something similar.

Problems which consist in proving a proposition p are going to be represented by the expression "(prove p)". Next, assuming also that the expression "(deduce q

[18] Our emphasis.

from Ω)" represents a basic kind of problem, where q is a proposition and Ω a set of hypotheses, the special case where Ω is empty constitutes a way for characterizing what is to prove a proposition p, that is:

(†) (prove p) ≡ (deduce p from ∅).

Thus, if it has been shown that (prove p) $\Vdash$ (prove q), then $\Vdash$ (prove p) ⊃ (prove q), by clause ($\supset^r$). This last means $\Vdash$ (deduce p from ∅) ⊃ (deduce q from ∅), according to (†). This corresponds to a transmission of demonstrability, similar to an admissible rule. Observe that it is distinct from $\Vdash$ (prove if p, then q) since this last means (deduce if p, then q from ∅) according to (†), and this last means (deduce q from p).[19]

Verbs commanding an action were used for talking about deductions and proofs since the expression of problems and solutions require practical propositions. Deductions and proofs become then just the trace of actions conveyed in the practical proposition. Two examples in Section 3.6 illustrate the point.

After examining in a few words what might be proving problems, we can deal with problem (5) given as an example by Kolmogorov. Recall that it is formulated as :"assuming that the number π has a rational expression, $\pi = m/n$, find a similar expression for the number e". Two distinct interpretations seem to be possible.

First, the problem can be understood as meaning: (prove $\pi = m/n$) ⊃ (find ...). This is a solvable problem, by the mere fact that (prove $\pi = m/n$) $\Vdash \bot$. Euclidean Geometry contains conditional problems where the condition is a proving problem. *Propositio* I.22 is an example. The construction of the triangle as required depends on supposing the provability of a certain relation between the three given straight-lines, A, B and C. Any of them must be shorter than the sum of the other two. The mere supposition that three straight-lines are in this relation does not guarantee that the procedure can be carried adequately.

Second possible interpretation, (5) could also be understood in a more general sense as being a problem to be solved under certain conditions. That π is not a rational number is a fact. Nonetheless, nothing forbidden us of supposing contrary to the fact that π were equal to m/n for m and n rational numbers. However, to derive a contradiction from this supposition is to misunderstand the point of the counterfactual. The condition in (5) could be interpreted as a counterfactual supposition. And, if this were what Kolmogorov intended, then the meaning of (5) would be different from the meaning considered in the precedent paragraph. It would be equivalent to state the following problem: find natural numbers x and y such that $(x/y)^{im/n} = -1$, supposing m and n to be natural numbers and i to be the imaginary number.

[19] The action of deduction can be defined by reflecting inside the object language the content of the focus clauses for Reduction Semantics. They can be characterized relative to the deduction problem as follows:

1. (deduce p and q from Ω) ≡ (deduce p from Ω) ∧ (deduce q from Ω)
2. (deduce p or q from Ω) ≡ (r)(((deduce r from Ω, p) ∧ (deduce r from Ω, q)) ⊃ (deduce r from Ω))
3. (deduce if p, then q from Ω) ≡ (deduce q from Ω, p)
4. (deduce absurd from Ω) ≡ for any atomic p (deduce p from Ω).

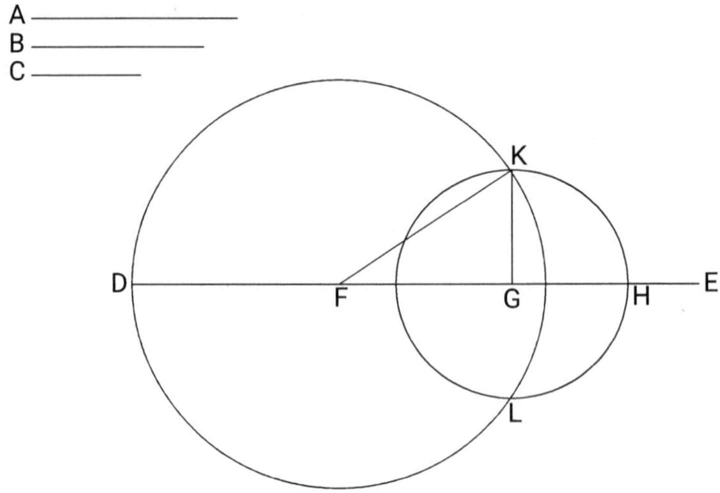

Fig. 2 Prop. I.22

We claim that a second kind of conditional problems should be considered in a calculus of problems, one in which the condition is not itself a problem, not even a proving problem. The condition involves just a description of a putative situation. There are related examples in Euclidean Geometry. One is the correctness-proof for the solution to the problem of finding the center of a circle in *Propositio* III.1. It starts by a contrary to fact supposition: that the point F is not the center. This supposition is the beginning of a *reductio* proof. It is not a supposition of having proved that F is not the center, even because exactly the opposite had been just showed by the construction. We come back to this issue in Section 4.2 when discussing the relation between hypotheses and problems.

Apparently, Kolmogorov did not consider the second interpretation above with respect to example (5) or, at least, he assumed it to be resolved with the first interpretation. And since our subject here is his calculus of problems, we put the alternative interpretation aside for the time being.

Next we turn to the analysis of solutions in ancient geometry. That is, we proceed to the second task concerning the adequacy question relative to Reduction Semantics.

3.4 Problems in ancient geometry

Now turning to Book I of Euclid's *Elements*, let's proceed with the epistemological analysis of problems on the basis of Reduction Semantics. What should be expected from it is a certain homogeneity in treating *Propositiones*, Postulates and Common Notions as problems.

Traditionally, logic has been used to offer an analysis, even if partial, of the proof steps and the structure of Euclidean Geometry. Some believe that since Aristotle, at least, proofs are taken to be discursive sequences of assertions or of declarative sentences. But this perspective does not correspond either with what we read in the three first Postulates or many other *Propositiones*, in particular the first three in Book I. Hence, it is a striking fact that the logical analysis through declarative sentences of Euclid's *Elements of Geometry* starts with an inadequate concept. When Hilbert (1899) came to light things changed, but then all *Propositiones* became declarative statements. Points, lines and planes became three distinct system of things related by declarative axioms.

Together with other authors, we think that it is possible to provide a logical analysis of ancient geometry maintaining its original formulation with all richness of its *Propositiones*. From our perspective it is Kolmogorov's merit to have devised an approach that can be fruitfully applied to ancient geometry, although it remained largely non-explored. We clearly mean his problem interpretation. So, next, the problem interpretation is employed as an epistemological tool for the analysis of the *Elements* via Reduction Semantics.

One attempt in such direction can be found in von Plato and Mäenpää (1990). The authors develop an investigation of Euclidean Geometry based on Martin-Löf's Type Theory assuming it to be an elucidation of that interpretation. They claim that the first three Postulates can be assimilated to constructive functions. According to them, (*ibid.*,p.281):

> The construction postulates lay down the permitted means of producing finite straight-lines, [. . .]. The functionality of postulates suggests a way of rendering them into the general pattern of natural deduction rules used in intuitionistic type theory. Its inferences may be viewed as functions from premises to conclusions. This proof functionality is explicitly recorded in proof objects, that is, in the objects given in the left side of judgements of the form a:A. It is judgements, not propositions, which figures as premises and conclusion in an inference rule.

This interpretation is further discussed and criticized in Naibo (2018). More recently, a deeper exegesis of Euclidean Geometry partially based on von Plato and Maenpaä's approach has been proposed by Sidoli (2018). But Sidoli does not focus the concept of problems or its elucidation. Also, together with von Plato and Maenpaä, he assumes that points are parameters of constructions. But there are reasons to disagree with them.

Postulate I.1 is interpreted by von Plato and Maenpaä as a constructive function producing a straight-line once two points are given as parameters, similar to a natural deduction rule. But it is doubtful that straight-lines depend on points as parameters in order to be produced. There are at least two reasons for doubting.

First, because the production of a straight-line can just start in a place whatever not determined beforehand and it can stop in a point not determined with antecedence. This is the case of the starting act of resolution of problem III.1 which consists in finding the center of a given circle as in Figure 1. This act is that of producing chord AB at random for the given circle ABC. It is a non-deterministic act. If conceived as a constructive function, it is doubtful that points A and B were previously established when Postulate I.1 was supposed to be applied. And nothing hinders one to have picked two points whatever over the circumference before drawing the line, similar to what was done in the resolution of *Propositio* I.9: *let the point D have been taken at random on AB* .... Most striking yet is that, after the chord is drawn, the perpendicular over the middle point *D* in the chord is raised and, although belonging to the circumference since the beginning, point *C* cannot be pinpointed before the perpendicular has been produced in Figure 1.

Second, Postulate I.2 — ... *to produce a finite straight-line continuously in a straight-line* — does not even mention points, much less a stopping point.

A mental image could be of some help here. Think about the points as being mainly the end or the start of a drawing/production. This is in accordance with Definition I.3. "Ends", as such, have no parts. Also, without an actual drawing there is no starting or stopping points. The first three Postulates strongly suggest such a drawing or action, although in the case of Postulate I.3 the center and the *radius* have to be provided before.

The three beginning *Propositiones* of Book I are naturally read as problems not as theorems. Their resolutions show how to fulfill what is being asked or commanded: a construction. Actually, the resolution text in the *Propositiones* contains two principal parts among others: *kastaskeue* and *apodeixis*. The *kastaskeue* contains a recipe or solution for the construction problem. It describes certain actions to be effected according to the Postulates. To draw a circle, to draw another circle, to draw a straight line, to draw another straight-line. The *apodeixis* contains a proof that this solution really produces what is demanded in the *Propositio*. Once I.1 — ... *to construct an equilateral triangle on a given finite straight-line* — has been resolved and the solution has been proven correct, it can be used for obtaining a solution for I.2 — ... *to place a straight-line equal to a given straight-line at a given point (as an extremity)*. Therefore, the *Propositio* that was formerly interpreted as *a problem* is now used as part of a *solution* for a new problem.

The change in perspective means that the distinction between problems and solutions is superficial and it depends of a certain history of accomplishments. First an action is seen as problematic. Then a procedure for effecting the action is offered, and the problematic action now just becomes a piece of knowledge that can be employed in the solution of another problem.

Proving problems were briefly described in the previous section and they must be considered as part of Kolmogorov's calculus of problems. Hence, they are part of the language of problems considered in Reduction Semantics.

Propositiones similar to I.47 — i.e., Pythagoras' — are theorems. Their proofs also contain *kataskeue* and *apodeixis*. When proving I.47, similar kinds of problems are to be solved again: first, a construction and, second, a proving problem. In fact both are intertwined and, although proofs in modern reconstructions are composed of assertions, the original geometrical proofs are never exclusively discursive since they involve practical problems: those involved in the construction.

We claim that Reduction Semantics fits the *Propositiones* of Book I the way they were formulated. Before the analysis goes on, it is necessary to effect an examination of Postulates and Common Notions within the perspective proposed.

3.5 Common Notions and Postulates

Common Notions are generic in the sense that they apply to different elements of geometry: lines, angles, triangles, etc. For example, in Common Notion I.2 — *if equal things are added to equal things then the wholes are equal* — the word "things" is a parametric word, which suggests that it can be interpreted as a schematic deductive rule. The same holds for Common Notions I.1 to I.4. Common Notion I.5 is an schematic axiom — *the whole is greater than the part.* Common Notions are either a proving problem assumed to be solved, as in I.5, or they are basic reduction rules as in all other cases. For example, I.2 can be rendered for things A, B and C as: $\Vdash$ (prove if $(A = B)$, then $(A + C = B + C)$). From it follows: (prove $A = B$) $\Vdash$ (prove $A + C = B + C$).

Postulates are the fundamental principles governing geometry. The first three Postulates of Book I are of a practical nature since they are about actions: ... *to produce a straight-line from any point to any point*; ... *to produce a finite straight-line continuously in a straight-line*; ... *to produce a circle with any center and radius*. We claim that they can easily and fairly be understood as problems, even if a very simple one.[20] But they are problems of a special kind. They are supposed to be solvable. This supposition does not require to suppose together any description of how to proceed or which tools should be employed. The alternative of assuming them to be solved seems to be stronger and unreasonable if it involves to suppose an infinity of acts to have been effected. The first three Postulates are rendered as:

I.1 $\Vdash$ (draw [any] finite straight-line XY), where X and Y are the two extremities of the line and they might be determined *a posteriori*;
I.2 for any straight-line AB: $\Vdash$ (extend AB into AX), where X might also be determined *a posteriori*;
I.3 for any point A and any straight-line AB: $\Vdash$ (draw a circle of center A and radius AB).

Postulate I.5 involves the problem of determining when two straight lines are not parallel. According to Definition I.23, straight-lines are parallels if they do not meet each other when indefinitely extended in any direction. And when another straight-line crossing both form internal angles less than two right angles, then they are not parallels. That is, the condition of making angles less than two right is sufficient for showing that they meet on the side they are less than that by indefinitely extending such straight-lines. This Postulate is then a specific principle for straight-lines guaranteeing a solution to the construction problem of finding the meeting point of two straight-lines when a certain condition is fulfilled. This is indeed a construction problem, hence the concept of postulate as presented by Kant fits here perfectly. It becomes, for three straight-lines AB, CD and EF, such that EF cuts AB in the point G and cuts CD in the point H: (prove $BGH + DHG < 2R\angle$) $\Vdash$ (find the intersection point of AB and

[20] The greek word *êitêsthô* is the first occurring in Postulate I.1. It is an imoerative verb and it means "ask for", "demand", which is exactly what one says when stating a problem.

CD, by extending them from B and D).[21] This practical principle effects a selection of which surfaces are to be considered in geometry: only flat surfaces.

Finally, Postulate I.4 — *that all right angles are equal to one another* — is a declarative statement. It does not look like a practical principle. It can be seen as an answer to the problem: when two distinct right angles are equal? We assume it as a solved proving problem differing from Common Notions, since it is specific, not generic. It states that any angles falling under Definition I.10[22] are equal angles. Depending on the surface being considered, this is not a trivial matter. It holds only for homogeneous curved surfaces, not for conic surfaces, for example. It is rendered, for any two angles ABC and DEF: ⊩ (prove if (ABC is a $R\angle$ and DEF is a $R\angle$), then $ABC = DEF$).

The proofs of the *Propositiones* achieve a problem reduction: from a more complex problem to less complex problems until the bottom — the Postulated problems and the Common Notions, in principle. Postulates and Common Notions are supposed to be solvable. And, of course, it does not make sense to ask a correctness-proof for both. Rephrasing Kolmogorov (1932, p. 151), but avoiding any commitment with the existence of solutions, and keeping in mind that problem reduction involves a specific relation between solutions already pointed before:[23]

> If we can reduce ~~the solution of~~ problem b to ~~the solution of~~ problem a and ~~the solution of~~ problem c to ~~the solution of~~ problem b, then ~~the solution of~~ c can also be reduced to ~~the solution of~~ a.

3.6 Towards a general theory of problems, or how to solve it?

Kolmogorov's point of view plunged intuitionistic logic into a general theory of problems. Effectively, there are reasons to think that the schemes of solutions for geometrical problems offer a model of how to approach logic in terms of problems and their solutions, following the structure of Euclid's *Elements*.

Concerning a general theory of problems, Veloso (1984, p. 29) points out that problem decomposition is one of the main strategies for solving problems:[24]

> A common approach to solving a problem is to partition the problem into smaller parts, find the solutions for the parts, and then combine the solutions for the parts into a solution for the whole.

We claim that logical constants give the patterns according to which problems should be decomposed and solved. From our perspective, it was Kolmogorov who

---

[21] It must be reminded that Kolmogorov's problem interpretation is silent about conditional construction problems with propositional conditions, which would be our preferred solution for this case. See Section 4.2.

[22] Right-angles are defined as the angles formed by the incidence of two straight lines making equal adjacent angles.

[23] That is, when problem b reduces to problem a, any solution of a gives rise to a solution of b.

[24] The original is in Aho, Hopcroft, and Ullman (1975, p. 60).

first devised it in his problem interpretation. Together, reduction and decomposition are the two main general strategies for solving problems, and they form the core of Reduction Semantics. Decomposition and reduction are going to be illustrated next.

Let's consider now how the resolution of *Propositio* I.1 takes place.

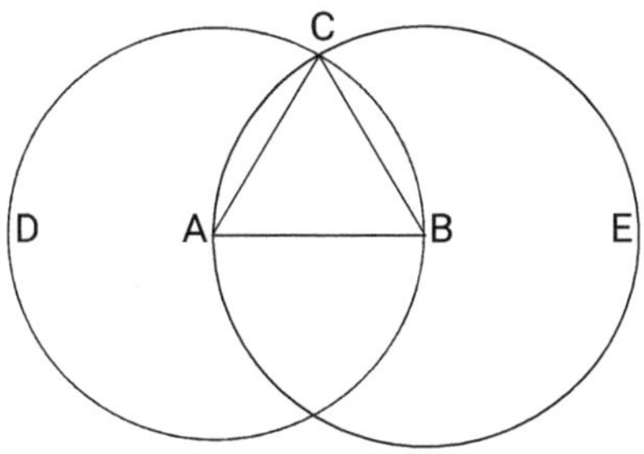

Fig. 3 Prop. I.1

The resolution of I.1 contains a construction solution part — to build the equilateral triangle over straight line AB — which is solved first; and it contains in the sequence a proving solution part — to prove for the three sides of the triangle that $AB = BC = AC$.

The construction part is done in two steps. First, the two circles of radius AB and BA are drawn/produced according to Postulate I.3, and they could be done in any order. Second, and only after the first part is done, the straight lines AC and BC are drawn/produced according to Postulate I.1, and they also could be done in any order. Drawing a circle and drawing a straight line are two problems that we assume to be solvable, since they are postulated. Conjunction ($\wedge$) is a natural way of composing actions when the order is irrelevant: (draw circle *BCD* of radius AB with center A) $\wedge$ (draw circle *ACE* of radius BA with center B). Conjunction is used again in the composition of the subsequent actions: (draw straight line AC) $\wedge$ (draw straight line BC). But, the next step requires the distinction of a before and an after, which we are going to represent by "$\rightsquigarrow$", a before-after conjunction:[25]

(‡) $\Vdash$ {(draw circle *BCD* of radius AB with center A) $\wedge$ (draw circle *ACE* of radius BA with center B)} $\rightsquigarrow$ {(draw straight line AC) $\wedge$ (draw straight line BC)}.[26]

At this point the construction problem is solved, a triangle was drawn/produced. What is the solution to the problem of drawing/producing triangle ABC? It is the composite action described in (‡). Since each action in (‡) is a problem considered to be solvable, then the whole complex structured action (‡) also shows the problem of drawing/producing an equilateral triangle to be solvable since Reduction Semantics explains how to understand each logical constant employed. Next, the solution of *Propositio* I.1 requires a verification showing that the triangle ABC is equilateral.

[25] The before-after conjunction is expressed in programming languages by the semicolon ";". But, as a logical symbol it might cause some confusion, reason why we do not use it.

[26] It can be read as follow: {(draw circle *BCD* of radius AB with center A) and (draw circle *ACE* of radius BA with center B)} and after {(draw straight line AC) and (draw straight line BC)}.

Proving that the triangle obtained is equilateral is the proving problem part. For proving problems, Postulates I.4 and I.5, the Common Notions, and the Definitions, all together, contribute to establish the basis of what is considered solvable. Hence the problem (prove $AB = BC = AC$) is decomposed (and solved) into:

(#) $\Vdash$ ((prove $AB = BC$) $\wedge$ (prove $AB = AC$)) $\rightsquigarrow$ (prove $BC = AC$)

Definition I.15 about circles establishes $\Vdash$ (prove $AB = BC$) as positively solvable, since both lines are radius of the same circle *ACE*. A similar reasoning holds for $\Vdash$ (prove $AB = AC$). In third place, from the two precedent equalities by using Common Notion I.1, $\Vdash$ (prove $BC = AC$) is established as positively solvable. That is, (#) makes explicit the decomposition of the problem (prove $AB = BC = AC$) into subproblems until reaching solvable subproblems. Additionally, the expression gives the trace of the proving procedure.[27] The expression (#) is the solution to the problem (prove $AB = BC = AC$) which turns now to be considered positively solvable.

The whole final expression with the solution of *Propositio* I.1 is:

$\Vdash$ {{(draw circle of radius AB with center A) $\wedge$ (draw circle of radius BA with center B)} $\rightsquigarrow$ {(draw straight line AC) $\wedge$ (draw straight line BC)}} $\rightsquigarrow$ {((prove $AB = BC$) $\wedge$ (prove $AB = AC$)) $\rightsquigarrow$ (prove $BC = AC$)}.

That is, the solution of a problem can be obtained by decomposition of a problem into subproblems until arriving to solvable subproblems and then composition of these subproblems already considered solved.[28]

The before-after conjunction is usually expressed in mathematics by means of function composition. If $f(x)$ and $g(x)$ are two functions, then $fog(x)$ represents the ordered application $f(g(x))$. Nonetheless, the before-after conjunction can be used when functions cannot be used. In the case of geometry, it is doubtful that drawing/extending a straight-line should be considered a function depending on previous determined parametric points, since the act of drawing/extending a straight-line might create, so to say, its starting and stopping points. That is, the actions corresponding to Postulate I.1 and I.2 cannot be identified with a function, and much less with a constructive function. As remarked, I.2 does not even mention an stopping point. Although powerful, the functional interpretation does not seem to be the best tool for effecting an epistemological analysis of ancient geometry.

After the example of a construction problem, let's examine a theorem *Propositio*, the proving problem I.15: *if two straight-lines cut one another then they make the vertically opposite angles equal to one another.* The proof starts by the following sentence: ($) *Let straight lines* AB *and* CD *cut one another at the point* E. It is

---

[27] All inferential relations can be read from this trace. Of course, this is not a natural deduction tree, but such a tree can be build on the basis of the trace given.

[28] One item might seem to be missing in the given solution: the action of determining the intersection point C. Notice that the Euclidean text contains a reference to point C in both circles drawn: circle *BCD* and circle *ACE*, thus apparently bypassing the issue, but being undeterministic about which intersection to consider, reminding that there are two of them. What is seen in the final solution is a reflection of the text in the *Elements*. If the action of determining the intersection point were added to the solution, we would be falsifying the original text.

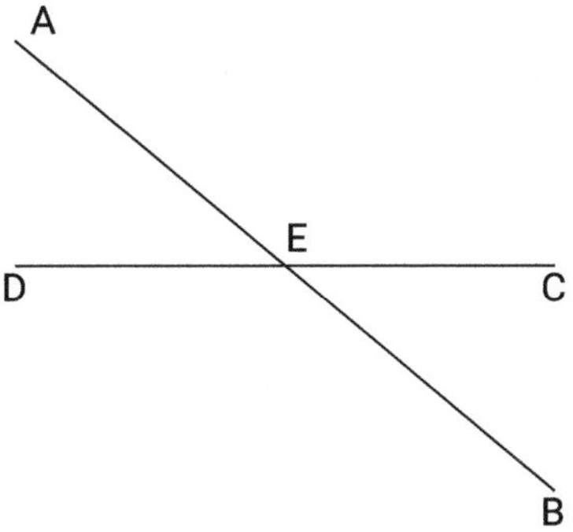

Fig. 4 Prop. I.15

required to show that angle *AEC* is equal to *DEB* (and *CEB* to *AED*). It is a conditional proving problem whose condition is: *if two straight-lines cut one another* .... The problem has to be formulated as: (deduce $AEC = DEB$ from [the supposition that] AE cuts CD). It cannot be formulated as (prove AE cuts CD) $\Vdash$ (prove $AEC = DEB$) because the antecedent problem does not seem to match the sentence (\$), "let ..." does not mean "suppose it has been proved that AE cuts CD". We come back to this point in Section 4.2 discussing the notions of hypothesis and assumption.

Suppose that the straight-line AE cuts CD. Hence, AE stands on the straight-line CD, making angles CEA and *AED*. The sum of the angles CEA and *AED* is thus equal to two right-angles according to *Propositio* I.13. That is, the problem $\Vdash$ (deduce $CEA + AED = 2R\angle$ from AE cuts CD) is solved. Also DE stands on *AB*, hence $\Vdash$ (deduce $AED + DEB = 2R\angle$ from AE cuts CD) is also solved by I.13 too. Next, by the Common Notion I.1, $\Vdash$ (deduce $CEA + AED = AED + DEB$ from AE cuts CD) is then solved. And next, subtracting *AED* from both sides, by Common Notion I.3, $\Vdash$ (deduce $CEA = DEB$ from AE cuts CD) is then solved. That is, $\Vdash$ (prove if AE cuts CD, then $CEA = DEB$). End of the proof. In this case, no construction is added to the given figure, so there is no *kataskeue*. The *apodeixis* covers the whole resolution of the proving problem and here is the final solution:

$\Vdash$ {{{(deduce $CEA+AED = 2R\angle$ from AE cuts CD) $\wedge$ (deduce $AED+DEB = 2R\angle$ from AE cuts CD)} $\rightsquigarrow$ (deduce $CEA + AED = AED + DEB$ from AE cuts CD)} $\rightsquigarrow$ (deduce $CEA = DEB$ from AE cuts CD)} $\wedge$ {... $\rightsquigarrow$ (deduce $CEB = DEA$ from AE cuts CD)}

It must be kept in mind that this expression is a rough statement of the solution in terms of actions. For example, the action of subtraction is not made explicit, although its use has been made clear. Also the deduction relating the conditions of I.13 and I.15 are not represented above. Solutions are communicated and as such the level of detail is variable and depends on the expertise expected from the audience, if they do not suffer of syntaxism.

As observed, *Propositio* I.15 starts by: *if two straight-lines cut one another* .... This is the statement of a condition that selects the situation to be considered and, of course, not every situation lies under this condition. Parallel lines do not, for example. But other configurations like that where with lines AB and FC in Figure 5 do not either, even if they are not parallels.

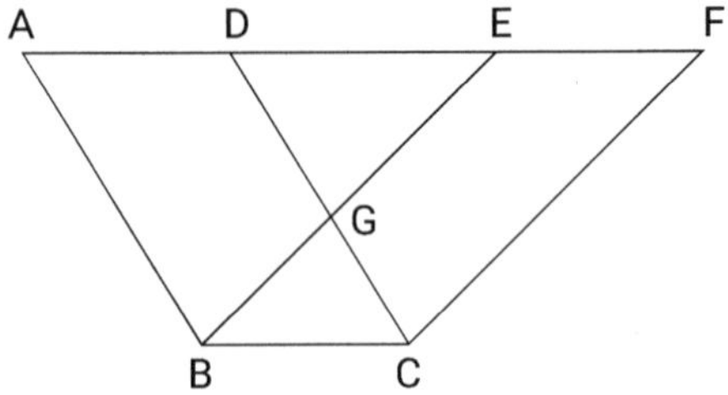

Fig. 5 Prop. I.35

Next, the condition is related to another condition, the condition in the formulation of *Propositio* I.13: *if a straight-line stood on a(nother) straight-line, ...*. In other words, any situation satisfying the identifying condition of *Propositio* I.15 has to be already a situation satisfying the identifying condition of *Propositio* I.13.

Now, some words about the before-after conjunction constant are required once both solutions in the two examples above employed it.

4 Logic and problems

4.1 Logical constants and algorithms

The question of what is a logical constant is deeply difficult and interesting. Kolmogorov's paper touches the heart of this question since he states that his calculus of problems should substitute intuitionistic logic. He also expressed he had hoped that the schemes of solutions of problems would become an important part in courses of logic.

There are distinct competing theories of what is logic. Many times, a decision of how to interpret logic already involves a decision of what is a logical constant. Logic has been assumed to be the science of logical truths.[29] However, it also has been assumed to be a science of formal deductions.[30] A third possibility is to conceive it as the central part of a general theory of problems or of a general theory on problems resolution. We think that this alternative differs from the previous two and, at the same time, it extends them once they are limited to proving problems. The third alternative deserves investigation, since the problem approach allows an homogeneous epistemological analysis of items in the history of mathematics the way they were

[29] See Gómez-Torrente (2019).

[30] See Došen (1989, p. 364).

formulated, without twists. After all, the basic sources for logic theorizing are the argumentative practices historically found mainly in mathematics.

Some might wonder if the before-after conjunction could not be defined by the other intuitionist constants. The formula $a \wedge (a \supset b)$ seems to be the candidate that comes to mind. That it is not can be realized when asking which of the two problems a or b should be taken as the "first" problem to be solved. This formula does not distinguish a first or a second element. Indeed, a problem like $a \wedge (a \supset b)$ can be settled by first solving problem b and next solving problem a, but this is unfaithful to the intuitive meaning of the before-after conjunction.

Examples of the use of before-after conjunction were given above. In the solution of *Propositio* I.1, the production of straight-lines AC and BC depends on having beforehand circles ACE and BCD as well as their intersection points. If the circles were not produced, there would be no intersection point C for drawing the two straight-lines. Point C is not determined by a constructive function over circles ACE and BCD as parameters. Only after the circles are produced, there will be two distinct intersection points. Any one can be picked as point C for drawing the triangle, it does not matter which. From this perspective, C is just a stopping point common to both circumferences produced in accordance with Postulate I.3.

The semantical clauses for the before-after conjunction are as follows:

In the repertoire:

($\rightsquigarrow^l$): $\Gamma, c \rightsquigarrow d \Vdash e \cong$ given any problem a ((first $a \Vdash c$, and after $a \Vdash d$) $\Rightarrow$ $\Gamma, a \Vdash e$)).

In the focus:

($\rightsquigarrow^r$): $\Gamma \Vdash c \rightsquigarrow d \cong$ (first $\Gamma \Vdash c$, and after $\Gamma \Vdash d$).

The before-after conjunction is not among the usual intuitionistic logical constants, neither among the classical. Function composition has many times been used for obtaining the before-after effect. We repeatedly pointed above why we prefer to consider it a logical constant for problems. Some basic actions do not fit in the role of functions.

More important, with the before-after conjunction the logical relations in the solution of a proving problem can be recovered since this kind of conjunction clearly distinguishes the temporal order in which the actions were resolved or established, thus allowing one to obtain the trace of the proof of an assertion. The following elimination rules partially state the inferential behaviour of the before-after conjunction: $c \rightsquigarrow d \vdash c$ and $c, c \rightsquigarrow d \vdash d$. Then, clearly, $c \rightsquigarrow d \vdash c \wedge (c \supset d)$. But the reverse is not correct as we already argued above.

The logicality of the before-after conjunction is a matter for debate. However, if Kolmogorov's problem interpretation of intuitionistic logic is regarded as a defensible position, then the absence of the before-after conjunction in the intuitionist description of the propositional logical constants would mean that this description is incomplete.

The before-after conjunction is at work everywhere in Euclidean Geometry. It corresponds to a sequential decomposition of a problem in view of producing its solution, one of the basic cases covered by the eternal strategist's *dictum*: divide and conquer. And since ancient geometry is a reference source for the concepts of proof and inference, any attempt at logic theorizing should keep a minimal *adequatio* with respect to what one can find in ancient texts.

Geometric proofs are the communication of how to solve certain problems, of which there are two important kinds: construction problems and proving problems. In general, the *kataskeue* contains the solution of a construction problem; the *apoidexis* contains the solution of a proving problem. But they are often mixed. In both cases what is communicated is a recipe, an algorithm. The communication of a geometric proof is then basically the communication of an algorithm. The structure of such communications has to be enough for understanding and undertaking the acts promoting evidence for a *Propositio*. Observe that the communication of an algorithm may sometimes hide evident or repetitive steps in order to make the communication shorter.

Some parts of the *Propositio* prepare or summarize the communication of the algorithm. The non linguisitc part is the accompanying diagram. It is not part of the algorithm. Diagrams exemplify the data that serve as the object modified, produced or examined in the algorithms being communicated, according to de Campos Sanz (2021). Usually, diagrams represent the final state of the drawing/production and in most cases it does not contain the traces of the actions comprised in the recipe as pointed by Sidoli (2018).

We next consider one particular item found on this argumentative practices: the use of hypotheses/suppositions in the solution/proof communications occurring in Euclid's *Propositiones*.

4.2 Hypotheses, assumptions and problems in geometric proofs

In the history of mathematics, problems seem to be by large the main theoretical concern, being later displaced by the assertion-theoremhood perspective. Since XIXth logic has become also driven by the assertion-theoremhood perspective. A valid inference is defined as a necessary relation holding among premiss-assertions and a conclusion-assertion. Nonetheless, when suppositions enter the picture things become delicate. Assumptions are sometimes conceived as a special kind of assertion for which we lack a proof and whose provability is then supposed.

In the context of construction problems, there is an exact correlate of assumptions conceived in the way just described. *Propositio* I.22 is a construction problem whose solution depends on supposing that a proving problem has been solved: *to construct a triangle from three straight-lines which are equal to three given. It is necessary for two taken together in any [way] to be greater than the remaining, . . .* Thus, not any trio of straight-lines fits the restriction. What is here being supposed is that a proving problem has been solved: the addition of any two straight-lines must be greater than the third remaining. Indeed, the solution described in the *kataskeue* of I.22 might not work properly if the straight-lines being used were not in the mentioned relation. Thus, here is an example of an assumption in the mould described, that is, the supposition of having a solution to a proving problem as the starting point for the resolution of I.22.

Nonetheless, there are many other cases in which a certain condition is supposed to hold without supposing that a proving problem has been solved. Take *Propositio* I.15: *if two straight-lines cut one another then they make the vertically opposite angles equal to one another.* The resolution of this proving problem starts with a simpler supposition: that they cut one another. It seems wrong to understand it as an assumption. The condition of cutting one another is the mere description of a possible situation. This situation could be produced in a myriad of different ways, it does not matter which. It even might be the case that we do not know how the straight-lines were produced. What matters is if we take the drawings as straight-lines and we suppose them to fall under the condition described, or not. The diagram in Figure 4 that comes together with the *Propositio* is just an exemplification of such a situation and, of course, the reader should suppose that it falls under the condition stated. In this sense, it also does not matter if the straight-lines drawn are not perfect straight-lines. It is enough to suppose that they are, in order to understand the proof establishing the equality of angles.

The solution for I.15 starts by noticing that the situation of cutting one another falls under the condition of *Propositio* I.13 which also describes a situation: *if a straight-line stood on a(nother) straight-line makes angles, it will certainly either make two right-angles, or (angles whose sum is) equal to two right-angles.* In both cases, no assertion is being made in the condition. The situations are merely considered and one is related to the other. In I.15, under the supposition that the situation is of cutting one another, it follows that it should also be a situation of standing on another, thus opening the way for using I.13 in the solution for I.15, as already described previously. Observe that there is no Definition, Common Notion, or Postulate, guaranteeing that any situation of cutting one another falls under the situation of standing on another. The comprehension of such a link seems to depend on perception and on intelect.

Concerning assumptions and hypotheses, it seems important to adopt a distinction. The concept of *assumption* should be reserved for those cases where either a proving problem is supposed to be solved or a construction problem is supposed to be solved. Assumptions are then hypotheses of a specific nature: those where we suppose the validity or the possession of a solution for a problem, I.22 being an example.

The concept of *hypothesis* is more general than that of assumption. There are cases where a situation is being supposed but this cannot be assimilated to the supposition that a proving problem or a construction problem is solvable. We claim that *Propositio* I.15 illustrates it. Hypotheses in a large sense are the supposition that a given situation falls under certain putative identifying conditions.

There are many examples of hypotheses in Euclid's text. For example, the proof of correctness for the solution of finding the center of the circle in *Propositio* III.1 starts by a counterfactual hypothesis: ... *I say that (point) F is the center of the [circle] ABC. For (if) not then, if possible, let G (be the center of the circle), ...* This hypothesis is the starting point of a *reductio* reasoning. That there should be a center *G* different of *F* in the circle just describes a situation that has been deduced from the counterfactual hypothesis and the fact that any circle has a center. The hypothesis does not seem an assumption, since this would be tantamount to suppose the possession of a construction showing that F is not the center of the circle or, at least, showing that a different G is the center of the circle. But what sense has such an assumption since the solution to the construction problem has just determined F as the center? The correctness-proof for F being the center seems to be in fact a proof of the statement that no other point inside the circle can be the center, established by a *reductio* proof.

Most of the hypotheses in Euclidean Geometry seem to be of this more general kind, not assumptions. If this is indeed the case, then the proving problem in which they appear involves a conditional proposition. The point certainly deserves a deeper investigation, but we leave it for future work.

5 Conclusion

Contrary to Kolmogorov's expectation, the calculus of problems has not become a substantial part of contemporary logic courses. Neither his problem interpretation has become widespreadly known in the community. If the above appreciation of both subjects is faithful, then we have reasons to regret that his expectations were deceived. Anyway, his work has been seminal for the development of Intuitionistic Type Theory, as stressed in Coquand (2007).

The above investigation examined the problem interpretation of intuitionistic logic and advanced Reduction Semantics as a way to further elucidate the conceptual structure in this interpretation, with some minor adjustments. Reduction Semantics brings together two main strategies for problem solving, thus giving the basics of general solution schemes. These strategies are reduction among problems and decomposition of problems. The reduction relation has been assumed as the basic semantical relation over which all patterns of problem composition-decomposition are characterized. These patterns happen to be the old well known logical constants, as Kolmogorov seems to have anticipated.

Reduction Semantics was also employed in an epistemological analysis of Euclidean Geometry. Besides construction problems, *Propositiones* like I.15 and many others are seen as proving problems. Again, Kolmogorov seems to have fully realized it, opening the door this way for considering all Euclid's *Propositiones* as problems by unifying construction problems and proving problems in the same approach.

Two points deserve to be stressed from the above investigation. First, intuitionistic logic becomes the core of a general theory of problems if we accept Kolmogorov's thesis that this logic is a calculus of problems. Second, this perspective leaves the way open to extend the usual set of intuitionist logical constants. The before-after conjunction was pointed as an example of a constant that cannot be defined employing the other constants. In the solution of any proving problem it allows one to keep trace of the reasoning.

What has been until now called proofs in Euclid's *Elements* emerge as the communications of a problem resolution. These must be seen as the transmission of an algorithm since virtually all *Propositiones* may be treated as problems, in view of Section 4.1. Diagrams just accompany the communication of the algorithms. They exemplify the objects being dealt with in the algorithm, similar to the role of syntactical data in the tape of a Turing machine, according to de Campos Sanz (2021).

Finally, the epistemological analysis of Euclidean Geometry has noticed that hypotheses are used in distinct roles inside *Propositiones*, the most general being that of identifying criteria for situations. They also appear as the supposition of having a solution. For this last case the word "assumption" has been reserved in order to distinguish it from other uses. These other uses include the case of counterfactual hypotheses, like that in the correctness proof of III.1. The existence of counterfactual hypotheses in Euclid's *Elements* highlights a noticeable fact. Historically, the language of mathematics is more rich than the current regimented languages with which logicians are used to work.

Acknowledgements We thank the anonymous referees and H. Oliveira for their valuable suggestions.

References

Aho, A. V., J. E. Hopcroft, and J. D. Ullman (1975). *The Design and Analysis of Computer Algorithms*. Reading: Addison-Wesley.

Coquand, T. (2007). Kolmogorov's contribution to intuitionistic logic. In: *Kolmogorov's Heritage in Mathematics*. Ed. by É. Charpentier, N. K. Nikolski, and A. Lesne. Heidelberg: Springer, 19–40.

de Campos Sanz, W. (2012). Kolmogorov e a lógica de problemas I. *Notae Philosophicae Scientiae Formalis* 1, 184–197.

– (2019). Hypo: a simple constructive semantics for intuitionistic sentential logic; soundness and completeness. In: *Proof-Theoretic Semantics: Assessment and Future Perspectives. Proceedings of the Third Tübingen Conference on Proof-Theoretic Semantics, 27–30 March 2019*, 153–178. URL: http://dx.doi.org/10.15496/publikation-35319.

– (2021). Euclidean machines: general theory of problems. In: *A Question is More Illuminating than an Answer. A Festschrift for Paolo A. S. Veloso*. Ed. by E. Haeusler, L. Pereira, and J. Viana. London: College.

Došen, K. (1989). Logical constants as punctuation marks. *Notre Dame Journal of Formal Logic* 30, 362–381.

Fitzpatrick, R. (2008). *Euclid's Elements of Geometry*. The Greek text of J. L. Heiberg (1883–1885) from *Euclidis Elementa*, edidit et Latine interpretatus est I. L. Heiberg, in aedibus B.G. Teubneri, 1883–1885; edited, and provided with a modern English translation, by Richard Fitzpatrick.

Gómez-Torrente, M. (2019). Logical Truth. In: *The Stanford Encyclopedia of Philosophy*. Ed. by E. N. Zalta. Spring 2019 edition. Metaphysics Research Lab, Stanford University.

Heyting, A. (1930). Die formalen Regeln der intuitionistischen Logik. In: *Sitzungsberichte der Preussischen Akademie der Wissenschaften. Physikalisch-Mathematische Klasse*, 42–56.

– (1956). *Intuitionism. An Introduction*. Amsterdam: North-Holland.

Hilbert, D. (1899). *The Foundations of Geometry*. La Salle, Illinois: Open Court 1950 english translation, 2ed.

Kant, I. (1992). The Hechsel Logic (in part). In: *Lectures on Logic*. Ed. by J. Young. The Cambridge Edition of the Works of Immanuel Kant. Cambridge University Press.

Kant, I. (1800/1904). *Logik*. Ed. by G. B. Jäsche. English translation by R. S. Hartman and W. Schwartz, *Logic*, Dover, N.Y. 1988. Leipzig: Felix Meiner.

Kolmogorov, A. (1932). Zur Deutung der intuitionistischen Logik. *Mathematische Zeitschrift* 35. Pagination from the English edition in Tikhomirov (1991), 58–65.

Lassalle-Casanave, A. (2019). *Por Construção de Conceitos – Em torno da Filosofia Kantiana da Matemática*. Rio de Janeiro: Puc-Rio.

Naibo, A. (2018). Constructivity and Geometry. In: *From Logic to Practice*. Ed. by G. et alli. Heidelberg: Springer, 123–161.

Sidoli, N. (2018). Use of constructions in Problems and Theorems in Euclid's Elements. *Archive for History of Exact Sciences* 72, 403–452.

Tikhomirov, V. M., ed. (1991). *Selected Works of A. N. Kolmogorov*. Vol. 1. Translated from the Russian by V. M. Volosov. Dordrecht: Kluwer.

Veloso, P. (1984). Aspectos de uma teoria geral de problemas. *Cadernos de História e Filosofia da Ciência* 7, 21–42.

von Plato, J. and P. Mäenpää (1990). The logic of Euclidean Construction Procedures. *Acta Phenica* 49, 275–293.

Disjunctive Syllogism without *Ex falso*

Luiz Carlos Pereira, Edward Hermann Haeusler and Victor Nascimento

Abstract The relation between *ex falso* and *disjunctive syllogism*, or even the justification of *ex falso* based on disjunctive syllogism, is an old topic in the history of logic. This old topic reappears in contemporary logic since the introduction of *minimal logic* by Johansson. The disjunctive syllogism seems to be part of our general non-problematic inferential practices and superficially it does not seem to be related to or to depend on our acceptance of the frequently disputable *ex falso* rule. We know that the acceptance of the *ex falso* is a sufficient condition for the acceptance of the disjunctive syllogism, but the interesting question is: is the *ex falso* a necessary condition for the acceptance of the disjunctive syllogism? The aim of the present paper is to discuss some possible ways to define systems that combines the preservation of the disjunctive syllogism with the rejection of the *ex falso*.

1 Introduction

The relation between *ex falso* and *disjunctive syllogism*, or even the justification of *ex falso* based on disjunctive syllogism, is an old topic in the History of Logic. This old topic reappears in contemporary logic since the introduction of *minimal logic* by Johansson. The disjunctive syllogism seems to be part of our general non-problematic inferential practices and superficially it does not seem to be related to or to depend on our acceptance of the *ex falso* rule; on the other hand, the general validity of the *ex falso* has been subjected to dispute. We know that the acceptance of the *ex falso* is

Luiz Carlos Pereira
Filosofia, PUC-Rio/UERJ/CNPq, Rio de Janeiro, Brazil, e-mail: luiz@inf.puc-rio.br

Edward Hermann Haeusler
Informatica, PUC-Rio, Rio de Janeiro, Brazil, e-mail: hermann@inf.puc-rio.br

Victor Nascimento
Filosofia, UERJ, Rio de Janeiro, Brazil, e-mail: victorluisbn@gmail.com

T. Piecha and K. F. Wehmeier (eds.), *Peter Schroeder-Heister on Proof-Theoretic Semantics*, Outstanding Contributions to Logic 29,
https://doi.org/10.1007/978-3-031-50981-0_6

a sufficient condition for the acceptance of the disjunctive syllogism, as the following simple derivation in an intuitionistic natural deduction system shows:

$$\dfrac{(A \vee B) \qquad [A]^1 \qquad \dfrac{\dfrac{[B]^2 \qquad \neg B}{\bot}\,{\neg}\mathrm{E}}{A}\,\bot_i}{A}\,\vee_{\mathrm{E}}\ 1,2$$

The interesting question is: is the *ex falso* a necessary condition for the acceptance of the disjunctive syllogism?

As it was said, the relation between *ex falso* and the disjunctive syllogism has a long history. A form of the disjunctive syllogism appears in Stoic Logic[1] as the fifth type of *undemonstrated* argument, "an argument which, having an exclusive disjunction and the contradictory of one of the disjuncts as premises, infers the other disjunct as its conclusion"[2]. Diogenes Laertius, in *Lives of Eminent Philosophers* (VII, 49), gives the following example:

Either it is day or it is night.
It is not night.
Therefore, it is day.

A medieval argument from the 12th century, attributed to William of Soissons, shows how to derive the *ex falso* from the disjunctive syllogism and other "non-problematic" rules. The argument in natural language is[3]:

> I wonder that certain men oppose the thesis that from a per se impossibility anything whatsoever follows . . .. For doesn't it follow that if Socrates is a man and not a man, then Socrates is a man, but if Socrates is a man, then Socrates is man or a stone. Therefore, if Socrates is a man and not a man, then Socrates is a man or a stone. But if Socrates is a man and Socrates is not a man, then Socrates is not a man. Therefore, if Socrates is a man and Socrates is not a man, then Socrates is a stone.

We can reconstruct this argument axiomatically[4] as:

1. $(A \wedge \neg A) \rightarrow A$ Conditional Simplification,
2. $(A \rightarrow (A \vee B))$ Conditional Addition,
3. $((A \wedge \neg A) \rightarrow (A \vee B))$ 1, 2, Transitivity,
4. $((A \wedge \neg A) \rightarrow \neg A)$ Conditional Simplification,
5. $((A \wedge \neg A) \rightarrow ((A \vee B) \wedge \neg A))$ 3, 4, Conditional Adjunction,
6. $(((A \vee B) \wedge \neg A) \rightarrow B)$ Conditional Disjunctive Syllogism,
7. $(A \wedge \neg A) \rightarrow B)$ 5, 6, Transitivity.

[1] It is worth noticing that, while stoic disjunction is exclusive, all the new systems examined in this paper use inclusive disjunctions. Even though changes are promoted in elimination rules, we are always allowed to use the standard rules of conjunction elimination and disjunction introduction to show that $A \wedge B \vdash A \vee B$.

[2] Benson Mates (1953), p. 73.

[3] See Martin (1986), p. 571.

[4] We prefer the axiomatic style here as it looks closer to the text.

This argument is considered a precursor to the argument known as *Lewis' argument*[5]:

(1) Assume $p \;\sim p$.

(2) $(1) ⥽ p$

If p is true and p is false, then p is true.

(3) $(1) ⥽ \sim p$

If p is true and p is false, then p is false.

(4) $(2) \,.\, ⥽ \,.\, p \vee q$

If, by (2), p is true, then at least one of the two, p and q, is true.

$(3) \,.\, (4)\!: ⥽ \,.\, q$

If, by (3), p is false, and by (4), at least one of the two, p and q, is true; then q must be true.

We can also easily show that the disjunctive syllogism axiom implies the *ex falso* theorem:

$$\dfrac{\dfrac{\dfrac{\dfrac{\dfrac{[B]^1}{(A \vee B)} \quad [\neg B]^2}{((A \vee B) \wedge \neg B)} \qquad (((A \vee B) \wedge \neg B) \to A)}{A}}{(B \to A)}\ 1}{(\neg B \to (B \to A))}\ 2$$

And the same result can be obtained if we add the *disjunctive-syllogism rule* (DS)

$$\dfrac{(A \vee B) \qquad \neg A}{B}\ \mathrm{DS}$$

to minimal logic[6]

$$\dfrac{\dfrac{\dfrac{\dfrac{[B]^1}{(A \vee B)} \quad [\neg B]^2}{A}\ \mathrm{DS}}{(B \to A)}\ {\to_{\mathrm{I}}}\ 1}{(\neg B \to (B \to A))}\ {\to_{\mathrm{I}}}\ 2$$

which allows a simple reconstruction of Soissons' argument in natural deduction:

---

[5] See Lewis and Langford (1959, p. 250).

[6] Rodolfo Ertola-Biraben called our attention to the fact that it is enough to add the following particular case $\mathrm{DS}_{\neg}$ of DS

$$\dfrac{(A \vee \neg A) \qquad \neg\neg A}{A}\ \mathrm{DS}_{\neg}$$

in order to obtain the full power of *ex-falso*.

$$\dfrac{\dfrac{\dfrac{(A \wedge \neg A)}{A}}{(A \vee B)} \qquad \dfrac{(A \wedge \neg A)}{\neg A}}{B}\,\text{DS}$$

It is interesting to observe that all these arguments and proofs are *not normal* in the proof-theoretical sense, as some occurrences of disjunctive formulas are both the conclusion of an introduction rule and the *major premise* of an application of the disjunctive syllogism, which has the shape of an elimination rule.

But are we really committed to the *ex falso* if we accept the disjunctive syllogism? Is it really necessary to resort to the *ex falso* in order to justify the *disjunctive syllogism*? Could we not try some sort of *admissibility argument* to justify the disjunctive syllogism?

2 An admissibility argument in minimal logic

An *admissibility* strategy was considered by Tim van der Molen in the paper "The Johansson/Heyting letters and the birth of minimal logic". In the very beginning of the paper we find the following interesting passage:

> The provability of Formula 4.41 $[(((A \wedge \neg A) \vee B) \rightarrow B)]$ in minimal logic is a desideratum because it stems from the disjunction property. The disjunction property is a property shared by all the usual intuitionistic formal systems. It states that if we can produce a proof of $(A \vee B)$, then we can also produce a proof of A or a proof of B. So, if $((A \wedge \neg A) \vee B)$ (the antecedent of 4.41) has been proved, then, by the disjunction property, $(A \wedge \neg A)$ is provable or B is. In a consistent system like minimal logic $(A \wedge \neg A)$ is not provable. Therefore, B (the consequent of 4.41) must be provable. This indicates that Formula 4.41 should hold in minimal logic. (van der Molen, 2016, p. 2)

The argument used by van der Molen has the form of an *admissibility argument*: in order to show that the rule

$$\frac{A_1 \quad \ldots \quad A_n}{B}$$

is admissible, we show that if $\vdash A_1, \ldots, \vdash A_n$, then $\vdash B$. If we try to apply this kind of *admissibility argument*[7] to the disjunctive syllogism we obtain:

[7] An alternative admissibility argument could be obtained by means of the so-called Dummett's fundamental assumption, according to which every proof (i.e., closed derivation) in intuitionistic logic — and *a fortiori* in minimal logic — can be reduced to a canonical proof (i.e., a closed derivation using an introduction rule in the last step). Both the disjunctive property and the consistency of minimal logic are immediate corollaries of applying the fundamental assumptions to $\vdash A \vee B$ and $\vdash \bot$ (due to the shape of the introduction rule for disjunction and to the absence of introduction rules for $\bot$). Then, since we must have either a proof of A or of B and (by assumption) we have a proof of $\neg A$, the fact that a proof of A could then lead us to a proof of $\bot$ and that it is impossible to obtain a proof of $\bot$ allows us to use a meta-application of the disjunctive syllogism and conclude $\vdash B$. (We would like to thank one of the anonymous reviewers for drawing our attention to this formulation of the admissibility argument).

1. Assume $\vdash (A \vee B)$ and $\vdash \neg A$;
2. By the disjunctive property, we have that $\vdash A$ or $\vdash B$;
3. But then, assuming that minimal logic is consistent, $\nvdash A$;
4. Thus, $\vdash B$.

Could then an intuitionist accept the disjunctive syllogism without accepting the general validity of *ex falso*? What would be Brouwer's own position concerning the *ex falso* and the disjunctive syllogism? According to van Atten, Brouwer would reject the irrestricted validity of the *ex falso*:

> In his dissertation from 1907, Brouwer gave an account of the hypothetical judgement that served him all his life. On that account, hypothetical judgements may in certain cases have false antecedents, but there is no justification of the general principle Ex Falso Sequitur Quodlibet. Neither is the familiar derivation of Ex Falso using the disjunctive syllogism acceptable on Brouwer's view of logic. A systematic conclusion, then, is that Brouwer's logic is a relevance logic. (van Atten, 2009, p. 123)

But again according to van Atten, a kind of "admissibility argument" in defense of the disjunctive syllogism can be attributed to Brouwer:

> The application of the disjunctive syllogism is not problematic either. For if $A \vee B$ is a description that applies to a mathematical construction, this means that we have a mathematical method that, when carried out, will show that the description A applies, or that the description B applies; a proof of $\neg B$ then simply tells us that the outcome of that method will be a proof of A. But then we also know that we would have obtained A as a description of the mathematical construction in question if no independent proof of $\neg B$ had been available to us. The disjunctive syllogism, then, accompanies the mathematical operation of leaving the construction described by A as is. (van Atten, 2009, p. 124)

From the way the argument is formulated, it is obvious that it seems circular: the last step of the argument is an explicit application of the very rule we are trying to justify, to wit, the *disjunctive syllogism*, even if only a *meta-application*! But is it necessary to understand this *meta*-application of disjunctive syllogism as dependent on a previous acceptance of ex falso? Let us consider the following scenario where we explore a comparison between our argument paths and trails we can follow in a promenade.

3 An informal account

Suppose that John is hiking in a forest and that at some point of the trail he finds a bifurcation point marked $A \vee B$. From this point, John could take the path marked by A or the path marked by B. But assume now that there is a sign (an extra information) $\neg B$ that indicates that the path B will lead to a dead-end (the $\perp$). In this case, the only path open to John is the path marked A. This situation can be graphically represented as:

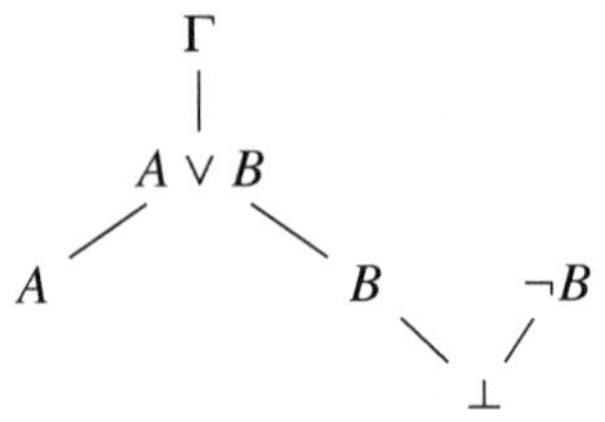

The sign ⊥ here indicates that John *can no longer go* along path B. In a certain sense, this scenario reminds an interesting passage in the third chapter of Brouwer's thesis where he says:

> 'But', the logician will retort, 'it might have happened that in the course of these reasonings a contradiction turned up between the newly deduced relations and those that had been kept in store. This contradiction, to be sure, will be observed as a logical figure, and this observation will be based upon the principium contradictionis.' To this I can reply: 'The words of your mathematical demonstration merely accompany a mathematical construction that is effected without words. At the point where you enounce the *contradiction*, I simply perceive that the construction no longer *goes*, that the required structure cannot be imbedded in the given basic structure.' And when I make this observation, I do no think of a principium contradictionis. (Brouwer, 1975, pp. 72–73)

The idea is that the traveller, as Brouwer says, *simply perceives that he can no longer consider path B, when he finds the contradiction*. It is true that the *ex falso* may be "secretly" used in the process, but our first impression is that this scenario would be acceptable to a *minimal logician*!

Just another point: this scenario also reminds Gentzen's interesting remark on the form of disjunction elimination. Gentzen says:

> In this example the tree form must appear somewhat artificial since it does not bring out the fact that it is after the enunciation of $X \vee (Y \vee Z)$ that we distinguish the cases X, Y and Z. (Gentzen's first example (1.1) on page 79 of Gentzen, 1969)

The point of this remark is that the fact that disjunction elimination has the form it has in the usual natural deduction systems is a kind of *artificial* adaptation to the general tree-form of derivations, but that, truly, the assumptions discharged come after the major premiss, as we do in some multiple-conclusion versions of natural deduction, as the following figure shows:

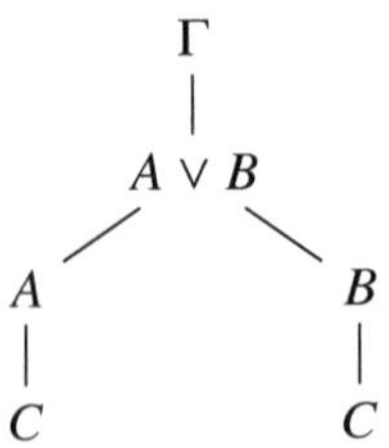

But if we consider a disjunction as a *branching point*, we need some kind of *synchronization mechanism* to bring paths together again!

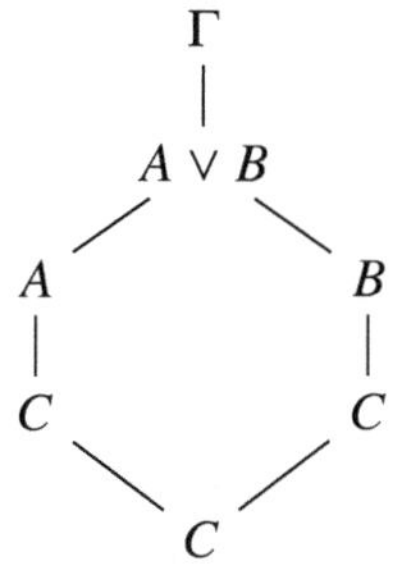

The point now is that the case of the disjunctive syllogism seems to require a new kind of *synchronization mechanism*, as the following figure shows:

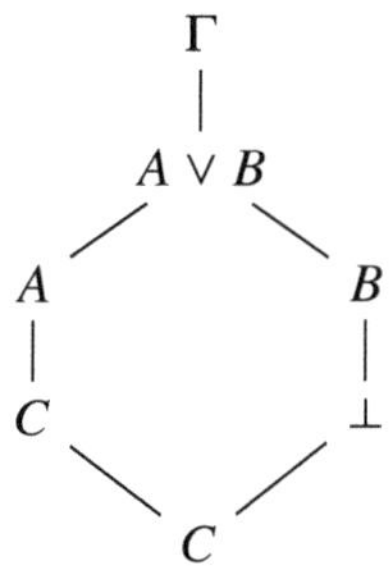

Maybe this is just an extravagant idea, but now what seemed to be an application of a disputable rule, the *ex falso*, is just a kind of *synchronization mechanism* required by disjunction. This idea applied to the disjunctive syllogism yields:

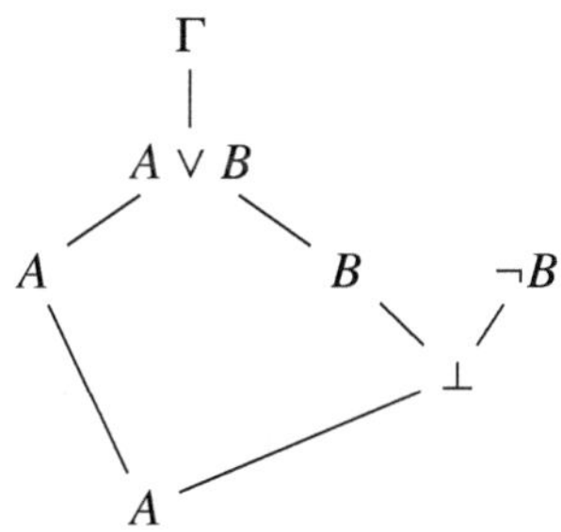

But as we shall see, this representation is not free of problems. In a certain sense, this representation suggests that we could go from the *end-point* $\bot$ to A and this path may hide an application of *ex falso*. Maybe a more faithful representation would be

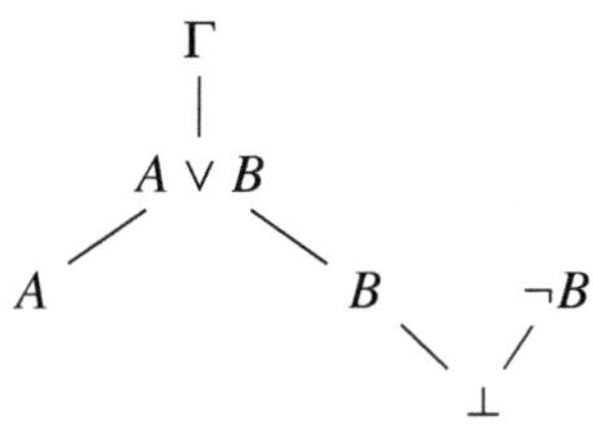

But this representation would inevitably leave us in the realm of multiple-conclusion systems.

4 The system $\mathsf{M}_\vee$

If we do not want to go *multiple-conclusion*, we could try to define a set of new disjunction eliminations ($\vee_\bot$-eliminations) as follows[8]:

1. $\vee_\bot$-elimination-1

$$\dfrac{\begin{matrix}\Gamma \\ \Pi \\ A \vee B\end{matrix} \quad \begin{matrix}[A]^m \\ \Pi_1 \\ C\end{matrix} \quad \begin{matrix}[B]^n \\ \Pi_2 \\ C\end{matrix}}{C}\; m,\, n$$

2. $\vee_\bot$-elimination-2

$$\dfrac{\begin{matrix}\Gamma \\ \Pi \\ A \vee B\end{matrix} \quad \begin{matrix}[A]^m \\ \Pi_1 \\ C\end{matrix} \quad \begin{matrix}[B]^n \\ \Pi_2 \\ \bot\end{matrix}}{C}\; m,\, n$$

3. $\vee_\bot$-elimination-3

$$\dfrac{\begin{matrix}\Gamma \\ \Pi \\ A \vee B\end{matrix} \quad \begin{matrix}[A]^m \\ \Pi_1 \\ \bot\end{matrix} \quad \begin{matrix}[B]^n \\ \Pi_2 \\ C\end{matrix}}{C}\; m,\, n$$

Let us the consider the natural deduction system $\mathsf{M}_\vee$ that is obtained from the propositional part of the Gentzen-Prawitz natural deduction system M for minimal logic through the replacement of the usual disjunction-elimination rule by this new set of disjunction eliminations ($\vee_\bot$-elimination-1, $\vee_\bot$-elimination-2 and $\vee_\bot$-elimination-3). It is clear that the system M is a proper subsystem of the new system $\mathsf{M}_\vee$ and that $\mathsf{M}_\vee$ is a subsystem of the propositional part of the Gentzen-Prawitz natural deduction system I for intuitionistic logic, but is $\mathsf{M}_\vee$ a proper subsystem of I? Is it possible to prove the full power of *ex falso*, to prove $(A \rightarrow (\neg A \rightarrow B)))$, in $\mathsf{M}_\vee$? If it is not possible, then the system $\mathsf{M}_\vee$ could be a good candidate to be an *intermediate* system, lying between the minimal system M and the intuitionistic system I. But consider now the following simple derivations:

$$\dfrac{\dfrac{\dfrac{\dfrac{[A]^2}{(A \vee B)} \quad \dfrac{[A]^3 \quad [\neg A]^1}{\bot} \quad [B]^4}{B}\vee\, 3,\, 4}{(\neg A \rightarrow B)}\rightarrow_{\mathrm{I}} 1}{(A \rightarrow (\neg A \rightarrow B))}\rightarrow_{\mathrm{I}} 2$$

[8] This modification of the disjunction-elimination rule was first proposed by Neil Tennant (1979).

$$\dfrac{\dfrac{\dfrac{\dfrac{[(B \land E)]^4}{B}}{(A \lor B)} \qquad \dfrac{\dfrac{[(A \lor B)]^3 \quad [A]^1 \quad \dfrac{[B]^2 \quad \neg B}{\bot}}{A}{\scriptstyle 1,\,2}}{((A \lor B) \to A)}{\scriptstyle 3}}{A}}{(B \land E) \to A}{\scriptstyle 4}$$

The first derivation is a correct non-normal proof of $(A \to (\neg A \to B))$ in $\mathsf{M}_\lor$ using the detour $A \lor B$; the second example is a correct derivation of $\{\neg B\} \vdash ((B \land E) \to A)$ using the *detour* $((A \lor B) \to A)$. How can we avoid these problematic derivations in $\mathsf{M}_\lor$?

The system $\mathsf{M}_\lor$ is clearly related to the intuitionistic relevant system IR defined by Tennant (1987). He recognizes that without further restrictions, the *ex falso* would be derivable, and he considers the following derivation:

$$\dfrac{\dfrac{A}{(A \lor B)}{\scriptstyle \lor I_1} \qquad \dfrac{[A]^1 \quad \neg A}{\bot}{\scriptstyle \neg E} \qquad [B]^2}{B}{\scriptstyle \lor E\ 1,\,2}$$

As we saw, Brouwer would have nothing against the use of the disjunctive syllogism in the derivation above; his qualms would be related to the *composition of derivations*:

> The problem is rather with the *composition* of these two inferences. The first inference requires that the mathematical construction being described is one for $\bot$; the second that it is one for A. As in general A and $\bot$ will not be equivalent descriptions, there is no general guarantee that when $\bot$ describes a mathematical construction, A describes it as well. This means that there is no guarantee that the linguistic figures in Lewis' argument accompany a mathematical procedure. (van Atten, 2009, p. 124)

What are "these two inferences" to which van Atten is referring? The recognition that there is a *composition problem* and that some restriction on the composition of derivations is needed is exactly what Tennant does: in any application α of an elimination rule, the major premiss of α cannot be the conclusion of an introduction rule. The derivations

$$\dfrac{A}{(A \lor B)}$$

$$\dfrac{(A \lor B) \qquad \dfrac{[A]^1 \quad \neg A}{\bot} \qquad [B]^2}{B}{\scriptstyle 1,\,2}$$

are correct, but the derivation (the result of the *composition*)

$$\dfrac{\dfrac{A}{(A \lor B)} \qquad \dfrac{[A]^1 \quad \neg A}{\bot} \qquad [B]^2}{B}{\scriptstyle 1,\,2}$$

is not.

We can certainly use Tennant's idea (forgetting everything about *relevance*) and impose *normality* by stipulation: only *normal derivations* are accepted as legitimate derivations. Obviously the system $\mathsf{M}_\vee$ is Tennant's system IR without the *relevance-restrictions*. As in the case of IR, we can say that:

1. The propositional part of the natural deduction system M for minimal logic is a subsystem of $\mathsf{M}_\vee$ (M satisfies normalization);
2. The disjunctive syllogism is derivable in $\mathsf{M}_\vee$; and
3. $\mathsf{M}_\vee$ does not have the *ex falso*.

It is true one could say that to impose *normality* is a too high price to pay in order to preserve the disjunctive syllogism without preserving the *ex falso*. In order to have a better understanding of what is happening with *detours* of the form $(A \vee B)$ in the derivation above, let us go back to our initial scenario where John is hiking in a trail. Suppose that John is hiking with a map and that John arrives at the same bifurcation point marked $(A \vee B)$. Let us suppose that path A with some extra information C leads to point E and that path B together with extra-information F leads to a point G, and that from the points E and G we can go to point H. We could try to represent this situation with the following figure:

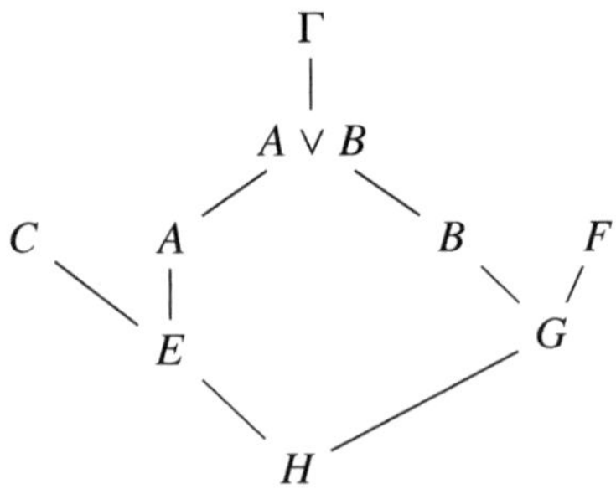

In the case of classical logic, if we are at point E, we can access point C, point F and all points in Γ: no *visibility/accessibility restrictions*. But in case of intuitionistic logic the situation is completely different: after the bifurcation point, *visibility/accessibility* restrictions are required. A more faithful representation of the situation is as follows:

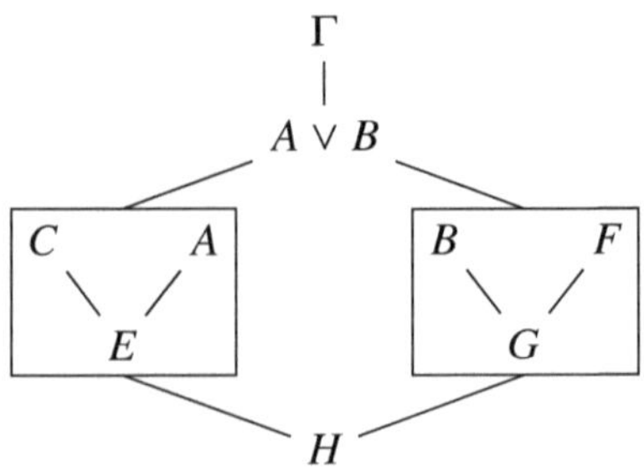

According to this new picture, at point E we have only access to what is inside its box (and the same holds for point G with respect to its box). In the next section we define a new system whose aim is to incorporate these *visibility/accessibility* restrictions.

5 The system $\mathsf{M}_\bot$

Let the natural deduction system $\mathsf{M}_\bot$ be obtained from the propositional part of the Gentzen-Prawitz natural deduction system M for minimal logic by the replacement of the usual disjunction elimination by the following new set of disjunction-elimination rules:

1. $\vee_\bot$-elimination-1

$$\dfrac{\begin{array}{c}\Gamma\\ \Pi\\ A \vee B\end{array} \quad \begin{array}{c}[A]^m\\ \Pi_1\\ C\end{array} \quad \begin{array}{c}[B]^n\\ \Pi_2\\ C\end{array}}{C}\ \vee_{\bot \text{E-1}}\ m,\ n$$

2. $\vee_\bot$-elimination-2

$$\dfrac{\begin{array}{c}\Gamma^*\\ \Pi\\ A \vee B\end{array} \quad \begin{array}{c}[A]^m\\ \Pi_1\\ C\end{array} \quad \begin{array}{c}[B]^n \quad \Gamma_2^*\\ \Pi_2\\ \bot\end{array}}{C}\ \vee_{\bot \text{E-2}}\ m,\ n,\ \Gamma^*,\ \Gamma_2^*$$

3. $\vee_\bot$-elimination-3

$$\dfrac{\begin{array}{c}\Gamma^*\\ \Pi\\ A \vee B\end{array} \quad \begin{array}{c}[A]^m \quad \Gamma_1^*\\ \Pi_1\\ \bot\end{array} \quad \begin{array}{c}[B]^n\\ \Pi_2\\ C\end{array}}{C}\ \vee_{\bot \text{E-3}}\ m,\ n,\ \Gamma^*,\ \Gamma_1^*$$

The notation Γ^* and Γ_i^* ($i = 1, 2$) indicates that the assumptions in Γ and in Γ_i ($i = 1, 2$) are *frozen*, i.e., that they cannot be discharged below the conclusion of the application of the corresponding $\vee_\bot$-elimination.[9]

The non-normal derivation of $(A \to (\neg A \to B))$ obtained in $\mathsf{M}_\vee$ is clearly not a correct derivation in $\mathsf{M}_\bot$: the application of disjunction elimination that was used is an application α of $\vee$-elimination-2 and the restriction demanded by the new $\vee_\bot$-elimination-2 rule is not satisfied in α, since the hypothesis $\neg A$ is discharged below the conclusion of α. The restrictions on Γ forbid the second problematic example given above. Although these problematic cases are not theorems of $\mathsf{M}_\bot$, we still have a non-normal derivation of $\{\neg A^{*2}, A^{*1}\} \vdash B$!

[9] The intuitionistic multiple succedent sequent calculus FIL defined in de Paiva and Pereira (2005) incorporates this idea by means of devices that control dependency relations between formulas in the antecedent and formulas in the succedent of a sequent.

$$\dfrac{\dfrac{A^{*1}}{(A \vee B)} \qquad \dfrac{[A]^1 \qquad \neg A^{*2}}{\bot} \qquad [B]^2}{B} \ \vee_{\bot E\text{-}3}\ 1,\ 2,\ A^{*1},\ \neg A^{*2}$$

In order to avoid these problematic derivability relations, let us examine the normalization problem for $\mathsf{M}_\bot$.

6 Normalization for $\mathbf{M}_\bot$

Let us assume now that our hiker when he arrives at the bifurcation point $(A \vee B)$ his map indicates that he should take path A. This situation can be pictured as:

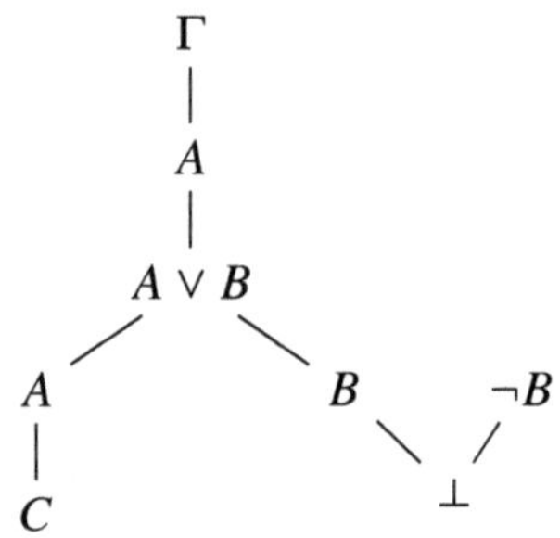

After taking the path marked by A, John's *promenade* looks as follows:

If John had found the indication to take path B, we would have the following picture:

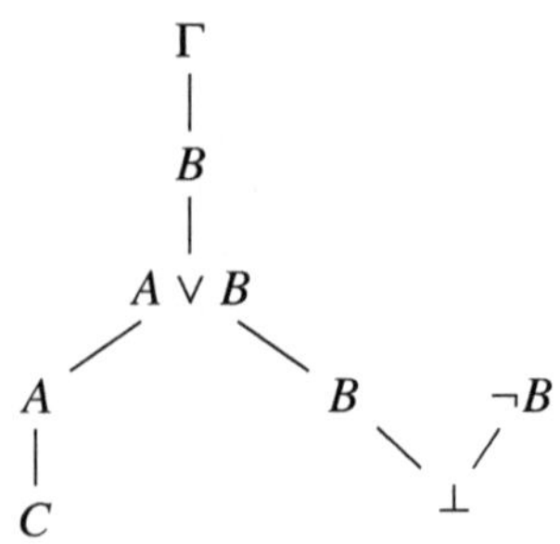

And after taking the path B, the situation is:

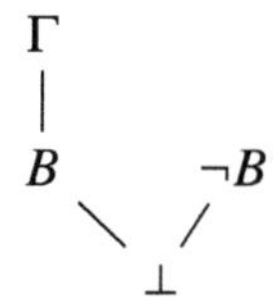

The new $\vee_\perp$ reductions corresponding to these figures are:

1. A derivation of the form

$$\begin{array}{c}\dfrac{\dfrac{\begin{matrix}\Gamma\\ \Pi\\ A\end{matrix}}{A \vee B} \quad \begin{matrix}[A]^m\\ \Pi_1\\ C\end{matrix} \quad \begin{matrix}[B]^n\\ \Pi_2\\ C_2\end{matrix}}{C}\,\vee_{\perp\text{E-1}}\ m,\ n\\ \Pi_3\end{array}$$

where C_2 is either C or $\perp$, reduces (as usual) to

$$\begin{matrix}\Gamma\\ \Pi\\ [A]\\ \Pi_1\\ [C]\\ \Pi_3\end{matrix}$$

2. A derivation of the form

$$\begin{array}{c}\dfrac{\dfrac{\begin{matrix}\Gamma\\ \Pi\\ B\end{matrix}}{A \vee B} \quad \begin{matrix}[A]^m\\ \Pi_1\\ C_1\end{matrix} \quad \begin{matrix}[B]^m\\ \Pi_2\\ C\end{matrix}}{C}\,\vee_{\perp\text{E-1}}\ m,\ n\\ \Pi_3\end{array}$$

where C_1 is either C or $\perp$, reduces (as usual) to

$$\begin{matrix}\Gamma\\ \Pi\\ [B]\\ \Pi_2\\ [C]\\ \Pi_3\end{matrix}$$

3. A derivation of the form

$$\begin{array}{c}\dfrac{\dfrac{\begin{matrix}\Gamma^*\\ \Pi\\ B\end{matrix}}{A \vee B} \quad \begin{matrix}[A]^m\\ \Pi_1\\ C\end{matrix} \quad \begin{matrix}[B]^n\\ \Pi_2\\ \perp\end{matrix}}{C}\,\vee_{\perp\text{E-2}}\ m,\ n,\ \Gamma^*\\ \Pi_3\end{array} \qquad \text{reduces to} \qquad \begin{matrix}\Gamma^*\\ \Pi\\ [B]\\ \Pi_2\\ \perp\end{matrix}$$

4. A derivation of the form

$$\dfrac{\dfrac{\begin{matrix}\Gamma^* \\ \Pi \\ A\end{matrix}}{A \vee B} \quad \begin{matrix}[A]^m \\ \Pi_1 \\ \bot\end{matrix} \quad \begin{matrix}[B]^n \\ \Pi_2 \\ C\end{matrix}}{\begin{matrix}C \\ \Pi_3\end{matrix}} \vee_{\bot\text{E-3}}\ m, n, \Gamma^* \qquad \text{reduces to} \qquad \begin{matrix}\Gamma^* \\ \Pi \\ [A] \\ \Pi_1 \\ \bot\end{matrix}$$

We could use here the strategy we used with respect to $\mathsf{M}_\vee$ and impose *normality* by stipulation: only *normal derivations* are accepted as legitimate derivations.

But we can also use a new strategy that was used by Prawitz in *Natural Deduction*. In the appendix A, on set theory, Prawitz introduces the notion of *quasi-derivation*:

> F $[F_c]$ is to be the system that is obtained from I $[C']$ by the addition of the λI- and λE-rule. A *quasi-deduction* in one of these systems is defined in the same way as a deduction was defined for Gentzen-type systems in general (Chapter I, §2). A deduction is then defined as a quasi-deduction that is normal. (Prawitz, 1965, pp. 94–95)

We could use this idea and define:

Definition 6.1 A *quasi-derivation* in $\mathsf{M}_\bot$ is a derivation as we usually define it. A *derivation* in $\mathsf{M}_\bot$ is a quasi-derivation that is *normal*.

We can now formulate the normalization theorem for $\mathsf{M}_\bot$ as follows:

Theorem 6.2 *Let* Π *be a quasi-derivation of* $\Gamma, \Delta^* \vdash A$. *Then, either* Π *reduces to a derivation* Π' *of* $\Gamma, \Delta^* \vdash A$ *or* Π *reduces to a derivation* Π'' *of* $\Gamma, \Delta^* \vdash \bot$.

The new normalization theorem establishes that a quasi-derivation will always take us either to a normal derivation of the same conclusion or to a normal derivation of $\bot$[10]. We can think of quasi-derivations in $\mathsf{M}_\bot$ as *deduction-maps* that may contain *detours*. Once we follow the map eliminating the detours (the *normalization guide*), we will either reach the marked goal or we will reach a *dead-end.*

Important remark: This idea of *deduction maps* cannot be applied to the system $\mathsf{M}_\vee$, since the system $\mathsf{M}_\vee$ does not satisfy the new normalization theorem. The following simple derivation is a counter-example to the new normalization theorem: This derivation

$$\dfrac{\dfrac{\dfrac{[A]^3}{(A \vee B)} \quad \dfrac{[A]^1 \quad \neg A}{\bot} \quad [B]^2}{B} \vee_{\bot\text{E-3}}\ 1, 2}{(A \to B)} \to_{\text{I}} 3$$

reduces to

$$\dfrac{A \quad \neg A}{\bot}$$

We immediately see that the number of open assumptions increases after the reduction (A was not open before the reduction).

[10] We can easily see that in some cases a quasi-derivation can be transformed into different derivations of $\bot$.

7 Conclusion

We have been examining two extensions of the propositional part of the Gentzen-Prawitz natural deduction system M for minimal logic, the systems $\mathsf{M}_\vee$ and $\mathsf{M}_\perp$, that satisfy *in a certain way* the disjunctive syllogism but that do not satisfy the *ex falso*.

The system $\mathsf{M}_\vee$ uses the rules $\vee$-elimination-1, $\vee$-elimination-2, and $\vee$-elimination-3, and *normality* is imposed by construction. In the system $\mathsf{M}_\vee$ we have the following results:

1. By means of the new rules for disjunction, we can easily show that

$$\{(A \vee B), \neg B\} \vdash_{\mathsf{M}_\vee} A.$$

2. Given that there is no restriction on the deduction theorem in $\mathsf{M}_\vee$, we also have

$$\vdash_{\mathsf{M}_\vee} (((A \vee B) \wedge \neg B) \rightarrow A).$$

3. The imposition of *normality by contruction* guarantees that

$$\nvdash_{\mathsf{M}_\vee} (\neg A \rightarrow (A \rightarrow B)).$$

4. $\{\neg A, A\} \nvdash_{\mathsf{M}_\vee} B$.
5. The $\perp$ works as the disjunction unity: $\vdash_{\mathsf{M}_\vee} ((A \vee \perp) \leftrightarrow A)$.
6. A general *composition of derivations* is lost with the imposition of *normality by contruction*. This feature of $\mathsf{M}_\vee$ does not seem to be faithful to our deductive practices, where composition seems absolutely natural, as in the case of proofs of theorems that use lemmas.

The system $\mathsf{M}_\perp$ uses the rules $\vee_\perp$-elimination-1, $\vee_\perp$-elimination-2, and $\vee_\perp$-elimination-3, and these rules impose a more strict control on *dependency relations* between assumptions and derived formulas. *Normality* is not imposed by construction, but it is used to define derivations: $\mathsf{M}_\perp$ works with the concept of *quasi-derivations* and defines derivations as quasi-derivations that are normal. The normalization theorem for $\mathsf{M}_\perp$ guarantees that every quasi-derivation Π of $\Gamma, \Delta^* \vdash A$ either reduces to a derivation Π' of $\Gamma, \Delta^* \vdash A$ or Π reduces to a derivation Π'' of $\Gamma, \Delta^* \vdash \perp$.

In the system $\mathsf{M}_\perp$ we have the following results:

1. $\{(A \vee B)^*, \neg B^*\} \vdash_{\mathsf{M}_\perp} A$. In the application of the new disjunction-elimination rule the assumptions $(A \vee B)$ and $\neg B$ must be *frozen*.
2. $\{(A \vee \perp)^*\} \vdash_{\mathsf{M}_\perp} A$.
3. $\nvdash_{\mathsf{M}_\perp} (((A \vee B) \wedge \neg B) \rightarrow A)$.
4. $\{\neg A, A\} \nvdash_{\mathsf{M}_\perp} B$.
5. $\nvdash_{\mathsf{M}_\perp} (\neg A \rightarrow (A \rightarrow B))$.
6. We can think of the quasi-derivations as *deduction plans* that may contain detours. The main idea now is that the normalization process will guide us through this plan and will lead us either to a derivation that ends with a certain goal, or it will lead us to a dead-end (possibly several dead-ends).

Obviously there is still a lot of work to be done: (1) a more detailed comparison between the systems $\mathsf{M}_\vee$ and $\mathsf{M}_\perp$; (2) a more detailed comparison with Tennant's systems[11,12]; (3) a deeper exploration of the idea of "freezing" hypotheses; and a more in-depth analysis of the proof-theory of $\mathsf{M}_\perp$.

Acknowledgements The work of Peter has been a great source of inspiration for us and this is especially true with respect to this paper: its original motivation was Peter's talk on "Paradoxes, local consequence, and the role and future of deductive logic", given at the Swedish Collegium for Advanced Study (SCAS), Uppsala. We would also like to thank Dag Prawitz, Valeria de Paiva, Wagner de Campos Sanz, Evandro Gomes, Abilio Rodrigues, Rodolfo Ertola-Biraben, Bogdan Dicher, Neil Tennant and the Tübingen group for important suggestions and criticisms. And finally, we would like to give a big thanks to the two anonymous referees for the extremely careful, detailed and generous reviews. Unfortunately, some important suggestions could not be incorporated into the revised version of our text, but they will certainly have to be taken into account in our future work. This research was supported by a CNPq project and by the projects CAPES/PRINT and CAPES/COFECUB.

References

Brouwer, L. E. J. (1975). *Collected Works 1. Philosophy and Foundations of Mathematics*. Ed. by A. Heyting. Amsterdam: North-Holland.

de Paiva, V. and L. C. Pereira (2005). A short note on intuitionistic propositional logic with multiple conclusions. *Manuscrito* 28, 317–329.

Gentzen, G. (1969). Investigations into logical deduction. In: *The Collected Papers of Gerhard Gentzen*. Ed. by M. Szabo. Amsterdam: North-Holland, 228–238.

Lewis, C. I. and C. H. Langford (1959). *Symbolic Logic*. New York: Dover Publications.

Martin, C. (1986). William's machine. *Journal of Philosophy* 83. Eighty-Third Annual Meeting of the American Philosophical Association, Eastern Division, 564–572.

Mates, B. (1953). *Stoic Logic*. Berkeley: University of California Press.

[11] Tennant (1987; 2017).

[12] After the text was ready and submitted, Prof. Bogdan Dicher called our attention to the many similarities between our results and those of yet another paper by Neil Tennant (1994). Both Tennant's extraction theorem and our notion of deduction maps heavily rely on procedures which are applied to derivations in order to produce either derivations with the same end formula or derivations of the absurdity constant. However, there are also many important differences between both approaches. Tennant imposes normal form by definition on derivations in his relevant intuitionistic logic, whereas our formulation allows derivations to contain maximal formulas which are later expunged by reduction procedures. Moreover, ours is a logic which stands strictly between minimal and intuitionistic logic, whereas the relation between Tennant's system and those logics is considerably more complex; though it allows one to prove all theorems of intuitionistic logic (even ones such as $\vdash \neg A \rightarrow (A \rightarrow B)$), it invalidates many of its deducibility relations (it can be shown that $\{\neg A, A\} \nvdash B$, for example). There also seem to be many other interesting differences and similarities between both approaches but, unfortunately, the late discovery of this relationship effectively prevented us from properly addressing the topic, and so further comparisons will have to be left to future work.

Prawitz, D. (1965). *Natural Deduction: A Proof-Theoretical Study*. Stockholm: Almqvist & Wiksell.
Tennant, N. (1979). Entailment and proofs. *Procedings of the Aristotelian Society* 79, 167–190.
– (1987). Natural deduction and sequent calculus for intuitionistic relevant logic. *Journal of Symolic Logic* 52, 665–680.
– (1994). Intuitionistic mathematics does not need ex falso quodlibet. *Topoi* 13, 127–133.
– (2017). *Core Logic*. Oxford: Oxford University Press.
van Atten, M. (2009). On the hypothetical judgement in the history of intuitionistic logic. In: *Logic, Methodology and Philosophy of Science. Proceedings of the Thirteenth International Congress*. Ed. by C. Glymour, W. Wei, and D. Westerståhl. London: College Publications, 122–136.
van der Molen, T. (2016). *The Johansson/Heyting letters and the birth of minimal logic*. Tech. rep. ILLC Publications, Technical Notes. URL: http://www.illc.uva.nl/Research/Publications/Reports/X-2016-04.text.pdf.

The Logicality of Equality

Andrzej Indrzejczak

Abstract The status of the equality predicate as a logical constant is problematic. In the paper we look at the problem from the proof-theoretic standpoint and survey several ways of treating equality in formal systems of different sorts. In particular, we focus on the framework of sequent calculus and examine equality in the light of criteria of logicality proposed by Hacking and Došen. Both attempts were formulated in terms of sequent calculus rules, although in the case of Došen it has a nonstandard character. It will be shown that equality can be characterised in a way which satisfies Došen's criteria of logicality. In the case of Hacking's approach the fully satisfying result can be obtained only for languages with a nonempty, finite set of predicate constants other than equality. Otherwise, cut elimination theorem fails to hold.

Key words: equality, logical constants, sequent calculus

1 Introduction

It is difficult to find serious applications of logic that do not use equality. Not only it is necessary for development of mathematical theories but it plays also an important role in philosophical applications. Yet it is problematic to show that it is a logical constant having a similar behaviour to undisputable cases like extensional connectives or quantifiers. In this paper we try to look at the problem from the proof-theoretic perspective and ask if it is possible to characterize equality by means of rules satisfying some of the proposed criteria of logicality. For simplicity's sake we restrict our considerations to classical first-order logic (FOL) although obtained results may be easily transmitted to intuitionistic logic (see concluding remarks). The criteria which will be examined with respect to equality are those proposed by Hacking (1979) and Došen (1989), and the framework for our considerations is provided by Gentzen's

Andrzej Indrzejczak
Department of Logic, University of Łódź, Poland, e-mail: andrzej.indrzejczak@filhist.uni.lodz.pl

T. Piecha and K. F. Wehmeier (eds.), *Peter Schroeder-Heister on Proof-Theoretic Semantics*, Outstanding Contributions to Logic 29,
https://doi.org/10.1007/978-3-031-50981-0_7

sequent calculus (SC). This is partly determined by the fact that both approaches to criteria of logicality were proposed in this framework, although in the case of Došen it was not a standard variant of SC. Moreover, the framework of SC seems to be particularly well suited for investigations concerning the problems of criteria for logical constants and in general for investigations in proof-theoretic semantics (see, e.g., Schroeder-Heister 2016).

It will be convenient to start our considerations with some general remarks concerning equality since the problem of its logical status begins with the proper understanding of what the equality predicate stands for. It is tacitly and commonly assumed that a binary predicate, usually symbolised as =, is introduced to formal languages as a characterization of the identity relation. In fact, the words 'identity' and 'equality' are often treated as synonymous by a majority of mathematicians and logicians (not excluding the author). Usually, it does not lead to any problems, but in the case where equality is itself an object of study, we should be more careful. Therefore, we prefer to follow here such authors like Manzano and Moreno (2017) or Kahle (2016), in keeping a strict distinction between identity and equality. Identity is a relation between objects and equality between terms. The former is a semantic relation that holds trivially only between an object and itself, whereas the latter is a syntactical relation which may hold between any terms of the language. It is natural to postulate that the equality predicate expresses in the language the identity of objects denoted by its arguments, however, there are serious problems hidden in such identification. First of all it can be even doubtful if identity is a genuine relation, and if so, if it should be represented by some binary predicate. Wittgenstein (1922) presented such a view in his rejection of the very symbol of equality from his language. In fact, Wittgenstein's view can be formally developed in an interesting way, as was shown by Hintikka (1956) and Wehmeier (2014).

Even if we follow a standard practice of treating identity as a relation, one must be aware that the kind of the correspondence between the equality predicate and the identity relation is not very strict. In model-theoretic terms identity is just a diagonal relation on the product of the domain of a model. But the equality predicate, as characterised in axiomatic systems of first-order logic (see Section 3) cannot express identity only. Even in the case of a language with a finite number of predicates, one can find nonstandard models in which axioms of equality do not characterise identity. It seems that the second-order logic (SOL) is better in this respect. Well, if we admit that the second-order logic is a genuine logic we can define identity in terms of equivalence and the second-order universal quantification, by means of Leibniz' axiom (see Section 3). But the second-order logic is expressive enough to capture Peano's Arithmetic so only the logicist position makes this argument unproblematic. Moreover, this holds only in the standard semantics for which SOL is not complete. If we take Henkin's generalised models to regain completeness, we can again find models where Leibniz' axiom does not determine identity (see, e.g., Manzano, 2005).

Despite the deficiency of equality as a definition of identity (in FOL in particular) it does make sense to check if equality itself may be conceived as a logical constant and this is our aim. In Section 2 we establish the notation and recall the basic information on sequent calculi and properties of rules which are important for this task. Several ways of formalising equality in axiomatic and natural deduction systems are surveyed in Section 3. Ways of dealing with equality in sequent calculi are discussed in a separate section. In Section 5 we recall criteria of logicality formulated by Hacking and check if equality can be formalised in a way conforming to these desiderata. It appears that most of the criteria hold for our proposal but cut elimination is sensitive to the kind of language under consideration. In Section 6 we describe Došen's approach and a nonstandard structural variant of sequent calculus adapted to its realization. It seems that equality formalised in this kind of a system satisfies fully the conditions for logical constant but only if equality is not the only predicate of the language. We finish with some remarks concerning further applications and possible generalizations of presented approach.

2 Preliminaries

The notation applied in the paper is mostly standard. φ, ψ, χ will represent arbitrary formulae built by means of $\neg, \wedge, \vee, \rightarrow, \forall, \exists$ from atomic formulae, i.e., predicates followed by a list of terms. Following Gentzen's custom, we distinguish between bound and free occurrences, reserving $x, y, z, \ldots$ for representing the former and $a, b, c, \ldots$ for the latter, usually called parameters. Nothing essential depends on this distinction, although it simplifies a definition of substitution for terms. Other terms, if any, will be constructed from function symbols of any arity. We will use f, g, h and metalevel θ^n for their representation. Arbitrary terms will be represented as $\tau_1, \tau_2, \ldots$. Predicates will also be divided into parameters (schematic symbols) and predicate constants of specific languages determining their signature. Predicates of both categories will be represented either by A, R or, in the metalevel, by π^n. Incidentally, X will be used for bound (predicate) variable of the second-order and. $\varphi(\tau)$ denotes a fomula having at least one occurrence of τ and $\varphi[x/\tau]$ the result of the correct substitution of τ for all free occurrences of x. $\Gamma, \Delta, \Sigma, \ldots$ represent finite multisets of formulae. Eventually $=$ will be used as a symbol of equality. In general, formulae of the form $\tau_1 = \tau_2$ are not counted as atomic, since $=$ is considered as a logical constant.

Following Church's (1956) terminology we distinguish the following language variants of FOLI (FOL with identity):

1. pure FOLI with predicate and term variables/parameters and equality as the only predicate constant;
2. applied FOLI with additional other predicate constants;
3. simple applied FOLI with no predicate parameters (only constants);
4. simple FOLI with no predicate parameters and constants except equality.

Thus pure FOLI is just a schematic version of FOL with equality whereas several cases of simple applied FOLIs are specific languages characterised by their signatures. For example the language of simple applied FOLI of set theory has two binary predicate constants: $=$ and $\in$. The cases of applied FOLI are of mixed character since in addition to constants they admit variables/parameters. This classification is essential for comparison of different possible characterisations of equality.

As our basic sequent calculus (SC) for classical logic FOL we will use a system which is essentially Gentzen's LK but with sequents built from multisets to avoid inessential complications. We also prefer to present all two-premiss rules in the multiplicative (or with independent contexts) version in contrast to Gentzen's original rules for $\wedge, \vee$. Again nothing essential hinges on this choice and other variants of SC can be also applied. The calculus consists of the following structural and logical rules:

$$(AX)\ \varphi \Rightarrow \varphi \qquad\qquad (Cut)\ \frac{\Gamma \Rightarrow \Delta, \varphi \quad \varphi, \Pi \Rightarrow \Sigma}{\Gamma, \Pi \Rightarrow \Delta, \Sigma}$$

$$(W\Rightarrow)\ \frac{\Gamma \Rightarrow \Delta}{\varphi, \Gamma \Rightarrow \Delta} \qquad\qquad (\Rightarrow W)\ \frac{\Gamma \Rightarrow \Delta}{\Gamma \Rightarrow \Delta, \varphi}$$

$$(C\Rightarrow)\ \frac{\varphi, \varphi, \Gamma \Rightarrow \Delta}{\varphi, \Gamma \Rightarrow \Delta} \qquad\qquad (\Rightarrow C)\ \frac{\Gamma \Rightarrow \Delta, \varphi, \varphi}{\Gamma \Rightarrow \Delta, \varphi}$$

$$(\neg\Rightarrow)\ \frac{\Gamma \Rightarrow \Delta, \varphi}{\neg\varphi, \Gamma \Rightarrow \Delta} \qquad\qquad (\Rightarrow\neg)\ \frac{\varphi, \Gamma \Rightarrow \Delta}{\Gamma \Rightarrow \Delta, \neg\varphi}$$

$$(\wedge\Rightarrow)\ \frac{\varphi, \psi, \Gamma \Rightarrow \Delta}{\varphi\wedge\psi, \Gamma \Rightarrow \Delta} \qquad\qquad (\Rightarrow\wedge)\ \frac{\Gamma \Rightarrow \Delta, \varphi \quad \Pi \Rightarrow \Sigma, \psi}{\Gamma, \Pi \Rightarrow \Delta, \Sigma, \varphi\wedge\psi}$$

$$(\vee\Rightarrow)\ \frac{\varphi, \Gamma \Rightarrow \Delta \quad \psi, \Pi \Rightarrow \Sigma}{\varphi\vee\psi, \Gamma, \Pi \Rightarrow \Delta, \Sigma} \qquad\qquad (\Rightarrow\vee)\ \frac{\Gamma \Rightarrow \Delta, \varphi, \psi}{\Gamma \Rightarrow \Delta, \varphi\vee\psi}$$

$$(\rightarrow\Rightarrow)\ \frac{\Gamma \Rightarrow \Delta, \varphi \quad \psi, \Pi \Rightarrow \Sigma}{\varphi\rightarrow\psi, \Gamma, \Pi \Rightarrow \Delta, \Sigma} \qquad\qquad (\Rightarrow\rightarrow)\ \frac{\varphi, \Gamma \Rightarrow \Delta, \psi}{\Gamma \Rightarrow \Delta, \varphi\rightarrow\psi}$$

$$(\forall\Rightarrow)\ \frac{\varphi[x/t], \Gamma \Rightarrow \Delta}{\forall x\varphi, \Gamma \Rightarrow \Delta} \qquad\qquad (\Rightarrow\forall)^1\ \frac{\Gamma \Rightarrow \Delta, \varphi[x/a]}{\Gamma \Rightarrow \Delta, \forall x\varphi}$$

where a is not in Γ, Δ, φ

$$(\exists\Rightarrow)\ \frac{\varphi[x/a], \Gamma \Rightarrow \Delta}{\exists x\varphi, \Gamma \Rightarrow \Delta} \qquad\qquad (\Rightarrow\exists)\ \frac{\Gamma \Rightarrow \Delta, \varphi[x/t]}{\Gamma \Rightarrow \Delta, \exists x\varphi}$$

where a is not in Γ, Δ, φ

Formulae displayed in the schemata are active whereas those in (possibly empty) multisets Γ, Δ are parametric (or form the context). In particular, a unique formula in the antecedent or succedent of the conclusion is the principal formula of the respective rule application whereas active formulae in the premisses are called side-formulae. The notion of a proof is standard, i.e., a tree labelled with sequents where each leaf is an axiom and edges are regulated by the rules. The height of a proof is the number of nodes in maximal branches. For stating criteria of logicality it is important to focus on some of the characteristic features of the logical rules. First of all they are rules of introduction of a constant, either to the antecedent or to the succedent of a sequent. Moreover, using a terminology of Wansing (1999) (see also Poggiolesi, 2011), we can observe that well-behaved rules have the following properties:

1. Separation: a rule for a constant should not exhibit any other constants in its schema.
2. Weak symmetry: each rule should either introduce a constant to the antecedent or to the succedent; if both rules for a constant are of such a kind, the calculus is (simply) symmetric with respect to this constant.
3. Weak explicitness: a constant should be present only in the conclusion; if only one occurrence of it is present a rule is (simply) explicit.
4. Subformula property: only formulae present in the conclusion and their subformulae (modulo substitution of terms) are present in the premisses.

Rules satisfying these properties are also called canonical by Avron (2001). In what follows we tacitly assume that candidates on rules characterising a logical constants should have these features or some reasonable generalizations of them. But these are considered only as necessary conditions, for sufficiency we will examine additional requirements formulated by Hacking and Došen.

3 Approaches to equality

As we remarked our main tool will be SC but it is profitable to recall first how equality was (and is) usually dealt with in the framework of other proof systems, in particular, in axiomatic or natural deduction systems (ND). In the philosophical considerations we can often find a reference to the traditional characterisation of equality due to Leibniz. This approach may be formally presented as a formula of the second-order logic (SOL) which we call LA (Leibniz Axiom):

$$\tau_1 = \tau_2 \leftrightarrow \forall X(X\tau_1 \leftrightarrow X\tau_2).$$

Commonly, the left-right implication is called the principle of indiscernibility of identiticals, whereas the converse is called the principle of identity of indiscernibles[1]. The latter principle implies immediately reflexivity of = whereas full LA is required to prove both symmetry and transitivity of = as implied by symmetry and transitivity of $\leftrightarrow$. Note also that LA may be weakened in two senses: (a) X may be restricted to atomic predicates (b) the rightmost equivalence may be replaced with implication. The restriction (b) is not independent from (a); LA is derivable from the weaker form (b) if we admit that not only predicates but complex formulae (of FOL) may be instantiated for X (see the proofs provided by Read, 2004 or Parlamento and Previale, 2019). This is perhaps not in conflict with the original intuitions of Leibniz since he seems to consider every context where respective terms may be exchanged salva veritate. On the other hand, in order to prove that equality is symmetric in restricted case (b) we must instantiate X with equalities. But this solution shows that LA in restricted form (b) cannot be treated as a definition of identity since it is either incomplete or circular. In the former case to obtain the full characteristics of identity

[1] Although some doubts may be raised against the correctness of the identification of these traditionally considered principles with this formula of SOL, see, e.g., Mates (1986).

we must add explicitly the condition of symmetry. Below we will show that using restricted form (b) of LA leads to other inadequacies as well.

On the other hand, restriction (a), i.e., restriction to atomic predicates as admissible instances of X, has some merits in the case of simple applied languages which was noted by Quine (1970). We can change the 'definiens' into conjunction of all possible cases. For example, if our language has one unary primitive predicate constant A and one binary R it takes the form:

$$\text{(LA}'\text{)} \qquad \tau_1 = \tau_2 \leftrightarrow ((A\tau_1 \leftrightarrow A\tau_2) \wedge \forall x((R\tau_1 x \leftrightarrow R\tau_2 x) \wedge (Rx\tau_1 \leftrightarrow Rx\tau_2)).$$

We already mentioned that this is not sufficient to obtain a real definition of identity in FOL but it can work as good stipulation of identity in the case of simple applied FOLI, i.e., languages with finite number of predicate constants.

If we restrict our considerations to FOL, a characterisation of equality in terms of LA is of no use in the case of pure or applied versions of language but still may have some heuristic value. In particular, on the ground of Hilbert systems one may distinguish two approaches which we call algebraic and Leibnizian. In the former, equality is characterised simply as a congruence on terms so we need to state first that it is an equivalence relation expressed by:

1. reflexivity axiom R: $\forall x(x = x)$;
2. symmetry axiom SYM: $\forall xy(x = y \rightarrow y = x)$
3. transitivity axiom TR: $\forall xyz(x = y \wedge y = z \rightarrow x = z)$.

This is enough for simple FOLI; in the case of simple applied versions of FOLI we must add two principles of congruence for every primitive atomic predicate and term:

1. Congruence of Predicates CP:

$$\forall x_1, \ldots, x_n, y_1, \ldots, y_n(x_1 = y_1 \wedge \ldots \wedge x_n = y_n \rightarrow (\pi^n(x_1, \ldots, x_n) \rightarrow \pi^n(y_1, \ldots, y_n)),$$

where π^n is n-argument predicate symbol.

2. Congruence of Terms CT:

$$\forall x_1, \ldots, x_n, y_1, \ldots, y_n(x_1 = y_1 \wedge \ldots \wedge x_n = y_n \rightarrow \theta^n(x_1, \ldots, x_n) = \theta^n(y_1, \ldots, y_n)),$$

where θ^n is n-argument function symbol.

This way of characterising equality is particularly elegant if we deal with simple applied first-order languages having only a small number of primitive predicate or function constants. It works better than characterization via LA′ since we obtain one axiom for every predicate instead of n equivalences for every n-ary predicate. Moreover, if we treat equality as a primitive atomic predicate it is not necessary for symmetry and transitivity of = to be explicitly added since they are provable by means of CP.

Authors dealing with pure or applied FOL, i.e., with predicate and function parameters usually prefer the latter approach which we called Leibnizian. It also requires reflexivity usually stated schematically as $\tau = \tau$ for every term, and the extensionality principle:

$$\text{(EP)} \qquad \forall xy(x = y \wedge \varphi[z/x] \rightarrow \varphi[z/y]),$$

where φ is arbitrary or atomic. The latter form is simpler to formulate since there is no problem with bound variables; moreover the general form is provable in extensional FOLI. In what follows we will keep the name EP for the version with arbitrary φ and call the version restricted to atoms Leibniz principle LP. One should note that it encodes one direction of LA, namely indiscernibility of identicals in weaker form (b), i.e., with $\leftrightarrow$ replaced by $\rightarrow$. It explains why we call this approach Leibnizian. One should also note that, contrary to ordinary custom, we defined EP (LP) by means of (correct) substitution of x, y for some free variable z. It is much more popular that it is characterised in terms of replacement:

$$\text{(EP}'\text{)} \qquad \forall xy(x = y \wedge \varphi \rightarrow \varphi[x//y]),$$

where $\varphi[x//y]$ denotes a replacement of some (not necessarily all) occurrences of x by y. It is perhaps intuitively more accessible but has some formal disadvantage since replacement is not an operation. To avoid this problem some authors define EP (LP) by means of a unique replacement:

$$\text{(EP}''\text{)} \qquad \forall xy(x = y \wedge \varphi(\ldots x \ldots) \rightarrow \varphi(\ldots y \ldots)),$$

where only one displayed occurrence of a variable (term) is taken into account. The last formulation is also simpler for arithmetization hence applied in the works dealing with Gödel's theorems. But it should be stressed that all these forms of characterization are equivalent. In particular, any possible application of EP$'$ is just a series of applications of EP$''$. Also SYM and TR are easily provable by any of these principles so it is not necessary to introduce them as primitive axioms.

Instead of REF we can find (for example in Tarski, 1941):

$$(\exists =) \qquad \exists x(x = \tau) \text{ where } x \text{ is not in } \tau.$$

This formula implies reflexivity by EP: $a = \tau \wedge a = \tau \rightarrow \tau = \tau$. In Mates (1965) the same characterization is applied but with universal closures of EP$'$ and $(\exists =)$.

There are also possible approaches which dispense with reflexivity, instead using just one formula which is equivalent to REF and EP. This approach is due to Wang (see Quine, 1966) and a dual axiom is due to Kalish and Montague (1957). On the ground of Hilbert system each one may be expressed by one axiom:

$$\exists x(x = \tau \wedge \varphi) \leftrightarrow \varphi[x/\tau], \text{ where } x \text{ is not free in } \tau$$

or

$$\forall x(x = \tau \rightarrow \varphi) \leftrightarrow \varphi[x/\tau], \text{ where } x \text{ is not free in } \tau.$$

In the framework of natural deduction (ND) the Leibnizian approach is prevalent although EP (LP) is usually presented as an inference rule of identity elimination:

(IE) $$\tau_1 = \tau_2, \varphi[x/\tau_1] \vdash \varphi[x/\tau_2],$$

whence reflexivity is treated as a zero-premiss rule of identity introduction:[2]

(II) $$\varnothing \vdash \tau = \tau.$$

Martin-Löf (1971) has shown that such a pair of rules is naturally induced for equality predicate in the general ND framework which satisfies normalization.

Kalish and Montague (1964) provide an ND system where the pair of rules is of the form:

(IE′) $\forall x(x = \tau \rightarrow \varphi) \vdash \varphi[x/\tau]$, where x is not free in τ,

(II′) $\varphi[x/\tau] \vdash \forall x(x = \tau \rightarrow \varphi)$, where x is not free in τ.

Both are clearly derived from the second axiom stated above.

There are also ND systems which operate not on formulae but on sequents of the form $\Gamma \vdash \varphi$, where Γ is a set (multiset, sequence) of active assumptions for φ[3]. In this setting rules for equality are formulated as follows:

(II″) $\Gamma \vdash \tau = \tau$,

(IE″) If $\Gamma \vdash \tau_1 = \tau_2$ and $\Delta \vdash \varphi[x/\tau_1]$, then $\Gamma, \Delta \vdash \varphi[x/\tau_2]$.

Finally let us point out the solution which is particularly important for our purposes. Read (2004) provides a rule of equality introduction as a proof construction rule of the form:

(RII) If $\Gamma, \varphi[x/\tau_1] \vdash \varphi[x/\tau_2]$, then $\Gamma \vdash \tau_1 = \tau_2$,
where φ is atomic and does not occur in Γ.

This rule is not sound in standard models for FOLI, however it is sound in so called Leibnizian models (see Read, 2004) and it may be shown that this class of models can equivalently characterise FOLI. Note that in the context of simple applied languages with a finite number of primitive predicates the corresponding result cannot be stated by means of such a rule; instead it may be stated by means of finite number of subproofs for each predicate constant. Something similar was proposed by Więckowski (2011) who provided a rule of the form:

(WRI) $$\varphi_1[x/\tau_1] \leftrightarrow \varphi_1[x/\tau_2], \ \dots \ \varphi_n[x/\tau_1] \leftrightarrow \varphi_n[x/\tau_2] \ \vdash \ \tau_1 = \tau_2.$$

Note that here we have an inference rule since instead of subproofs we have a finite number of premisses. The drawback of this rule is that another constant, namely the

[2] Some authors add also inference rules corresponding to SYM and TR but of course they are derivable.

[3] The fact that we deal with sequents not with formulae is sometimes hidden since Γ consists not of formulae but of numbers referring to lines where assumptions were stated — see, e.g., Suppes (1957), Lemmon (1965).

equivalence connective, is present so the rule is not separate (see the previous section). It could be changed into a separate rule (i.e., with displayed equality only) by replacing each premiss $\varphi_i[x/\tau_1] \leftrightarrow \varphi_i[x/\tau_2]$ with a pair of subproofs $\varphi_i[x/\tau_1] \vdash \varphi_i[x/\tau_2]$ and $\varphi_i[x/\tau_2] \vdash \varphi_i[x/\tau_1]$. Note that it does not directly corresponds to Read's rule — the latter should have an additional subproof leading from $\varphi[x/\tau_2]$ (as an assumption) to $\varphi[x/\tau_1]$. The reasons why Read dispenses with the second subproof was explained above; WRI is based on LA (or rather LA′) whereas RII on LA with nonrestricted instantiation for X which enables replacement of equivalences by implication. Read's solution may be criticised not only from technical but also from philosophical point of view (see, e.g., Griffiths 2014, or Klev 2019) but deserves carefull examination. In what follows we will check how it works in the setting of SC.

4 Equality in sequent calculi

SC provides a framework which not only easily accomodates all approaches described so far but allows for several other solutions. An interesting thing is that in SC we can characterise equality not only by local rules but globally in the following way:

$$(\text{SUB})\ \frac{\tau_1 = \tau_2, \Gamma[x/\tau_1] \Rightarrow \Delta[x/\tau_1]}{\tau_1 = \tau_2, \Gamma[x/\tau_2] \Rightarrow \Delta[x/\tau_2]} \qquad (\text{REF})\ \Rightarrow \tau = \tau$$

where $\Gamma[x/\tau]$ denotes a uniform substitution of x by τ in all elements of Γ. Such solution was first introduced by Kanger (1957) but also proposed by Wang (1960) in the version where substitution is made only in Δ; this apparently weaker version is in fact sufficient. Essentially the same solution was used among others, by Mints (1968), Došen (1989), Seligman (2001), in several variants (for example with $\tau_1 = \tau_2$ only in the conclusion). Usually (SUB) is introduced in two variants where in the second we have $\tau_2 = \tau_1$, but it is redundant. Similar approach was also applied by Schroeder-Heister (1994) in the formalization of the free equality investigated in the setting of logic programming.

This form of introduction of equality is global, since we can treat a sequent as expressing a whole proof at some stage. Hence to imitate the application of this rules in ND we should rewrite the whole derivation. So in fact it is global in comparison with the solution proposed in ND setting. In SC this is reduced to the operation performed on the sequent not on the whole derivation. Such an approach has obvious virtues. One may easily prove everything which is needed to show that it is adequate; we obtain a proof of LP immediately. Moreover, cut elimination holds for it (see, e.g., Seligman, 2001 for a constructive proof; the version of Kanger is just a cut-free variant of G3). It is also worth noting that it was the most influential approach in automated theorem proving based on SC[4]. However, it seems that this approach is not fully convincing as a way for justifying equality as a logical constant. Even in the form proposed by Došen (see Section 6) equality is presented as something which

[4] Degtyarev and Voronkov (2001) present it as the only SC-based approach to equality formalization in automated deduction.

affects the whole sequent and looks like something of slightly different character from other logical constants which are characterised by local rules.

Other approaches are of local character and can be divided according to the possible ways of formalizing theories in the framework of SC. Negri and von Plato (2001) described four approaches to this question:

1. Addition of axiomatic sequents $\Rightarrow \varphi$ for each axiom φ.
2. Addition of "mathematical basic sequents" which consist of atomic formulae.
3. Addition of all axioms as a context in the antecedents of all provable sequents.
4. Addition of new rules corresponding to axioms.

In every class we can find SC formalizations of equality. The first approach may be called the "naive" since it treats SC in the same way as Hilbert system and does not refer to its specific features. Not surprisingly such a solution obstructs the application of specific virtues of SC; in particular cut elimination cannot be proved.

The second approach may be seen as a refinement of the first and was already applied by Gentzen in his formalization of Peano arithmetic. Restricted to equality it leads to addition of two axiomatic (atomic) sequents of the form:

$$\Rightarrow \tau = \tau$$

$$\tau_1 = \tau_2, \varphi[x/\tau_1] \Rightarrow \varphi[x/\tau_2] \text{ with } \varphi \text{ atomic}$$

In Takeuti (1987) one may find variants of this approach and the proofs of some of its features. It is interesting that although cut elimination cannot be proved in general for such a system it may be proved in restricted form. Let us call inessential any cut in which the cut formula is an equality, otherwise cut is essential. For Takeuti's system it holds that all essential cuts are eliminable. Recently Parlamento and Previale (2019) proved an even stronger result showing that after an additional series of transformations cut can be eliminated from all proofs.

The third approach was also considered by Gentzen. Interestingly enough one can prove cut elimination for it but for any theorem φ of FOLI we do not obtain proofs of $\Rightarrow \varphi$ but of $\Gamma \Rightarrow \varphi$ where Γ is a collection of instances of REF and LP. Accordingly this approach does not provide an interesting tool for analysis of proofs.

There is a variant of this approach which in fact can be treated also as belonging to the last group. For each axiom φ an SC rule is postulated for elimination of this axiom, which is of the form:

$$\text{(AE)} \qquad \frac{\varphi, \Gamma \Rightarrow \Delta}{\Gamma \Rightarrow \Delta}$$

Such formalization of cut-free SC for FOLI is considered in Gallier (1986), where φ can be an instance of REF, CP or CT. Also Bell and Machover (1977) apply this approach in the tableau framework where it is represented just as a rule of introduction of suitable instances of REF, CT or CP on the branch. Note that in the context of SC this solution although applied in the cut-free version is in fact equivalent to addition of the special form of cut. This follows from the result that the cut elimination theorem is equivalent to the result showing eliminability of (AE) where φ is an arbitrary thesis (see Indrzejczak, 2017).

It seems that the most interesting approach, in particular for our purposes, is the last one. Gallier's solution is literaly speaking of this sort but not very satisfactory since (AE) is rather a trivial kind of rule mechanically applied to any formula. The subformula property does not hold and cut freeness is apparent, as we noted above. Generally speaking, it is not in any sense better than the first approach. What we need is the generation of genuine rules with active formulae in premisses and conclusions and satisfying possibly some welcome proof-theoretic properties like cut elimination or a reasonable form of the subformula property. Such solution which, in some specific form, was advocated by Negri and von Plato (2001) and applied to formalisations of several theories on the basis of SC, found many adherents. In particular, Troelstra and Schwichtenberg (1996) in the second edition of their well known textbook introduced this characterization of equality instead of the second one which was present in the first edition. Equality is characterised by means of two rules:

$$\text{(RE)}\ \frac{\tau_1 = \tau_2, \varphi[x/\tau_1], \varphi[x/\tau_2], \Gamma \Rightarrow \Delta}{\tau_1 = \tau_2, \varphi[x/\tau_1], \Gamma \Rightarrow \Delta} \qquad \text{(REP)}\ \frac{\tau = \tau, \Gamma \Rightarrow \Delta}{\Gamma \Rightarrow \Delta}$$

which are added to purely logical variant G3 . In general the specific features of Negri and von Plato's approach are connected with the fact that active formulae are atomic and occur only on one side of sequents. We will call this variant the one-sided approach; systems in Negri and Plato (2001) are in fact left- (or antecedent-)sided but in Negri and Plato (2011) right-(or succedent-)sided systems are also considered. Rules of this kind can safely be added without destroying all results concerning admissibility of structural rules, including cut. However, the rule-based approach to the characterisation of theories may be realised in many different ways, not necessarily as one-sided, despite its obvious virtues.

In order to put things in a systematic way we apply the following theorem (Indrzejczak, 2018b):

Theorem 4.1 (Rule-generation) *For any sequent* $\Gamma \Rightarrow \Delta$ *with* $\Gamma = \{\varphi_1, \ldots, \varphi_k\}$ *and* $\Delta = \{\psi_1, \ldots, \psi_n\}, k \geq 0, n \geq 0, k + n \geq 1$ *there are* $2^{k+n} - 1$ *equivalent rules captured by the general schema:*

$$\frac{\Pi_1, \Rightarrow \Sigma_1, \varphi_1, \ \ldots, \ \Pi_i \Rightarrow \Sigma_i, \varphi_i \qquad \psi_1, \Pi_{i+1} \Rightarrow \Sigma_{i+1}, \ \ldots, \ \psi_j, \Pi_{i+j} \Rightarrow \Sigma_{i+j}}{\Gamma^{-i}, \Pi_1, \ldots, \Pi_i, \Pi_{i+1}, \ldots, \Pi_{i+j} \Rightarrow \Sigma_1, \ldots, \Sigma_i, \Sigma_{i+1}, \ldots, \Sigma_{i+j} \Delta^{-j}}$$

where $\Gamma^{-i} = \Gamma - \{\varphi_1, \ldots, \varphi_i\}$ *and* $\Delta^{-j} = \Delta - \{\psi_1, \ldots, \psi_j\}$ *for* $0 \leq i \leq k,\ 0 \leq j \leq n$.

It should be stressed that the proof of this theorem requires only applications of axioms and cut (see Indrzejczak, 2018b). Informally, it shows that for any sequent we can provide different rules which are interderivable with it. Premisses of these rules are obtained either by deleting some formula from the antecedent of the respective sequent and putting it in the succedent (of the respective premiss), or conversely, by deleting a formula in the succedent and putting it into the antecedent of a premiss. The conclusion of such a rule is provided by what remains intact in the input sequent. Let us see what kind of rules can be generated on the basis of LP (or EP if we wish), expressed as a sequent (1=) $\tau_1 = \tau_2, \varphi[x/\tau_1] \Rightarrow \varphi[x/\tau_2]$. We obtain the following equivalent rules:

$$(2=)\ \frac{\varphi[x/\tau_2], \Gamma \Rightarrow \Delta}{\tau_1 = \tau_2, \varphi[x/\tau_1], \Gamma \Rightarrow \Delta} \qquad (3=)\ \frac{\Gamma \Rightarrow \Delta, \varphi[x/\tau_1]}{\tau_1 = \tau_2, \Gamma \Rightarrow \Delta, \varphi[x/\tau_2]}$$

$$(4=)\ \frac{\Gamma \Rightarrow \Delta, \tau_1 = \tau_2}{\varphi[x/\tau_1], \Gamma \Rightarrow \Delta, \varphi[x/\tau_2]} \qquad (5=)\ \frac{\Gamma \Rightarrow \Delta, \tau_1 = \tau_2 \quad \Pi \Rightarrow \Sigma, \varphi[x/\tau_1]}{\Gamma, \Pi \Rightarrow \Delta, \Sigma, \varphi[x/\tau_2]}$$

$$(6=)\ \frac{\Gamma \Rightarrow \Delta, \tau_1 = \tau_2 \quad \varphi[x/\tau_2], \Pi \Rightarrow \Sigma}{\varphi[x/\tau_1], \Gamma, \Pi \Rightarrow \Delta, \Sigma}$$

$$(7=)\ \frac{\Gamma \Rightarrow \Delta, \varphi[x/\tau_1] \quad \varphi[x/\tau_2], \Pi \Rightarrow \Sigma}{\tau_1 = \tau_2, \Gamma, \Pi \Rightarrow \Delta, \Sigma}$$

$$(8=)\ \frac{\Gamma \Rightarrow \Delta, \tau_1 = \tau_2 \quad \Pi \Rightarrow \Sigma, \varphi[x/\tau_1] \quad \varphi[x/\tau_2], \Lambda \Rightarrow \Theta}{\Gamma, \Pi, \Lambda \Rightarrow \Delta, \Sigma, \Theta}$$

Each of them may be used to express LP as a rule of SC, and in fact all were used for that. For example, Negri and von Plato (2001) applied (2=) (but with the repetition of active formulae in the premiss to save admissibility of contraction — see the rule (RE) above) and Manzano (1999) prefers its dual form (3). Some authors, for example Parlamento and Previale (2019) used both (2=) and (3=) although it is redundant. Note also that both (2) and (3) may be seen as special forms of (SUB) (with singular antecedent and succedent containing only φ). (4) was applied by Reeves (1987), although in the framework of tableaux and in an apparently different way. He considers rules modelled not on LP but on CP, so the general schema would be rather:

$$(\text{CP})\quad \frac{\Gamma_1 \Rightarrow \Delta_1, \tau_1 = \tau_1' \quad \dots \quad \Gamma_n \Rightarrow \Delta_n, \tau_n = \tau_n'}{\varphi(\tau_1, \dots \tau_n), \Gamma_1, \dots \Gamma_n \Rightarrow \Delta_1, \dots \Delta_n, \varphi(\tau_1', \dots \tau_n')}$$

and similarly for CT. Moreover, it deals with tableaux so the rules are branching downwards and there are no sequents but formulae as nodes.

Indrzejczak (2019) used (5=), whereas Baaz and Leitsch (2011) both (5=) and (6=) which is its dual. Nagashima (1966) used (7=) and Indrzejczak (2018a) (8=).

Let us discuss these rules in the light of properties required from well-behaved SC rules. Although all these rules are separate most of them are rather not satisfactory with respect to other features. Only (2=), (3=) and (7=) are weakly symmetric and explicit, in the sense that they may be treated as equality introduction rules. Together with REF treated as 0-premiss rule we may even say that such pair is symmetric. Only (2=) and (3=) satisfy the subformula property. In the remaining cases this property holds in the generalised sense: in any (cut-free) proof either subformulae of the proven sequent or atomic formulae occur.

What with cut elimination? Before we answer this question it should be established also in what form the reflexivity of equality is represented in the system. On the basis of the rule generation theorem, the only nontrivial rule (except axiomatic sequent REF) is the above mentioned:

$$\text{(REP)} \qquad \frac{\tau = \tau, \Gamma \Rightarrow \Delta}{\Gamma \Rightarrow \Delta},$$

which is used by Negri and von Plato (2001) (but also by Nagashima, 1966 and Gallier, 1986), whereas other authors prefer axiomatic sequents. However this is not an introduction but rather elimination rule. Of course we can think also of making use of Kalish and Montague (1964) solution from their ND system. Let us consider the possibility of application of the rule generation theorem to a sequent:

$$\forall x(x = \tau \to \varphi) \Rightarrow \varphi[x/\tau], \text{ where } x \text{ is not free in } \tau$$

corresponding to the reflexivity axiom. It may be expressed as a rule in the following ways:

$$(1)\ \frac{\varphi[x/\tau], \Gamma \Rightarrow \Delta}{\forall x(x = \tau \to \varphi), \Gamma \Rightarrow \Delta} \qquad (2)\ \frac{\Gamma \Rightarrow \Delta, \forall x(x = \tau \to \varphi)}{\Gamma \Rightarrow \Delta, \varphi[x/\tau]}$$

$$(3)\ \frac{\Gamma \Rightarrow \Delta, \forall x(x = \tau \to \varphi) \qquad \varphi[x/\tau], \Pi \Rightarrow \Sigma}{\Gamma, \Pi \Rightarrow \Delta, \Sigma}$$

Moreover, variants (2) and (3) may be improved in a way which dispenses with other constants:

$$(2')\ \frac{a = \tau, \Gamma \Rightarrow \Delta, \varphi[x/a]}{\Gamma \Rightarrow \Delta, \varphi[x/\tau]} \qquad (3')\ \frac{a = \tau, \Gamma \Rightarrow \Delta, \varphi[x/a] \qquad \varphi[x/\tau], \Pi \Rightarrow \Sigma}{\Gamma, \Pi \Rightarrow \Delta, \Sigma}$$

where a is not free in τ, Γ, Δ.

Such a rule is closer to the ordinary way of defining rules in SC setting since it is separated in a sense that no other constant is present in the schema. On the other hand, it is an elimination, not an introduction rule, similarly as (REP).

The last option for a nontrivial rule expressing reflexivity of equality is Read's rule from ND presented in the previous section. It may be used also in SC framework for formalization of reflexivity. It looks like this:

$$\frac{\varphi[x/\tau_1], \Gamma \Rightarrow \Delta, \varphi[x/\tau_2]}{\Gamma \Rightarrow \Delta, \tau_1 = \tau_2},$$

where φ is a predicate not present in Γ, Δ.

Such a rule is considered also by Parlamento and Previale (2019) whereas Restall (2020) considers a rule that is closer to Więckowski's solution (in fact he considers a stronger rule — see the next section):

$$(\Rightarrow=) \qquad \frac{\varphi[x/\tau_1], \Gamma \Rightarrow \Delta, \varphi[x/\tau_2] \qquad \varphi[x/\tau_2], \Gamma \Rightarrow \Delta, \varphi[x/\tau_1]}{\Gamma \Rightarrow \Delta, \tau_1 = \tau_2}$$

where φ is a predicate not present in Γ, Δ.

These two rules seem to be the best choice from the syntactical point of view, since they are nontrivial equality introduction rules. Moreover they are separate, weakly symmetric (together with (2=), (3=) or (7=) even symmetric) and explicit. They also satisfy the subformula property in the generalised sense. Note also that for any simple applied language with n primitive predicates, these rules may be replaced with rules which do not refer to fresh predicate parameters. Consider a language having only n unary predicates, then a suitable rule will have just n (Read's variant) or $2n$ (Restall's variant) premisses. Of course in the case of languages having k-ary predicate constants (for $k > 1$) the situation is more complicated since for every such predicate we must have k- (Read variant) or $2k$-premisses which take into account all positions of τ_1, τ_2 as arguments of this predicate, exactly as in Quine's counterpart of LA.

Now let us consider which combinations of rules yield cut-free SC. At first we consider only how rules (2=)-(8=) behave in this respect together with (REF) or (REP). It is well known, as shown in Negri and Plato (2001), that SC with (REP) is cut free. In fact, also (2=) allows for cut elimination in LK again only with (REP). A similar situation holds for (6=). On the other hand (5=), (7=) and (8=) provide cut-free LK independently of the choice of (REP) or (REF). For systems with (3=) and (4=) it is not clear if cut elimination can be constructively proved. Consider the following:

$$\dfrac{(3=)\,\dfrac{\Gamma \Rightarrow \Delta, a = c}{a = b, \Gamma \Rightarrow \Delta, b = c} \qquad \dfrac{\Pi \Rightarrow \Sigma, Ab}{b = c, \Pi \Rightarrow \Sigma, Ac}\,(3=)}{a = b, \Gamma, \Pi \Rightarrow \Delta, \Sigma, Ac}\,(\text{Cut})$$

It is neither possible to reduce the height of this cut or the complexity of cut-formula in the standard manner. A similar counterexample may be easily provided for (4=). Of course one can easily notice that in such cases the problem is connected with the fact that equalities are allowed as instances of φ in schemata of the respective rules. Of course if we think of rules for equality not only satisfying some desirable syntactic criteria for logicality, but also as being in a sense definitions of this constant, it would be desirable to restrict instances of φ to atomic formulae other than equalities. But then neither symmetry nor transitivity of equality can be proved. In fact, this holds for all seven rules considered in connection with (REF) or (REP).

Although most of the systems are cut-free, taking into account other properties, the only reasonable candidate for our aim is LK with (REF) and (7=). In the remaining combinations at least one rule is not a rule of equality introduction. Still (REF) is also rather a poor candidate for our aim. So eventually we should take (7) and ($\Rightarrow =$) (or its one-premiss version due to Read) as the pair of rules which, at least at the first sight, look better. In the next section we examine LK with such a pair of rules in the light of criteria of logicality proposed by Hacking.

5 Hacking's criterion

Hacking (1979) based his considerations on the criteria of logicality on the standard form of SC with canonical logical and structural rules[5]. He did not consider equality

[5] Although he is using sequents built from finite sets so contraction rules are dispensable.

but, as we will show, his analysis may be applied to this constant. In fact, Hacking follows closely Gentzen's suggestions concerning rules as possible definitions of constants. Hence rules of this kind have to satisfy all the properties we discussed in the preceding section with special attention paid to the subformula property. As we noticed, the two rules we have chosen satisfy this property only in the generalised sense but it seems that this generalisation is reasonable. However, this is not enough. In order to satisfy Hacking's requirements of logicality it should be possible to show three elimination theorems for a system consisting of such rules:

1. every axiom built from complex formulae must be eliminated in favor of atomic axioms;
2. every application of weakening rules introducing complex formulae must be eliminated in favor of weakening made on atomic formulae;
3. applications of cut must be eliminated.

The last one is Gentzen's famous Hauptsatz, whereas the remaining ones have a slightly weaker character since they do not postulate the complete elimination of purely logical axioms or weakening but only their reduction to the atomic level. In fact, if we consider purely logical versions of SC, i.e., without primitive structural rules, like G3 [6], then atomic axioms are present as primitive and all these requirements may be presented in a more uniform way as respective admissibility results. However, we opt for LK as our basis and check how it behaves when enriched with equality rules.

We start with SC for pure and applied FOLI. As we concluded the preceding section the only reasonable candidates are (7=) as the antecedent introduction rule, and Read's or Restall's rule as the succedent introduction rule. Accordingly we consider two variants, and in both (7=) will be taken (possibly in two symmetric versions) as the antecedent introduction rule. LKI1 is LK with:

$$(1{=}{\Rightarrow}) \qquad \frac{\Gamma \Rightarrow \Delta, \varphi[a/\tau_1] \qquad \varphi[a/\tau_2], \Pi \Rightarrow \Sigma}{\tau_1 = \tau_2, \Gamma, \Pi \Rightarrow \Delta, \Sigma}$$

$$(1{\Rightarrow}{=}) \qquad \frac{\varphi[a/\tau_1],\ \Gamma \ \Rightarrow\ \Delta,\ \varphi[a/\tau_2]}{\Gamma \Rightarrow \Delta, \tau_1 = \tau_2}$$

where in the latter rule φ is an atomic predicate not in Γ, Δ (in antecedent introduction rules it is an arbitrary atomic formula).

Whereas in LKI2 we have instead:

$$(1{=}{\Rightarrow}) \qquad \frac{\Gamma \Rightarrow \Delta, \varphi[a/\tau_1] \qquad \varphi[a/\tau_2], \Pi \Rightarrow \Sigma}{\tau_1 = \tau_2, \Gamma, \Pi \Rightarrow \Delta, \Sigma}$$

$$(2{=}{\Rightarrow}) \qquad \frac{\Gamma \Rightarrow \Delta, \varphi[a/\tau_2] \qquad \varphi[a/\tau_1], \Pi \Rightarrow \Sigma}{\tau_1 = \tau_2, \Gamma, \Pi \Rightarrow \Delta, \Sigma}$$

---

[6] See, e.g., Troelstra and Schwichtenberg (1996) or Negri and von Plato (2001).

$$(2{\Rightarrow}{=})\quad \frac{\varphi[a/\tau_1], \Gamma \Rightarrow \Delta, \varphi[a/\tau_2] \quad \varphi[a/\tau_2], \Pi \Rightarrow \Sigma, \varphi[a/\tau_1]}{\Gamma, \Pi \Rightarrow \Delta, \Sigma, \tau_1 = \tau_2}$$

where in the latter rule φ is an atomic predicate not in Γ, Δ.

The distinction between LKI1 and LKI2 follows from the way we treat equalities. If they are treated as atomic formulae, LKI1 is sufficient; otherwise we need LKI2. The key point is how to prove symmetry in both systems; in the former we have the following proof:

$$(1{=}{\Rightarrow})\ \dfrac{(1{\Rightarrow}{=})\ \dfrac{A\tau_1 \Rightarrow A\tau_1}{\Rightarrow \tau_1 = \tau_1} \qquad \tau_2 = \tau_1 \Rightarrow \tau_2 = \tau_1}{\tau_1 = \tau_2 \Rightarrow \tau_2 = \tau_1}$$

where $\varphi[a/\tau_i]$ is $a = \tau_1[a/\tau_i]$.

In LKI2 it looks like that:

$$(2{\Rightarrow}{=})\ \dfrac{(2{=}{\Rightarrow})\ \dfrac{A\tau_2 \Rightarrow A\tau_2 \quad A\tau_1 \Rightarrow A\tau_1}{\tau_1 = \tau_2, A\tau_2 \Rightarrow A\tau_1} \qquad \dfrac{A\tau_1 \Rightarrow A\tau_1 \quad A\tau_2 \Rightarrow A\tau_2}{\tau_1 = \tau_2, A\tau_1 \Rightarrow A\tau_2}\ (1{=}{\Rightarrow})}{\tau_1 = \tau_2 \Rightarrow \tau_2 = \tau_1}$$

Both systems are complete and it is not difficult to extend them to cover complex terms. However suitable rules of the antecedent introduction for equality must be modified in the way which ensures derivability of CT. It is only necessary to require that in two atomic formulae which are side formulae in both premisses the respective occurrences of terms being arguments of principal formula, may appear not only as arguments of this predicate but also as arguments of complex terms being its arguments. Under this generalised understanding a proof of the simplest case of CT in LKI2 for some unary operation f looks like this:

$$(2{\Rightarrow}{=})\ \dfrac{(1{=}{\Rightarrow})\ \dfrac{Af\tau_1 \Rightarrow Af\tau_1 \quad Af\tau_2 \Rightarrow Af\tau_2}{\tau_1 = \tau_2, Af\tau_1 \Rightarrow Af\tau_2} \qquad \dfrac{Af\tau_2 \Rightarrow Af\tau_2 \quad Af\tau_1 \Rightarrow Af\tau_1}{\tau_1 = \tau_2, Af\tau_2 \Rightarrow Af\tau_1}\ (2{=}{\Rightarrow})}{\tau_1 = \tau_2 \Rightarrow f\tau_1 = f\tau_2}$$

There are other solutions which work as well as LKI2 but are simpler. A close inspection of proofs needed to prove completeness shows that it is also possible to obtain two variants of LKI1 keeping the proviso concerning equalities in LKI2 (i.e., that they are not treated as atoms being possible instances of the antecedent introduction rule). Instead of one 2-premiss rule (2⇒=) we can add two one-premiss rules: It is enough either to use both rules of the antecedent introduction, as in LKI2, or to add a symmetric version of (1⇒=) (corresponding to the right premiss of (2⇒=)):

$$(1{\Rightarrow}{=}')\quad \frac{\varphi[a/\tau_2],\ \Gamma \Rightarrow \Delta,\ \varphi[a/\tau_1]}{\Gamma \Rightarrow \Delta, \tau_1 = \tau_2}$$

(with the same proviso concerning fresh φ). It may be easily checked that suitable proofs of symmetry and transitivity may be obtained from the proofs in LKI2 stated above by deleting derivations of one branch. This provides an adequate SC for FOL with equality and simplifies many proofs since the branching factor is lower.

The last thing concerns the question how these variants of SC fare with respect to Hacking's criteria of logicality. We have noticed above that the subformula property is generalised but in the sense which is acceptable. As for eliminability conditions one may easily check that the first two hold for all stated versions of LKI. What with cut elimination? First of all note that in this respect LKI1 is not a suitable system. Consider the case where the cut formula is an equality which is a principal formula of the last applied rule, and moreover side formulae of (1=⟹) application were equalities:

$$\dfrac{\dfrac{\varphi(\tau_1), \Gamma \Rightarrow \Delta, \varphi(\tau_2)}{\Gamma \Rightarrow \Delta, \tau_1 = \tau_2} \qquad \dfrac{\Pi \Rightarrow \Sigma, \tau_1 = \tau_3 \qquad \tau_2 = \tau_3, \Pi \Rightarrow \Sigma}{\tau_1 = \tau_2, \Pi \Rightarrow \Sigma}}{\Gamma, \Pi \Rightarrow \Delta, \Sigma}$$

In this case, there is no possibility of making a reduction either on the height or on the degree (complexity) of cut formula.

Perhaps LKI2, or modified versions of LKI1 with additional rule but excluding equality as atomic, are better. Again consider the case where the cut formula is an equality which is a principal formula of the last applied rule:

$$\dfrac{\dfrac{\varphi(\tau_1), \Gamma \Rightarrow \Delta, \varphi(\tau_2) \qquad \varphi(\tau_2), \Gamma \Rightarrow \Delta, \varphi(\tau_1)}{\Gamma \Rightarrow \Delta, \tau_1 = \tau_2} \qquad \dfrac{\Pi \Rightarrow \Sigma, \psi(\tau_1) \qquad \psi(\tau_2), \Pi \Rightarrow \Sigma}{\tau_1 = \tau_2, \Pi \Rightarrow \Sigma}}{\Gamma, \Pi \Rightarrow \Delta, \Sigma}$$

(The argument for the right premiss deduced by (2=⟹) is symmetric). We assume that all atomic formulae have complexity 0 but equalities have complexity 1 since = is treated on a par with other logical constants. In order to apply induction on the complexity of the cut formula we should however first unify side formulae in the premisses of both applications of equality rules. The situation is analogous to the case of reduction made when the cut formula is introduced by quantifier rules. In this case we first apply substitution to fresh parameter of the premiss of ($\Rightarrow \forall$) (or ($\exists\Rightarrow$)) and then we can make a reduction by making cut on the premisses. In the present case, if a similar kind of substitution is possible, the following reduction of cut degree would do:

$$\dfrac{\dfrac{\dfrac{\Pi \Rightarrow \Sigma, \psi(\tau_1) \qquad \psi(\tau_1), \Gamma \Rightarrow \Delta, \psi(\tau_2)}{\Pi, \Gamma \Rightarrow \Sigma, \Delta, \psi(\tau_2)} \qquad \psi(\tau_2), \Pi \Rightarrow \Sigma}{\Gamma, \Pi, \Pi \Rightarrow \Delta, \Sigma, \Sigma}}{\Gamma, \Pi \Rightarrow \Delta, \Sigma}$$

where in the middle sequent $\psi(\tau_1), \psi(\tau_2)$ were substituted for unique occurrences of $\varphi(\tau_1), \varphi(\tau_2)$ in the leftmost premiss of the previuos figure. The problem is that no corresponding result for substitution of fresh atomic formulae can be proved in the presence of rules for antecedent introduction. Consider the following example which shows the source of the problem:

$$\dfrac{\tau_3 = \tau_2, \varphi(\tau_1), \Gamma_1 \Rightarrow \Delta_1, \varphi(\tau_2) \qquad \dfrac{\varphi(\tau_2) \Rightarrow \varphi(\tau_2) \qquad \varphi(\tau_3), \Gamma_2 \Rightarrow \Delta_2, \varphi(\tau_1)}{\tau_3 = \tau_2, \varphi(\tau_2), \Gamma_2 \Rightarrow \Delta_2, \varphi(\tau_1)}}{\tau_3 = \tau_2, \Gamma \Rightarrow \Delta, \tau_1 = \tau_2}$$

Assuming that it is a proof of the left premiss of the considered cut application we cannot change φ for ψ since the middle top sequent will not be an axiom.

This shows that in case of pure and applied FOLI our solution does not satisfy the most important condition in the list of logicality criteria provided by Hacking. Since it is not applicable to simple FOLI there is only one possibility — that it works for simple applied versions of FOLI. Certainly it works for all languages having only one unary predicate constant. In this case all the rules of LKI2 (or variants of LKI1) have the same shape but with $\varphi(\tau)$ being always $A\tau$ and with no side condition for the succedent introduction rules. All proofs are intact, moreover, the problem connected with the proof of cut elimination does not hold since there is only one atomic formula in both premisses of cut, i.e., both $\varphi(\tau)$ and $\psi(\tau)$ are $A\tau$ and the reduction of cut degree holds. What with languages having richer signature? Following Quine's recipe mentioned in Section 3 we must suitably modify both rules of succedent introduction: In the case of (1⇒=) for each n-ary predicate constant we must use n premisses with respective term as an argument in all positions, for (2⇒=) we must introduce $2n$ of such premisses. For example suitable form of (1⇒=) for the language with unary A and binary R which was considered in Section 3 for the illustration of LA′ looks like that:

$$\frac{A\tau_1, \Gamma_1 \Rightarrow \Delta_1, A\tau_2 \qquad R\tau_1 a, \Gamma_2 \Rightarrow \Delta_2, R\tau_2 a \qquad Ra\tau_1, \Gamma_3 \Rightarrow \Delta_3, Ra\tau_2}{\Gamma \Rightarrow \Delta, \tau_1 = \tau_2}$$

where a is not in $\Gamma, \Delta, \tau_1, \tau_2$.

Rules for the antecedent introduction remain intact but φ is now an instance of an arbitrary predicate constant with at least one occurrence of suitable term. Such versions of SC with succedent introduction rules having no fixed number of premisses but relative to the signature are not very satisfactory. In particular, from the proof-search point of view in the case of richer signature the branching factor is too big to use them in practice. However, we are not concerned here with practical application but with satisfiability of theoretical desiderata and from this point of view this solution works. In particular, cut elimination can be proved since every application of (2⇒=) (and (1⇒=) in variants of LKI1) has always among its premisses the one which has identical side formula as the ones occuring in actual premisses of (1=⇒) or (2=⇒).

6 Došen's criterion

In this section we focus on the criterion of logicality proposed by Došen. It can be seen as a successful refinement of a proposal due to Popper (1947a; 1947b) concerned with some conception of proof-theoretical semantics which was however not articulated in a satisfactory way. Popper tried to characterize constants by means of inferential definitions which yield double-valid rules characterising constants of the form:

$$(\to)\ \frac{\varphi, \chi \Rightarrow \psi}{\chi \Rightarrow \varphi \to \psi} \qquad\qquad (\wedge)\ \frac{\varphi, \psi \Rightarrow \chi}{\varphi \wedge \psi \Rightarrow \chi}$$

$$(\vee)\ \frac{\varphi \Rightarrow \chi \quad \psi \Rightarrow \chi}{\varphi \vee \psi \Rightarrow \chi} \qquad (\neg)\ \frac{\varphi,\ \psi\ \Rightarrow\ \chi}{\neg\chi, \psi\ \Rightarrow\ \neg\varphi}$$

Popper's project was criticised by Kleene, Curry, and many others, however it was convincingly shown by Schroeder-Heister (1984) (see also Schroeder-Heister, 2006 and Binder and Piecha, 2021) that his works contain an interesting proposal for a slightly weaker plan of providing criteria for being a logical constant. Such an enterprise was undertaken much later by Došen, first in his doctoral thesis and then in Došen (1989). It seems that if we do not use them as a way for establishing a variant of proof theoretic semantics but only as an independent criterion of logicality it works well. In fact, the opinions on the very nature of the relationships between criteria of logicality and proof-theoretic semantics differ strongly. Despite of Došen's own remarks that in his proposal rules are not intended as definitions of constants[7] it may be treated as a good framework even for developing proof-theoretic semantics. The idea of the application of double-line rules as definitional rules reappeared in such significantly different frameworks like Koslow's structural approach to logics (see Koslow, 1992), Sambin's Basic Logic (see Sambin, Battilotti, and Faggian, 2000), or categorical logic (see Maruyama, 2016). Moreover, this strong interpretation of Došen's proposal is provided independently by Gratzl and Orlandelli (2017) and by Restall (2019). In particular, the former work proposes an interesting explanation of the reasons for such a choice in terms of harmony. Since the 1960s, harmony was treated as crucial for the explanation of proper rules for defining logical constants, and a lot of work was offered in which a clarification of this notion was provided[8]. Harmony is in general understood as a kind of balance between two kinds of rules, introduction and elimination in ND, or antecedent and succedent introduction in SC. However, this notion is explicated in many different ways. Gratzl and Orlandelli (2017) proposed an explanation of harmony as a kind of deductive equilibrium which was first proposed by Tennant (2010). Their approach may be seen as an improvement of Tennant's solution in the sense that it is purely local, i.e., an analysis of a constant in terms of rules is independent of what other constants are already present in the language. Moreover, in contrast to other approaches to harmony it allows for the unique determination of one kind of rules from the other kind and vice versa.

Došen's system described below serves as an exemplification of his theory of criteria of logicality. The starting point of analysis of logical constants is the conviction that logic is the science of formal proofs. Hence basic formal proofs are of purely structural character, i.e., where only structural rules were applied[9]. It follows that an expression is logical iff it is analysable in purely structural terms. As he emphasized in the title of his paper — logical constants are punctuation marks. An analysis should satisfy three conditions:

[7] In particular, the proposed criteria do not necessarily satisfy the criteria of eliminability and non-creativity required from the well-stated definitions.

[8] See, e.g., Schroeder-Heister (2012), Poggiolesi (2011) or Kürbis (2019).

[9] It is in a sense a development of Hertz's programme (see Hertz, 1929).

1. any sentence (sequent) with constant to be analysed should be equivalent to a sentence (sequent) without it;
2. the analysis must be adequate (i.e., sound and complete);
3. the constant must be uniquely characterised.

His rules provide such an analysis in the sense that on the one side we have only structural sequents, i.e., with no constant displayed. According to Došen, in order to claim that an expression is a logical constant it is necessary to find such a double-valid rule which after addition to structural rules allows for obtaining a full characterisation of this constant. Došen (1989) proposed a structural version of LK, which we will call SLK, in the language without negation but with $\bot$ and $\top$ but we introduce a rule for negation instead. In this system the set of structural rules is primitive and not eliminable. Every constant is characterised by means of only one, but double-line (i.e., invertible) rule:

$$(\to)\ \frac{\varphi,\Gamma\Rightarrow\Delta,\psi}{\Gamma\Rightarrow\Delta,\varphi\to\psi}\qquad(\wedge)\ \frac{\Gamma\Rightarrow\Delta,\varphi\quad\Gamma\Rightarrow\Delta,\psi}{\Gamma\Rightarrow\Delta,\varphi\wedge\psi}\qquad(\neg)\ \frac{\varphi,\Gamma\Rightarrow\Delta}{\Gamma\Rightarrow\Delta,\neg\varphi}$$

$$(\vee)\ \frac{\varphi,\Gamma\Rightarrow\Delta\quad\psi,\Gamma\Rightarrow\Delta}{\varphi\vee\psi,\Gamma\Rightarrow\Delta}\qquad(\forall)^1\ \frac{\Gamma\Rightarrow\Delta,\varphi[x/a]}{\Gamma\Rightarrow\Delta,\forall x\varphi}\qquad(\exists)^1\ \frac{\varphi[x/a],\Gamma\Rightarrow\Delta}{\exists x\varphi,\Gamma\Rightarrow\Delta}$$

where in the latter two rules a is not in Γ, Δ, φ. In each case in addition to the rule of introduction we have also a rule of elimination if we read the rule upside down. Every rule is then a counterpart of a suitable equivalence characterising the respective constant within Scott's theory of consequence relations (see Scott, 1974). In what follows we will use notation ($\downarrow\!\to$) and ($\uparrow\!\to$) for suitable halves of the rule for implication and similarly for other constants.

The first condition is obviously satisfied for all rules. The second one can be shown by providing a proof of equivalence with some standard version of SC which is known to be adequate; our version of LK is sufficient. Since one half of each of Došen's rules corresponds to a suitable introduction rule such a proof amounts in principle to demonstration of derivability of the remaining rules by means of structural rules only. For example if we take his rule for implication, then ($\downarrow\!\to$) is exactly our ($\Rightarrow\to$) and the following shows derivability of ($\to\Rightarrow$) by means of ($\uparrow\!\to$):

$$\dfrac{\Gamma\Rightarrow\Delta,\varphi \qquad \dfrac{(\uparrow\to)\ \dfrac{\varphi\to\psi\Rightarrow\varphi\to\psi}{\varphi,\varphi\to\psi\Rightarrow\psi}\qquad \psi,\Pi\Rightarrow\Sigma}{\varphi,\varphi\to\psi,\Pi\Rightarrow\Sigma}\,(Cut)}{\varphi\to\psi,\Gamma,\Pi\Rightarrow\Delta,\Sigma}\,(Cut)$$

whereas the converse derivability goes like that:

$$\dfrac{\Gamma\Rightarrow\Delta,\varphi\to\psi\qquad \dfrac{\varphi\Rightarrow\varphi\qquad\psi\Rightarrow\psi}{\varphi\to\psi,\varphi\Rightarrow\psi}\,(\to\Rightarrow)}{\varphi,\Gamma\Rightarrow\Delta,\psi}\,(Cut)$$

Note that both demonstrations of derivability are structural and moreover in addition to the respective logical rules they use only axioms and cut. Such demonstration of interderivability is sufficient to show that rules of LK are harmonious in the sense explained by Gratzl and Orlandelli (2017). One may easily check that the remaining halves of Došen's rules are also interderivable with $(\wedge\Rightarrow)$, $(\neg\Rightarrow)$, $(\Rightarrow\vee)$, $(\forall\Rightarrow)$ and $(\Rightarrow\exists)$. However, in the case of interderivability of $(\wedge\Rightarrow)$ and $(\Rightarrow\vee)$ with $(\uparrow\wedge)$ and $(\downarrow\vee)$ we must additionally use contraction and weakening. Derivations are still structural but some authors tend to be careful with that and admit as fully satisfactory analyses only those where axioms and cut are the only rules (see, e.g., Gratzl and Orlandelli, 2017). However, it is easy to provide a remedy. In fact, Došen's rules were based on the original Gentzen's rules for LK which are additive and for these rules interderivability requires only axioms and cut. On the other hand we have chosen all rules to be multiplicative. If we change Došen's rules for conjunction and disjunction into:

$$(\wedge)\ \frac{\varphi, \psi,\ \Gamma \Rightarrow\ \Delta}{\overline{\varphi \wedge \psi, \Gamma \Rightarrow \Delta}} \qquad\qquad (\vee)\ \frac{\Gamma \Rightarrow\ \Delta, \varphi, \psi}{\overline{\Gamma \Rightarrow \Delta, \varphi \vee \psi}}$$

we can provide interderivability proofs for our version of LK by means of axioms and cut only.

On the other hand note that showing derivability of $(\forall\Rightarrow)$ and $(\Rightarrow\exists)$ in structural variant requires additional structural rule of substitution:

$$(SUB)\qquad \frac{\Gamma \Rightarrow \Delta}{\Gamma[a/\tau] \Rightarrow \Delta[a/\tau]}$$

which is necessary to enable unrestricted instantiation (modulo correct substitution) of terms in these two rules.

The last condition, i.e., uniqueness may be demonstrated as the provability for each constant $*$ of two sequents: $*\varphi \Rightarrow \star\varphi$ and $\varphi \Rightarrow *\varphi$, where $\star$ is a notational variant having the same rule. Suitable proofs are trivial in this setting (although not in general — see, e.g., Došen, 1985).

How can equality be characterised in this framework? In fact, Došen proposed a rule which is of global character:

$$(=)\qquad \frac{\Gamma[a/\tau_1] \Rightarrow \Delta[a/\tau_1]}{\overline{\tau_1 = \tau_2, \Gamma \Rightarrow \Delta}}$$

In Section 2 we explained why such kind of rules cannot be treated as providing a criterion of logicality. What kind of local rules can be used instead? The obvious candidates are double versions of the rules for succedent introduction which we examined in the last section:

$$(1=)\qquad \frac{\varphi[a/\tau_1],\ \Gamma \Rightarrow\ \Delta,\ \varphi[a/\tau_2]}{\overline{\Gamma \Rightarrow \Delta, \tau_1 = \tau_2}}$$

$$(2=)\qquad \frac{\varphi[a/\tau_1], \Gamma \Rightarrow \Delta, \varphi[a/\tau_2] \qquad \varphi[a/\tau_2], \Gamma \Rightarrow \Delta, \varphi[a/\tau_1]}{\overline{\Gamma \Rightarrow \Delta, \tau_1 = \tau_2}}$$

where in both rules φ is atomic predicate not in Γ, Δ. The first one is an obvious SC version of Read's rule and the second of Restall's one. Let us consider a pure or applied FOLI as formalised by Došen's SLK in two variants: SLK1 (with (1=)) and SLK2 (with (2=)). We start with the latter, moreover we add, similarly as in LKI2, the proviso that equalities are not atomic predicates, so only predicate parameters (and other predicate constants in applied version) are admitted as instances of φ. This system satisfies the first condition so we must check the second, i.e., adequacy. Both directions of (2=) are cases of (2⇒=) and (4=) respectively so SLK2 is sound. Of course we can also easily provide demonstrations of the interderivability of (1=↑) and (2=↑) with (1=⇒) and (2=⇒) which is enough to show that the equality rules of LKI1 and LK2 are harmonious in the sense of Gratzl and Orlandelli (2017). To show that SLK (in both versions) is complete we must be able to prove the reflexivity axiom and LP which are immediate:

$$(\downarrow 2{=})\ \frac{A\tau \Rightarrow A\tau \qquad A\tau \Rightarrow A\tau}{\Rightarrow \tau = \tau}$$

$$(\uparrow 2{=})\ \frac{\tau_1 = \tau_2 \Rightarrow \tau_1 = \tau_2}{\tau_1 = \tau_2, \varphi[a/\tau_1] \Rightarrow \varphi[a/\tau_2]}$$

where φ is any atomic predicate (parameter or constant).

However, it is not sufficient. Since equalities cannot be instances of φ we must provide also proofs of symmetry and transitivity for =:

$$\begin{array}{l} (2{=}\uparrow) \\ (2{=}\downarrow) \end{array} \frac{\dfrac{\tau_1 = \tau_2 \Rightarrow \tau_1 = \tau_2}{\tau_1 = \tau_2, A\tau_2 \Rightarrow A\tau_1} \qquad \dfrac{\tau_1 = \tau_2 \Rightarrow \tau_1 = \tau_2}{\tau_1 = \tau_2, A\tau_1 \Rightarrow A\tau_2}\ (2{=}\uparrow)}{\tau_1 = \tau_2 \Rightarrow \tau_2 = \tau_1}$$

for transitivity we derive:

$$\begin{array}{l} (2{=}\uparrow) \\ (\mathrm{Cut}) \end{array} \frac{\dfrac{\tau_1 = \tau_2 \Rightarrow \tau_1 = \tau_2}{\tau_1 = \tau_2, A\tau_1 \Rightarrow A\tau_2} \qquad \dfrac{\tau_2 = \tau_3 \Rightarrow \tau_2 = \tau_3}{\tau_2 = \tau_3, A\tau_2 \Rightarrow A\tau_3}\ (2{=}\uparrow)}{\tau_1 = \tau_2, \tau_2 = \tau_3, A\tau_1 \Rightarrow A\tau_3}$$

and

$$\begin{array}{l} (2{=}\uparrow) \\ (\mathrm{Cut}) \end{array} \frac{\dfrac{\tau_2 = \tau_3 \Rightarrow \tau_2 = \tau_3}{\tau_2 = \tau_3, A\tau_3 \Rightarrow A\tau_2} \qquad \dfrac{\tau_1 = \tau_2 \Rightarrow \tau_1 = \tau_2}{\tau_1 = \tau_2, A\tau_2 \Rightarrow A\tau_1}\ (2{=}\uparrow)}{\tau_1 = \tau_2, \tau_2 = \tau_3, A\tau_3 \Rightarrow A\tau_1}$$

which together by (2=↓) yield $\tau_1 = \tau_2, \tau_2 = \tau_3 \Rightarrow \tau_1 = \tau_3$.

It is important to note that in both proofs the use of both premisses of (2=) is essential. It is not possible to provide a proof of symmetry and transitivity of = in SLK1 with the same proviso. On the other hand, if we admit equalities as possible instances of φ, SLK1 appears also to be an adequate formalization of pure or applied FOLI, similarly as LKI1. Proofs of symmetry and transitivity look like that:

$$\begin{array}{l} (1{=}\downarrow) \\ (\mathrm{Cut}) \end{array} \frac{\dfrac{A\tau_1 \Rightarrow A\tau_1}{\Rightarrow \tau_1 = \tau_1} \qquad \dfrac{\tau_1 = \tau_2 \Rightarrow \tau_1 = \tau_2}{\tau_1 = \tau_1, \tau_1 = \tau_2 \Rightarrow \tau_2 = \tau_1}\ (1{=}\uparrow)}{\tau_1 = \tau_2 \Rightarrow \tau_2 = \tau_1}$$

where in the application of $(1{=}\uparrow)$ on the right $\varphi[a/\tau_i]$ is $a = \tau_1[a/\tau_i]$.

$$\dfrac{\tau_1 = \tau_2 \Rightarrow \tau_1 = \tau_2 \qquad \dfrac{\tau_2 = \tau_3 \Rightarrow \tau_2 = \tau_3}{\tau_1 = \tau_2, \tau_2 = \tau_3 \Rightarrow \tau_1 = \tau_3}\,(1{=}\uparrow)}{\tau_1 = \tau_2, \tau_2 = \tau_3 \Rightarrow \tau_1 = \tau_3}\,(\mathrm{Cut})$$

where in the application of $(1{=}\uparrow)$ on the right $\varphi[a/\tau_i]$ is $\tau_1 = a[a/\tau_i]$. Since the proof of LP is correct also with (1=) and the proof of reflexivity by (1=) is involved in the proof of symmetry above we are done. Also the proofs of uniqueness are directly obtainable in both versions of SLK. The above proofs show also that simple FOLI may be also formalised by means of SLK1; it is enough to change in the proof of reflexivity $A\tau \Rightarrow A\tau$ for $\tau = \tau \Rightarrow \tau = \tau$.

We leave the problem of formalization of simple applied versions of FOLI in SLK — it may follow the way described in the preceding section. The only difference is that now such many-premiss rules are treated as double valid. In particular, in the case of the language with just one unary predicate we can use just (2=) without proviso that φ is fresh.

There is no problem with the treatment of complex terms in the structural variant provided we will treat substitution of terms in the same way as in the systems of the preceding section. For illustration we can show how to provide a proof of CT for a binary operation f of the form: $\forall xyzv(x = y \wedge z = v \rightarrow fxz = fyv)$

We can prove:

$$(2{=}\downarrow)\,\dfrac{(2{=}\uparrow)\,\dfrac{a = b \Rightarrow a = b}{a = b, Afac \Rightarrow Afbc} \qquad \dfrac{a = b \Rightarrow a = b}{a = b, Afbc \Rightarrow Afac}\,(2{=}\uparrow)}{a = b \Rightarrow fac = fbc}$$

In a similar way we prove $c = d \Rightarrow fbc = fbd$. Since $fac = fbc, fbc = fbd \Rightarrow fac = fbd$ is provable as an instance of transitivity we obtain $a = b, c = d \Rightarrow fac = fbd$ by two applications of cut, and then the result by $(\wedge), (\rightarrow), (\forall)$. This proof generalizes for any n-ary operation.

We finish this section with the remark that similarly as in the case of other constants, there is another possibility of characterizing equality by means of invertible rule. We can use the antecedent-based rule:

$$(=3) \qquad \dfrac{\Gamma \Rightarrow \Delta, \varphi[a/\tau_1], \varphi[a/\tau_2] \qquad \varphi[a/\tau_1], \varphi[a/\tau_2], \Gamma \Rightarrow \Delta}{\tau_1 = \tau_2, \Gamma \Rightarrow \Delta}$$

This rule has an advantage of having no side condition on φ except that it is atomic. However, showing that it satisfies Došen's criteria is slightly more involved and we do not pursue this task here.

7 Final comments

The results of our analysis do not provide a decisive answer to the problem that was posed concerning logicality of equality. They show that the status of equality as a logical constant is dependent not only on the criteria which are taken into account but also sensitive to the version of the language to which equality is appended. In this respect the structural variant of Došen works better. All criteria of logicality hold for pure, applied and simple applied variants of FOLI introduced in Section 6. Moreover, invertible rules of this calculus allow us to show that standard introduction rules from Section 5 are harmonious in the sense advocated by Gratzl and Orlandelli (2017). However, in the latter case, i.e., SC realizing Hacking's approach, it is striking that the most important condition, namely cut eliminability, fails for the pure and applied version of FOLI. On the other hand it can be proved for many other sequent formalizations of FOLI but with non-canonical equality rules. Such rules however cannot be considered as proof-theoretic characterisations of a logical constant. Note also that simple FOLI does not satisfy the criteria of logicality in either formalization. Neither cut elimination holds for LKI variants, nor the first condition of Došen's analysis holds if equality is a sole predicate.

Our analysis was provided in the setting of standard SC and its slightly nonstandard variant but still based on the standard notion of a sequent. One may ask if the application of some other generalised setting may show in a more decisive way that equality satisfies the criteria of logicality. For example, in Hacking (1979) modal operators are taken as a negative example since in the setting of standard SC they cannot be characterised by means of canonical rules. But one can find more satisfying solutions on the ground of generalised formalisations. For example, in the setting of display logic one can provide rules for modal operators satisfying Hacking's demands (see, e.g., Wansing, 1999).

One possible generalization which works for equality can be based on using sequents with terms occuring on a par with formulae. The idea of formal systems with terms treated as fully fledged elements of deductions is not new. For the first time it was introduced by Jaśkowski (1934) in his first system of ND. Quite recently the idea was independently undertaken and developed in the setting of ND by Textor (2017) and Gazzari (2019). In the framework of SC it was developed by Restall (2019) and in Indrzejczak (2021). It seems that in this slightly generalised setting not only equality but also several kinds of term-forming operators may be formalised in a way important for further development of proof-theoretic semantics.

In this study we have restricted considerations to classical FOLI but the results we obtained may be easily extended to Intuitionistic version by restricting sequents to single-succedent ones. It is an open problem how to adapt this kind of analysis to other predicates of similar character applied in nonclassical logics.

Acknowledgements I am greatly indebted to Nils Kürbis and the referees of this paper for many valuable remarks which helped to improve the final version. The results reported in this paper are supported by the National Science Centre, Poland (grant number: DEC-2017/25/B/HS1/01268).

References

Avron, A. and I. Lev (2001). Canonical propositional gentzen-type systems. In: *Proceedings of IJCAR'01*. Vol. 2083. LNCS, 529–543.

Baaz, M. and A. Leitsch (2011). *Methods of Cut-Elimination*. Springer.

Bell, J. L. and M. Machover (1977). *A Course in Mathematical Lgic*. Amsterdam: North-Holland.

Binder, D. and T. Piecha (2021). Popper on quantification and identity. In: *Karl Popper's Science and Philosophy*. Ed. by Z. Parusnikova and D. Merrit. To appear. Springer.

Church, A. (1956). *Introduction to Mathematical Logic*. Vol. I. Princeton: Princeton University Press.

Degtyarev, A. and A. Voronkov (2001). Equality reasoning in sequent-based calculi. In: *Handbook of Automated Reasoning vol I*. Ed. by A. Robinson and A. Voronkov. Elsevier, 611–706.

Došen, K. (1985). Sequent-systems for modal logic. *Journal of Symbolic Logic* 50, 149–159.

– (1989). Logical constants as punctuation marks. *Notre Dame Journal of Formal Logic* 30, 362–381.

Gallier J., H. (1986). *Logic for Computer Science*. New York: Harper and Row.

Gazzari, R. (2019). The calculus of natural calculation. In: *Proof-Theoretic Semantics: Assesment and Future Perspectives. Proceedings of the Third Tübingen Conference of Proof-Theoretic Semantics*. Ed. by T. Piecha and P. Schroeder-Heister. Tübingen, 123–139.

Gratzl, N. and E. Orlandelli (2017). Double-line harmony in sequent calculi. In: *The Logica Yearbook 2016*. Ed. by P. Arazim and T. Lavicka. College Publications, 157–171.

Griffiths, O. (2014). Harmonious rules for identity. *The Review of Symbolic Logic* 7, 499–510.

Hacking, I. (1979). What is logic? *Journal of Philosophy* 76, 285–319.

Hertz, P. (1929). Über axiomensysteme für beliebige satzsysteme. *Mathematische Annalen* 101, 457–514.

Hintikka, J. (1956). Identity, variables and impredicative definitions. *Journal of Symbolic Logic* 21, 225–245.

Indrzejczak, A. (2017). Tautology elimination, cut elimination and s5. *Logic and Logical Philosophy* 26, 461–471.

– (2018a). Cut-free modal theory of definite descriptions. In: *Advances in Modal Logic 12*. Ed. by G. B. et al. College Publications, 387–406.

– (2018b). Rule-generation theorem and its applications. *The Bulletin of the Section of Logic* 47, 265–281.

– (2019). Fregean description theory in proof-theoretical setting. *Logic and Logical Philosophy* 28, 137–155.

– (2021). A novel approach to equality. *Synthese* 199, 4749–4774.

Jaśkowski, S. (1934). On the rules of suppositions in formal logic. *Studia Logica* 1, 5–32.

Kahle, R. (2016). Towards a proof-theoretic semantics of equalities. In: *Advances in Proof-theoretic Semantics*. Ed. by T. Piecha and P. Schroeder-Heister. Springer, 153–160.

Kalish, D. and R. Montague (1957). Remarks on descriptions and natural deduction. *Archiv. für Mathematische Logik und Grundlagen Forschung* 3, 65–73.

– (1964). *Logicical Techniques of Formal Reasoning*. New York.

Kanger, S. (1957). *Provability in Logic*. Stockholm: Almqvist & Wiksell.

Klev, A. (2019). The harmony of identity. *Journal of Philosophical Logic* 48, 867–884.

Koslow, A. (1992). *A Structuralist Theory of Logic*. Cambridge: Cambridge University Press.

Kürbis, N. (2019). *Proof and Falsity. A Logical Investigation*. Cambridge: Cambridge University Press.

Lemmon, E. J. (1965). *Beginning Logic*. London: Nelson.

Manzano, M. (1999). *Model Theory*. Oxford: Oxford University Press.

– (2005). *Extensions of First-Order Logic*. Cambridge: Cambridge University Press.

Manzano, M. and M. C. Moreno (2017). Identity, equality, nameability and completeness. *The Bulletin of the Section of Logic* 46, 169–196.

Martin-Löf, P. (1971). Hauptsatz for the intuitionistic theory of iterated inductive definitions. In: *Proceedings of the Second Scandinavian Logic Symposium*. Ed. by J. E. Fenstad. North-Holland.

Maruyama, Y. (2016). Categorical harmony and paradoxes in proof-theoretic semantics. In: *Advances in Proof-Theoretic Semantics*. Ed. by T. Piecha and P. Schroeder-Heister. Springer, 95–114.

Mates, B. (1965). *Elementary Logic*. Oxford: Oxford University Press.

– (1986). *The Philosophy of Leibniz. Metaphysics and Language*. Oxford: Oxford University Press.

Mints, G. (1968). Some calculi of modal logic. *Trudy Mat. Inst. Steklov* 98. in Russian, 88–111.

Nagashima, T. (1966). An extension of the craig-schütte interpolation theorem. *Annals of the Japan Association for the Philosophy of Science* 3, 12–18.

Negri, S. and J. von Plato (2001). *Structural Proof Theory*. Cambridge: Cambridge University Press.

– (2011). *Proof Analysis. A Contribution to Hilbert's Last Problem*. Cambridge: Cambridge University Press.

Parlamento, F. and F. Previale (2019). *The elimination of atomic cuts and the semishortening property for Gentzen's sequent calculus with equality*. on-line first. The Review of Symbolic Logic.

Poggiolesi, F. (2011). *Gentzen Calculi for Modal Propositional Logic*. Springer.

Popper, K. (1947a). Logic without assumptions. *Proceedings of the Aristotelian Society* 47, 251–292.

– (1947b). New foundations for logic. *Mind* 56.

Quine, W. V. (1966). *Set Theory and its Logic*. Harvard University Press.

– (1970). *Philosophy of Logic*. Prentice Hall.

Read, S. (2004). Identity and harmony. *Analysis* 64, 113–119.

Reeves, S. V. (1987). Adding equality to semantic tableau. *Journal of Automated Reasoning* 3, 225–246.

Restall, G. (2019). Generality and existence 1: quantification and free logic. *The Review of Symbolic Logic* 12, 354–378.

– (2020). Assertions, denials, questions, answers and the common ground. URL: http://consequently.org/presentation.

Sambin, G., G. Battilotti, and C. Faggian (2000). Basic logic: reflection, symmetry, visibility. *Journal of Symbolic Logic* 65, 979–1013.

Schroeder-Heister, P. (1984). Popper's theory of deductive inference and the concept of a logical constant. *History and Philosophy of Logic* 5, 79–110.

– (1994). Definitional reflection and the completion. In: *Extensions of Logic Programming. Fourth International Workshop, St. Andrews, Scotland, April 1993, Proceedings*. Ed. by R. Dyckhoff. Vol. 798. Berlin/Heidelberg/New York: Springer Lecture Notes in Artificial Intelligence, 333–347.

– (2006). Popper's structuralist theory of logi. In: *Karl Popper: A Centenary Assesment, vol III: Science*. Ed. by I. Jarvie, K. Milford, and D. Miller. Ashgate Publishing: Aldershot, 17–36.

– (2012). Proof-theoretic semantics. In: *Stanford Encyclopedia of Philosophy*. URL: https://plato.stanford.edu/entries/proof-theoretic-semantics/.

– (2016). Open problems in proof-theoretic semantics. In: *Advances in Proof-theoretic Semantics*. Ed. by T. Piecha and P. Schroeder-Heister. Springer, 253–283.

Scott, D. (1974). Rules and derived rules. In: *Logical Theory and Semantical Analysis*. Ed. by S. Stenlund, 147–161.

Seligman, J. (2001). Internalization: the case of hybrid logics. *Journal of Logic and Computation* 11, 671–689.

Suppes, P. (1957). *Introduction to Logic*. Princeton: Van Nostrand.

Takeuti, G. (1987). *Proof Theory*. Amsterdam: North-Holland.

Tarski, A. (1941). *Introduction to Logic*. Oxford University Press.

Tennant, N. (2010). Harmony in a sequent setting. *Analysis* 70, 462–468.

Textor, M. (2017). Towards a neo-brentanian theory of existence. *Philosophers' Imprint* 17, 1–20.

Troelstra, A. S. and H. Schwichtenberg (1996). *Basic Proof Theory*. Oxford: Oxford University Press.

Wang, H. (1960). Toward mechanical mathematics. *IBM Journal of Research and Development* 4, 2–22.

Wansing, H. (1999). *Displaying Modal Logics*. Dordrecht: Kluwer Academic Publishers.

Wehmeier, K. (2014). How to live without identity – and why. *Australasian Journal of Philosophy* 90, 761–777.

Więckowski, B. (2011). Rules for subatomic derivations. *Review of Symbolic Logic* 4, 219–236.

Wittgenstein, L. (1922). *Tractatus Logico-Philosophicus*. Brace and Co., New York: Harcourt.

Eight Rules for Implication Elimination

Michael Arndt

Abstract Eight distinct rules for implication in the antecedent for the sequent calculus, one of which being Gentzen's standard rule, can be derived by successively applying a number of cuts to the logical ground sequent $A \to B, A \Rightarrow B$. A naive translation into natural deduction collapses four of those rules onto the standard implication elimination rule, and the remaining four rules onto the general elimination rule. This collapse is due to the fact that the difference between a formula occurring in the succedent of a premise of a sequent calculus rule and that formula instead occurring in the antecedent of the conclusion cannot be adequately expressed in the framework of natural deduction. In contrast to this, the difference between a formula occurring in the succedent of the conclusion of a sequent calculus rule and that formula instead occurring in the antecedent of a premise corresponds exactly to the distinction between the standard implication elimination rule and its general counterpart. This incongruity can be remedied by introducing a notational facility that requires of the particular premise of a rule to which it is attached to be an assumption, i.e., to prevent it from being the conclusion of another rule application. Applying this facility to implication elimination results in eight distinct rules that correspond exactly to the eight sequent calculus rules. These eight rules are presented and discussed in detail. It turns out that a natural deduction calculus (for positive implication logic) that employs the rule corresponding to the standard left implication rule of the sequent calculus as well as a rule for explicit substitution can be seen as a natural deduction style sequent calculus.

1 Introduction

In a previous article we presented eight sequent calculus rules for implication in the antecedent (left implication) and discussed their properties (Arndt, 2019). That

---

Michael Arndt
Department of Computer Science, University of Tübingen, Germany,
e-mail: arndt@cs.uni-tuebingen.de

T. Piecha and K. F. Wehmeier (eds.), *Peter Schroeder-Heister on Proof-Theoretic Semantics*, Outstanding Contributions to Logic 29,
https://doi.org/10.1007/978-3-031-50981-0_8

$$\frac{}{A \to B, A \Rightarrow B}\ (\to\mathrm{L}_{000})$$

$$\frac{\Gamma \Rightarrow A}{A \to B, \Gamma \Rightarrow B}\ (\to\mathrm{L}_{010})$$

$$\frac{\Gamma \Rightarrow A \to B}{A, \Gamma \Rightarrow B}\ (\to\mathrm{L}_{100})$$

$$\frac{\Gamma_1 \Rightarrow A \to B \qquad \Gamma_2 \Rightarrow A}{\Gamma_1, \Gamma_2 \Rightarrow B}\ (\to\mathrm{L}_{110})$$

$$\frac{B, \Gamma \Rightarrow \Lambda}{A \to B, A, \Gamma \Rightarrow \Lambda}\ (\to\mathrm{L}_{001})$$

$$\frac{\Gamma_1 \Rightarrow A \qquad B, \Gamma_2 \Rightarrow \Lambda}{A \to B, \Gamma_1, \Gamma_2 \Rightarrow \Lambda}\ (\to\mathrm{L}_{011})$$

$$\frac{\Gamma_1 \Rightarrow A \to B \qquad B, \Gamma_2 \Rightarrow \Lambda}{A, \Gamma_1, \Gamma_2 \Rightarrow \Lambda}\ (\to\mathrm{L}_{101})$$

$$\frac{\Gamma_1 \Rightarrow A \to B \qquad \Gamma_2 \Rightarrow A \qquad B, \Gamma_3 \Rightarrow \Lambda}{\Gamma_1, \Gamma_2, \Gamma_3 \Rightarrow \Lambda}\ (\to\mathrm{L}_{111})$$

Fig. 1 The eight inference rules for intuitionistic left implication.

investigation was motivated by Schroeder-Heister's argument for a 'conceptually more elementary and more plausible' rule (Schroeder-Heister 2010; 2011) against the backdrop of our own research into the origins of structural reasoning as promoted by Paul Hertz (1922; 1923; 1929) as well as von Plato's recent documentation (2003a) of the development of the sequent calculus from Hertz' sentence calculus. From this historical vantage point both the conventional rule (→L) as well as Schroeder-Heister's alternative rule (→L)° are derivable in a purely structural sequent calculus from the *logical ground sequent* $A \to B, A \Rightarrow B$ by applications of the cut rule. In such a calculus logical ground sequents are the main way of meaningfully introducing logical symbols into sequents.[1]

Since this particular logical ground sequent consists of three formulae, any number of which can be used as cut formula(e), there are eight possible rules that can be derived in this manner. Figure 1 presents the intuitionistic variants of those eight rules that will be relevant for this investigation. The intuitionistic restriction being that Λ (where applicable) must not contain more than one formula.[2] Each one of the

---

[1] Weakening and rules with an additive flavour can always be used to introduce arbitrary logical symbols.

[2] For each of these, the corresponding classical rule is easily restored by adding (where applicable) Δ_1 to each premise's succedent that is A and Δ_2 to each premise's succedent that is $A \to B$ while adding the same to the conclusion, as well as dropping the restriction on Λ.

rules is systematically labelled ($\rightarrow$L$_{xyz}$), where $x, y, z \in \{0, 1\}$. The indices x, y, z correspond to the three formulae $A \rightarrow B$, A and B occurring in the ground sequent, and the number specifies for each of these formulae whether it is used as a cut formula in the derivation of the respective rule or not. To begin with, if $x = 0$ then the formula $A \rightarrow B$ *is not* used as a cut formula in the derivation of that rule and, consequently, occurs in the conclusion of that rule as an antecedent formula; on the other hand, if $x = 1$ then the formula $A \rightarrow B$ *is* used as a cut formula in the derivation of that rule, which means that it can no longer occur in its conclusion; moreover, the derived rule must have a premise in which $A \rightarrow B$ occurs as a succedent formula. In the same manner the value of y specifies the occurrences of formula A. The value of z plays a similar role with regard to the occurrences of formula B, except that "antecedent" has to be exchanged by "succedent" in the characterization and vice versa. Thus rule ($\rightarrow$L$_{xyz}$) has exactly $x + y + z$ premises and $3 - (x + y + z)$ formulae of the ground sequent occur in its conclusion.

For example, consider the derivation of the standard sequent calculus rule ($\rightarrow$L$_{011}$) from the logical ground sequent, which is employed as the logical axiom (lgs). Each of the formulae A and B of the ground sequent are the cut formulae in subsequent applications of cut. Note that the resulting derived rule is irrespective of the order in which the cuts are applied.

$$\dfrac{\dfrac{\Gamma_1 \Rightarrow A \qquad \dfrac{}{A \rightarrow B, A \Rightarrow B}\,(\text{lgs})}{A \rightarrow B, \Gamma_1 \Rightarrow B}\,(\text{cut}) \qquad B, \Gamma_2 \Rightarrow \Lambda}{A \rightarrow B, \Gamma_1, \Gamma_2 \Rightarrow \Lambda}\,(\text{cut})$$

Rule ($\rightarrow$L$_{000}$) is simply the logical axiom that has the ground sequent itself as its conclusion, since none of its formulae are used as cut formulae. In each of the rules ($\rightarrow$L$_{100}$), ($\rightarrow$L$_{010}$) and ($\rightarrow$L$_{001}$), one of the formulae in the ground sequent is cut, resulting in a single premise rule each, where the premise contains the respective formula in the appropriate position and the conclusion contains the remaining formulae in the positions in which they occur in the ground sequent. Schroeder-Heister's alternative ($\rightarrow$L)° is the rule ($\rightarrow$L$_{010}$). In each rule ($\rightarrow$L$_{011}$), ($\rightarrow$L$_{101}$) and ($\rightarrow$L$_{110}$), only one of the formulae in the ground sequent is not cut and the other two are the cut formulae of two applications of cut, thereby resulting in a two premise rule. Rule ($\rightarrow$L$_{011}$) is the standard sequent calculus rule. Finally, the rule ($\rightarrow$L$_{111}$) of three premises is the result of cutting all three of the formulae that occur in the ground sequent. All of these rules and their respective properties are thoroughly discussed in the precursor article (Arndt, 2019).

As a notational convenience for the purpose of talking about several rules, the wildcard symbol "$*$" for x, y or z admits either value, and we employ the same notation when talking about implication elimination rules later in this article.

It is the purpose of the present investigation to transfer the observations that were made with regard to these sequent calculus rules to the calculus of natural deduction. This immediately poses a number of questions:

1. As natural deduction is fundamentally a calculus based on rules, what kind of object could correspond to the logical ground sequent that was the starting point in our preceding work?

2. How does the specific use of cut in the derivation of those eight rules from the logical ground sequent translate into natural deduction, where cut is not a rule but a meta-theoretical operation?
3. Since the natural deduction inference rule for eliminating implications has two variants, namely the standard elimination rule and the general elimination rule, do these correspond to any particular two of those eight rules, and if so, how can there be other (significantly different) inference rules than these two?

The key to answering these questions lies in two fundamental modifications of the calculus of natural deduction. Firstly, in an effort to delaminate the logical and the compositional aspects of natural deduction rules, we add a structural rule of *explicit composition* to the calculus that governs the assembly of derivations.[3] This rule is related to rules of explicit substitution or delayed substitution that was first introduced for the λ-calculus (Abadi, Cardelli, Curien, and Lévy, 1991) and then investigated for its logical properties by, e.g. Espírito Santo (2007) and Pfenning (2007). However, in those cases it was studied for λ-calculi or sequent-style natural deduction systems with term labels, whereas we will provide a straightforward natural deduction rule in standard notation. We will employ this rule for the purpose of composing (i.e., constructing) derivations ab initio.[4] Secondly, we will introduce *proudness markers* that prevent premises thus marked from being the conclusion of any rule application. A suitably marked variant of standard implication elimination will correspond to the logical ground sequent. These two modifications together form an adequate basis for the translations of the eight rules for left implication into natural deduction.

This article is structured as follows. Section 2 recalls Schroeder-Heister's argument for suggesting his alternative sequent calculus rule $(\to L)^{\circ}$, based on an observation how derivations are composed in natural deduction, thus motivating the introduction of an explicit composition rule. It also addresses the matter of preventing the composition of derivations, which results in the notion of proudness markers. In Section 3 both additions to the calculus of natural deduction are used to distinguish eight rules for implication elimination and to demonstrate their interderivability. The properties of these eight rules with regard to their usefulness are discussed in Sections 4 and 5. Section 6 widens the discussion from a merely technical matter to questions of improving the relationship between Gentzen's calculi as well as the role of directionality in the calculus of natural deduction. Section 7 closes with a summary and a brief outlook.

[3] The name of the rule is taken from one of our earlier articles (Arndt and Tesconi, 2014), where it is used as the single generalized composition principle in a sequent calculus famework.

[4] Explicit composition is thus used in the same specific manner as the cut rule is used in the precursor article to obtain sequent calculus rules from ground sequents.

$$\frac{\Gamma \Rightarrow A \qquad B, \Delta \Rightarrow C}{A \rightarrow B, \Gamma, \Delta \Rightarrow C}\,(\rightarrow\mathrm{L}) \qquad\qquad \frac{\Gamma \Rightarrow A}{A \rightarrow B, \Gamma \Rightarrow B}\,(\rightarrow\mathrm{L})^{\circ}$$

Fig. 2 Standard left introduction (→L) and Schroeder-Heister's (→L)°.

2 Composing derivations

Schroeder-Heister's suggestion of a new rule (→L)° to replace the standard sequent calculus rule is based on his interpretation of *implications-as-rules* in the calculus of natural deduction (Schroeder-Heister 1984; 2010; 2011). According to this view, an implication $A \rightarrow B$ codifies the fact that it is possible to derive the formula B from an assumption A, which is represented by a new "defined" rule that has A as a premise and B as a conclusion.[5] Thus, the standard implication elimination rule is replaced by a generic schema for rule application that has the implication, which codifies the rule in question, as a parameter:

$$\frac{A \rightarrow B \qquad A}{B} \qquad\qquad A \rightarrow B\,\frac{A}{B}$$

The schema reads as follows: "B can be inferred from A under the assumption that there is a rule $A \rightarrow B$ that admits this inference." Schroeder-Heister observes that, while this formal reinterpretation of *implications-as-rules* is easy to implement in natural deduction, the standard inference rule of the sequent calculus (→L) that corresponds to implication elimination does not suggest such a reading. While the rule obviously concludes an implication as part of the antecedent, i.e., introduces a hypothetical implication, it is clearly not about utilizing this assumption for the purpose of deriving a formula B. This is why Schroeder-Heister suggests a new inference rule (→L)° for the sequent calculus, which makes it possible to conclude, on the basis that A is the consequence of certain formulae, that formula B must be the consequence of those formulae if the additional hypothesis $A \rightarrow B$ is introduced at the same time. Figure 2 presents both sequent calculus rules.

There are two ways of looking at the relationship between the rules (→L) and (→L)°. In order to see what is taking place inside of (→L), we can derive the rule on the basis of (→L)° by means of a subsequent cut on formula B:

$$\dfrac{\dfrac{\begin{matrix}\mathcal{D}_1\\ \Gamma \Rightarrow A\end{matrix}}{\Gamma, A \rightarrow B \Rightarrow B}\,(\rightarrow\mathrm{L})^{\circ} \qquad \begin{matrix}\mathcal{D}_2\\ B, \Delta \Rightarrow C\end{matrix}}{\Gamma, A \rightarrow B, \Delta \Rightarrow C}\,(\mathrm{cut})$$

In other words, if we give primacy to (→L)°, then the standard rule is easily derived by performing a cut on the succedent formula B in the conclusion of that rule and the hypothetical formula B in the antecedent of another sequent. Thus, the consequence

[5] Indeed, Schroeder-Heister uses a considerably more involved natural deduction system that employs the notation $A \Rightarrow B$ (not to be mistaken as a sequent arrow) for rules of higher order, from which implications $A \rightarrow B$ can be obtained through introduction rules and vice versa.

$$(\to\mathrm{L})^\circ\;\dfrac{\begin{array}{c}\mathcal{D}_1\\ \Gamma \Rightarrow A\end{array}}{\Gamma, A \to B \Rightarrow B} \qquad \rightsquigarrow \qquad \begin{array}{c}\mathcal{D}_1^*\\ A \to B \;\dfrac{A}{B}\end{array} \qquad (1)$$

$$(\to\mathrm{L})\;\dfrac{\begin{array}{c}\mathcal{D}_1\\ \Gamma \Rightarrow A\end{array} \qquad \begin{array}{c}\mathcal{D}_2\\ \Delta, B \Rightarrow C\end{array}}{\Gamma, \Delta, A \to B \Rightarrow C} \qquad \rightsquigarrow \qquad \begin{array}{c}\mathcal{D}_1^*\\ A \to B \;\dfrac{A}{B}\\ \mathcal{D}_2^*\\ C\end{array} \qquad (2)$$

Fig. 3 $(\to\mathrm{L})^\circ$, $(\to\mathrm{L})$ and their translations into natural deduction with rules.

formula B features merely as an intermediate result that is subsequently related to a corresponding hypothesis in another sequent. Alternatively, if we give primacy to the standard rule, then $(\to\mathrm{L})^\circ$ is trivially derivable by choosing a suitable instance of the structural axiom for the right premise:

$$\dfrac{\begin{array}{c}\mathcal{D}_1\\ \Gamma \Rightarrow A\end{array} \qquad \dfrac{}{B \Rightarrow B}\,(\mathrm{id})}{A \to B, \Gamma \Rightarrow B}\,(\to\mathrm{L})$$

Both analyses convey the same picture, namely that in a certain sense too much is taking place within an application of $(\to\mathrm{L})$. Not only does it encode the information that when a formula A is given (as the consequence of certain other formulae), the assumption of an implication $A \to B$ allows us to simultaneously proceed to a new consequence B; it furthermore proceeds to use that newly obtained formula in order to create a link to some other consequence relation in which B figures as a hypothesis, something which in itself quite obviously has no immediate connection to an intuitive understanding of implication.[6] This second effect can only be prevented by providing as second premise the trivial consequence relation that relates B back to itself. It is apparent that a more natural approach is to employ the rule $(\to\mathrm{L})^\circ$, which expresses the core feature of implication, and to use the cut rule to explicitly create a link to another consequence relation.[7]

Figure 3 relates the two sequent calculus rules under discussion to the calculus of natural deduction that additionaly uses a rule-based inference schema.[8] What was just observed for the sequent calculus rules is immediately reflected in their translations. In (1), the application of a rule in natural deduction merely extends the derivation $\mathcal{D}_1^*$

[6] Note that we are talking about implication in an intuitionistic setting. In the classical case, instead of talking of hypotheses and consequences, a semantic argument about the truth values of the various formulae in a sequent immediately establishes $(\to\mathrm{L})$ as the ideal rule.

[7] This immediately leads to the issue of non-eliminability of cuts, which, from a technical perspective, is less than desirable. Schroeder-Heister (2011) argues that these cuts are unproblematic, and we systematically discussed the question of analyticity of cuts (Arndt, 2019).

[8] Following the usual conventions for writing down schematic derivations in natural deduction with the purpose of providing more elegant presentations, we omit the parametric sets of hypotheses Γ and Δ in the translation.

of A, which exemplifies Schroeder-Heister's paradigm of *implications-as-rules*. On the other hand, in (2) the implication furthermore establishes as a link between two derivations $\mathcal{D}_1^*$ and $\mathcal{D}_2^*$, which corresponds to the paradigm of *implications-as-links*.

At this point it is important to be mindful of the fact that natural deduction is fundamentally a unidirectional calculus in which derivations are constructed from the top downward. The prevalent notation for substitution suggests that any derivation can simply be plugged into another one if the conclusion of the former matches an assumption of the latter. Prawitz' assemblage of formal definitions states clearly that a derivation is constructed inductively from other derivations by applying a rule whose premises match the end formulae of those derivations, thereby combining them into a new, single derivation whose end formula is the conclusion of that rule (Prawitz, 1965). Every rule of natural deduction represents a tree constructor which has one or more derivation trees as its argument(s) and yields a new derivation tree whose immediate subtrees are those arguments, arranged according to where their end formulae occur in the premises of the rule.[9] Thus, the operative part of a derivation tree is always its end formula, which is the conclusion of its bottommost rule application. In contrast to this, neither the substitution of derivations for assumptions nor the linking of derivations on certain formulae are operations that are part of the calculus proper. They are meta-theoretical operations that are formally defined by recursively deconstructing a given derivation up to its leafs, some or one of which are then replaced by the derivation that is the substituens, after which the deconstructed derivation is reassembled around the substituted part(s).

Consequently, the difference between (1) and (2) in Figure 3 is a substantial one. The former is merely the application of a rule, i.e., an operation within the calculus (albeit in a somewhat extended calculus), whereas the latter adds to this an operation that, strictly speaking, lies outside of the calculus proper. This observation certainly underscores Schroeder-Heister's case that (1) is philosophically more fundamental than (2).

2.1 Explicit composition

It is quite easy to restate (2) in a manner that is rule-based. The key is to use a rule that merely states the intention that some derivation is to be substituted for an assumption in another derivation. This corresponds to the explicit substitution of terms into terms in the λ-calulus (Espírito Santo, 2007; Pfenning, 2007), whose typing rule is closely related to the cut rule. As Figure 4 demonstrates, it is quite straightforward to translate an application of the cut rule into natural deduction by using well-known translation principles without having to perform a substitution. The left premise of cut, in which the cut formula A occurs in the succedent, is translated as a premise A of the new natural deduction rule that is the end formula of a derivation with assumptions Γ obtained from a translation of $\mathcal{D}_1$. The right premise of cut has the cut formula A

[9] Of course, discharge must be seen as a side effect of such a tree constructor.

$$(\text{cut}) \frac{\begin{matrix} \mathcal{D}_1 \\ \Gamma \Rightarrow A \end{matrix} \quad \begin{matrix} \mathcal{D}_2 \\ A, \Delta \Rightarrow C \end{matrix}}{\Gamma, \Delta \Rightarrow C} \quad \rightsquigarrow \quad \frac{\begin{matrix} \mathcal{D}_1^* \\ A \end{matrix} \quad \begin{matrix} [A]^n \\ \mathcal{D}_2^* \\ C \end{matrix}}{C} (\text{ec}{:}n)$$

Fig. 4 The translation of cut yields explicit composition.

occurring in its antecedent and the succedent is a parametric formula C, which is translated as a second premise C that is the end formula of a derivation obtained from a translation of $\mathcal{D}_2$ that has A as an assumption beside Δ. Corresponding to the conclusion of cut in which the antecedents of the premises are joined and the succedent is the parametric formula C, the conclusion of the new rule repeats the parametric C of its second premise. However, this C is independent of the assumption A, which may therefore be discharged. We call the resulting rule for the natural deduction setting *explicit composition*:

$$\frac{A \quad \begin{matrix} [A] \\ C \end{matrix}}{C} (\text{ec})$$

The same rule was employed by Prawitz (2015) under the name "substitution" to delay substitutions resulting from contractions of maximal implications for the purpose of retaining the overall structure of derivations under normalization. Of course, it is possible to provide a contraction step that executes the substitution:

$$\frac{\begin{matrix} \mathcal{D}_1 \\ A \end{matrix} \quad \begin{matrix} [A]^1 \\ \mathcal{D}_2 \\ C \end{matrix}}{C} (\text{ec}{:}1) \quad \triangleright \quad \begin{matrix} \mathcal{D}_1 \\ A \\ \mathcal{D}_2 \\ C \end{matrix}$$

In the standard calculus of natural deduction, a composition of derivation $\mathcal{D}_1$ with end formula A and derivation $\mathcal{D}_2$ that has an assumption A is only expressible by means of the meta-theoretical substitution operation. In contrast to that, rule (ec) is a proper inference rule of an extended calculus, i.e., a tree constructor, that allows the composition of $\mathcal{D}_1$ and $\mathcal{D}_2$ into a single derivation. Since the substitution occurs in the contractum, the redex itself can also be considered to be a form of "delayed substitution" or "explicit substitution", which is what its sequent-style relative has been called by other authors (Espírito Santo, 2007; Pfenning, 2007).

Considering that (ec) is neither an introduction nor an elimination rule, any derivation that contains one or more applications of (ec) cannot be considered to be in normal form. A maximal formula occurrence is the conclusion of an introduction rule that is also the major premise of an elimination rule, thereby being the result of and thus representing a detour in a derivation, and a redex consists of the corresponding succession of the introduction and the elimination rules. This new rule (ec) is rather extraordinary as it is essentially a self-contained redex. However, the formula A is not a maximal occurrence of a formula. Rather, via both of its occurrences, it redirects the derivation $\mathcal{D}_1$ with this end formula to $\mathcal{D}_2$, which depends on its assumption.

The idea of a rule that redirects the flow of a derivation from one of its premises to the assumption, upon which another premise depends, is not new. The standard disjunction elimination rule ($\vee$E) uses this feature, the difference being that the redirection is divided upon the assumptions of two derivations. In this rule the disjunction $A \vee B$ is eliminated by a case distinction on its subformulae. However, as this is not expressible in the single conclusion setting of natural deduction, it has to be executed indirectly by separately considering two derivations, one depending on A and the other on B, and bringing them together by having reached the same end formula C in both derivations. Thus, C can summarily be concluded, thereby eliminating the disjunction, and this occurrence of C is no longer dependent on A nor on B.

$$\frac{A \vee B \quad \begin{array}{c}[A]\\ C\end{array} \quad \begin{array}{c}[B]\\ C\end{array}}{C}\ (\vee\mathrm{E})$$

This rule does not relate its major premise to a formula in the conclusion that has to bear any relation to the disjunction or any of its subformulae. Instead, the conclusion of ($\vee$E) is a parametric device to collect (and contract) the side premises. The logical content thus resides in the setting of assumptions, and the only additional effect an application of the rule has is to discharge those assumptions.

The fundamental idea is that of anticipating the possible aftermath of the application of an inference rule before that rule is applied. In other words, knowing that the application of a certain rule has some formula, say A, as its conclusion, instead of actually applying the rule and continuing the construction of the subsequent derivation, that subsequent derivation is instead constructed beforehand with A as an assumption. The end formula of that derivation, say C, is added as an additional premise of the rule in question, and the conclusion of the rule is changed from A to C. Thus, the modified rule not only has the proper premises of the original rule, it also has another premise with the sole purpose of anticipating and collecting the aftermath of the application of the original rule.[10] This idea, which was conceived by Gentzen for the purpose of allowing for an elimination rule for disjunction in a calculus that only admits single conclusions, was first suggested by Prawitz (1979), subsequently elaborated by Schroeder-Heister (1984), and later by von Plato (2001) in the form of *general elimination rules*. The explicit substitution rule (ec) epitomizes the principle underlying this idea.

[10] While it is easily possible to add additional premises or to indicate extensive changes in the structure of derivations, due to the emphasis that natural deduction places on the singular conclusions of inference rules and singular end formulae of derivations, this kind of shifting of the conclusion of a rule into an assumption on a premise requires the introduction of a parametric formula C. This is due to the way natural deduction implicitly addresses structural matters. The corresponding modification in the sequent calculus would simply result in an empty succedent, which could subsequently be weakened into a singleton succedent C.

Similarly, the notation $[A]$ does not necessarily refer to a singular assumption A, but can refer to multiple such assumptions or even none at all. We will not attempt an exegesis of such structural matters at this point.

$$\frac{A \to B \quad A}{B}\,(\to\mathrm{E}) \qquad\qquad \frac{A \to B \quad A \quad \begin{array}{c}[B]\\ C\end{array}}{C}\,(\to\mathrm{gE})$$

Fig. 5 Standard and general elimination rules for implication.

$$(\to\mathrm{L})\ \frac{\begin{array}{c}\mathcal{D}_1\\ \Gamma \Rightarrow A\end{array} \quad \begin{array}{c}\mathcal{D}_2\\ \Delta, B \Rightarrow C\end{array}}{\Gamma, \Delta, A \to B \Rightarrow C} \quad \rightsquigarrow \quad \frac{A \to B \quad \begin{array}{c}\mathcal{D}_1^*\\ A\end{array} \quad \begin{array}{c}[B]^n\\ \mathcal{D}_2^*\\ C\end{array}}{C}\,(\to\mathrm{gE}{:}n)$$

$$(\to\mathrm{L})^\circ\ \frac{\begin{array}{c}\mathcal{D}_1\\ \Gamma \Rightarrow A\end{array}}{\Gamma, A \to B \Rightarrow B} \quad \rightsquigarrow \quad \frac{A \to B \quad \begin{array}{c}\mathcal{D}_1^*\\ A\end{array}}{B}\,(\to\mathrm{E})$$

Fig. 6 $(\to\mathrm{L})^\circ$, $(\to\mathrm{L})$ and their translations into natural deduction.

2.2 Standard and general elimination rules

The standard elimination rule and the general elimination rule for implication are presented in Figure 5. The similarity of these rules is rather obvious due to the fact that both of them have premises $A \to B$ and A. It is immediately apparent that the instantiation of B for C in ($\to$gE) yields ($\to$E), as the discharge can be applied to the premise itself, thereby rendering it inert:

$$\frac{A \to B \quad A \quad [B]}{B}\,(\to\mathrm{gE})$$

Conversely, the rule ($\to$gE) can be (schematically) derived by applying (ec) to the conclusion of ($\to$E):

$$\frac{\dfrac{A \to B \quad A}{B}\,(\to\mathrm{E}) \qquad \begin{array}{c}[B]\\ C\end{array}}{C}\,(\mathrm{ec})$$

This is the same phenomenon that we have observed for the relationship between the rules ($\to$L) and ($\to$L)° in the sequent calculus, with the discharged premise B playing the role of the trivial sequent and (ec) playing the role of the cut rule.

Based on our observation about the composition of derivations in natural deduction, we will tentatively consider ($\to$gE) as the translation of ($\to$L) as well as ($\to$E) as the translation of ($\to$L)°, as depicted in Figure 6. It should be noted that this does not counter Schroeder-Heister's view of considering ($\to$L) as representing the paradigm of *implications-as-links*. Rather than the usual linking by means of substitution, we rather utilize linking by explicit composition, in which the hidden conclusion B appears as (reference to) the assumption B in another subderivation.

This analysis of composition accounts for the difference in the translations of sequent calculus rules ($\rightarrow$L$_{010}$) and ($\rightarrow$L$_{011}$), i.e., how the consequence formula B of the logical ground sequent $A \rightarrow B, A \Rightarrow B$ is treated under translation. In one case the formula B is retained in the conclusion of the rule application, in the other case it is mentioned as an assumption in another derivation. In both cases a derivation is extended in the direction of the conclusion of the rule application.

3 Preventing composition of derivations

At this point one could content oneself with the observation that any one of the four rules ($\rightarrow$L$_{**0}$) translates into ($\rightarrow$E), and that any one of the other four rules ($\rightarrow$L$_{**1}$) translates into ($\rightarrow$gE), since it is the last digit in the index that determines whether formula B occurs in the succedent of the conclusion of these rules, which corresponds to ($\rightarrow$E), or in the antecedent of a premise, which corresponds to ($\rightarrow$gE). In the usual natural deduction setting these are the two known alternative rules governing the elimination of implication.

The translations depicted in Figure 6 already exhibit the key to a somewhat more fine grained analysis, however. For both rules ($\rightarrow$E) and ($\rightarrow$gE) have premises $A \rightarrow B$ and A, but only the premise A is mentioned as the end formula of a derivation $\mathcal{D}_1^*$, the translation of a sequent calculus derivation $\mathcal{D}_1$, whereas the main premise $A \rightarrow B$ *stands proud* as an assumption in the sense of Tennant (1992). We propose that the former, i.e., the usual understanding of premises of natural deduction rules, is the image of the translation of a succedent formula in a premise of a sequent calculus rule, whereas the latter is the image of the translation of an antecedent formula in the conclusion of a sequent calculus rule. In order to give an adequately differentiated account of the different roles of premises $A \rightarrow B$ and A, we have to find a mechanism that awards the power to make premises of natural deduction rules unavailable for the construction (composition) of derivations. It is instructive for this purpose to review how inference rules give rise to derivations in natural deduction.

3.1 Constructing derivations from inference rules

Gentzen (1935) himself did not provide a procedure for the construction of derivations, instead mererly giving the following characterization:

> (3.2) A *proof figure*, called a *derivation* for short, consists of a number of formulae (at least one), which combine to form inference figures in the following way: Each formula is the lower formula of at most one inference figure; each formula (with the exception of exactly one: the *endformula*) is the upper formula of at least one inference figure; and the system of inference figures is non circular [. . .].
> (3.3) [. . .] A derivation is in *tree form*, if each one of its formulae is upper formula of *at most* one inference figure. [. . .] We shall have to treat only of derivations in tree form.

Thus, inference rules essentially serve as tools for the local verification that a given figure is indeed a derivation by correlating them to certain premises (upper formulae) and conclusions (lower formulae).

A formal method for the construction of derivations is formulated by Prawitz in his seminal work on natural deduction (Prawitz, 1965). For this purpose, he introduces the noteworthy distinction between *inference rules* and *deduction rules*. The former essentially express the immediate logical relationships between certain (schematic) premises and certain (schematic) conclusions. He accounts for the additional allowance for discharging certain assumptions that some of the inference rules express by calling those *"improper"* in distinction to the *"proper"* inference rules that only address the relationships between certain premises and a conclusion. What is presumably improper about certain rules is the fact that they express logical relationships in an indirect manner. The immediacy of the proper inference rules is depicted in the usual notation by the fact that the respective formulae are only separated by a horizontal stroke. The premise or premises of a rule are written immediately above this stroke, and the conclusion is recorded immediately below it. On the other hand, the improper inference rules make it necessary to talk about formulae that lie beyond this immediate relation between premises and conclusions. In order to avoid having to refer directly to derivations, Prawitz formulates *deduction rules* for the improper inference rules. Instead of simply relating one or more premise formulae to a conclusion formula, a deduction rule relates one or more general kind of premises, namely pairs of assumptions and a formula dependent on them, to a more general form of conclusion, also a pair of assumptions, suitably obtained from the assumptions of the premises, and a dependent formula. For example, the deduction rule that is obtained for the rule ($\vee$E) is given as $\langle\langle\Gamma_1, A \vee B\rangle, \langle\Gamma_2, C\rangle, \langle\Gamma_3, C\rangle/\langle\Delta, C\rangle\rangle$[11] with the side condition that $\Delta = \Gamma_1 \cup (\Gamma_2 - \{A\}) \cup (\Gamma_3 - \{B\})$.[12] Finally, based on the proper inference rules and the deduction rules (which replace the improper inference rules for this purpose), Prawitz describes a procedure for the composition of new derivations from given ones, based on their end formulae and the assumptions those depend on. For a proper inference rule this is done by identifying the premise(s) of a suitable instance of the rule with the end formula(e) of the given derivation(s) and producing the conclusion of the rule as new end formula as well as using (the union of) the assumptions as the formulae on which this end formula depends. For the deduction rules, this is done by simultaneously identifying the assumptions of given derivations as well as their end formulae with a premise pair of a suitable instance of the derivation rule, and by producing as conclusion a pair consisting of an appropriately modified set of assumptions and the new end formula.

We are recalling this procedure in such detail not because it illustrates the effort

[11] For clarity, we use a slightly modified notation. Prawitz uses another comma to separate the conclusion pair from the premise pairs instead of the slash "/".

[12] It is quite apparent that this deduction rule could with only small modifications be formulated in a sequent-style natural deduction format as follows:

$$\frac{\Gamma_1 \vdash A \vee B \qquad \Gamma_2 \vdash C \qquad \Gamma_3 \vdash C}{\Gamma_1, \Gamma_2 - \{A\}, \Gamma_3 - \{B\} \vdash C}\ (\vee\text{E})$$

$$(1) \qquad \frac{A \to B \quad A}{B}\,(\to\mathrm{E}) \qquad\qquad \frac{A \lor B \quad \begin{matrix}[A]\\ C\end{matrix} \quad \begin{matrix}[B]\\ C\end{matrix}}{C}\,(\lor\mathrm{E})$$

$$(2) \qquad \frac{A \to B \quad A}{B}\,(\to\mathrm{E}) \qquad\qquad \frac{\begin{matrix}\vdots\\ A \lor B\end{matrix} \quad \begin{matrix}[A]\\ \vdots\\ C\end{matrix} \quad \begin{matrix}[B]\\ \vdots\\ C\end{matrix}}{C}\,(\lor\mathrm{E})$$

$$(3) \qquad \frac{A \to B \quad A}{B}\,(\to\mathrm{E}) \qquad\qquad \frac{A \lor B \quad \begin{matrix}[A]\\ \vdots\\ C\end{matrix} \quad \begin{matrix}[B]\\ \vdots\\ C\end{matrix}}{C}\,(\lor\mathrm{E})$$

$$(4) \qquad \frac{\begin{matrix}\vdots\\ A \to B\end{matrix} \quad \begin{matrix}\vdots\\ A\end{matrix}}{B}\,(\to\mathrm{E}) \qquad\qquad \frac{\begin{matrix}\vdots\\ A \lor B\end{matrix} \quad \begin{matrix}[A]\\ \vdots\\ C\end{matrix} \quad \begin{matrix}[B]\\ \vdots\\ C\end{matrix}}{C}\,(\lor\mathrm{E})$$

Fig. 7 Four styles of displaying inference rules.

that goes into explaining how inference rules give rise to deductions. Instead, the curious distinction that Prawitz makes between proper inference rules and deduction rules inhabit an interesting position in the different manners in which the inference rules of natural deduction are understood and, consequently, presented with regard to their role in the construction of derivations.

Figure 7 compares four styles of displaying the rules of natural deduction calculi. Style (1) is the notation used by Gentzen to express *inference figure schemata*. They contain no direct reference to derivations whatsoever. While this style is adopted by Prawitz (who calls them *inference rules*), he subsequently distinguishes between proper and improper inference rules and introduces deduction rules as a more detailed explication of the improper inference rules for the purpose of describing a procedure for constructing derivations. Style (2) is a modification of the former that we propose to account for the difference that results from this distinction Prawitz made. His deduction rules are represented in a manner more suitable for a comparison with other styles.[13] It is quite remarkable that the main formula $A \lor B$ of the derivation rule for the improper inference rule ($\lor$E) is understood in relation to a derivation, whereas this is not the case for the main formula $A \to B$ in the proper inference rule ($\to$E). Style (3), perhaps most prominently used by van Dalen (1980), has become the standard notation for displaying rules of natural deduction. It uses what he calls *derivation*

[13] It must be noted that the dots do not signify derivations yet, but merely the relation between certain assumption formulae and another formula depending on those, and, moreover, the parametric assumptions are not mentioned nor is the manner in which the forming of the union and the subtraction of the explicit assumptions occurs.

rules that double as demonstrating immediate inferences as well as addressing matters of assumption discharge that are only properly explicable in the derivational setting. What is rather remarkable in this case is the fact that derivational composition on the "proper" premises, i.e., premises that do not depend on assumptions that may be discharged, is not indicated (although it is implicitly assumed), whereas it is explicitly indicated that premise formulae C are the end formulae of two derivations depending on respective assumptions, of which those having certain forms may be discharged. Thus, this style is hardly an improvement on (1), as there it is at least apparent that both the premises with discharge markers and those without have to be explained in terms of how derivations are to be composed. Style (4), the uniform representation of the idea that all of the premises of a deduction rule can figure as end formulae of derivations, is also used to depict the rules of the calculus. Noteworthy in this regard are the books by Girard, Lafont, and Taylor (1989) and by Troelstra and Schwichtenberg (1996)[14]. This style of notation is occasionally employed by authors, who use style (1) and (3) for stating the rules of the calculus, when talking about actual applications of the deduction rules in concrete but otherwise unspecified derivations.

In the literature it is usually assumed to be understood by the reader (or it is occasionally briefly made explicit) how the rules are to be read and how they are to be used for the purpose of constructing derivations. However, in view of the question how the composition of derivations is to be prevented on certain premises, the notational insignificancy of employing dots to signify the possibility of an occurrence of a derivation on top of premises, acquires some importance. For the purpose of distinguishing which premises are available to continue derivations from those that are not, we would prefer to use style (4), i.e., to write vertical dots above premises to indicate that they can be used to continue a derivation with that end formula, and to purposefully *not* write vertical dots above premises that must stand proud. However, we suspect that, due to the frequent use of style (3) in the presentation of rules and of style (4) for concrete derivations in the literature, this would lead to some confusion.

3.2 Proudness markers

We thus adopt style (1) when stating inference rules, and we introduce as an additional notational measure a *proudness marker* that can be applied to individual premises as long as they do not also carry discharge information.[15] This proudness marker has the form of a dotted line above a premise in an inference rule. It signifies that this premise must not be used to continue a derivation, i.e. that it can only figure as an assumption in the derivation.

[14] Here, derivations are written as parameters $\mathcal{D}, \mathcal{D}', \ldots, \mathcal{D}_1, \mathcal{D}_2, \ldots$ instead of vertical dots.

[15] If we were to mark a premise with discharge information to stand proud, the premise itself would be the only assumption that could be discharged. Hence, the proudness marker would trivialize every possible application of the rule by either forcing an empty discharge or by having a proud assumption discharged immediately.

Take as an example the inference rule obtained from standard implication elimination by marking the left (major) premise to stand proud:

$$\frac{\overset{\cdots\cdots\cdot}{A \to B} \quad A}{B}\,(\to\mathrm{E})'$$

As this premise can never be the continuation of a derivation, i.e., this occurrence of $A \to B$ is designated as an assumption in any derivation resulting from the application of this rule.

There are two quite different ways in which the proudness marker can be understood. The first interpretation remains firmly within the standard conception of natural deduction. According to this the proudness marker merely expresses the requirement that the premise must be an assumption, i.e., a trivial derivation that is already present as the rule is applied. The second interpretation has it that, given a derivation with end formula A, in order to conclude the formula B by means of this rule, an ad hoc assumption of the major premise $A \to B$ is made. While this differs from the first interpretation perhaps only in the sequence of the steps comprising the rule application, it is quite easy to conflate this sequence and understand it as having an almost bidirectional quality in the sense that not only does the application of the rule produce a conclusion, but it also introduces a new assumption at the same time.[16]

It is the second interpretation that grants the extra conceptual space in which suitably different implication elimination inference rules can be explained and maintained. The standard implication elimination rule is used to extend two derivations, one with end formula $A \to B$ and one with end formula A, to a new derivation with end formula B that depends on the union of those assumptions. The rule $(\to\mathrm{E})'$ above can be used to extend a derivation of A from assumptions to a derivation of B with the additional assumption $A \to B$. The variant with both premises marked to stand proud can be understood as playing the role of a *ground derivation* that expresses that a new derivation of a formula B from two new assumptions $A \to B$ and A can be stated.

4 Eight rules and their relationships

We can now combine the notational facility of restricting the composition on either one of its premises by means of proudness markers together with the power of using explicit composition to preempt the subsequent application of rules to the conclusion of the standard implication elimination rule. As each of these modifications can be applied to the two premises or the conclusion independently of one another, this results in eight different rules. Figure 8 displays all of the possible combinations of assigning proudness markers to the premises $A \to B$ and A in both the standard rule and the general elimination rule.

[16] These two interpretations correspond to the difference in the sequent calculus between the implication occurring as both antecedent and succedent in an identity axiom in a premise of a rule $(\to\mathrm{L}_{1**})$ versus that formula being introduced as an antecedent formula in the conclusion in a rule $(\to\mathrm{L}_{0**})$. The details of this will be developed shortly.

$$\frac{\overset{\cdots\cdots}{A \to B} \quad \overset{\cdots}{A}}{B}\ (\to\mathrm{E}_{000}) \qquad \frac{\overset{\cdots\cdots}{A \to B} \quad \overset{\cdots}{A} \quad \overset{[B]}{C}}{C}\ (\to\mathrm{E}_{001})$$

$$\frac{\overset{\cdots\cdots}{A \to B} \quad A}{B}\ (\to\mathrm{E}_{010}) \qquad \frac{\overset{\cdots\cdots}{A \to B} \quad A \quad \overset{[B]}{C}}{C}\ (\to\mathrm{E}_{011})$$

$$\frac{A \to B \quad \overset{\cdots}{A}}{B}\ (\to\mathrm{E}_{100}) \qquad \frac{A \to B \quad \overset{\cdots}{A} \quad \overset{[B]}{C}}{C}\ (\to\mathrm{E}_{101})$$

$$\frac{A \to B \quad A}{B}\ (\to\mathrm{E}_{110}) \qquad \frac{A \to B \quad A \quad \overset{[B]}{C}}{C}\ (\to\mathrm{E}_{111})$$

Fig. 8 The eight inference rules for implication elimination.

4.1 Classification

Each inference rule is systematically labelled $(\to\mathrm{E}_{xyz})$, where $x, y, z \in \{0, 1\}$. The indices x, y, z correspond to the schematic formulae $A \to B, A$ and B. These numbers specify for each of these formulae whether it is available for the purpose of composition with other derivations. While $A \to B$ and A are the premises of the standard elimination rule, B is its conclusion. Thus, "composition" means something quite different in these two cases.[17]

If $x = 0$ then the premise $A \to B$ must stand proud in any application of that rule and is thus not usable for the purpose of continuing derivations with that end formula through this premise. If $x = 1$ then the premise $A \to B$ may be used to continue a derivation that has this implication as its end formula. The value of y specifies the role of premise A in the same manner. The value of z plays a rather more striking role with regard to formula B, as it determines the format of the rule. If $z = 0$ then the formula B is the conclusion of the two premise standard elimination rule. Since any derivation resulting from an application of this rule has B as its end formula and can thus be further attached to subsequent rules applications, this value for z does not preclude composition as such in contrast to what the value 0 for x and y specifies for formulae $A \to B$ and A. If $z = 1$ then the formula B does not feature as the conclusion of the respective rule but as an assumption to be discharged in the additional premise of the general elimination rule. As mentioned before, when talking about groups of rules, the symbol "$*$" for x, y or z can admit either value.

[17] While there seems to be a difference between what "composition" means at this point, we shall see later on that we can render composition in all three cases by explicit composition.

4.2 Understanding the rules

So far the new rules are little more than the product of a combinatorial exercise. While the juxtaposition of the standard elimination rule versus the general elimination rule has been discussed in the literature with some benefit, the addition of the proudness marker might appear to be something of an ad hoc device. Tennant's result that every derivation in natural deduction can be transformed into a derivation in *exposed normal form* of the same end formula in which all of the major premises of elimination rules stand proud (Tennant, 1992) gives some pertinence to the idea of restricting major premises of the usual inference rules. However, being able to block the minor premise of an elimination rule from being the result of a derivation does not seem to have a meaningful application. We will proceed to consider possible justifications for the various new rules by elaborating on their differences.

The new rule ($\rightarrow E_{000}$) is the standard elimination rule that requires both of its premises to stand proud. The rule is apparently not particularly useful, since it introduces two assumptions, $A \rightarrow B$ and A, just for the purpose of obtaining the conclusion B. A more efficient way of obtaining B is to simply assume B itself (unless, of course, some dependency from $A \rightarrow B$ or A is to be subsequently utilized). Since the rule cannot be used to continue a derivation, the whole figure can only occur as a topmost application in a derivation. Thus, the rule is nothing more than a static schema relating the assumption of an implication and the assumption of its implicational antecedent to the consequence of its succedent, as expressed by the derivability statement $A \rightarrow B, A \vdash B$. If we consider a formal framework for deducing such derivability statements for natural deduction, this would correspond to the following axiom schema that immediately delivers statements of this form:[18]

$$\overline{A \rightarrow B, A \vdash B}$$

In contrast to this, rule ($\rightarrow E_{110}$), the standard implication elimination rule that can be used to extend derivations through both of its premises, can by no means be reduced to something as static as a consequence statement. It has to be expressed in the manner of a derivation rule that states that, if both $A \rightarrow B$ and A are derivable from certain assumptions, then B, dependent on the union of these assumptions, can be concluded. Thus, rule ($\rightarrow E_{110}$) clearly expresses a certain relationship of derivability statements that could be written as a rule of inference for such statements as follows:

$$\frac{\Gamma_1 \vdash A \rightarrow B \qquad \Gamma_2 \vdash A}{\Gamma_1, \Gamma_2 \vdash B}$$

Note that the derivability statement $A \rightarrow B, A \vdash B$ mentioned above is easily obtained in the special case that both $A \rightarrow B$ and A are indeed assumptions, (albeit assumptions that are made independently of the subsequent application of the rule):

[18] As we will take the step of relating the left implication rules of the sequent calculus to the new elimination rules shortly, we refrain from introducing labels for these ad hoc inference rules. However, we expressly talk about derivability in natural deduction as opposed to the syntactic notion of sequent, which is why we use "⊢" instead of "⇒" here.

$$\frac{A \to B \vdash A \to B \quad A \vdash A}{A \to B, A \vdash B}$$

The new rules ($\to E_{010}$) and ($\to E_{100}$) are positioned between these two rules, because they both continue a derivation through one of their premises while simultaneously introducing a single new assumption. These are the respective inference rules for derivability statements:

$$\frac{\Gamma \vdash A}{A \to B, \Gamma \vdash B} \qquad\qquad \frac{\Gamma \vdash A \to B}{A, \Gamma \vdash B}$$

Finally, as we have already discussed, the difference between rules ($\to E_{**0}$), the standard format, and ($\to E_{**1}$), the general elimination format, lies in the fact that the former ends with the conclusion B whereas the latter presumes that another derivation with end formula C from the assumption B is considered for the application of the rule. This consideration of an additional derivation has the effect on the respective inference rule for derivability statements of adding the statement $B, \Gamma_i \vdash C$ as another premise while replacing the original consequence B in the resulting inference rule by the parametric C:

$$\frac{\ldots}{\ldots \vdash B} \quad \rightsquigarrow \quad \frac{\ldots \quad B, \Gamma_i \vdash C}{\ldots \vdash C}$$

The induced inference rules for derivability statements show that a rule ($\to E_{xyz}$) composes $x + y + z$ derivations into a new derivation. As we have just illustrated, every index that is 1 expresses the power to compose some derivation that is assumed to be given by the application of the rule, whereas every index that is 0 renders the corresponding premise inactive in the case of x and y, or abstains from preempting the subsequent continuation of the derivation in the form of the general elimination format in the case of z. It is thus apparent that from a compositional point of view the differences between the eight inference rules are certainly noteworthy. It is perhaps unexpected that this is not only the case for the two formats of inference rules that are strikingly different, the standard and the general elimination rules. The significantly distinct derivational aspects of the implication elimination inference rules which only differ in whether certain premises are marked to stand proud or not are of the same rank as the difference between those two formats, since they determine how many derivations are to be composed through an application of the respective inference rule.

By way of the ad hoc inference rules for derivability statements it has also become obvious how the eight sequent calculus rules of Figure 1 are related to the eight implication elimination rules in Figure 8. Each rule ($\to L_{xyz}$) corresponds to the rule ($\to E_{xyz}$) in the sense that (i) there is an appropriate implication elimination rule to be used for the translation of derivations from the sequent calculus into natural deduction, and (ii) the derivational character of each one of the natural deduction rules is described in a sequent-style format by the respective sequent calculus rule.[19]

[19] Note that the sequent calculus rules are formulated with the usual intuitionistic constraint of $|\Lambda| \leq 1$ for arbitrary succedents, whereas the natural deduction rules use a parametric C which may be the absurdity constant. Replacing Λ by C and allowing absurdity in the sequent calculus yields the exact correspondence.

Whereas the latter correspondence is well known from the literature in the form of sequent-style natural deduction rules, using ($\to L_{110}$) for standard implication eliminiation and ($\to L_{111}$) for the general elimination rule, the reverse is less prevalent. Indeed, while there are many interesting studies about the translations from the sequent calculus into natural deduction and the role of the cut rule therein (Zucker, 1974; Negri and von Plato, 2001; von Plato, 2003b; Tesconi, 2011), they are all based on the translation of standard sequent calculus rule ($\to$L), i.e., rule ($\to L_{011}$) in our classification, to either the standard or the general elimination rule. It is apparent that it is actually rule ($\to E_{011}$), the general implication elimination rule in which the premise $A \longrightarrow B$ stands proud, that is its proper correspondent, since the standard left implication rule of the sequent calculus introduces the implication into the antecedent of its conclusion.

4.3 Interderiving the rules

The sequent calculus rules ($\to L_{xyz}$) have been shown to be interderivable by means of cut in our companion article (Arndt, 2019). We will now demonstrate a similar interderivability of the eight implication elimination rules with explicit composition (ec) playing the role of cut. Just as ($\to L_{000}$) is the rule (axiom) from which all other rules can be derived, the inference rule ($\to E_{000}$), the standard elimination rule with both premises standing proud, lies at the heart of this exercise. This means that we consider as fundamental the standard elimination rule that expresses the immediate relationship between the relevant formulae while being derivationally inactive. All of the rules ($\to E_{xyz}$) are derivable from this rule by means of explicit composition. By means of (ec) the inactivity of the components of rule ($\to E_{000}$) can be selectively removed in order to obtain any kind of compositionality format required.

This matter is complicated by the fact that we don't want to make the easy assumption that the order of the premises in a rule is irrelevant. The left premise of rule ($\to E_{000}$) is $A \longrightarrow B$, and its right premise is A. An application of (ec) to the conclusion B with regard to composition formula A discharges that right premise, but in the resulting derivation the new assumption A ends up on the left of the premise $A \longrightarrow B$:

$$\dfrac{A \qquad \dfrac{\overset{\cdots\cdots}{A \longrightarrow B} \qquad \overset{\cdots}{[A]^1}}{B}\,(\to E_{000})}{B}\,(\text{ec}{:}1)$$

The rule thus derived is not ($\to E_{010}$), since that rule has the premises occurring in reverse order. The way to remedy this unfortunate complication is to consider a companion rule (ce) to explicit composition[20] that has its premises reversed:

[20] Read this rule as "composition explicite" (French) or "composición explícita" (Spanish).

$$\frac{A \quad \overset{[A]}{C}}{C}\,(\mathrm{ec}) \qquad\qquad \frac{\overset{[A]}{C} \quad A}{C}\,(\mathrm{ce})$$

It is apparent that using this new rule for any composition on formula A retains the relative position of the premise with regard to the other (the left) premise $A \to B$ of rule ($\to$E$_{000}$).

The derivations of all of the rules are shown in Figure 9. They are to be understood as derivation schemas rather than concrete derivatons. This is particularly obvious and relevant in the cases of the rules ($\to$E$_{**1}$), in which the right premise C of the final application of (ec) is decorated with the information that assumptions B can be discharged. This is simply due to the fact that this premise has never been instantiated in that application of (ec).

Rule ($\to$E$_{000}$) is listed for the sake of completeness. The remaining three variants of the standard elimination rule ($\to$E$_{**0}$) are obtained by applying one or both versions of explicit composition to the conclusion B in ($\to$E$_{000}$). For the rule ($\to$E$_{010}$), the second premise of (ce) is an arbitrary assumption A, and, consequently, the premise A of ($\to$E$_{000}$) can be discharged. Similarly, for rules ($\to$E$_{1*0}$), composition rule (ec) is applied to an arbitrary assumption $A \to B$, and the corresponding implication as premise of ($\to$E$_{000}$) can be discharged. Alternatively to the manner it is derived in Figure 9, rule ($\to$E$_{110}$) can be obtained by applying the variants of explicit composition in reverse order, i.e., by first applying (ce), thereby addressing A, and then applying (ec), which takes care of $A \to B$.

In the cases of rules ($\to$E$_{xy1}$) that have the general elimination format it is easy to see that they can be derived from the respective rule ($\to$E$_{xy0}$) already discussed by a final application of (ec). As this application of (ec) introduces the premise C with discharge decoration $[B]$ as a right premise, it cannot be exchanged with the application of (ce) that introduces a new assumption A on the right hand side, as this would result in an inference rule that has C with discharge decoration $[B]$ as its second and A as its third premise, which is not the usual format for the general implication elimination rule. Thus, in the derivations of rules ($\to$E$_{*11}$), the application of (ce) with regard to A must always come before the application of (ec) with regard to B. In contrast to this, the application of (ec) with regard to $A \to B$ can occur before or after the application of (ec) with regard to B.

The various ways of deriving the rules ($\to$E$_{***}$) from ($\to$E$_{000}$) are shown in Figure 10. Each arrow corresponds to an application of (ec) in the cases with the labels $A \to B$ or B or to an application of (ce) in the cases with the label A. Notable are the omissions of arrows from ($\to$E$_{001}$) to ($\to$E$_{011}$) and from ($\to$E$_{101}$) to ($\to$E$_{111}$) reflecting the fact that an application of (ce) as composition on A after the application of (ec) as composition on B would result in a general elimination rule in which the second and third premises are reversed.[21] On the other hand, the subdiagram for rules ($\to$E$_{**0}$) commutes, since the application of (ec) to premise $A \to B$ and the

[21] It should be noted that it is possible to force the issue and derive ($\to$E$_{111}$) from ($\to$E$_{101}$) by extraordinarily applying (ec) instead of (ce) with regard to A followed by applying (ec) with regard to $A \to B$. However, this tweak cannot be used to derive ($\to$E$_{011}$) from ($\to$E$_{001}$), as in both of these $A \to B$ must stand proud.

$(\to E_{000})$:

$$\dfrac{\overset{\vdots}{A \to B} \qquad \overset{\vdots}{A}}{B}\,(\to E_{000})$$

$(\to E_{010})$:

$$\dfrac{\dfrac{\overset{\vdots}{A \to B} \qquad \overset{\vdots}{[A]^1}}{B}\,(\to E_{000}) \qquad A}{B}\,(\mathrm{ce}{:}1)$$

$(\to E_{100})$:

$$\dfrac{A \to B \qquad \dfrac{\overset{\vdots}{[A \to B]^1} \qquad \overset{\vdots}{A}}{B}\,(\to E_{000})}{B}\,(\mathrm{ec}{:}1)$$

$(\to E_{110})$:

$$\dfrac{\dfrac{A \to B \qquad \dfrac{\overset{\vdots}{[A \to B]^1} \qquad \overset{\vdots}{[A]^2}}{B}\,(\to E_{000})}{B}\,(\mathrm{ec}{:}1) \qquad A}{B}\,(\mathrm{ce}{:}2)$$

$(\to E_{001})$:

$$\dfrac{\dfrac{\overset{\vdots}{A \to B} \qquad \overset{\vdots}{A}}{B}\,(\to E_{000}) \qquad \begin{matrix}[B]^1 \\ C\end{matrix}}{C}\,(\mathrm{ec}{:}1)$$

$(\to E_{011})$:

$$\dfrac{\dfrac{\dfrac{\overset{\vdots}{A \to B} \qquad \overset{\vdots}{[A]^1}}{B}\,(\to E_{000}) \qquad A}{B}\,(\mathrm{ce}{:}1) \qquad \begin{matrix}[B]^2 \\ C\end{matrix}}{C}\,(\mathrm{ec}{:}2)$$

$(\to E_{101})$:

$$\dfrac{\dfrac{A \to B \qquad \dfrac{\overset{\vdots}{[A \to B]^1} \qquad \overset{\vdots}{A}}{B}\,(\to E_{000})}{B}\,(\mathrm{ec}{:}1) \qquad \begin{matrix}[B]^2 \\ C\end{matrix}}{C}\,(\mathrm{ec}{:}2)$$

$(\to E_{111})$:

$$\dfrac{\dfrac{\dfrac{A \to B \qquad \dfrac{\overset{\vdots}{[A \to B]^1} \qquad \overset{\vdots}{[A]^2}}{B}\,(\to E_{000})}{B}\,(\mathrm{ec}{:}1) \qquad A}{B}\,(\mathrm{ce}{:}2) \qquad \begin{matrix}[B]^3 \\ C\end{matrix}}{C}\,(\mathrm{ec}{:}3)$$

Fig. 9 Deriving the eight rules from $(\to E_{000})$.

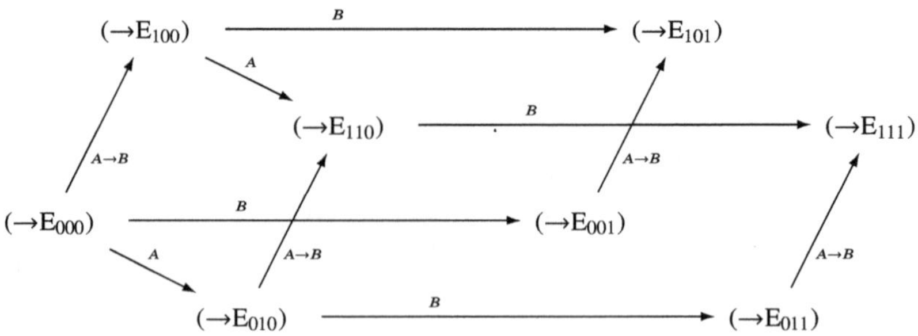

Fig. 10 Derivability of the rules by (ec) and (ce).

application of (ce) to premise A do not interfere with each other and can thus be transposed. To illustrate this, compare the derivation given for ($\to E_{110}$) in Figure 9 to the following one:

$$\dfrac{A \to B \qquad \dfrac{\dfrac{[A \to B]^2 \quad [A]^1}{B}\,(\to E_{000}) \qquad A}{B}\,(\text{ce:1})}{B}\,(\text{ec:2})$$

Furthermore, the subdiagram for rules ($\to E_{*0*}$) commutes, as does the subdiagram for rules ($\to E_{*1*}$), as the two applications of (ec), one with regard to premise $A \to B$ and the other with regard to the conclusion B, can be transposed without affecting the order of the premises. Compare (as the most involved example) the following derivation to the one provided for ($\to E_{111}$) in Figure 9:

$$\dfrac{A \to B \qquad \dfrac{\dfrac{\dfrac{[A \to B]^3 \quad [A]^1}{B}\,(\to E_{000}) \qquad A}{B}\,(\text{ce:1}) \qquad \overset{[B]^2}{C}}{C}\,(\text{ec:2})}{C}\,(\text{ec:3})$$

Note that after the transposition of the two applications of (ec) the conclusion of the composition on $A \to B$ is no longer B but the parametric formula C.

5 Employing the new rules in the calculus

Having established the uniqueness and interderivability of the eight rules for implication elimination (two of which being the standard and the general elimination rules) by means of explicit composition, it is necessary to discuss the usefulness and

$$\dfrac{\dfrac{\dfrac{\dfrac{[(p \land q) \to r]^3 \qquad \dfrac{[p]^2 \quad [q]^1}{(p \land q)}\,(\land\mathrm{I})}{r}\,(\to\mathrm{E}_{110})}{q \to r}\,(\to\mathrm{I}{:}1)}{p \to (q \to r)}\,(\to\mathrm{I}{:}2)}{((p \land q) \to r) \to (p \to (q \to r))}\,(\to\mathrm{I}{:}3)$$

Fig. 11 A formula derivable with $(\to\mathrm{E}_{110})$ but not with $(\to\mathrm{E}_{000})$.

suitability of these rules both as alternatives to the known rules as well as in relation to one another for the purpose of generating derivations and proofs in the calculus.

The fact that all of the rules can be obtained from $(\to\mathrm{E}_{000})$ by means of the rules (ec) and (ce) apparently informs us that such a discussion is only meaningful in a calculus in which those composition rules are not available. For this reason we begin by considering independently the effect of replacing the standard rule for implication elimination with one of the other eight rules. Following these considerations we will propose a calculus of compositional natural deduction in which only the *logical ground inference rule* $(\to\mathrm{E}_{000})$ is used together with the composition rules.[22]

5.1 One rule replacing standard implication elimination

The relationship between the standard rules and the general elimination rules in natural deduction has been thoroughly investigated (Schroeder-Heister 1982; 2010; von Plato, 2001; Read, 2010). Specifically, it has been argued that the general elimination rules are the ideal format for deliberations on the meaning of the logical constants, as they can be thought of as being immediately justified by the corresponding introduction rules by means of an inversion principle.[23] One important observation in connection with these investigations is the fact that a calculus in which the standard elimination rule $(\to\mathrm{E}_{110})$ is replaced by the general elimination rule $(\to\mathrm{E}_{111})$ proves exactly the same formulae as the standard calculus.

In contrast to this, it is quite apparent that neither one of the rules $(\to\mathrm{E}_{*0*})$ is a suitable replacement for the standard elimination rule. Figure 11 shows a derivation of the formula $((p \land q) \to r) \to (p \to (q \to r))$ in which the minor premise of the standard implication elimination is the conjunction $p \land q$, the conclusion of an application of the rule $(\land\mathrm{I})$. It is inconceivable how its subformulae, p and q, could be made available for the purpose of deriving the formula $p \to (q \to r)$ when only a rule $(\to\mathrm{E}_{*0*})$ is available that requires its minor premise to stand proud. Consequently,

[22] For this purpose, we will initially look at minimal logic, but will argue that inference rules for other logical connectives can also be replaced by logical ground inferences, thereby obtaining a uniform format for logical inference rules.

[23] Unfortunately, this immediacy is not evident in the case of implication, at least not without resorting to a higher level calculus (Schroeder-Heister, 1982).

$$\dfrac{\dfrac{\dfrac{\dfrac{\dfrac{[p\to(q\to r)]^3 \quad [p]^1}{q\to r}\,(\to E_{110}) \qquad [q]^2}{r}\,(\to E_{110})}{(p\to r)}\,(\to I{:}1)}{q\to(p\to r)}\,(\to I{:}2)}{(p\to(q\to r))\to(q\to(p\to r))}\,(\to I{:}3)$$

Fig. 12 A formula derivable with $(\to E_{110})$ but not with $(\to E_{010})$.

$$\dfrac{\dfrac{\dfrac{\dfrac{[p\to(q\to r)]^5 \qquad [p]^3 \qquad \dfrac{[q\to r]^2 \quad [q]^4 \quad [r]^1}{r}\,(\to E_{011}{:}1)}{r}\,(\to E_{011}{:}2)}{(p\to r)}\,(\to I{:}3)}{q\to(p\to r)}\,(\to I{:}4)}{(p\to(q\to r))\to(q\to(p\to r))}\,(\to I{:}5)$$

Fig. 13 Deriving the formula with $(\to E_{011})$.

any calculus that replaces the standard rule by either one of the four rules $(\to E_{*0*})$ must be incomplete.

Only two more rules, namely $(\to E_{010})$ and $(\to E_{011})$, have to be considered. These are the two rules in which the major premise but not the minor premise must stand proud, where the former has the standard format and the latter the general elimination format. The challenge for rules $(\to E_{0**})$ is the successive decomposition of implications that are nested to the right.

Figure 12 demonstrates that this is easily accomplished by means of the standard rule, in which the major premise may be a derived formula. Successive applications of that elimination rule eventually retrieve the succedent of the innermost implication. Clearly, this cannot be accomplished by rule $(\to E_{010})$. While a first application of this rule to the proud major premise $p \to (q \to r)$ yields the formula $q \to r$, this in turn no longer stands proud and can thus not be used as premise for a second application. On the other hand, while an application of this rule to a proud premise $q \to r$ yields the required formula r, it is not feasable to combine these two derivations into a single one. Consequently, the end formula of Figure 12 cannot be derived by means of rule $(\to E_{010})$, thereby rendering a calculus using this rule of implication elimination incomplete.

On the other hand, the general elimination variant $(\to E_{011})$ of this rule does allow for the composition of these two subderivations via the third premise. Figure 13 demonstrates that the consequence formula r of the first application of this rule becomes the parametric formula in the second application that really composes the two subderivations by referencing formula $q \to r$ as an assumption. This simple procedure can be repeated as required to decompose arbitrarily deep nestings of implications.[24] Indeed, this observation is a consequence of Tennant's more general result mentioned earlier that all derivations in a natural deduction calculus that employs general elimination rules[25], in which all major premises must stand proud for all connectives, are already in normal form.

To summarize, only three of the eight rules for implication elimination are suitable for the purpose of a complete calculus, only one of which is new. Apart from the standard elimination rule ($\rightarrow E_{110}$) and the general elimination rule ($\rightarrow E_{111}$), only the general elimination rule ($\rightarrow E_{011}$) with proud major premise retains completeness. Any choice of the former two does not restrict compositionality of the premises. Quite to the contrary, the general elimination rule even allows for the pre-empting of subsequent rules applications by enabling the composition of a third derivation that depends on the formula which is the conclusion of the standard rule. Interestingly enough, it is this general feature that renders completeness possible in the case of the proud rule ($\rightarrow E_{011}$). In contrast to this enabling feature of the general format, proudness markers in general significantly restrict the usefulness of the respective rules. Most significantly, the requirement of a proud minor premise prohibits any derivation of a complex implicational antecedent, which leaves the four rules ($\rightarrow E_{*0*}$) so significantly impaired as to practically rendering them useless. The requirement of a proud major premise without the redeeming feature of the general rule format is similarly disabling, thereby rendering rule ($\rightarrow E_{010}$) useless. It is to be expected that restricting the built-in means of composing derivations by means of proudness markers should have a debilitating effect on the calculus employing the respective rule for the elimination of implication. For this reason the positive result for the rule ($\rightarrow E_{011}$) is what stands out in this review. We will return to this important observation and expand on it in the following discussion.

5.2 A calculus for logical ground inferences and composition

The situation changes significantly if rules for explicit composition, (ec) as well as (ce), are available in the calculus. Since all of the rules ($\rightarrow E_{***}$) are derivable from ($\rightarrow E_{000}$) alone by means of explicit compositions, this rule in itself codifies all pertinent logical relationships between the relevant (schematic) formulae for the delineation how implication is to be used. Consequently, in such a setting the matter of composing derivations can be regarded completely independently from the purely logical considerations. As far as implication elimination is concerned, this results in *logical ground inferences*, i.e., "seedling" instances of ($\rightarrow E_{000}$), that can subsequently

---

[24] This procedure cannot be used to circumvent the problem encountered with rules in which the minor premise must stand proud, since the general elimination format adds a third premise that is via its assumption related to the conclusion of the standard rule (the succedent of the implication), not its minor premise (the antecedent of the implication).

[25] Tennant calls the general elimination format "parallel forms".

be composed with other derivations by means of explicit composition.[26] That is to say that the elimination of implication is never applied as a rule to conclusions of derivations but always occurs as an instance of ($\rightarrow E_{000}$) with premises standing proud. It is instructive to become aware of the fact that the application of a rule consists of two acts that can actually be viewed as independent of one another. This amounts to nothing less than the distinction between the logical aspect of the rule on the one hand, that which is expressed by the logical ground inference, and its structural or procedural aspect, which allows the integration of said logical aspect into a continuing flow of other inferences. Such a view is something of a contrast to the usual manner in which presentations of the calculus of natural deduction cast reasoning itself as progressions of eliminations and introductions of logical symbols. Thus, the suggested use of logical ground inferences together with explicit composition could serve as a starting point for a systematic fine-grained analysis of the notion of reasoning in natural deduction.

Unfortunately, the generalization of this approach to the entire calculus of natural deduction meets the substantial difficulty that not all logical inference rules can be reduced to logical ground inferences. We have concerned ourselves only with implication elimination, and a similar analysis can be made for most of the inference rules of natural deduction. The notable exceptions are the elimination rules for disjunction and existential quantification as well as the introduction rule of implication and the classical absurdity rule. The two elimination rules are already in the general elimination format, which quite apparently cannot be undone.[27] While that still leaves the possibility to add proudness markers to the major premises of those rules, the idea of an undifferentiated generality has thereby been shown to be impossible. Implication introduction only has a single premise that allows the discharge of an assumption. Even if we were to allow a proudness marker for premises that allow for discharge, the only implications derivable would be trivial tautological instances. Moreover, while (ec) can be used on the conclusion of implication introduction, it merely results in a general introduction rule, something we have not considered in the context of this investigation at all.[28] The classical absurdity rule has a format quite similar to that of implication introduction. The premise of that rule must be the absurdity constant, whereas the application of the rule allows the discharge of a negated formula.

[26] This situation is similar to the one we described for the sequent calculus that uses the logical ground sequents (Arndt, 2019). The implication left rule can be replaced by the ground sequent $A \rightarrow B, A \Rightarrow B$ that expresses how implication is to be used. The cut rule must be used to relate such statements of logical relationships to other sequents. Thus, purely logical relationships are expressed horizontally, i.e. within logical ground sequents, whereas matters of reasoning are encoded in the structural rules of the calculus.

[27] In the case of disjunction elimination, one would have to introduce a dual conclusion logical ground inference, and in the case of existential elimination, a logical ground inference would violate the Eigenvariable condition.

[28] What is remarkable in connection with implication introduction, however, is the fact that it is closely related to explicit composition. Consider the following two contractions, which proceed from a maximal occurrence of an implication in the derivation on the left to its usual contractum in the derivation on the right:

Therefore, demanding absurdity to stand proud renders this rule a severely restricted variant of the intuitionistic absurdity rule.

Moreover, there are several useful properties of the natural deduction calculus that are countered or at least made more complicated by an approach that delaminates the inference rules. For one, a single application of standard implication elimination has to be simulated by one instance of ($\to E_{000}$) and one application of each (ec) and (ce), so derivations of the same end formula obviously involve more rules applications. More serious, however, is the fact that maximal occurrences of implications are no longer easily identifiable, since the major premise in ($\to E_{000}$) stands proud. Maximality becomes a relational property between the simple premise of explicit composition and an assumption that its other premise depends on. Consequently, the meta theory of normalizability becomes considerably more involved.[29]

6 Discussion

The professed aim of this article was to give an exhaustive list of all the inference rules that can be considered to express the idea of eliminating an implication in the calculus of natural deduction. The starting point was a previous comparison of eight variants of inference rules for implication in the antecedent in the sequent calculus. Despite the fact that the only alternative to the standard implication elimination rule that has been seriously discussed in the literature is the general elimination rule, the guiding idea was that it should somehow be possible to formulate eight distinct implication elimination inference rules in accordance to the left rules obtained in the sequent calculus.

Prawitz (1965) observed that the (intuitionistic) sequent calculus can be understood

---

$$\dfrac{\dfrac{\begin{array}{c}[A]^1\\ \mathcal{D}_1\\ C\end{array}}{A \to C}\,(\to\text{I:}1) \qquad \begin{array}{c}\mathcal{D}_2\\ A\end{array}}{C}\,(\to\text{E}) \quad \rhd \quad \dfrac{\begin{array}{c}[A]^1\\ \mathcal{D}_1\\ C\end{array} \qquad \begin{array}{c}\mathcal{D}_2\\ A\end{array}}{C}\,(\text{ce:}1) \quad \rhd \quad \begin{array}{c}\mathcal{D}_2\\ A\\ \mathcal{D}_1\\ C\end{array}$$

We already demonstrated on page 246 that the derivation on the right is the contractum of an application of (ec), and it is naturally also the contractum of (ce). Thus, while an introduced implication $A \to C$ expresses the fact that it is possible to derive C from the assumption(s) A, explicit composition further states that if, in addition to that derivation, A is given independently, then C can be reobtained. In other words, while ($\to$I) codifies (certain aspects of) a derivation into a formula, explicit composition uses that derivation and a single instance of its discharged assumption(s) to move on to its conclusion. Formally, the premise A of (ec) becomes the antecedent subformula of the implication in ($\to$I):

$$\dfrac{A \qquad \begin{array}{c}[A]\\ C\end{array}}{C}\,(\text{ec}) \qquad\qquad \dfrac{\begin{array}{c}[A]\\ C\end{array}}{A \to C}\,(\to\text{I})$$

[29] This is not necessarily a disqualifying feature. The theory of λ-calculi with explicit substitution has become a prolific field because of, not despite of, these rather fruitful complexities.

$$\dfrac{\begin{matrix} \mathcal{D}_1 & \mathcal{D}_2 \\ \Gamma \Rightarrow A & B, \Delta \Rightarrow C \end{matrix}}{A \to B, \Gamma, \Delta \Rightarrow C}\,(\to\text{L}) \quad \rightsquigarrow \quad \dfrac{A \to B \quad \begin{matrix} \Gamma \\ \mathcal{D}_1^* \\ A \end{matrix} \quad \begin{matrix} [B]^n, \Delta \\ \mathcal{D}_2^* \\ C \end{matrix}}{C}\,(\to\text{E'}{:}n)$$

Fig. 14 The translation of (→L) into natural deduction.

as a meta-calculus for natural deduction in the sense that its rules can be read as derivability statements for deductions. Every premise in a sequent calculus rule corresponds to some natural deduction derivation of its succedent formula from the assumptions given in its antecedent. The application of the rule in natural deduction combines those given derivations into one corresponding to the conclusion of the sequent calculus rule. The new end formula must thus be the succedent formula, and the assumptions of the new derivation are simply the combination of the assumptions of the given derivations. If any specific formula mentioned in the antecedent of a premise does not reoccur in the antecedent of the conclusion, this assumption must be marked as discharged in the resulting natural deduction derivation.[30] Remarkably, this naive characterization does not cover the correspondence between the main formulae of the left rules and the main formulae of elimination rules. Since such a formula does not occur in a premise of the sequent calculus rule but is instead introduced into the antecedent of the conclusion, a corresponding derivation in natural deduction must introduce that formula as an additional assumption that stands proud.

The translation for a schematic application of rule (→L) based on this interpretation is depicted in Figure 14. The key to an accurate correspondence between the rules in the sequent calculus and rules in natural deduction must thus be to take the different treatments of main formula and side formula seriously. The former yields a new assumption, whereas the latter results in an ordinary premise that can be used to continue a given derivation with that end formula. For the rule (→L) the main formula is $A \to B$ and the relevant side formula is A. Both correspond to premises in the natural deduction rule, but the former must be marked to stand proud in order to guarantee that it must be an assumption of the derivation. Thus, the natural deduction rule that is obtained by translating the standard left implication rule of the sequent calculus in this manner is rule (→E$_{011}$):

$$\dfrac{\overset{\cdots\cdots\cdot}{A \to B} \quad A \quad \begin{matrix} [B] \\ C \end{matrix}}{C}\,(\to\text{E}_{011})$$

As we discussed in the previous section, replacing the standard implication elimination with most of the new rules would render the calculus incomplete. This would be a

[30] An exact correspondence would have to address various structural issues, such as whether antecedents should rather be considered as sequences (without an exchange rule) than as multisets, and whether the mentioning of a formula in the antecedent of a premise refers to a single formula occurrence or to several such occurrences that are to be removed in the conclusion rather than by the application of an independent contraction rule.

rather disappointing result, were it not for the exception of this rule ($\rightarrow E_{011}$), which deserves to be called *proud implication elimination*. Taking this rule seriously grants the opportunity to readdress several issues regarding the calculus of natural deduction.

6.1 Improving the relationship between Gentzen's calculi

Due to the manner in which this new rule was obtained, a natural deduction calculus using proud implication elimination exhibits an intimate correspondence to the sequent calculus without the cut rule (as far as minimal implication logic is concerned). This apparent fact holds the key for a fruitful reconsideration of the relationship between Gentzen's two calculi.

Translations between the calculi have been unwieldy from the very beginning (Gentzen, 1935), and the exact nature of the complications have been discussed in textbooks and specialised treatises (Negri and von Plato, 2001; von Plato, 2003b). The difficulties preventing a one-to-one translation in as far as they pertain to the meta theories of the two calculi have been succinctly characterized by Tesconi (2011):

> [One translation] maps normal to non cut-free derivations, because elimination rules are translated to left-side rules with the help of the cut rule; [...] whereas it is necessarily the case that [the converse translation] maps cut-free on normal derivations, it does not necessarily map cuts on instances of non-normality, because composition – by means of which the cut rule is translated – of normal derivations may or may not preserve normality.

Both of these problems are due to the fact that major premises of elimination rules in natural deductions can be arbitrarily derived formulae, i.e., that derivations can pass through such premises. Consequently, the complexity (the number of logical constants) of formulae encountered along a branch passing downward through major premises can increase or decrease. In contrast to this, since all of the logical rules of the sequent calculus are introduction rules, the complexity of formulae can only increase through their application. The elimination of a major premise must therefore be mimicked by the introduction of that formula into the antecedent of a sequent. The fact that this very premise may be obtained as a conclusion of a subderivation in the natural deduction derivation can only be rendered in the sequent calculus by a subsequent cut of that formula with a sequent that corresponds to that subderivation in which the formula appears in the succedent.

The fact that the structural rule of cut, which is the object of the sequent calculus' meta theory, is required to render the utterly basic act of applying an elimination rule to build up a derivation in the calculus of natural deduction is astounding. Consequently, the meta theories of the two calculi are egregiously misaligned. The *Hauptsatz* of the sequent calculus revolves around the fact that in any derivation all instances of the cut rule, a proper rule of the calculus, can be eliminated. On the other hand, normalization considers the removal of maximal formula occurrences by means of substitution, a procedure that is a meta operation on derivations.

In a natural deduction calculus with proud implication elimination, the continuation of derivations through major premises is impossible. This restriction prevents the

$$\dfrac{\begin{array}{c}\mathcal{D}_1\\ A \to B\end{array} \quad \begin{array}{c}\mathcal{D}_2\\ A\end{array} \quad \begin{array}{c}[B]^n\\ \mathcal{D}_3\\ C\end{array}}{C}\,(\to\mathrm{gE}{:}n) \quad \leadsto \quad \dfrac{\begin{array}{c}\mathcal{D}_1\\ A \to B\end{array} \quad \dfrac{\begin{array}{c}\cdots\cdots\cdots\\ [A \to B]^1\end{array} \quad \begin{array}{c}\mathcal{D}_2\\ A\end{array} \quad \begin{array}{c}[B]^n\\ \mathcal{D}_3\\ C\end{array}}{C}\,(\to\mathrm{E}_{011}{:}n)}{C}\,(\mathrm{ec}{:}1)$$

Fig. 15 Recovering (→gE) by means of (ec).

formation of maximal implicational formula occurrences, and, in the case of minimal implicational logic, it guarantees normality of derivations.[31] Nonetheless, a rich meta theory is still feasible. What is required is the addition of the rules of explicit substitution (ec) and its variant (ce) to the calculus, which take on the roles of enabling redirections that can circumvent the proudness requirement. Figure 15 reiterates how an application of the general elimination rule can be rendered by an application of rule ($\to E_{011}$), which is followed by an explicit substitution of the proud premise $A \to B$.

While an in-depth discussion of the impact that these modifications have on the translations between natural deduction and the sequent calculus is a task left for another place, it should already be apparent from this presentation that they can certainly help to significantly improve the relationship between the two calculi. What is also apparent is that such a modification rather changes the character of the calculus of natural deduction. Having been the showcase of the inferentialist theory of meaning, the addition of a rule for explicit composition could be downplayed as a merely technical matter that changes the meta theory with the effect of somewhat dulling the shine of the calculus in view of that theory.

6.2 Bidirectional natural deduction

Natural deduction with its emphasis on the relationship between premises and conclusions (the latter of which can be rendered as succedent formulae in a sequent style notation) is a fundamentally asymmetrical calculus. While assumptions are explicitly stated, they retreat into the background as ever new applications of rules remove them further and further from the current end formula (the conclusion of the last rule application), and the only possible purpose to revisit any of them is to finally discharge them. The most significant conceptual divide between the two Gentzen calculi lies in the fact that, in stark contrast to natural deduction, the sequent calculus is symmetrical in that both antecedent and succedent are fundamentally treated in the same manner.[32] Indeed, to achieve the suppositiously "natural" relationship between

[31] According to Tennant's result, systematically using proud elimination rules for the other logical constants guarantees universal normality of derivations.

[32] At first glance the conceptual divide appears to be the fact that in the sequent calculus the (horizontal) direction relating hypotheses to consequence(s) is decoupled from the (vertical)

any numbner of assumptions and a single assertive conclusion in the sequent calculus, its sequents have to be artificially restricted to singleton succedents.

When considering a possible evolution of calculi from the entirely assertion based Frege-Hilbert style calculus[33] towards calculi which give increasingly more significance to the role of assumptions, natural deduction is clearly only the first step. For, while it allows derivations to be dependent on assumptions and implicitly develops consequence relations between assumptions and the current end formula, the assumptions remain unaltered until discharged. Clearly, this is the exercise of *assertive reasoning* on the basis of initial assumptions. The full capacity of reasoning, including *assumptive reasoning*, in which an inference step can consider some (momentarily) static conclusion in relation to changing assumptions, is only realized in the sequent calculus.

This touches upon a matter that has been discussed by Schroeder-Heister. In his discussion of a proper basis of proof-theoretic semantics (Schroeder-Heister, 2009), he describes four features that are taken as integral to the standard paradigm of proof-theoretic semantics:

(α) the primacy of closed over open proofs,
(β) the placeholder view of assumptions,
(γ) the transmission view of consequence,
(δ) the global view of proofs.

About these he states:

> These four features are intimately connected to the model of natural deduction as its formal background. This holds especially for (β) and (δ), which specify (α) and (γ), respectively.

Schroeder-Heister criticises these commitments that are imposed simply by the formal framework of natural deduction, which he calls its *unidirectionality*. His suggestion of an alternative paradigm is based on his preference of local reasoning over global reasoning. Seeing that realized in the sequent calculus, he proceeds to formulate a natural deduction rule for implication elimination that embraces and introduces into natural deduction the *bidirectionality* that is inherent in the sequent calculus:

---

direction of reasoning, whereas the two are aligned in natural deduction. The sequent calculus explicitly derives sequents, i.e., syntactic representations of consequence relations between sets of formulae. On the other hand, the rules of natural deduction appear to be formulated for the purpose of deriving individual formulae, since every rule of the calculus locally relates one or more premises to a conclusion. However, most rules have the significant side effect of modifying the global dependency of that conclusion on assumptions, either by means of allowing the discharge of certain assumptions or by means of joining the assumptions on which the premises depended. Thus, every new conclusion obtained by the application of an inference rule has to be seen in relation to the collection (usually in a set or multiset) of still undischarged assumptions. Indeed, it is easy to record the dependencies of conclusions on assumptions explicitly in a sequent-style manner, which most notably results in redundant accounting of assumptions. Consequently, a derivation in natural deduction is simply a highly compactified representation of onion layers of consequence relationships.

[33] The deduction theorem, which states that $\Gamma, A \vdash B$ iff $\Gamma \vdash A \rightarrow B$ and only through this admits derivations based on assumptions, constitutes the meta theory of those calculi.

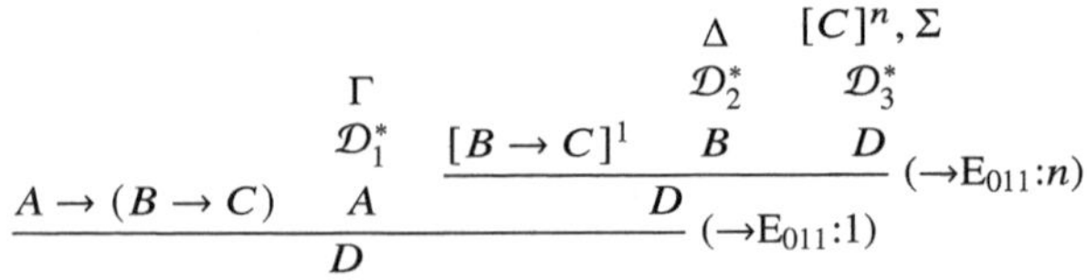

Fig. 16 Two successive applications of ($\rightarrow$E$_{011}$).

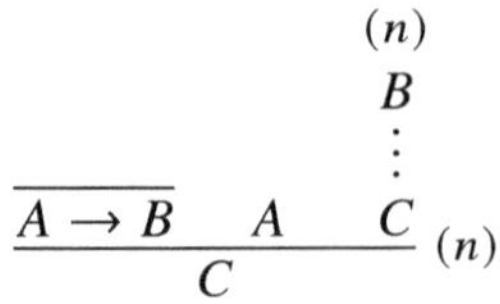

The bar above the major premise expresses "the crucial modification that the major premiss must now be an assumption, i.e., must occur in top position". Thus, Schroeder-Heister had already singled out proud implication elimination, the rule that is ($\rightarrow$E$_{011}$) in our terminology, as the philosophically most promising alternative to the standard rule for the purpose of developing a more general approach to proof theoretic semantics.

It should be noted, however, that "bidirectionality" must not be understood in the literal sense of directly applying rules upwards, i.e., by drawing an inference stroke above some undischarged assumption and placing the relevant premises as new assumptions on top of it. Instead, such an assumption must be discharged by a regular (downwards) application of the respective proud inference rule, thereby utilizing its capability to preempt and explicitly compose a derivation that would ordinarily follow the application of the corresponding standard rule. Figure 16 depicts two successive applications of proud implication elimination. It is apparent that the major premise $B \rightarrow C$ of the first rule application remains in topmost position even after the second application of the same rule, although it gets discharged in the process. Nonetheless, by means of this discharge and the positioning of a new major premise the second application effectively replaces the previous assumption $B \rightarrow C$ by the new assumption $A \rightarrow (B \rightarrow C)$.[34]

7 Summary and outlook

We had set ourselves the somewhat technical task of translating the eight alternative rules for left implication in the sequent calculus into natural deduction. The fact that natural deduction does not encompass a cut rule, which is used to derive those alternatives in the sequent calculus, made the task somewhat intractable. The solution was to introduce an explicit composition rule into the calculus of natural deduction

[34] It remains to be seen whether this apparently quite cumbersome and indirect notion of bidirectionality amounts to anything useful beyond a mere proof of principle.

(indeed, two such rules with flipped premises were required). The possibility to suspend derivations following the application of a rule in natural deduction allowed for the derivation of the general implication elimination rule from the standard one. However, this enhancement only served to distinguish four of the original alternatives (namely those that would translate into the standard elimination rule) from four others (those that would translate into the general elimination rule). Furthermore, a mechanism was required that restricts the application of rules in such a way that certain premises must be guaranteed to stand proud as assumptions. Assigning these proudness markers to any one of the two premises of either the standard or the general implication elimination rule resulted in eight distinct natural deduction rules reflecting the character of the eight left implication rules of the sequent calculus. An analysis of the new natural deduction rules revealed that apart from the standard and the general implication elimination rules, only one of the new rules retains completeness of the calculus, namely the rule of proud implication elimination. It is this rule that should be considered as the definitive translation of the standard left implication rule in the sequent calculus. As such, it not only significantly improves the translations of sequent calculus derivations into natural deduction, it also introduces the paradigm of bidirectionality or assumptive reasoning into the framework of natural deduction.

The obvious next step is a comprehensive investigation of bidirectional natural deduction employing proud elimination rules for all logical constants as well as the rules of explicit composition.

Acknowledgements Supported by the DFG project "Paul Hertz and his foundations of structural proof theory" (DFG AR 1010/2-1).

References

Abadi, M., L. Cardelli, P.-L. Curien, and J.-J. Lévy (1991). Explicit substitutions. *Journal of Functional Programming* 1/4, 375–416.

Arndt, M. (2019). Eight inference rules for implication. *Studia Logica*.

Arndt, M. and L. Tesconi (2014). Principles of explicit composition. In: *Second Pisa Colloquium in Logic, Language and Epistemology, Proceedings*. Ed. by E. Moriconi. ETS, 19–67.

Avron, A. (1990). Gentzenizing Schroeder-Heister's natural extension of natural deduction. *Notre Dame Journal of Formal Logic* 31, 127–135.

Dyckhoff, R. (1987). Implementing a simple proof assistant. In: *Workshop on Programming for Logic Teaching*. Ed. by J. Derrick and H. A. Lewis. Centre for Theoretical Computer Science, University of Leeds, 49–59.

Espírito Santo, J. (2007). Completing Herbelin's programme. In: *Typed Lambda Calculi and Applications, 8th International Conference, Proceedings*, 118–132.

Gentzen, G. (1933). Über die Existenz unabhängiger Axiomensysteme zu unendlichen Satzsystemen. *Mathematische Annalen* 107, 329–350.

Gentzen, G. (1935). Untersuchungen über das logische Schließen. I-II. *Mathematische Zeitschrift* 39, 176–210, 405–431.
Girard, J.-Y., Y. Lafont, and P. Taylor (1989). *Proofs and Types*. Cambridge University Press.
Hertz, P. (1922). Über Axiomensysteme für beliebige Satzsysteme. I. Teil. Sätze ersten Grades. *Mathematische Annalen* 87, 246–269.
– (1923). Über Axiomensysteme für beliebige Satzsysteme. II. Teil. Sätze höheren Grades. *Mathematische Annalen* 89, 76–100.
– (1929). Über Axiomensysteme für beliebige Satzsysteme. *Mathematische Annalen* 101, 457–514.
Negri, S. and J. von Plato (2001). *Structural Proof Theory*. Cambridge University Press.
Pfenning, F. (2007). On a logical foundation for explicit substitutions. In: *Typed Lambda Calculi and Applications*. Ed. by S. R. Della Rocca. Springer.
Prawitz, D. (1965). *Natural Deduction. A Proof-Theoretic Study*. Almqvist & Wiksell.
– (1979). Proofs and the meaning and completeness of the logical constants. In: *Essays on Mathematical and Philosophical Logic*. Ed. by J. Hintikka, I. Niiniluoto, and E. Saarinen. Reidel, 25–40.
– (2015). A note on how to extend Gentzen's second consistency proof to a proof of normalization for first order arithmetic. In: *Gentzen's Centenary: The Quest for Consistency*. Springer.
Read, S. (2010). General-elimination harmony and the meaning of the logical constants. *Journal of Philosophical Logic* 39, 557–576.
Restall, G. (2014). Normal proofs, cut free derivations and structural rules. *Studia Logica* 102, 1143–1166.
Schroeder-Heister, P. (1982). Logische Konstanten und Regeln. Zur Deutung von Aussagenoperatoren. *Conceptus* 16, 45–59.
– (1984). A natural extension of natural deduction. *Journal of Symbolic Logic* 49, 1284–1300.
– (2009). Sequent calculi and bidirectional natural deduction: on the proper basis of proof-theoretic semantics. In: *The Logica Yearbook 2008*. Ed. by M. Peliš. College Publications, 245–259.
– (2010). Generalized elimination inferences, higher-level rules, and the implications-as-rules interpretation of the sequent calculus. In: *Advances in Natural Deduction*. Ed. by E. H. Haeusler, L. C. Pereira, and V. de Paiva. Springer.
– (2011). Implications-as-rules vs. implications-as-links: an alternative implication-left schema for the sequent calculus. *Journal of Philosophical Logic* 40, 95–101.
Smullyan, R. M. (1968). Analytic cut. *Journal of Symbolic Logic* 33, 560–564.
Tennant, N. (1992). *Autologic*. Edinburgh University Press.
– (2002). Ultimate normal forms for parallelized natural deductions. *Logic Journal of the IGPL* 10, 299–337.
Tesconi, L. (2010). Some not so obvious remarks about the cut rule. In: *First Pisa Colloquium in Logic, Language and Epistemology*. Ed. by C. Marletti. ETS, 116–132.

Troelstra, A. S. and H. Schwichtenberg (1996). *Basic Proof Theory*. Vol. 43. Cambridge Tracts in Theoretical Computer Science. Cambridge University Press.
van Dalen, D. (1980). *Logic and Sructure*. Springer.
von Plato, J. (2001). Natural deduction with general elimination rules. *Archive for Mathematical Logic* 40/7, 541–567.
– (2003a). Gentzen's proof systems: byproducts in a work of genius. *Bulletin of Symbolic Logic* 18/3, 313–367.
– (2003b). Translations from natural deduction to sequent calculus. *Mathematical Logic Quarterly* 49, 435–443.
Zimmermann, E. (2007). Substructural logics in natural deduction. *Logic Journal of the IGPL* 15, 211–232.
Zucker, J. (1974). The correspondence between cut-elimination and normalization. *Annals of Mathematical Logic* 7, 1–112.

Focusing Gentzen's LK Proof System

Chuck Liang and Dale Miller

Abstract Gentzen's sequent calculi LK and LJ are landmark proof systems. They identify the structural rules of weakening and contraction as notable inference rules, and they allow for an elegant statement and proof of both cut elimination and consistency for classical and intuitionistic logics. Among the undesirable features of those sequent calculi is that their inferences rules are low-level and frequently permute over each other. As a result, large-scale structures within sequent calculus proofs are hard to identify. In this paper, we present a different approach to designing a sequent calculus for classical logic. Starting with Gentzen's LK proof system, we examine the *proof search* meaning of his inference rules and classify those rules as involving either *don't care nondeterminism* or *don't know nondeterminism*. Based on that classification, we design the *focused* proof system LKF in which inference rules belong to one of two phases of proof construction depending on which flavor of nondeterminism they involve. We then prove that the cut rule and the general form of the initial rule are admissible in LKF. Finally, by showing that the inference rules for LK are all admissible in LKF, we can give a relative completeness proof for LKF provability with respect to LK provability. We shall also apply these properties of the LKF proof system to establish other meta-theoretic properties of classical logic, including Herbrand's theorem.

Key words: sequent calculus, Gentzen's LK, focused proof system, LKF, polarization, cut elimination

Chuck Liang
Department of Computer Science, Hofstra University, Hempstead, NY, United States of America, e-mail: cscccl@hofstra.edu

Dale Miller
Inria & LIX/Ecole Polytechnique, Palaiseau, France, e-mail: dale.miller@inria.fr

T. Piecha and K. F. Wehmeier (eds.), *Peter Schroeder-Heister on Proof-Theoretic Semantics*, Outstanding Contributions to Logic 29,
https://doi.org/10.1007/978-3-031-50981-0_9

1 Introduction

In his attempt to prove the *Hauptsatz* (cut elimination) for both intuitionistic and classical logics, Gentzen (1935) moved away from natural deduction to the sequent calculus. The sequent calculus allowed him to introduce the structural rules of weakening and contraction: their use on the right-hand side of sequents was fundamental to capturing both classical and intuitionistic logics in one framework. If we are only interested in proving cut elimination and consistency, then the sequent calculus, as Gentzen presented it, is a great tool. If, however, we wish to apply logic and proof theory to, say, computation, then Gentzen's sequent calculus has some significant problems: we discuss four such problems in Section 2.

In earlier work (Liang and Miller, 2009), we have presented the *focused proof system* LJF as an improved version of Gentzen's sequent system LJ for intuitionistic logic. Such focused proof systems have been used to give a foundation to logic programming (Miller, 1989; Miller, Nadathur, Pfenning, and Scedrov, 1991), model checking (Heath and Miller, 2019), and term representation (Herbelin, 1995; Scherer, 2016).

This paper examines a *focused* version of the LK sequent calculus proof system, called LKF. The key properties of LKF — cut elimination and relative completeness for LK — have been proved elsewhere (Liang and Miller 2009; 2011) by using complex and indirect arguments involving linear logic (Girard, 1987), a focused proof system for linear logic due to Andreoli (1992), and the focused proof system LJF. Here, we present LKF from first principles: we make no use of intuitionistic or linear logics nor the meta-theory of other proof systems. Additionally, proof transformations here, including those for cut elimination, should be easier to formalize in, say, proof assistants than the more abstract arguments used elsewhere (including in Liang and Miller, 2011).

After presenting the LK inference rules, we describe some of the shortcomings of that proof system in Section 2. In Section 3, that criticism of LK motivates the design of LKF. We then prove the following results about LKF.

1. The cut rule in LKF is admissible in (cut-free) LKF (Section 4).
2. While the initial rule in LKF is limited to atomic formulas, the general form of the initial rule is admissible (Section 5).
3. The rules of LK are admissible in LKF (Section 7).

Taken together, these results prove that LKF is complete for LK. A similar proof outline for proving the relative completeness of focused proof systems has been used by Laurent (2004) for linear logic, by Chaudhuri, Pfenning, and Price (2008) for an intuitionistic version of linear logic, and by Simmons (2014) for a propositional intuitionistic logic. The proofs of these meta-theoretic results for LKF rely almost exclusively on tedious arguments about the permutability of inference rules. One of the design goals for LKF has been to build a calculus that can be used directly to prove other proof-theoretic results without the need to involve such tedious permutation arguments. We illustrate this principle by proving the admissibility of cut in cut-free LK (Section 9.1) and by proving Herbrand's theorem (Section 9.3): both proofs do not explicitly involve permutation arguments.

STRUCTURAL RULES

$$\frac{\Gamma, B, B \vdash \Delta}{\Gamma, B \vdash \Delta}\; cL \qquad \frac{\Gamma \vdash \Delta, B, B}{\Gamma \vdash \Delta, B}\; cR \qquad \frac{\Gamma \vdash \Delta}{\Gamma, B \vdash \Delta}\; wL \qquad \frac{\Gamma \vdash \Delta}{\Gamma \vdash \Delta, B}\; wR$$

IDENTITY RULES

$$\frac{}{B \vdash B}\; init \qquad \frac{\Gamma \vdash \Delta, B \quad \Gamma, B \vdash \Delta'}{\Gamma, \Gamma' \vdash \Delta, \Delta'}\; cut$$

INTRODUCTION RULES

$$\frac{\Gamma, B_i \vdash \Delta}{\Gamma, B_1 \wedge B_2 \vdash \Delta}\; \wedge L \qquad \frac{\Gamma \vdash \Delta, B \quad \Gamma \vdash \Delta, C}{\Gamma \vdash \Delta, B \wedge C}\; \wedge R \qquad \frac{}{\Gamma \vdash \Delta, t}\; tR$$

$$\frac{\Gamma, B \vdash \Delta \quad \Gamma, C \vdash \Delta}{\Gamma, B \vee C \vdash \Delta}\; \vee L \qquad \frac{}{\Gamma, f \vdash \Delta}\; fL \qquad \frac{\Gamma \vdash \Delta, B_i}{\Gamma \vdash \Delta, B_1 \vee B_2}\; \vee R$$

$$\frac{\Gamma \vdash \Delta, B \quad \Gamma, C \vdash \Delta'}{\Gamma, \Gamma', B \supset C \vdash \Delta, \Delta'}\; \supset L \qquad \frac{\Gamma, B \vdash \Delta, C}{\Gamma \vdash \Delta, B \supset C}\; \supset R$$

$$\frac{\Gamma, [s/x]B \vdash \Delta}{\Gamma, \forall x.B \vdash \Delta}\; \forall L \qquad \frac{\Gamma \vdash \Delta, [y/x]B}{\Gamma \vdash \Delta, \forall x.B}\; \forall R \qquad \frac{\Gamma, [y/x]B \vdash \Delta}{\Gamma, \exists x.B \vdash \Delta}\; \exists L \qquad \frac{\Gamma \vdash \Delta, [s/x]B}{\Gamma \vdash \Delta, \exists x.B}\; \exists R$$

Fig. 1 The rules for LK. In the $\forall R$ and $\exists L$ rules, the variable y is not free in the conclusion. In the $\wedge L$ and $\vee R$ rules, $i \in \{1, 2\}$. In $\forall L$ and $\exists R$, s is a first-order term.

2 The LK proof system

Formulas for first-order classical logic are defined as follows. Atomic formulas are of the form $P(t_1, \ldots, t_n)$, where $n \geq 0$, P is a predicate of arity n, and $t_1, \ldots, t_n$ is a list of first-order terms. Formulas are built from atomic formulas using both the logical connectives $\wedge$, t, $\vee, f, \supset$ as well as the two first-order quantifiers $\forall$ and $\exists$. We shall assume the usual treatment of bound variables and substitutions: in particular, the expression $[s/x]B$ denotes the result of performing a capture-avoiding substitution of term s for all free occurrences of the variable x in the formula B.

Figure 1 presents the LK sequent proof calculus of Gentzen (1935). Inference rules are between *sequents* which are pairs of multisets of formula, formally written with an infix $\vdash$. The rules there are divided into introduction rules, structural rules, and *identity rules*. Note that the rules in this latter group, namely the *init* and the *cut* rules, require checking that two occurrences of a formula, here B, on different sides of a sequent or in different sequents, have the same identity (e.g., are equal). Note also that the first important results about the LK sequent calculus imply that completeness is maintained if almost all of the identity rules are eliminated: one must only retain occurrences of the *init* rule where B is atomic.

The main differences between the proof system in Figure 1 and Gentzen's presentation of LK are the following.

1. In Gentzen's system, contexts are lists of formulas, and the exchange rule, which allowed two adjacent formulas to be swapped, was used. In Figure 1, contexts (Γ and Δ) are multisets of formulas, and the exchange rule is not used.

2. Gentzen did not have the logical units for true and false while here they are explicitly written as t and f: they also have associated inference rules.
3. Gentzen's system contained negation as a primitive connective while we shall treat it as an abbreviation: in particular, $\neg B$ is defined to be $B \supset f$.

For this paper, we shall make the following distinction between proof and derivation. By *proof*, we mean a tree structure of inference rules and sequents such that all premises are closed, in the sense that the inference rules at the leaves have zero premises (such as the initial rule). By *derivation*, we mean a similar tree structure of inference rules and sequents, but we do not assume that all leaves are closed: derivations can have unproved premises.

Gentzen's sequent calculus was designed to support the proof of cut elimination (for both classical and intuitionistic logics). As we suggested in the introduction, sequent calculus is difficult to apply in a number of application areas. We describe four major shortcomings of the LK sequent calculus.

2.1 The collision of cut and the structural rules

Consider the following instance of the cut rule.

$$(\dagger) \qquad \frac{\Gamma \vdash C \qquad \Gamma', C \vdash B}{\Gamma, \Gamma' \vdash B}\ \mathit{cut}.$$

If the right premise is proved by a left-contraction rule from $\Gamma', C, C \vdash B$, then cut elimination proceeds by permuting the *cut* rule to the right premises, yielding the derivation

$$\cfrac{\cfrac{\Gamma \vdash C \qquad \cfrac{\Gamma \vdash C \qquad \Gamma', C, C \vdash B}{\Gamma, \Gamma', C \vdash B}\ \mathit{cut}}{\Gamma, \Gamma, \Gamma' \vdash B}\ \mathit{cut}}{\Gamma, \Gamma' \vdash B}\ cL.$$

(An inference figure written with double lines indicates possibly several applications of the rules listed as its justification.) In the intuitionistic variant of the sequent calculus, it is not possible for the occurrence of C in the left premise of $(\dagger)$ to be contracted since two formulas are not allowed on the right of the sequent arrow. If the cut inference in $(\dagger)$ takes place in the classical proof system LK, it is possible that the left premise is the conclusion of a contraction applied to $\Gamma \vdash C, C$. In that case, cut elimination can also proceed by permuting the cut rule to the left premise.

$$\cfrac{\cfrac{\cfrac{\Gamma \vdash C, C \qquad \Gamma', C \vdash B}{\Gamma, \Gamma' \vdash C, B}\ \mathit{cut} \qquad \Gamma', C \vdash B}{\Gamma, \Gamma', \Gamma' \vdash B, B}\ \mathit{cut}}{\Gamma, \Gamma' \vdash B}\ cL,\ cR$$

Thus, in LK, it is possible for both occurrences of C in (†) to be contracted and, hence, the elimination of cut is nondeterministic since the cut rule can move to both the left and right premises.

Such nondeterminism in cut elimination is even more pronounced when we consider the collision of the cut rule with weakening. Consider the derivation (taken from Girard, Taylor, and Lafont, 1989, Appendix B).

$$\dfrac{\dfrac{\dfrac{\Xi_1}{\vdash B}}{\vdash C, B}\,wR \qquad \dfrac{\dfrac{\Xi_2}{\vdash B}}{C \vdash B}\,wL}{\dfrac{\vdash B, B}{\vdash B}\,cR}\,cut$$

cut elimination here can yield either Ξ_1 or Ξ_2: thus, nondeterminism arising from weakening can lead to completely different proofs of B. This kind of example does not occur in the intuitionistic (single-sided) version of the sequent calculus.

These problems with cut elimination and the structural rules were noted in Danos, Joinet, and Schellinx (1997) and by Lafont in Girard, Taylor, and Lafont (1989). Lafont concludes that in order to avoid this problem with cut elimination, one can choose from among two solutions: either make the sequent calculus asymmetric (leading to intuitionistic logic where the structural rules are not available on the right) or forbid all structural rules (leading to linear logic where structural rules are not available on the left and right). It is possible, however, to remain in classical logic by employing a third solution that uses both *polarization* and *focused proof systems*. Such an approach was proposed by Girard (1991) in his LC proof system and by Danos, Joinet, and Schellinx (1997) in their LK^{η} proof system. In this paper, we present the LKF proof system, which is also based on the notions of polarization and focusing. As we shall see, the problems with the nondeterminism in cut elimination caused by the use of structural rules in classical logic disappear in LKF for two reasons. First, weakening will be allowed only in the initial rules of LKF where it cannot cause problems with cut elimination. Second, a cut takes place between two *polarized formulas* of opposite *polarity* and, in LKF, contraction is only applied to positive formulas.

2.2 Permutations of inference rules

A dominating feature of sequent calculus proofs in LK is that many pairs of inference rules permute over each other (Kleene, 1952). For example, when an occurrence of $\supset L$ is below $\forall R$, as in the derivation

$$\dfrac{\Gamma_1 \vdash B, \Delta_1 \qquad \dfrac{\Gamma_2, C \vdash [y/x]D, \Delta_2}{\Gamma_2, C \vdash \forall x.D, \Delta_2}\,\forall R}{\Gamma_1, \Gamma_2, B \supset C \vdash \forall x.D, \Delta_1, \Delta_2}\,\supset L,$$

the order of these two rules can be switched to form the derivation

$$\dfrac{\dfrac{\Gamma_1 \vdash B, \Delta_1 \quad \Gamma_2, C \vdash [y/x]D, \Delta_2}{\Gamma_1, \Gamma_2, B \supset C \vdash [y/x]D, \Delta_1, \Delta_2} \supset L}{\Gamma_1, \Gamma_2, B \supset C \vdash \forall x.D, \Delta_1, \Delta_2} \forall R.$$

Similarly, the following two deviations are such that permuting the inference rules in one derivation yields the other derivation.

$$\dfrac{\dfrac{\Gamma, B_i, C_j \vdash \Delta}{\Gamma, B_i, C_1 \wedge C_2 \vdash \Delta}}{\Gamma, B_1 \wedge B_2, C_1 \wedge C_2 \vdash \Delta} \qquad \dfrac{\dfrac{\Gamma, B_i, C_j \vdash \Delta}{\Gamma, B_1 \wedge B_2, C_j \vdash \Delta}}{\Gamma, B_1 \wedge B_2, C_1 \wedge C_2 \vdash \Delta}$$

If one is trying to find structure in sequent calculus proofs, then it is likely that both of these pairs of derivations should be identified in some way.

The existence of such permutations of inference rules suggests that uncovering structures in proofs will always be disturbed by the possibilities of such shallow rearrangements of inference rules. For such reasons, people have often argued that the "essence" of proof structures is better captured in some radically different proof systems, such as, for example, expansion trees (Miller, 1987), proof nets (Girard, 1987; Laurent, 2011), and atomic flows (Guglielmi and Gunderson, 2008). In this paper, we also replace Gentzen-style sequent calculus with something else, namely LKF, but this time, that replacement will still resemble sequent calculus but with more structure added to both sequents and inference rules.

An introduction rule of LK is *invertible* if whenever there is an LK proof of its conclusion, there are LK proofs of the premises. When attempting to build a proof from the bottom-up, invertible rules can always be applied without losing provability. If an introduction rule is not invertible, it is *non-invertible*. The LK introduction rules can be classified as follows: the invertible rules are $\wedge R$, tR, $\vee L, fL$, $\supset R$, $\forall R$, $\exists L$ while the non-invertible rules are $\wedge L$, $\vee R$, $\supset L$, $\forall L$, $\exists R$. Note that every connective has an invertible introduction rule on one side of the $\vdash$, and every occurrence of the corresponding introduction rule on the other side is non-invertible. (This last statement is vacuously true for t and f since they have zero introduction rules on the left and right, respectively.) Observing the invertibility of introduction rules allows us to give some structure to the permutation of inference rules. In particular, an invertible rule above any other rule can always be permuted down. Furthermore, two non-invertible rules, one above the other, can always be permuted as well.

We make one additional observation: if an occurrence of a non-atomic formula on the left or right of a sequent can be the consequence of an invertible rule, that formula occurrence never needs to have a structural rule applied to it. For example, the contraction-left rule never needs to be applied to a disjunction since the disjunction-left rule is invertible.

These three observations about invertible and non-invertible rules — the left-right duality regarding invertibility; the permutations involving invertible and non-invertible rules; and the connection between invertible rules and the structural rules — will all be made explicit in the design of the LKF proof system.

$$\frac{\Gamma, B_1, B_2 \vdash \Delta}{\Gamma, B_1 \wedge B_2 \vdash \Delta}\ \wedge L \qquad \frac{\Gamma \vdash \Delta}{\Gamma, t \vdash \Delta}\ tL \qquad \frac{\Gamma \vdash \Delta, B \quad \Gamma' \vdash \Delta', C}{\Gamma, \Gamma' \vdash \Delta, \Delta', B \wedge C}\ \wedge R \qquad \frac{}{\cdot \vdash t}\ tR$$

$$\frac{\Gamma, B \vdash \Delta \quad \Gamma', C \vdash \Delta'}{\Gamma, \Gamma' B \vee C \vdash \Delta, \Delta'}\ \vee L \qquad \frac{}{f \vdash \cdot}\ fL \qquad \frac{\Gamma \vdash \Delta, B_1, B_2}{\Gamma \vdash \Delta, B_1 \vee B_2}\ \vee R \qquad \frac{\Gamma \vdash \Delta}{\Gamma \vdash \Delta, f}\ fR$$

Fig. 2 The introduction rules for conjunction, disjunction, and their units using multiplicative instead of additive rules.

2.3 Additive and multiplicative rules and connectives

The LK rules that have two premises can be classified as either *additive*, in which case the side formulas (Γ, Δ) are the same in the conclusion as well as in both premises, or *multiplicative*, in which case the side formulas in the premises (Γ, Δ and Γ', Δ') are accumulated to form the side formulas in the conclusion. Of the four inference rules in Figure 1 with two premises, the *cut* rule and the implication-left rule are multiplicative while the disjunction-left rule and the conjunction-right rule are additive.

Consider the alternative inference rules in Figure 2 for conjunction and disjunction. The rules in that figure with two premises are multiplicative. We can make the following observations.

1. The rules in Figure 2 are inter-admissible, in the sense of preserving the provability of sequents, with those for the same connectives given in Figure 1. Establishing that fact requires using the structural rules of weakening and contraction.
2. The $\wedge R$ rule in Figure 1 is invertible while the corresponding $\wedge R$ rule in Figure 2 is not invertible. Similarly, the $\vee R$ rule in Figure 1 are not invertible while the corresponding $\vee R$ rule in Figure 2 is invertible.
3. If we are keen to separate the roles of structural rules from cut elimination, then we should not mix the various rules in Figures 1 and 2. For example, if we replace the $\wedge L$ rule in Figure 1 with $\wedge L$ in Figure 2, then the proof that a cut of a conjunction can be eliminated will necessarily use a structural rule.

Although Gentzen used the additive rules for conjunction and disjunction, there are reasons to admit other choices. For example, it is a popular choice to select the invertible right introduction rules for both conjunction and disjunction, which means selecting the additive conjunction and the multiplicative disjunction. Ketonen introduced such a variant of Gentzen's original calculus and used it to give "a strikingly elegant proof of completeness" (von Plato, 2012). People working in automated theorem proving often use the invertible rules since it simplifies implementations of proof search. In particular, it is possible to define one-sided sequent systems for classical logic in such a way that all (right) introduction rules are invertible except for the existential introduction rule. As a result, proof search algorithms can limit backtracking to only the treatment of existential quantifiers.

The LKF proof system contains both the additive and multiplicative versions of conjunction and disjunction (and their units).

2.4 The need for synthetic inference rules

Our final criticism of LK is that its inference rules are too small, especially for applications involving theories. For example, assume that we are working with a theory (a set of assumptions) that has an axiom that declares that the binary predicate *path* is transitive: that is, that the theory contains the formula

$$\forall x \forall y \forall z\ (path(x,y) \supset path(y,z) \supset path(x,z)).$$

If that formula is invoked in an LK proof, there will be a minimum of five introduction rules involved in that invocation. That seems unfortunate since it is more natural to view that formula as denoting one of the following inference rules.

$$\frac{\Gamma \vdash \Delta,\ path(x,y) \quad \Gamma \vdash \Delta,\ path(y,z)}{\Gamma \vdash \Delta, path(x,z)}$$

or

$$\frac{path(x,y),\ path(y,z),\ path(x,z), \Gamma \vdash \Delta}{path(x,y),\ path(y,z), \Gamma \vdash \Delta}$$

These *synthetic rules* would be a more appropriate way to invoke the transitivity axiom. Such synthetic rules have been addressed before in the literature, particularly as a back-chaining inference rule (Hallnäs and Schroeder-Heister, 1990; Miller, Nadathur, Pfenning, and Scedrov, 1991) or as a forward-chaining inference rule (Negri and von Plato, 1998). One of the immediate applications of LKF is as a formal framework for computing and justifying the addition of such synthetic inference rules to LK.

3 The LKF proof system

The LKF proof system does not deal with formulas but with *polarized formulas*: these are built from atomic formulas and negated atomic formulas (collectively called literals), and *polarized logical connectives* as well as the first-order quantifiers $\forall$ and $\exists$. The polarized logical connectives come in two flavors: the *positive connectives* are f^+, $\vee^+$, t^+, $\wedge^+$, and $\exists$ while the *negative connectives* are t^-, $\wedge^-$, f^-, $\vee^-$, and $\forall$.

Literals are also assigned a polarity as follows. An *atomic bias assignment* is a function $\delta(\cdot)$ that maps atomic formulas to the set of two tokens $\{+,-\}$: if $\delta(A)$ is $+$ then that atomic formula is positive and if $\delta(A)$ is $-$ then that atomic formula is negative. We extend $\delta(\cdot)$ to literals by setting $\delta(\neg A)$ to be the opposite polarity of $\delta(A)$. We may ask that all atomic formulas are positive, that they are all negative, or we can mix polarity assignments. In particular, the atomic bias assignment $\delta^+(\cdot)$ assigns all atoms the positive polarity while $\delta^-(\cdot)$ assigns all atoms the negative polarity. We shall often suppress explicit reference to atomic bias assignments, assuming that they have been specified and fixed. The only restriction we impose on atomic bias assignments is that they are stable under substitution: that is, for all atomic formulas

A and every first-order variable x and term s, then $\delta(A) = \delta([x/s]A)$. This restriction is equivalent to saying that the value of $\delta(\cdot)$ is determined by the predicate that is the top-level symbol of A: that is, if A and A' are two atoms formed with the same predicate, then $\delta(A) = \delta(A')$.

A polarized formula is *positive* if it is a positive literal or its top-level connective or quantifier is positive (i.e, it is of the form $A \wedge^+ B$, $A \vee^+ B$, $\exists x.A$, t^+ or f^+); similarly, a polarized formula is *negative* it is a negative literal or its top-level connective or quantifier is negative (i.e, it is of the form $A \wedge^- B$, $A \vee^- B$, $\forall x.A$, t^- or f^-).

Polarized formulas are in *negation normal form* (nnf), meaning that there are no occurrences of implication $\supset$ and that the negation symbol $\neg$ has only atomic scope. When the negation symbol $\neg$ is used with the non-atomic polarized formulas of LKF, we shall view it as the following function that transforms that polarized formula to its De Morgan dual.

Definition 3.1 The negation symbol $\neg$ is defined as the following function when applied to non-atomic polarized formulas.

1. $\neg\neg a = a$ for atomic formula a
2. $\neg(A \wedge^+ B) = \neg A \vee^- \neg B, \quad \neg(A \vee^- B) = \neg A \wedge^+ \neg B$
3. $\neg(A \vee^+ B) = \neg A \wedge^- \neg B, \quad \neg(A \wedge^- B) = \neg A \vee^+ \neg B$
4. $\neg\exists x.A = \forall x.\neg A, \quad \neg\forall x.A = \exists x.\neg A$

It is easily shown that $\neg\neg A = A$ for all polarized formulas A. Clearly, negation is treated differently between unpolarized formulas (where it is an abbreviation for "implies false") and polarized formulas (where it computes the De Morgan dual).

The sequent calculus LKF for polarized formulas is presented in Figure 3: this presentation is a simplification of our original presentation given in Liang and Miller (2009). This proof system uses one-sided sequents, but of two varieties, namely, $\vdash \Gamma \Uparrow \Theta$ and $\vdash A \Downarrow \Theta$, where Γ is a multiset of polarized formulas, Θ is a set of polarized formulas, and A is a single polarized formula. The up and down arrows separate sequents into two *zones*: the zone on the right of the arrows (written using Θ) is called the *storage* for that sequent. In notation such as $\vdash \Gamma, \Gamma' \Uparrow \Theta, \Theta'$, the multiset Γ, Γ' represents the multiset sum of Γ and Γ' while the set Θ, Θ' represents the union of the two sets Θ and Θ': it is, of course, possible for Θ and Θ' to share a non-empty intersection. When moving a collection of polarized formulas from the left of the $\Uparrow$ into storage, we coerce multisets into sets in the obvious way. Note that by inspection, the storage of the sequent in the conclusion of an inference rule is always a subset of storage of the sequents in the premises. We say that the polarized formula B has an LKF proof if the sequent $\vdash B \Uparrow \cdot$ has a proof using the inference rules from Figure 3.

Before completing the details of the LKF proof system, we informally describe its relationship to Gentzen's LK proof system. Just as polarized formulas are essentially regular formulas with some additional structure added (the + and − annotations), LKF sequents are essentially LK sequents with additional structure. That extra structure is the establishment of two zones within a sequent, namely, the storage zone and the non-storage zone. Additionally, LKF sequents come in two kinds, as is witnessed by the use of either $\Uparrow$ or $\Downarrow$. Thus, if one wishes to relate LKF sequents to Gentzen's original sequents, one only needs to forget this additional structure. In particular, the

ASYNCHRONOUS INTRODUCTION RULES

$$\frac{}{\vdash t^-, \Gamma \Uparrow \Theta}\ t^- \qquad \frac{\vdash A, \Gamma \Uparrow \Theta \quad \vdash B, \Gamma \Uparrow \Theta}{\vdash A \wedge^- B, \Gamma \Uparrow \Theta}\ \wedge^- \qquad \frac{\vdash \Gamma \Uparrow \Theta}{\vdash f^-, \Gamma \Uparrow \Theta}\ f^- \qquad \frac{\vdash A, B, \Gamma \Uparrow \Theta}{\vdash A \vee^- B, \Gamma \Uparrow \Theta}\ \vee^-$$

$$\frac{\vdash [y/x]B, \Gamma \Uparrow \Theta}{\vdash \forall x.B, \Gamma \Uparrow \Theta}\ \forall$$

SYNCHRONOUS INTRODUCTION RULES

$$\frac{}{\vdash t^+ \Downarrow \Theta}\ t^+ \qquad \frac{\vdash A \Downarrow \Theta \quad \vdash B \Downarrow \Theta}{\vdash A \wedge^+ B \Downarrow \Theta}\ \wedge^+ \qquad \frac{\vdash B_i \Downarrow \Theta}{\vdash B_1 \vee^+ B_2 \Downarrow \Theta}\ \vee^+,\ i \in \{1,2\} \qquad \frac{\vdash [s/x]B \Downarrow \Theta}{\vdash \exists x.B \Downarrow \Theta}\ \exists$$

INITIAL, STORE, RELEASE, AND DECIDE RULES

$$\frac{}{\vdash p \Downarrow \neg p, \Theta}\ init \qquad \frac{\vdash \Gamma \Uparrow Q, \Theta}{\vdash Q, \Gamma \Uparrow \Theta}\ store \qquad \frac{\vdash N \Uparrow \Theta}{\vdash N \Downarrow \Theta}\ release \qquad \frac{\vdash P \Downarrow P, \Theta}{\vdash \cdot \Uparrow P, \Theta}\ decide$$

Fig. 3 The inference rules for LKF. Here, P is a positive polarized formula and p is a positive literal; N is a negative polarized formula and Q is a positive polarized formula or negative literal. The rule for $\forall$ has the usual eigenvariable restriction: y is not free in any polarized formula in the concluding sequent.

arrows $\Uparrow$ and $\Downarrow$ can be replaced by a comma and all the polarization annotations on polarized formulas can be deleted.

We borrow the terminology *asynchronous* and *synchronous* rules from Andreoli (1992). A derivation composed only of asynchronous introduction rules (see Figure 3) and the *store* rule will be called an *asynchronous phase*, and a derivation composed only of synchronous introduction rules and the *init* rule will be called a *synchronous phase*. The sequents in an asynchronous phase all involve $\Uparrow$-sequents while the sequents in a synchronous phase all involve $\Downarrow$-sequents. An LKF proof is composed of alternations of these two kinds of phases. In particular, the *decide* rule connects a synchronous phase above its premise with an asynchronous phase below its conclusion, and the *release* rule connects an asynchronous phase above its premise with a synchronous phase below its conclusion.

The asynchronous phase can be used to encapsulate what is often called *don't care nondeterminism*. That is, if we consider the asynchronous phase as a large scale inference rule having a sequent of the form $\vdash N \Uparrow \Theta$ as its conclusion and sequents of the form $\vdash \cdot \Uparrow \Theta'$ as its premises, then that large scale rule is independent of the sequence of rule applications within the asynchronous phase (see Lemma 3.4). On the other hand, the synchronous phase is a sequence of applications of inference rules with choices (particularly for the $\vee^+$ and $\exists$ introduction rules), and different choices will yield different synchronous phases: such phases, therefore, capture *don't know nondeterminism*.

While the weakening and contraction rules are not explicitly given in LKF, both of these rules occur implicitly. The *decide* rule does an implicit contraction on the polarized formula P: hence, the only polarized formulas contracted in an LKF proof are positive polarized formulas. The *init* and the t^+ rules do implicit weakening on the polarized formulas in Θ: thus weakening is available for positive polarized formulas and negative literals. Thus, a negative, non-literal polarized formula is never weakened nor contracted: such polarized formulas are treated *linearly*, in the sense of linear logic (Girard, 1987).

Polarized formulas in the storage zone play two different roles in proof search. With the *decide* rule, a positive polarized formula in the storage is *simultaneously* contracted and made available to introduction rules. On the other hand, with the *init* rule, a negative literal in storage is available to end the proof. No other kind of polarized formula will occur in storage.

The four binary logical connectives of LKF — $\vee^+$, $\vee^-$, $\wedge^+$, $\wedge^-$ — can be classified using three different attributes: positive or negative; additive or multiplicative; and conjunctive or disjunctive. By fixing any two of these attributes, the third attribute is uniquely determined. For example, a connective that is both additive and positive must be the disjunction $\vee^+$. Note also that the De Morgan dual of a logical connective (in the sense of Definition 3.1) flips its polarity and conjunctive/disjunctive status but does not change its additive/multiplicative status. The introduction rule for $\wedge^+$ looks additive since the storage Θ in the conclusion and the premises are all the same. The essential multiplicative character of $\wedge^+$ is not apparent in this proof system in which there can be only one focused polarized formula in a sequent. In Section 10, we present a *multifocused* version of LKF, and in that enlarged setting, it will be clear that $\wedge^+$ is, in fact, a multiplicative connective.

The proof system for LKF given in Figure 3 has no cut rule; thus the proofs built using the rules in Figure 3 are cut-free proofs. Cut-rules for LKF and a cut-elimination theorem will be presented in the next section.

Let B be a polarized formula and let $\check{B}$ be the *depolarized* version of B: that is, $\check{B}$ is the unpolarized formula that results from removing the superscript + and – from the logical connectives in B. Since B is in negation normal form, the formula $\check{B}$ might have occurrences of negated atomic formulas, say $\neg A$, and these should be seen as abbreviations for $A \supset f$. Depolarizing a multiset or set of polarized formulas Γ is the set $\check{\Gamma}$ resulting from depolarized the formulas in Γ.

Theorem 3.2 (Soundness of LKF**)** *Let B be a polarized formula and let Γ and Θ be a multiset and set, respectively, of polarized formulas. If $\vdash \Gamma \Uparrow \Theta$ is provable in* LKF *then $\vdash \check{\Gamma}, \check{\Theta}$ is provable in* LK. *If $\vdash B \Downarrow \Theta$ is provable in* LKF *then $\vdash \check{B}, \check{\Theta}$ is provable in* LK.

Proof This theorem can be proved by a straightforward mutual induction on the structure of (cut-free) LKF proofs. Most cases of this mutual induction are straightforward. For example, the introduction rule for $\vee^+$ in LKF corresponds to the introduction rule for $\vee$ in LK, while the introduction rule for $\vee^-$ in LKF corresponds to the multiplicative version of the introduction rule for $\vee$ in Figure 2. The *init* rule in LKF corresponds, however, to the following LK derivation.

$$\dfrac{\dfrac{\overline{p \vdash p}\ init}{p \vdash p, f, \check{\Theta}}\ wR}{\vdash p, p \supset f, \check{\Theta}}\ \supset R$$

Finally, *decide* in LKF corresponds to the *cR* rule, and *store* and *release* do not contribute to the LK proof. □

The converse of this soundness theorem is more challenging to prove: we shall state and prove such completeness as Theorem 8.4 in Section 8. (Every time we mention completeness theorems in this paper, we shall mean *relative completeness* with respect to another proof system: we will not use the model theory notion of validity in this paper.) In anticipation of that result, we state a version of that completeness theorem here. Let B be a first-order polarized formula, let $\delta(\cdot)$ be any atomic bias assignment, and let C be the unpolarized formula $\check{B}$. If C is provable in LK (in the sense that $\vdash C$ is provable in LK) then B is provable in LKF. A consequence of this completeness theorem is the following: if C is an unpolarized formula that is provable in LK, then for every polarized formula B (and atomic bias assignment) such that $\check{B}$ is C, then B has an LKF proof. Note that if there are n occurrences of propositional connectives in C, there are 2^n polarized formulas B such that $\check{B} = C$. Clearly, polarization does not affect provability, but it can have a large impact on the structure of (focused) proofs.

We now state two properties about (cut-free) LKF proofs.

Lemma 3.3 (Admissibility of Weakening) *If* $\vdash \Gamma \Uparrow \Theta$ *and* $\vdash A \Downarrow \Theta$ *are (cut-free) provable and if* Θ' *is a set of positive polarized formulas and negative literals then* $\vdash \Gamma \Uparrow \Theta, \Theta'$ *and* $\vdash A \Downarrow \Theta, \Theta'$ *are also provable.*

This lemma is proved easily by induction on the structure of proofs. The proof further shows that weakening also does not affect the structure of proofs in that the same inference rules are applied at each step.

The following lemma captures the fact that the asynchronous phase of inference rules can deal with don't-care-nondeterminism: any polarized formula to the left of the $\Uparrow$ can be selected to be processed first.

Lemma 3.4 *If there is a (cut-free) proof of* $\vdash A, \Gamma \Uparrow \Theta$ *then there is a (cut-free) proof that ends with either an introduction of* A *or a store rule on* A.

Proof This lemma holds because the asynchronous introduction rules permute over each other in such a way that the same premises remain. The formal proof of this lemma is by induction on the sum of the sizes of formulas in Γ. The size of a formula is the number of occurrences of literals, connectives, and quantifiers in the formula. In particular, A and $\neg A$ are of the same size. In the base case, Γ is empty, and the result is trivial. For the inductive case, let $\Gamma = B, \Gamma'$ and assume that the sequent $\vdash A, B, \Gamma' \Uparrow \Theta$ is the conclusion of an inference rule ρ which is either an introduction or *store* on B. We then proceed to show that the ρ rule can be permuted above the introduction or *store* of A. There are several cases to consider.

Case: A and B are both either positive formulas or negative literals. In this case, ρ is *store* on B with premise $\vdash A, \Gamma' \Uparrow \Theta, B$. By inductive hypothesis on the smaller

Γ', the next rule above must be a *store* on A, with premise $\vdash \Gamma' \Uparrow \Theta, A, B$. But clearly we can switch the order of the two *store* rules:

$$\dfrac{\dfrac{\vdash \Gamma' \Uparrow \Theta, A, B}{\vdash B, \Gamma' \Uparrow \Theta, A}\ \mathit{store}}{\vdash A, B, \Gamma' \Uparrow \Theta}\ \mathit{store}$$

Case: A is a positive formula or negative literal and B is a non-literal negative formula. In this case, we consider the structure of B. For example, if B is $B_1 \vee^- B_2$, then the premise of ρ is $\vdash A, B_1, B_2, \Gamma' \Uparrow \Theta$. Since the size of B_1, B_2, Γ' is smaller than the size of $B_1 \vee^- B_2, \Gamma'$, the inductive hypothesis provides a proof where the rule above ρ is the *store* rule applied to A with premise $\vdash B_1, B_2, \Gamma' \Uparrow \Theta, A$. Starting from that sequent, we can switch the *store* and $\vee^-$ rules, resulting in

$$\dfrac{\dfrac{\vdash B_1, B_2, \Gamma' \Uparrow \Theta, A}{\vdash B_1 \vee^- B_2, \Gamma' \Uparrow \Theta, A}\ \vee^-}{\vdash A, B_1 \vee^- B_2, \Gamma' \Uparrow \Theta}\ \mathit{store}$$

The cases of B is t^-, $B_1 \wedge^- B_2$, $\forall x.B'$ and f^- are similar.

Case: B is a positive formula or negative literal and A is a non-literal negative formula. This case is analogous to the above case. We illustrate with the case that A is $A_1 \wedge^- A_2$. Since the ρ rule is *store* on B, its premise is $\vdash A_1 \wedge^- A_2, \Gamma' \Uparrow \Theta, B$. By the inductive hypothesis, the next rule above is the introduction for $\wedge^-$:

$$\dfrac{\dfrac{\vdash A_1, \Gamma' \Uparrow \Theta, B \quad \vdash A_2, \Gamma' \Uparrow \Theta, B}{\vdash A_1 \wedge^- A_2, \Gamma' \Uparrow \Theta, B}\ \wedge^-}{\vdash A_1 \wedge^- A_2, B, \Gamma' \Uparrow \Theta}\ \mathit{store}.$$

These rules can be permuted to yield the desired form

$$\dfrac{\dfrac{\vdash A_1, \Gamma' \Uparrow \Theta, B}{\vdash A_1, B, \Gamma' \Uparrow \Theta}\ \mathit{store} \qquad \dfrac{\vdash A_2, \Gamma' \Uparrow \Theta, B}{\vdash A_2, B, \Gamma' \Uparrow \Theta}\ \mathit{store}}{\vdash A_1 \wedge^- A_2, B, \Gamma' \Uparrow \Theta}\ \wedge^-.$$

Case: A and B are both non-literal negative polarized formulas. There are several cases to consider, but they are all similar. For example, if A and B are $A_1 \vee^- A_2$ and $B = B_1 \vee^- B_2$, respectively, and the last rule introduces B, we just need to show that the two $\vee^-$-introductions permute over each other, which follows easily from the fact that both proofs can be constructed from the common premise of $\vdash A_1, A_2, B_1, B_2, \Gamma' \Uparrow \Theta$. In the case where A is $A_1 \vee^- A_2$ and B is $B_1 \wedge^- B_2$, introducing $B_1 \wedge^- B_2$ results in the premises $\vdash A_1 \vee^- A_2, B_1, \Gamma' \Uparrow \Theta$ and $\vdash A_1 \vee^- A_2, B_2, \Gamma' \Uparrow \Theta$, both of which have a smaller inductive measure, which allows us to assume that the next rule above will introduce $A_1 \vee^- A_2$ and we can therefore build the proof

$$\frac{\vdash A, \Gamma \Uparrow \Theta \quad \vdash \neg A, \Gamma' \Uparrow \Theta'}{\vdash \Gamma, \Gamma' \Uparrow \Theta, \Theta'}\ cut_u \qquad \frac{\vdash A \Downarrow \Theta \quad \vdash \neg A, \Gamma' \Uparrow \Theta'}{\vdash \Gamma' \Uparrow \Theta, \Theta'}\ cut_f$$

$$\frac{\vdash \Gamma \Uparrow \Theta, P \quad \vdash \neg P, \Gamma' \Uparrow \Theta'}{\vdash \Gamma, \Gamma' \Uparrow \Theta, \Theta'}\ dcut_u \qquad \frac{\vdash B \Downarrow \Theta, P \quad \vdash \neg P \Uparrow \Theta'}{\vdash B \Downarrow \Theta, \Theta'}\ dcut_f$$

Fig. 4 The Cut Rules of LKF. Here, A and B are arbitrary polarized formulas and P is a positively polarized formula.

$$\frac{\dfrac{\vdash A_1, A_2, B_1, \Gamma' \Uparrow \Theta \quad \vdash A_1, A_2, B_2, \Gamma' \Uparrow \Theta}{\vdash A_1, A_2, B_1 \wedge^- B_2, \Gamma' \Uparrow \Theta}\ \wedge^-}{\vdash A_1 \vee^- A_2, B_1 \wedge^- B_2, \Gamma' \Uparrow \Theta}\ \vee^-.$$

The remaining cases are treated similarly. □

Definition 3.5 We say that a (cut-free) proof of $\vdash A, \Gamma \Uparrow \Theta$ is *eager* with respect to A if the last inference rule introduces A or is a *store* rule on A. We say that the proof is *delayed* with respect to A if either

1. Γ is empty, or
2. the last inference rule does not introduce A, is not a *store* rule on A, and each immediate subproof above $\vdash A, \Gamma \Uparrow \Theta$ is also delayed with respect to A.

In other words, a proof is delayed with respect to A if A is only subject to an introduction or *store* rule on A when it appears in a conclusion of the form $\vdash A \Uparrow \Theta$. Note also that a proof of $\vdash A \Uparrow \Theta$ is both eager and delayed with respect to A.

Lemma 3.4 implies that a proof can be transformed into either the eager or the delayed form.

4 Cut Elimination for LKF

Given that LKF has two kinds of sequents and each of these has two zones for holding polarized formulas, we introduce in Figure 4 a total of four cut rules in order to state and prove the cut-elimination theorem for LKF. The cut_u rule (called the *unfocused* cut rule) applies only to $\Uparrow$-sequents while the cut_f rule (called the *focused* cut rule) involves one $\Downarrow$-sequent. Both of those cut rules also have a "delayed" version in which one of the occurrences of the polarized cut formula is "locked" in storage.

It is important to note that in the delayed cuts, the polarized cut formula P is positive and not a negative literal: in particular if P were a negative literal in the $dcut_f$ rule and if $B = \neg P$ then $dcut_f$ is not admissible since focusing on a positive literal requires the proof to end in an initial rule.

A simple observation shows that the cut-rules in Figure 4 do not suffer the collision problems mentioned in Section 2.1. As we noted in the previous section, only positive polarized formulas are contracted (by the *decide* rule) in LKF proofs: as a result, exactly one of the pair of polarized formulas A and $\neg A$ involved in a cut rule will be positive, and only one of them can be contracted. Similarly, weakening only appears within the *init* rule in LKF proofs and, as a result, the problematic case involving weakening also disappears.

The general strategy for proving cut elimination in LKF extended with these cut rules is familiar: we reduce cuts to "key cases" in which the polarized cut formula is principal in both premises. The proof proceeds by simultaneous induction over the permutabilities of all four cuts. The inductive measure is the lexicographical ordering consisting of the size of the polarized cut formula followed by the sum of the heights of the subproofs above the cut. We apply the procedure to the topmost cuts first, thus assuming that the cuts to be reduced have cut-free subproofs.

Lemma 3.4 is used to simplify the cut-elimination proof. However, the application of this lemma for proof transformation may affect the height of proofs (because of the t^- rule). These transformations must be applied carefully to preserve the inductive measure. For the cut-elimination proof, we further require that the following conditions be placed on the cut rules.

1. In cut_u, the subproof of the premise with the positive cut formula must be *eager* with respect to the cut formula; the subproof of the premise with the negative cut formula must be *delayed* with respect to the cut formula.
2. In $dcut_u$, the subproof of the premise with the negative cut formula must be *delayed* with respect to the cut formula.
3. In cut_f, the subproof of the sequent $\vdash \neg A, \Gamma' \Uparrow \Theta'$, where $\neg A$ is the cut formula, must be *eager* with respect to $\neg A$ regardless of the polarity of A.

The third requirement may appear inconsistent with the others when $\neg A$ is negative in cut_f: however, the transition from cut_u or $dcut_u$ to a cut_f only occurs when the cut formula is decomposed into subformulas, which reduces the stronger inductive measure. For the $dcut_f$ rule, the subproof above the negative cut formula $\neg P$ can be considered both eager and delayed with respect to $\neg P$ because it is the only formula to the left of $\Uparrow$. By Lemma 3.4, any proof can be transformed into the required forms so that the reducibility of the restricted cuts also implies the reducibility of the unrestricted versions. In other words, before the application of any cut, we can always apply Lemma 3.4 to assume that the subproofs are in the required forms. The cut-elimination arguments will show that all restrictions are preserved when any of the four cut rules are permuted to other cut rules.

We detail the permutation of each of the four cuts. We sometimes do not repeat cases that are obvious, and we generally ignore the quantifiers as the first-order quantifiers add nothing to the argument: their treatment is completely standard.

4.1 Permutations of cut_u

The cut_u rule has the general form, repeated here for convenience:

$$\frac{\vdash A, \Gamma \Uparrow \Theta \qquad \vdash \neg A, \Gamma' \Uparrow \Theta'}{\vdash \Gamma, \Gamma' \Uparrow \Theta, \Theta'}\ cut_u.$$

Assume without loss of generality that A is positive and, therefore, $\neg A$ is negative. It is also required that the left subproof above cut_u is *eager* with respect to the positive A, i.e., it ends in a *store* rule on the cut formula A. Furthermore, the right subproof above the negative cut formula $\neg A$, is required to be *delayed* with respect to $\neg A$. These assumptions mean that this cut can be transformed immediately into a $dcut_u$:

$$\frac{\dfrac{\vdash \Gamma \Uparrow \Theta, A}{\vdash A, \Gamma \Uparrow \Theta}\ store \qquad \vdash \neg A, \Gamma' \Uparrow \Theta'}{\vdash \Gamma, \Gamma' \Uparrow \Theta, \Theta'}\ cut_u \quad \longrightarrow \quad \frac{\vdash \Gamma \Uparrow \Theta, A \quad \vdash \neg A, \Gamma' \Uparrow \Theta'}{\vdash \Gamma, \Gamma' \Uparrow \Theta, \Theta'}\ dcut_u.$$

Clearly the restriction on the *delayed* form of the subproof above the negative cut formula $\neg A$ is preserved for the $dcut_u$ rule. The inductive measure is reduced by the smaller height of the left subproofs above the cut.

4.2 Permutations of $dcut_u$

The delayed, unfocused $dcut_u$ rule has the form

$$\frac{\vdash \Gamma \Uparrow \Theta, P \qquad \vdash \neg P, \Gamma' \Uparrow \Theta'}{\vdash \Gamma, \Gamma' \Uparrow \Theta, \Theta'}\ dcut_u$$

where the cut formula P is positive. It is required that the subproof above the right premise is *delayed* with respect to the cut formula $\neg P$. These cuts are permuted to the point where P is selected for focus, at which point the cut transforms into a combination of cut_f and $dcut_f$. In other words, the "goal" or "target" of all permutations of $dcut_u$ is to be able to apply the following transformation when the left premise of the $dcut_u$ is the *decide* rule.

$$\frac{\dfrac{\vdash P \Downarrow \Theta, P}{\vdash \cdot \Uparrow \Theta, P} \qquad \vdash \neg P \Uparrow \Theta'}{\vdash \cdot \Uparrow \Theta, \Theta'}\ dcut_u \quad \longrightarrow \quad \frac{\dfrac{\vdash P \Downarrow \Theta, P \quad \vdash \neg P \Uparrow \Theta'}{\vdash P \Downarrow \Theta, \Theta'}\ dcut_f \qquad \vdash \neg P \Uparrow \Theta'}{\vdash \cdot \Uparrow \Theta, \Theta'}\ cut_f.$$

In the transformed proof, the upper $dcut_f$ has subproofs of lesser height measure, while the lower cut_u is a *key case* cut where the cut formula is principal in both subproofs. That is, cut-free proofs for $\vdash P \Downarrow \Theta, \Theta'$ and $\vdash \neg P \Uparrow \Theta'$ must both end with the cut formulas P and $\neg P$ subject to an inference rule. The key-case cuts immediately decompose into cuts on subformulas of a smaller size than P (or reduces completely by weakening in case of P being a positive literal). Thus, the inductive measures of both cuts are reduced.

Note that the *eager* restriction on the right subproof above cut_f is trivially preserved since $\neg P$ is the only polarized formula on the left of $\Uparrow$.

All other permutations of $dcut_u$ make progress toward this case. We organize these permutations into two stages.

The first stage performs permutations over inference rules in the right subproof of $dcut_u$. The right subproof above $dcut_u$ ends in $\vdash \neg P, \Gamma' \Uparrow \Theta'$. We permute $dcut_f$ until it has such a right subproof with an empty Γ'. The fact that this subproof is *delayed* with respect to $\neg P$ means that if it ends in a conclusion $\vdash \neg P, B, \Gamma' \Uparrow \Theta'$ we can assume that the last rule either introduces B or is a *store* rule on B (and not on $\neg P$). There are many subcases depending on the form of B:

Case: B is a positive polarized formula or negative literal. In this case, the rule above is a *store* on B, resulting in the following permutation.

$$\dfrac{\vdash \Gamma \Uparrow \Theta, P \quad \dfrac{\vdash \neg P, \Gamma' \Uparrow \Theta', B}{\vdash \neg P, B, \Gamma' \Uparrow \Theta'}\ store}{\vdash B, \Gamma, \Gamma' \Uparrow \Theta, \Theta'}\ dcut_u \quad \longrightarrow \quad \dfrac{\dfrac{\vdash \Gamma \Uparrow \Theta, P \quad \vdash \neg P, \Gamma' \Uparrow \Theta', B}{\vdash \Gamma, \Gamma' \Uparrow \Theta, \Theta', B}\ dcut_u}{\vdash B, \Gamma, \Gamma' \Uparrow \Theta, \Theta'}\ store$$

The "delayed" restriction on the right subproof above $dcut_u$ is preserved by definition: an immediate subproof of a delayed proof is also delayed. This property applies similarly to all subsequent cases.

Case: B is $B_1 \vee^- B_2$. In this case, we can transform

$$\dfrac{\vdash \Gamma \Uparrow \Theta, P \quad \dfrac{\vdash \neg P, B_1, B_2, \Gamma' \Uparrow \Theta'}{\vdash \neg P, B_1 \vee^- B_2, \Gamma' \Uparrow \Theta'}\ \vee^-}{\vdash B_1 \vee^- B_2, \Gamma, \Gamma' \Uparrow \Theta, \Theta'}\ dcut_u$$

into the following derivation.

$$\dfrac{\dfrac{\vdash \Gamma \Uparrow \Theta, P \quad \vdash \neg P, B_1, B_2, \Gamma' \Uparrow \Theta'}{\vdash B_1, B_2, \Gamma, \Gamma' \Uparrow \Theta, \Theta'}\ dcut_u}{\vdash B_1 \vee^- B_2, \Gamma, \Gamma' \Uparrow \Theta, \Theta'}\ \vee^-$$

Case: B is $B_1 \wedge^- B_2$. In this case, we can transform

$$\dfrac{\vdash \Gamma \Uparrow \Theta, P \quad \dfrac{\vdash \neg P, B_1, \Gamma' \Uparrow \Theta' \quad \vdash \neg P, B_1, \Gamma' \Uparrow \Theta'}{\vdash \neg P, B_1 \wedge^- B_2, \Gamma' \Uparrow \Theta'}\ \wedge^-}{\vdash B_1 \wedge^- B_2, \Gamma, \Gamma' \Uparrow \Theta, \Theta'}\ dcut_u$$

into the following derivation.

$$\dfrac{\dfrac{\vdash \Gamma \Uparrow \Theta, P \quad \vdash \neg P, B_1, \Gamma' \Uparrow \Theta'}{\vdash B_1, \Gamma, \Gamma' \Uparrow \Theta, \Theta'}\ dcut_u \quad \dfrac{\vdash \Gamma \Uparrow \Theta, P \quad \vdash \neg P, B_2, \Gamma' \Uparrow \Theta'}{\vdash B_2, \Gamma, \Gamma' \Uparrow \Theta, \Theta'}\ dcut_u}{\vdash B_1 \wedge^- B_2, \Gamma, \Gamma' \Uparrow \Theta, \Theta'}\ \wedge^-$$

The other cases for B are proved similarly. This stage ends when the right subproof concludes with a sequent of the form $\vdash \neg P \Uparrow \Theta'$.

The second stage performs permutation over inference rules in the left subproof of $dcut_u$. The cases of asynchronous introduction rules are analogous to the cases demonstrated above and are equally straightforward. Generally speaking, the permutation of cut above introduction rules is always straightforward. The important cases to point out are the *decide*, *release*, and *store* rules. A *store* rule ending the left subproof is also a trivial case because it cannot affect the cut formula. The interesting case is when the left subproof ends in the form $\vdash \cdot \Uparrow \Theta, P$. The rule above this sequent must be *decide*. There are two cases depending on whether or not the polarized formula selected for focus is the cut formula P or not. If it is not the cut formula but, say, another formula Q, then we can permute inference rules as follow.

$$\dfrac{\dfrac{\vdash Q \Downarrow Q, \Theta, P}{\vdash \cdot \Uparrow Q, \Theta, P}\,\mathit{decide} \qquad \vdash \neg P \Uparrow \Theta'}{\vdash \cdot \Uparrow Q, \Theta, \Theta'}\,\mathit{dcut}_u \quad \longrightarrow \quad \dfrac{\dfrac{\vdash Q \Downarrow Q, \Theta, P \qquad \vdash \neg P \Uparrow \Theta'}{\vdash Q \Downarrow Q, \Theta, \Theta'}\,\mathit{dcut}_f}{\vdash \cdot \Uparrow Q, \Theta, \Theta'}\,\mathit{decide}$$

If the polarized formula selected for focus is P, then we have reached the targeted transition to key-case cuts as already described above.

4.3 Permutations of dcut_f

The general form of dcut_f is

$$\dfrac{\vdash B \Downarrow \Theta, P \qquad \vdash \neg P \Uparrow \Theta'}{\vdash B \Downarrow \Theta, \Theta'}\,\mathit{dcut}_f$$

with P positive. This cut permutes over synchronous introduction rules until reaching an *init* or *release* rule on its left premise, at which point the cut will transition to a dcut_u with lower subproofs:

$$\dfrac{\dfrac{\vdash B \Uparrow \Theta, P}{\vdash B \Downarrow \Theta, P}\,\mathit{release} \qquad \vdash \neg P \Uparrow \Theta'}{\vdash B \Downarrow \Theta, \Theta'}\,\mathit{dcut}_f \quad \longrightarrow \quad \dfrac{\dfrac{\vdash B \Uparrow \Theta, P \qquad \vdash \neg P \Uparrow \Theta'}{\vdash B \Uparrow \Theta, \Theta'}\,\mathit{dcut}_u}{\vdash B \Downarrow \Theta, \Theta'}\,\mathit{release}$$

Besides the cases of initial rules, all other permutations of dcut_f make progress towards this case. Since $\neg P$ is the only polarized formula to the left of $\Uparrow$, the "delayed" requirement of dcut_u is trivially met. The right-side subproof with the negative cut formula stays intact during these permutations. We consider two cases where B is a positive polarized formula: the other cases are treated similarly. If B is a positive literal, then $\vdash B \Downarrow \Theta, P$ must be the conclusion of an initial rule. Since P is also positive, it must be the case that $B^{\perp} \in \Theta$. Thus $\vdash B \Downarrow \Theta, \Theta'$ is also the conclusion of an initial rule. If B is $B_1 \vee^{+} B_2$, then we have the following transformation (here, i is 1 or 2):

$$\dfrac{\dfrac{\vdash B_i \Downarrow \Theta, P}{\vdash B_1 \vee^{+} B_2 \Downarrow \Theta, P}\,\vee^{+} \qquad \vdash \neg P \Uparrow \Theta'}{\vdash B_1 \vee^{+} B_2 \Downarrow \Theta, \Theta'}\,\mathit{dcut}_f \quad \longrightarrow \quad \dfrac{\dfrac{\vdash B_i \Downarrow \Theta, P \qquad \vdash \neg P \Uparrow \Theta'}{\vdash B_i \Downarrow \Theta, \Theta'}\,\mathit{dcut}_f}{\vdash B_1 \vee^{+} B_2 \Downarrow \Theta, \Theta'}\,\vee^{+}$$

4.4 Permutations of cut_f

The cut_f rule has the general form

$$\frac{\vdash A \Downarrow \Theta \qquad \vdash \neg A, \Gamma' \Uparrow \Theta'}{\vdash \Gamma' \Uparrow \Theta, \Theta'}\ cut_f.$$

It is required that the subproof above the unfocused sequent $\vdash \neg A, \Gamma' \Uparrow \Theta'$ is *eager* with respect to $\neg A$.

If A is negative, then the left subproof above cut_f must be the conclusion of a *release* rule, and the cut permutes to a cut_u with shorter subproofs:

$$\frac{\vdash A \Uparrow \Theta \qquad \vdash \neg A, \Gamma' \Uparrow \Theta'}{\vdash \Gamma' \Uparrow \Theta, \Theta'}\ cut_u.$$

As for the restrictions on cut_u, $\neg A$ must be positive if A is negative, so the subproof above the positive cut formula stays *eager* with respect to that polarized formula, and the subproof above $\vdash A \Uparrow \Theta$ is trivially *delayed* above the negative cut formula.

If A is positive, then the left subproof above cut_f must be either *init* or an introduction of the cut formula A. We illustrate three cases below: the other cases are similar.

1. If A is a positive literal p then the left premise of cut_f, namely $\vdash p \Downarrow \Theta$, is the conclusion of an initial rule with $\neg p \in \Theta$. The other, *eager* subproof of $\vdash \neg p, \Gamma' \Uparrow \Theta'$, must end in a *store* rule on $\neg p$, with premise $\vdash \Gamma' \Uparrow \Theta', \neg p$. But since $\neg p \in \Theta$, the provability of $\vdash \Gamma' \Uparrow \Theta, \Theta'$ follows from weakening.

2. If A is $A_1 \vee^+ A_2$, then $\neg A$ is $\neg A_1 \wedge^- \neg A_2$. This key case requires transforming the derivation

$$\frac{\dfrac{\vdash A_i \Downarrow \Theta}{\vdash A_1 \vee^+ A_2 \Downarrow \Theta}\ \vee^+ \qquad \dfrac{\vdash \neg A_1, \Gamma' \Uparrow \Theta' \quad \vdash \neg A_2, \Gamma' \Uparrow \Theta'}{\vdash \neg A_1 \wedge^- \neg A_2, \Gamma' \Uparrow \Theta'}\ \wedge^-}{\vdash \Gamma' \Uparrow \Theta, \Theta'}\ cut_f$$

into the derivation

$$\frac{\vdash A_i \Downarrow \Theta \qquad \vdash \neg A_i, \Gamma' \Uparrow \Theta'}{\vdash \Gamma' \Uparrow \Theta, \Theta'}\ cut_f.$$

The inductive measure is reduced by the size of the cut formulas. Here we can apply Lemma 3.4 to the subproof above $\vdash \neg A_i, \Gamma' \Uparrow \Theta'$ so that it becomes *eager* with respect to (each) $\neg A_i$ without regard to how the transformation might affect the height of proofs because the lexicographical inductive measure is still reduced. This argument similarly applies to the other key cases.

3. if A is $A_1 \wedge^+ A_2$ then $\neg A$ is $\neg A_1 \vee^- \neg A_2$ and the proof is transformed as follows:

$$\frac{\dfrac{\vdash A_1 \Downarrow \Theta \quad \vdash A_2 \Downarrow \Theta}{\vdash A_1 \wedge^+ A_2 \Downarrow \Theta}\ \wedge^+ \qquad \dfrac{\vdash \neg A_1, \neg A_2, \Gamma' \Uparrow \Theta'}{\vdash \neg A_1 \vee^- \neg A_2, \Gamma' \Uparrow \Theta'}\ \vee^-}{\vdash \Gamma' \Uparrow \Theta, \Theta'}\ cut_f$$

$$\downarrow$$

$$\dfrac{\vdash A_2 \Downarrow \Theta \qquad \dfrac{\vdash A_1 \Downarrow \Theta \qquad \vdash \neg A_1, \neg A_2, \Gamma' \Uparrow \Theta'}{\vdash \neg A_2, \Gamma' \Uparrow \Theta, \Theta'}\ cut_f}{\vdash \Gamma' \Uparrow \Theta, \Theta'}\ cut_f$$

The two cuts introduced are both on smaller cut formulas compared to the original cut: the inductive hypothesis is first applied to the upper cut to obtain a cut-free proof, then to the lower one.

With these permutation results in hand, we can now prove the cut-admissibility theorem for LKF.

Theorem 4.1 *The rules cut_u, cut_f, $dcut_u$ and $dcut_f$ are admissible in* LKF.

Proof The formal proof is a nested induction argument: first on the number of cuts in each proof, the second on the lexicographical measure for each cut. The corresponding procedure is: select a top-most cut with cut-free subproofs and apply Lemma 3.4 so that the subproofs satisfy the requirements concerning the *eager* and *delayed* properties. Then apply the transformations to reduce the cut. Apply this procedure repeatedly until all cuts are eliminated. □

5 Admissibility of the general *init* rule

The initial rule of LKF requires A to be a literal in order to prove the sequent $\vdash A \Downarrow \neg A, \Theta$. Just as important as the admissibility of cut is the admissibility of the more general form of *init*: that is, the sequent $\vdash A, \neg A \Uparrow \Theta$ is provable for every polarized formula A. For unfocused sequent calculus, the proof of this result is straightforward because of the perfect duality between the introduction rules for dual logical connectives. In particular, assuming that A is negative, apply its (invertible) introduction rule followed by the introduction rule for $\neg A$ (reading rules from conclusion to premises). The induction hypothesis can then be applied directly to the premises. In a focused setting, however, the proof becomes more difficult since multiple asynchronous or synchronous connectives are introduced in a single phase. To solve this problem, we introduce the following relation, which was also used in Liang and Miller (2011).

Definition 5.1 Let $\uparrow$ be the binary relation between a polarized formula and multisets of polarized formulas defined inductively as follows.

1. $A \uparrow \{A\}$ if A is a positive polarized formula or negative literal.
2. $f^- \uparrow \{\}$.
3. $A \vee^- B \uparrow \Phi, \Phi'$ if $A \uparrow \Phi$ and $B \uparrow \Phi'$.
4. $A \wedge^- B \uparrow \Phi$ if $A \uparrow \Phi$ or $B \uparrow \Phi$.
5. $\forall x.A \uparrow \Phi$ if $A \uparrow \Phi$.

Clearly each such Φ contains only positive polarized formulas and negative literals. Note that the polarized formulas t^- and $A \vee^- t^-$ are not $\uparrow$-related to any multiset of polarized formulas.

The following lemmas establish the properties of the asynchronous and synchronous phases in a form that allows us to derive the admissibility of the general *init* rule.

Lemma 5.2 *For all polarized formulas* A*, multisets of polarized formulas* Γ*, and sets of polarized formulas* Θ*, if* $\vdash \Phi, \Gamma \Uparrow \Theta$ *is provable for all* Φ *such that* $A \uparrow \Phi$*, then* $\vdash A, \Gamma \Uparrow \Theta$ *is also provable.*

Proof The proof is by induction on the size of A. If a polarized formula A is not $\uparrow$-related to any multiset of polarized formulas then we say that $\uparrow$ is *undefined* for A. Note that if $\uparrow$ is undefined for A then the lemma implies that $\vdash A, \Gamma \Uparrow \Theta$ is provable.

1. If A is a positive polarized formula or negative literal, the property is trivial since only $A \uparrow \{A\}$ holds and Φ contains only A.
2. If A is the constant f^-, then the property holds by the f^- rule.
3. If A is the constant t^-, then $\vdash t^-, \Gamma \Uparrow \Theta$ is provable by the rule for t^-.
4. Let A be the polarized formula $B \wedge^- C$. If $\uparrow$ is undefined for A, then it is undefined for B and for C, and the inductive hypothesis states that $\vdash B, \Gamma \Uparrow \Theta$ and $\vdash C, \Gamma \Uparrow \Theta$ are provable. Otherwise, if $\vdash \Phi, \Gamma \Uparrow \Theta$ is provable for all Φ such that $A \uparrow \Phi$, then it is provable for all Φ such that $B \uparrow \Phi$ or $C \uparrow \Phi$. The inductive hypothesis yields the provability of both $\vdash B, \Gamma \Uparrow \Theta$ and $\vdash C, \Gamma \Uparrow \Theta$. In either case, the $\wedge^-$ rule yields a proof of $\vdash B \wedge^- C, \Gamma \Uparrow \Theta$.
5. Let A be the polarized formula $B \vee^- C$. Assume that $\vdash \Phi, \Gamma \Uparrow \Theta$ is provable for all Φ such that $B \vee^- C \uparrow \Phi$. This assumption is equivalent to assuming that $\vdash \Phi', \Phi'', \Gamma \Uparrow \Theta$ is provable for all Φ' and Φ'' such that $B \uparrow \Phi'$ and $C \uparrow \Phi''$. Now assume that $B \uparrow \Phi'$ and $C \uparrow \Phi''$ hold. By the above hypothesis, we have $\vdash \Phi', \Phi'', \Gamma \Uparrow \Theta$ is provable. By the inductive hypothesis applied to B, we know that $\vdash B, \Phi'', \Gamma \Uparrow \Theta$ is provable and by the inductive hypothesis applied to C, we know that $\vdash B, C, \Gamma \Uparrow \Theta$ is provable. If $\uparrow$ is undefined for either B or C, we reach the same conclusion. The $\vee^-$ rule thus yields a proof of $\vdash B \vee^- C, \Gamma \Uparrow \Theta$.
6. Let A be the polarized formula $\forall x.B$ and assume that x is not free in Γ, Θ. If $A \uparrow \Phi$ then $B \uparrow \Phi$. If $\uparrow$ is undefined for A then it is also undefined for B. In either case the inductive hypothesis states that if $\vdash \Phi, \Gamma \Uparrow \Theta$ is provable for all Φ such that $B \uparrow \Phi$, then $\vdash B, \Gamma \Uparrow \Theta$ is provable. The property is established by applying the $\forall$ rule. □

The next lemma connects the synchronous phase with the $\uparrow$-relation.

Lemma 5.3 *For all polarized formulas* A *and multisets of polarized formulas* Φ*, if* $A \uparrow \Phi$ *then* $\vdash \neg A \Downarrow \Phi$ *is provable.*

Proof The proof proceeds by induction on the size of A, which is the same as the size of $\neg A$.

1. If A is t^-, then the property holds vacuously.
2. If A is a negative literal then the property holds by the initial rule *init*.
3. If A is f^-, the property holds by the rule for t^+.

4. If A is $B \wedge^- C$ then $\neg A$ is $\neg B \vee^+ \neg C$. Assuming that $A \uparrow \Phi$ then either $B \uparrow \Phi$ or $C \uparrow \Phi$. Assume without loss of generality that $B \uparrow \Phi$: by inductive hypothesis $\vdash \neg B \Downarrow \Phi$ is provable. Thus, $\vdash \neg B \vee^+ \neg C \Downarrow \Phi$ is provable using the $\vee^+$ rule.

5. If A is $B \vee^- C$ then $\neg A$ is $\neg B \wedge^+ \neg C$. Assume that $A \uparrow \Phi, \Phi'$ such that $B \uparrow \Phi$ and $C \uparrow \Phi'$. By the inductive hypotheses, we know that $\vdash \neg B \Downarrow \Phi$ and $\vdash \neg C \Downarrow \Phi'$ are provable. Apply weakening (Lemma 3.3) to both sequents and we get that $\vdash \neg B \Downarrow \Phi, \Phi'$ and $\vdash \neg C \Downarrow \Phi, \Phi'$ are provable. Thus $\vdash \neg B \wedge^+ \neg C \Downarrow \Phi, \Phi'$ is provable using the $\wedge^+$ rule.

6. If A is $\forall x.B$ then $\neg A$ is $\exists x.\neg B$. If $A \uparrow \Phi$ then $B \uparrow \Phi$. By inductive hypothesis we have $\vdash \neg B \Downarrow \Phi$ and by the $\exists$ rule, we have $\vdash \exists x.\neg B \Downarrow \Phi$.

7. If A is a positive polarized formula, then the inductive hypothesis also applies to the proper subformulas of $\neg A$, which is negative and of the same size as A. Thus if $\neg A \uparrow \Phi$ then the cases above show that $\vdash A \Downarrow \Phi$ is provable. By weakening $\vdash A \Downarrow A, \Phi$ is also provable, and we can form the derivation

$$\dfrac{\dfrac{\vdash A \Downarrow A, \Phi}{\vdash \cdot \Uparrow A, \Phi}\ \mathit{decide}}{\vdash \Phi \Uparrow A}\ \mathit{store}$$

where a sequence of *store* rules are applied to the positive polarized formulas and negative literals in Φ. This holds for all Φ such that $\neg A \uparrow \Phi$, so by Lemma 5.2, $\vdash \neg A \Uparrow A$ is provable, and by applying the *release* rule, we have a proof of $\vdash \neg A \Downarrow A$. This establishes the property for positive A for which only $A \uparrow \{A\}$ holds. □

The following theorem states the admissibility of the general form of the *init* rule.

Theorem 5.4 $\vdash A, \neg A \Uparrow \cdot$ *is provable for all polarized formulas A.*

Proof Assume without loss of generality that A is positive. Then $A \uparrow \{A\}$ and Lemma 5.3 states that $\vdash \neg A \Downarrow A$ is provable. Since $\neg A$ is negative, this sequent must be the conclusion of a release rule in a cut-free proof, so $\vdash \neg A \Uparrow A$ is provable. Applying the store rule on A to this sequent gives a proof of $\vdash A, \neg A \Uparrow \cdot$. □

6 Generalized invertibility

The following results about the invertibility of the negative introduction rules is now easily proved using the admissibility of cut. The following corollary is the converse of Lemma 5.2.

Corollary 6.1 *If* $\vdash A, \Gamma \Uparrow \Theta$ *is provable and* $A \uparrow \Phi$, *then* $\vdash \Phi, \Gamma \Uparrow \Theta$ *is provable.*

Proof Given the assumption $A \uparrow \Phi$, Lemma 5.3 implies that the sequent $\vdash \neg A \Downarrow \Phi$ is provable. Using a cut rule, we therefore have the following proof.

$$\dfrac{\dfrac{\vdash A, \Gamma \Uparrow \Theta \quad \vdash \neg A \Downarrow \Phi}{\vdash \Gamma \Uparrow \Theta, \Phi}\ cut_f}{\vdash \Phi, \Gamma \Uparrow \Theta}\ store.$$

The final result follows from the admissibility of cut (Theorem 4.1). □

From the generalized invertibility property and Lemma 5.2, we can derive the invertibility of the individual asynchronous introduction rules.

Lemma 6.2 *The introduction rules for the negative connectives are invertible; i.e., the provability of the conclusion of each rule implies the provability of all of its premises.*

Proof First, consider the case for $\vee^-$. Assume that $\vdash A \vee^- B, \Gamma \Uparrow \Theta$ is provable and assume that A is $\uparrow$-related to exactly the multisets $\Phi^1_A, \ldots, \Phi^n_A$ and that B is $\uparrow$-related to exactly $\Phi^1_B, \ldots, \Phi^m_B$, where $n, m \geq 0$. By the definition of $\uparrow$, we know that $A \vee^- B \uparrow \Phi^i_A \Phi^k_B$ for each i and k such that $1 \leq i \leq n$ and $1 \leq k \leq m$. (Note that if either n or m is 0 then this statement is vacuously true.) Corollary 6.1 implies that $\vdash \Phi^i_A \Phi^k_B, \Gamma \Uparrow \Theta$ is provable. By Lemma 5.2, this means that $\vdash A, B, \Gamma \Uparrow \Theta$ is provable.

To consider the case for $\wedge^-$ assume that $\vdash A \wedge^- B, \Gamma \Uparrow \Theta$ is provable and (as above) A is $\uparrow$-related to $\Phi^1_A, \ldots, \Phi^n_A$ and B is $\uparrow$-related to $\Phi^1_B, \ldots, \Phi^m_B$, where $n, m \geq 0$. Then $A \wedge B \uparrow \Phi^i_A$ for each i such that $1 \leq i \leq n$ and $A \wedge B \uparrow \Phi^k_B$ for each k such that $1 \leq k \leq m$. By Corollary 6.1, this implies that $\vdash \Phi^i_A, \Gamma \Uparrow \Theta$ is provable for each i such that $1 \leq i \leq n$ and $\vdash \Phi^k_B, \Gamma \Uparrow \Theta$ is provable for each k such that $1 \leq k \leq m$. By Lemma 5.2, $\vdash A, \Gamma \Uparrow \Theta$ and $\vdash B, \Gamma \Uparrow \Theta$ are provable.

The cases for t^- and $\forall$ are similar and omitted. □

Given Lemmas 5.2 and 5.3, we often use the following *argument schema* to establish the provability of $\vdash A_1, \ldots, A_n, \Gamma \Uparrow \Theta$: If $\uparrow$ is undefined for any A_i then Lemma 5.2 already shows that the sequent is provable. Otherwise, assume that for each $i \in \{1, \ldots, n\}$ there is an n_i greater than or equal to 1 such that A_i is $\uparrow$-related to exactly $\Phi^1_i, \ldots, \Phi^{n_i}_i$. Show that for each possible selection of $\Phi^{k_1}_1, \ldots, \Phi^{k_n}_n$, the sequent $\vdash \Gamma \Uparrow \Theta, \Phi^{k_1}_1, \ldots, \Phi^{k_n}_n$ is provable. Then $\vdash A_1, \ldots, A_n, \Gamma \Uparrow \Theta$ is provable by Lemma 5.2 plus enough applications of the *store* rule to move each member of $\Phi^{k_i}_i$ to the left side of $\Uparrow$. Furthermore, if Γ consists of a single positive polarized formula P (P can also be in Θ with Γ empty) and $\vdash P \Downarrow P, \Theta, \Phi^{k_1}_1, \ldots, \Phi^{k_n}_n$ is provable, then using the *decide* rule

$$\dfrac{\vdash P \Downarrow P, \Theta, \Phi^{k_1}_1, \ldots, \Phi^{k_n}_n}{\vdash \cdot \Uparrow P, \Theta, \Phi^{k_1}_1, \ldots, \Phi^{k_n}_n}\ decide$$

the provability $\vdash A_1, \ldots, A_n, P \Uparrow \Theta$ also follows from Lemma 5.2 and the *store* rule. The provability of the focused sequent above *decide* often follows from Lemma 5.3.

7 Returning to LK

In this section, we show how the unfocused LK proof system can be faithfully captured within LKF. We do this in three steps: (1) we translate the two-sided proof system LK into a one-sided system; (2) we show that a more general form of contraction is admissible in LKF; and (3) we prove that the unfocused introduction rules of (the one-side version of) LK are admissible in LKF. As a consequence, LKF is complete for LK.

Gentzen's original version of LK used the additive versions of conjunction and disjunction, namely $\wedge^-$ and $\vee^+$, while his implication $\supset$ was multiplicative. Gentzen himself noted (Gentzen, 1935, Remark 2.4) that LK is 'dual' in the sense that the left and right inference rules are symmetrical except for $\supset$. In LKF, the multiplicative connective $\vee^-$ can be used to encode $A \supset B$ into $\neg A \vee^- B$: hence, $\neg(A \supset B)$ is encoded as $A \wedge^+ \neg B$. As a result, we can remove implications and negated implications by mapping them to these multiplicative connectives.

Definition 7.1 The LK *-polarization* $(\cdot)^{\pm}$ *of classical formulas* is defined as follows (recall that the negation of polarized formulas is given in Definition 3.1):

1. For any atomic polarized formula a, $a^{\pm} = a$ and $(\neg a)^{\pm} = \neg a$.
2. $(A \wedge B)^{\pm} = A^{\pm} \wedge^- B^{\pm}$; $(A \vee B)^{\pm} = A^{\pm} \vee^+ B^{\pm}$; $t^{\pm} = t^-$; $f^{\pm} = f^+$;
3. $(A \supset B)^{\pm} = \neg A^{\pm} \vee^- B^{\pm}$

We also assume that all atomic polarized formulas are polarized positively.

Figure 5 contains the inference rules for LKi, a sequent calculus intermediate between LK and LKF in the sense that it is a one-sided sequent calculus that contains polarized formulas but it is not focused. An LK sequent $A_1, \ldots, A_n \vdash B_1, \ldots, B_m$ is represented in this setting as $\vdash \neg A_1{}^{\pm}, \ldots, \neg A_n{}^{\pm}, B_1{}^{\pm}, \ldots, B_m{}^{\pm}$. Each inference rule of LK is translated directly into this setting: replace each sequent in the premises and conclusion of the rule with their one-sided, polarized versions. Left-introductions rules on A_i are thus represented as one-sided introduction rules on $\neg A_i{}^{\pm}$.

Theorem 7.2 *Let $n, m \geq 0$ and let $A_1, \ldots, A_n, B_1, \ldots, B_m$ be unpolarized formulas. If the sequent $\vdash \neg A_1{}^{\pm}, \ldots, \neg A_n{}^{\pm}, B_1{}^{\pm}, \ldots, B_m{}^{\pm} \Uparrow \cdot$ is provable in* LKF *then the sequent $A_1, \ldots, A_n \vdash B_1, \ldots, B_m$ is provable in* LK.

Proof Note that an LKF proof of $\vdash \neg A_1{}^{\pm}, \ldots, \neg A_n{}^{\pm}, B_1{}^{\pm}, \ldots, B_m{}^{\pm} \Uparrow \cdot$ can easily be translated to an LKi proof of $\vdash \neg A_1{}^{\pm}, \ldots, \neg A_n{}^{\pm}, B_1{}^{\pm}, \ldots, B_m{}^{\pm}$. Such an LKi proof can then be converted to a proof of the two-sided sequent $A_1, \ldots, A_n \vdash B_1, \ldots, B_m$ in LK. In this later transformation, when the multiplicative connectives $\vee^-$ and $\wedge^+$ are introduced in the LKi proof, implications are introduced on the right or left in the LK proof. □

We shall now proceed to prove that the rules of LKi are admissible in LKF by presenting new admissible LKF rules derived from the LKi rules. When naming the new admissible LKF rules, we will add parentheses around the name of the LKi rule. For example, the *init* rule of LKi yields the admissible LKF rule

STRUCTURAL RULES AND IDENTITY RULES

$$\frac{\vdash \Delta, B, B}{\vdash \Delta, B}\ cR \qquad \frac{\vdash \Delta}{\vdash \Delta, B}\ wR \qquad \frac{}{\vdash B, \neg B}\ init \qquad \frac{\vdash B, \Delta \quad \vdash \neg B, \Delta'}{\vdash \Delta, \Delta'}\ cut$$

INTRODUCTION RULES

$$\frac{}{\vdash t^-, \Delta}\ t^- \qquad \frac{\vdash A, \Theta \quad \vdash B, \Theta}{\vdash A \wedge^- B, \Theta}\ \wedge^- \qquad \frac{}{\vdash t^+}\ t^+ \qquad \frac{\vdash A, \Theta \qquad \vdash B, \Theta'}{\vdash A \wedge^+ B, \Theta, \Theta'}\ \wedge^+$$

$$\frac{\vdash B_i, \Theta}{\vdash B_1 \vee^+ B_2, \Theta}\ \vee^+ \qquad \frac{\vdash \Theta}{\vdash f^-, \Theta}\ f^- \qquad \frac{\vdash A, B, \Theta}{\vdash A \vee^- B, \Theta}\ \vee^-$$

$$\frac{\vdash \Delta, [y/x]B}{\vdash \Delta, \forall x.B}\ \forall \qquad \frac{\vdash \Delta, [s/x]B}{\vdash \Delta, \exists x.B}\ \exists$$

Fig. 5 The rules for LKi. In the $\forall$ rule, the variable y is not free in the conclusion. In the $\vee^+$ rule, $i \in \{1, 2\}$.

$$\frac{}{\vdash B, \neg B, \Gamma \Uparrow \cdot}\ (init).$$

The admissibility of (*init*) follows immediately from Theorem 5.4. The admissibility of (wR), namely,

$$\frac{\vdash \Delta \Uparrow \Theta}{\vdash B, \Delta \Uparrow \Theta, \Theta'}\ (wR)$$

follows from Lemma 3.3 and a simple induction on the structure of B. We delay the proof of the admissibility of the LKi *cut* rule until Section 9.1. We now proceed to prove the admissibility of contraction and the introduction rules of LKi.

Unlike LK and LKi, LKF does not include explicit rules for contraction. In LKF, the rule of contraction is only applied to positive polarized formulas and only within the *decide* rule. We now show that contraction for *all* polarized formulas is admissible in LKF.

Lemma 7.3 *The following rule is admissible in* LKF *for all polarized formulas A.*

$$\frac{\vdash A, A, \Gamma \Uparrow \Theta}{\vdash A, \Gamma \Uparrow \Theta}\ (cR)$$

Proof Assume that $\vdash A, A, \Gamma \Uparrow \Theta$ has an LKF proof. Using Lemma 3.4, we can assume that this proof is eager for the first occurrence of A. If A is a positive polarized formula or negative literal, then the only rule that can be applied to it is *store*, which means that the sequent $\vdash A, \Gamma \Uparrow A, \Theta$ has an LKF proof. Again, this sequent has a proof eager for A and, thus, must be proved by the *store* rule, which implies that $\vdash \Gamma \Uparrow A, \Theta$ has an LKF proof. By using that sequent as the premise of the *store* rule we have an LKF proof of $\vdash A, \Gamma \Uparrow \Theta$.

Consider the cases where A is a non-literal negative polarized formula. The case where A is t^- is immediate. The case where A is f^- follows using Lemma 6.2 twice. If A is $B \vee^- C$ then, using Lemmas 3.4 and 6.2 twice, it is the case that $\vdash B, B, C, C, \Gamma \Uparrow \Theta$ is provable. The result follows by using the inductive assumption twice along with the $\vee^-$ rule. If A is $B \wedge^- C$ then, using Lemmas 3.4 and 6.2 twice, it is the case that both $\vdash B, B, \Gamma \Uparrow \Theta$ and $\vdash C, C, \Gamma \Uparrow \Theta$ are provable. The result follows by using the inductive assumption twice along with the $\wedge^-$ rule. Finally, the case where A is universally quantified is similar and omitted here. □

From the results in the preceding sections, we can show the admissibility of the unfocused introduction rules (corresponding to the rules of LKi) in LKF.

Theorem 7.4 (Admissibility of unfocused introduction rules) *All the introduction rules of* LKi *are admissible in* LKF.

Proof Throughout this proof, we use the admissibility of cut combined with the argument schema outlined at the end of Section 6.

The $\vee^+$-introduction rule for LKi is admissible in LKF in the form

$$\frac{\vdash B_i, \Gamma \Uparrow \Theta}{\vdash B_1 \vee^+ B_2, \Gamma \Uparrow \Theta}\ (\vee^+)$$

for $i \in \{1, 2\}$. Admissibility follows from using the admissibility of the cut_u rule in the derivation

$$\frac{\vdash B_i, \Gamma \Uparrow \Theta \qquad \vdash \neg B_i,\ B_1 \vee^+ B_2 \Uparrow \cdot}{\vdash B_1 \vee^+ B_2, \Gamma \Uparrow \Theta}\ cut_u.$$

To show the provability of the right premise above the cut we apply the argument schema of Section 6. Let $\neg B_i \uparrow \Phi^1, \ldots, \neg B_i \uparrow \Phi^n$ be an exhaustive list of multisets of polarized formulas $\uparrow$-related to $\neg B_i$, for $n \geq 0$. If $n = 0$ then the sequent is provable by Lemma 5.2. Otherwise, n is positive. For each Φ^k ($k \in 1, \ldots, n$), construct the following subproof

$$\dfrac{\dfrac{\dfrac{\vdash B_i \Downarrow B_1 \vee^+ B_2, \Phi^k}{\vdash B_1 \vee^+ B_2 \Downarrow B_1 \vee^+ B_2, \Phi^k}\ \vee^+}{\vdash \cdot \Uparrow B_1 \vee^+ B_2, \Phi^k}\ decide}{\vdash B_1 \vee^+ B_2, \Phi^k \Uparrow \cdot}\ store$$

The provability of the top sequent follows from Lemma 5.3 and the provability of $\vdash \neg B_i,\ B_1 \vee^+ B_2 \Uparrow \cdot$ follows from all such subproofs by Lemma 5.2.

The $\wedge^+$-introduction rule for LKi is admissible in LKF in the form

$$\frac{\vdash A, \Gamma \Uparrow \Theta \quad \vdash B, \Gamma \Uparrow \Theta}{\vdash A \wedge^+ B, \Gamma \Uparrow \Theta}\ (\wedge^+).$$

This rule is also justified using the admissibility of cut_u as follows.

$$\dfrac{\vdash B,\Gamma \Uparrow \Theta \qquad \dfrac{\vdash A,\Gamma \Uparrow \Theta \quad \vdash \neg A,\neg B, A\wedge^{+} B \Uparrow \cdot}{\vdash \neg B, A\wedge^{+} B,\Gamma \Uparrow \Theta}\ cut_u}{\dfrac{\vdash A\wedge^{+} B,\Gamma,\Gamma \Uparrow \Theta}{\vdash A\wedge^{+} B,\Gamma \Uparrow \Theta}\ (cR)}\ cut_u$$

The provability of the top right sequent uses the argument schema described above: let $\neg A \uparrow \Phi^1_{\neg A}, \ldots, \neg A \uparrow \Phi^n_{\neg A}$ and $\neg B \uparrow \Phi^1_{\neg B}, \ldots, \neg B \uparrow \Phi^m_{\neg B}$ be exhaustive lists of multiset of set related to $\neg A$ and $\neg B$, respectively. If either n or m is 0, then the sequent is already provable. Otherwise for each pair $\Phi^i_{\neg A}, \Phi^k_{\neg B}$ construct the subproof

$$\dfrac{\dfrac{\dfrac{\vdash A \Downarrow A\wedge^{+} B, \Phi^i_{\neg A}, \Phi^k_{\neg B} \qquad \vdash B \Downarrow A\wedge^{+} B, \Phi^i_{\neg A}, \Phi^k_{\neg B}}{\vdash A\wedge^{+} B \Downarrow A\wedge^{+} B, \Phi^i_{\neg A}, \Phi^k_{\neg B}}\ \wedge^{+}}{\vdash \cdot \Uparrow A\wedge^{+} B, \Phi^i_{\neg A}, \Phi^k_{\neg B}}\ decide}{\vdash A\wedge^{+} B, \Phi^i_{\neg A}, \Phi^k_{\neg B} \Uparrow \cdot}\ store.$$

The provability of the top sequents follows from Lemma 5.3 and from these subproofs the provability of $\vdash \neg A, \neg B, A\wedge^{+} B \Uparrow \cdot$ follows by Lemma 5.2.

To prove the admissibility of the introduction of $\exists$, we similarly rewrite

$$\dfrac{\vdash A[s/x],\Gamma \Uparrow \Theta}{\vdash \exists x.A,\Gamma \Uparrow \Theta}\ (\exists) \quad\longrightarrow\quad \dfrac{\vdash A[s/x],\Gamma \Uparrow \Theta \quad \vdash \neg A[s/x], \exists x.A \Uparrow \cdot}{\vdash \exists x.A,\Gamma \Uparrow \Theta}\ cut_u$$

The provability of the right premise again uses the argument schema of Section 6: let $\neg A[s/x] \uparrow \Phi^1, \ldots, \neg A[s/x] \uparrow \Phi^n$ be the exhaustive list of multisets that are $\uparrow$-related to $\neg A[s/x]$. If $n = 0$, then the premise is already provable. Otherwise, for each Φ^i we have

$$\dfrac{\dfrac{\dfrac{\vdash A[s/x] \Downarrow \exists x.A, \Phi^i}{\vdash \exists x.A \Downarrow \exists x.A, \Phi^i}\ \exists}{\vdash \cdot \Uparrow \exists x.A, \Phi^i}\ decide}{\vdash \exists x.A, \Phi^i \Uparrow \cdot}\ store$$

from which the provability of $\vdash \neg A[s/x], \exists x.A \Uparrow \cdot$ follows.

The LKi introduction rule for t^+ yields the following admissible rule, which can be justified by the associated LKF derivation.

$$\dfrac{}{\vdash t^+,\Gamma \Uparrow \Theta}\ (t^+) \quad\longrightarrow\quad \dfrac{\dfrac{\dfrac{\dfrac{}{\vdash t^+ \Downarrow t^+}\ t^+}{\vdash \cdot \Uparrow t^+}\ decide}{\vdash t^+ \Uparrow \cdot}\ store}{\vdash t^+,\Gamma \Uparrow \Theta}\ (wR)$$

The negative introduction rules already apply on the left side of $\Uparrow$. Thus every unfocused inference rule can be emulated on the left side $\Uparrow$, and the completeness of LKF with respect to the intermediate LKi, and to the original LK is therefore established.

Theorem 7.5 (Weak completeness of LKF) *If the sequent $A_1, \ldots, A_n \vdash B_1, \ldots, B_m$ is provable in* LK *then the sequent* $\vdash \neg A_1{}^{\pm}, \ldots, \neg A_n{}^{\pm}, B_1{}^{\pm}, \ldots, B_m{}^{\pm} \Uparrow \cdot$ *is provable in* LKF.

We have labeled this theorem as "weak completeness" since it states that if an unpolarized formula is provable in LK, then there is *some* polarization of that formula (namely $(\cdot)^{\pm}$) which is provable in LKF. Theorem 8.4 in the next section is a stronger version of the completeness theorem since it states that *every* polarization of an unpolarized theorem is provable in LKF.

8 Choosing the polarization of formulas

We are now able to prove that every polarization of a formula provable in LK is provable in LKF. Formally, we say that the polarized formula B (together with an atom bias assignment $\delta(\cdot)$) is a *polarization* of C if $\check{B}$ is C.

We write $A \equiv B$ to mean that both $\vdash \neg A, B \Uparrow \cdot$ and $\vdash \neg B, A \Uparrow \cdot$ are provable. We first show that the positive and negative versions of each connective are equivalent.

Lemma 8.1 *For every pair of polarized formulas A and B, it is the case that* $A \vee^{+} B \equiv A \vee^{-} B$ *and* $A \wedge^{+} B \equiv A \wedge^{-} B$.

Proof To prove the first equivalence, we need proofs of $\vdash \neg A \wedge^{-} \neg B,\ A \vee^{-} B \Uparrow \cdot$ and $\vdash \neg A \wedge^{+} \neg B,\ A \vee^{+} B \Uparrow \cdot$ The first of these is straightforward given the admissibility of the general initial rule. The provability of the second sequent is equally simple given the admissibility of the *unfocused* introduction rules shown in Section 7, as demonstrated by the following derivation.

$$\dfrac{\dfrac{\vdash \neg A,\ A \Uparrow \cdot}{\vdash \neg A,\ A \vee^{+} B \Uparrow \cdot}\ (\vee^{+}) \qquad \dfrac{\vdash \neg B,\ B \Uparrow \cdot}{\vdash \neg B,\ A \vee^{+} B \Uparrow \cdot}\ (\vee^{+})}{\vdash \neg A \wedge^{+} \neg B,\ A \vee^{+} B \Uparrow \cdot}\ (\wedge^{+})$$

Showing $A \wedge^{+} B \equiv A \wedge^{-} B$ is similar, and the equivalences between the positive and negative versions of the units are straightforward. □

Definition 8.2 Let $\circ$ represent one of the binary connectives $\vee^{-}$, $\vee^{+}$, $\wedge^{-}$, or $\wedge^{+}$ and let F be a syntactic variable ranging over arbitrary polarized formulas. Let S range over *subformula contexts* which are defined inductively by

$$S \;=\; [\cdot] \mid S \circ F \mid F \circ S \mid \exists x.S \mid \forall x.S.$$

Here, $[\cdot]$ is a constant denoting a primitive subformula context. The notation $S[A]$ denotes the polarized formula that results from replacing $[\cdot]$ in S with A.

Theorem 8.3 *Let S be a subformula context. If* $A \equiv B$ *then* $S[A] \equiv S[B]$.

Proof We prove the general property: if $\vdash \neg A, B \Uparrow \cdot$ is provable then for any subformula context S, $\vdash \neg S[A], S[B] \Uparrow \cdot$ is also provable.

The proof of this property essentially repeats the arguments for eliminating the generalized initial rule. However, instead of replicating Lemmas 5.2 and 5.3, we can take advantage of the admissibility of unfocused rules for the positive connectives.

We proceed by induction on S. In the base case, $S = [\cdot]$ and the property is immediate. If, instead, $S = F \vee^- S'$ then $S[A] = F \vee^- S'[A]$, and $\neg S[A] = \neg F \wedge^+ \neg S'[A]$: we construct

$$\dfrac{\dfrac{\vdash \neg F, F, S'[B] \Uparrow \cdot \qquad \vdash \neg S'[A], S'[B], F \Uparrow \cdot}{\vdash \neg F \wedge^+ \neg S'[A],\ F, S'[B] \Uparrow \cdot}\ (\wedge^+)}{\vdash \neg F \wedge^+ \neg S'[A],\ F \vee^- S'[B] \Uparrow \cdot}\ \vee^-$$

The left premise follows from the general initial rule admissibility and the right premise is provable by inductive hypothesis (plus weakening). All the other cases are proved similarly. □

Theorem 8.4 (Strong completeness of LKF**)** *Let C be an unpolarized formula that is provable in* LK *and let B be a polarization of C. Then B is provable in* LKF*.*

Proof Let C be an unpolarized formula that is provable in LK and let B be a polarized version of C and let $\delta(\cdot)$ be any atomic bias assignment. By weak completeness (Theorem 7.5), we know that $C^{\pm}$ is provable in LKF. Since the only difference between $C^{\pm}$ and B are polarized formulas is that the + and − signs on logical connectives might be different and, by construction, the atoms in C are all given positive bias. Using the equivalences in Lemma 8.1 and Theorem 8.3, we can conclude that B is provable, assuming that all atoms are positively biased.

What remains to be shown is that provability is preserved by imposing the atomic bias assignment $\delta(\cdot)$. Translating a proof with a negative atom a into one where a is considered positive is the same as translating a proof with $\neg a$ considered positive to one where $\neg a$ is considered negative, so we only need to show one direction of the translation. Assume that a is considered negative in a proof. A strategy for reconstructing the proof where a is considered positive is to use *delays* together with cut. In particular, we define the polarized formula B^{δ} as the result of replacing every occurrence of a in B with $a \vee^- f^-$ (and therefore every occurrence of $\neg a$ by $\neg a \wedge^+ t^+$). The strategy is to show that every proof of $\vdash B \Uparrow \cdot$ with a considered negative corresponds to a proof of $\vdash B^{\delta} \Uparrow \cdot$ with a considered positive. Then by the cut rule

$$\dfrac{\vdash B^{\delta} \Uparrow \cdot \qquad \vdash \neg B^{\delta}, B \Uparrow \cdot}{\vdash B \Uparrow \cdot}\ cut_u$$

we derive a proof of B without delays and with a considered positive. We can easily generalized the proof of a single formula to the proof of a sequent since (by invertibility) a multiset $\{B_1, \ldots, B_n\}$ is equivalent to $B_1 \vee^- B_2 \ldots \vee^- B_n$.

The rules that may have a literal as principal formula are *store*, *release*, *decide*, and *init*. We show how each rule is emulated in a proof of B^{δ}:

1. Both a and $\neg a$ can be subject to a store, in which case the emulations are as follow.

$$\dfrac{\vdash \Gamma \Uparrow a, \Theta}{\vdash a, \Gamma \Uparrow \Theta}\ \mathit{store} \quad \longrightarrow \quad \dfrac{\dfrac{\dfrac{\vdash \Gamma \Uparrow a, \Theta}{\vdash a, \Gamma \Uparrow \Theta}\ \mathit{store}}{\vdash a, f^-, \Gamma \Uparrow \Theta}\ f^-}{\vdash a \vee^- f^-, \Gamma \Uparrow \Theta}\ \vee^-$$

$$\dfrac{\vdash \Gamma \Uparrow \neg a, \Theta}{\vdash \neg a, \Gamma \Uparrow \Theta}\ \mathit{store} \quad \longrightarrow \quad \dfrac{\vdash \Gamma \Uparrow \neg a \wedge^+ t^+, \Theta}{\vdash \neg a \wedge^+ t^+, \Gamma \Uparrow \Theta}\ \mathit{store}$$

Thus, in a proof of B^δ, a will appear on the right side of $\Uparrow$ and $\Downarrow$ as a but $\neg a$ will appear as $\neg a \wedge^+ t^+$.

2. The *release* rule is applicable when a is considered negative and is still applicable to $a \vee^- f^-$ when a is considered positive. Since a is a literal, the only rule that can apply above *release* is *store*.

$$\dfrac{\dfrac{\vdash \cdot \Uparrow a, \Theta}{\vdash a \Uparrow \Theta}\ \mathit{store}}{\vdash a \Downarrow \Theta}\ \mathit{release} \quad \longrightarrow \quad \dfrac{\dfrac{\dfrac{\dfrac{\vdash \cdot \Uparrow a, \Theta}{\vdash a \Uparrow \Theta}\ \mathit{store}}{\vdash a, f^- \Uparrow \Theta}\ f^-}{\vdash a \vee^- f^- \Uparrow \Theta}\ \vee^-}{\vdash a \vee^- f^- \Downarrow \Theta}\ \mathit{release}$$

3. In the *init* rule, a is negative: it is emulated as indicated.

$$\dfrac{}{\vdash \neg a \Downarrow a, \Theta}\ \mathit{init} \quad \longrightarrow \quad \dfrac{\dfrac{\dfrac{\dfrac{\dfrac{}{\vdash a \Downarrow \neg a, a, \Theta}\ \mathit{init}}{\vdash \cdot \Uparrow \neg a, a, \Theta}\ \mathit{decide}}{\vdash \neg a \Uparrow a, \Theta}\ \mathit{store}}{\vdash \neg a \Downarrow a, \Theta}\ \mathit{release} \qquad \dfrac{}{\vdash t^+ \Downarrow a, \Theta}\ t^+}{\vdash \neg a \wedge^+ t^+ \Downarrow a, \Theta}\ \wedge^+$$

4. Finally, when a is considered negative, the *decide* rule can only be applied to $\neg a$, and must be preceded from above by an *init*, and so is emulated as follows

$$\dfrac{\vdash \neg a \Downarrow \neg a, a, \Theta}{\vdash \cdot \Uparrow \neg a, a, \Theta}\ \mathit{decide} \quad \longrightarrow \quad \dfrac{\vdash \neg a \wedge^+ t^+ \Downarrow \neg a \wedge^+ t^+, a, \Theta}{\vdash \cdot \Uparrow \neg a \wedge^+ t^+, a, \Theta}\ \mathit{decide}$$

The proof of the remaining premise is easy to find.

Finally, to show that $\vdash \neg B^\delta, B \Uparrow \cdot$ is provable with a considered positive, we induct on the structure of B:

1. If B is a or $\neg a$, consider the following derivations.

$$\dfrac{\dfrac{\dfrac{\dfrac{\dfrac{}{\vdash a \Downarrow a, \neg a}\ init}{\vdash \cdot \Uparrow a, \neg a}\ decide}{\vdash a, \neg a \Uparrow \cdot}\ store}{\vdash a, f^-, \neg a \Uparrow \cdot}\ f^-}{\vdash a \vee^- f^-, \neg a \Uparrow \cdot}\ \vee^- \qquad \dfrac{\dfrac{\vdash \neg a \wedge^+ t^+ \Downarrow \neg a \wedge^+ t^+, a}{\vdash \cdot \Uparrow \neg a \wedge^+ t^+, a}\ decide}{\vdash \neg a \wedge^+ t^+, a \Uparrow \cdot}\ store$$

The proof of $\vdash \neg(a \vee^- f^-), a \Uparrow \cdot$ on the right is preceded from above by the same subproof as in the imitation of *decide* for $\neg a \wedge^+ t^+$.

2. If B is $C \vee^- D$, we apply the admissible unfocused rules to simplify the proof:

$$\dfrac{\dfrac{\vdash \neg C^\delta, C, D \Uparrow \cdot \quad \vdash \neg D^\delta, C, D \Uparrow \cdot}{\vdash \neg C^\delta \wedge^+ \neg D^\delta, C, D \Uparrow \cdot}\ (\wedge^+)}{\vdash \neg C^\delta \wedge^+ \neg D^\delta, C \vee^- D \Uparrow \cdot}\ \vee^-$$

The premises are provable by inductive hypotheses and by weakening.

3. If B is $C \vee^+ D$:

$$\dfrac{\dfrac{\vdash \neg C^\delta, C \Uparrow \cdot}{\vdash \neg C^\delta, C \vee^+ D \Uparrow \cdot}\ (\vee^+) \quad \dfrac{\vdash \neg D^\delta, D \Uparrow \cdot}{\vdash \neg D^\delta, C \vee^+ D \Uparrow \cdot}\ (\vee^+)}{\vdash \neg C^\delta \wedge^- \neg D^\delta, C \vee^+ D \Uparrow \cdot}\ \wedge^-$$

The premises are provable by inductive hypotheses.

4. The cases of $C \wedge^+ D$ and $C \wedge^- D$ are symmetrical to the above. The cases of $\exists$ and $\forall$ are also similar and cases where a does not appear in B follows directly from the admissibility of the general initial rule. □

Pimentel, Nigam, and Neto (2016) give a similar analysis of how changing the polarity of atoms within the intuitionistic focused proof system LJF (Liang and Miller, 2009) affects the structure of such proofs.

9 Four applications of LKF

Part of the motivation for developing the LKF proof system is that its meta-theory should help in proving other proof-theoretic results about first-order classical logic. To support this claim, we present four applications of LKF.

9.1 The admissibility of *cut* in LK

We can prove that the admissibility of cut holds for LK given that we have proved cut-admissibility for the more complex proof system LKF. While it is no surprise that this can be done, it is reassuring to see that that result for LK follows so directly from the results for LKF.

Theorem 9.1 *The cut rule for* LK *is admissible in the cut-free fragment of* LK.

Proof Assume that the sequents $\Gamma \vdash \Delta, B$ and $\Gamma', B \vdash \Theta'$ have cut-free LK-proofs. By the weak completeness of LKF (Theorem 7.5), the sequents $\vdash \neg(\Gamma)^{\pm}, (\Delta)^{\pm}, B^{\pm} \Uparrow \cdot$ and $\vdash \neg(\Gamma')^{\pm}, \neg(B^{\pm}), (\Delta')^{\pm} \Uparrow \cdot$ both have (cut-free) LKF proofs. By the admissibility of cut for LKF (Theorem 4.1), we know that $\vdash \neg(\Gamma)^{\pm}, \neg(\Gamma')^{\pm}, (\Delta)^{\pm}, (\Delta')^{\pm} \Uparrow \cdot$ has a (cut-free) LKF proof. Finally, by Theorem 7.2, we know that $\Gamma, \Gamma' \vdash \Delta, \Delta'$ has a cut-free LK proof. □

9.2 Synthetic inference rules

Following up on the suggestion in Section 2.4, we show how to define larger-scale, synthetic inference rules using the LKF proof system.

A sequent of the form $\vdash \cdot \Uparrow \Theta$ is called a *border sequent*. The only LKF proof rule that can have a border sequent as a conclusion is the *decide* rule.

Definition 9.2 (Synthetic inference rule) A *synthetic inference rule* is an inference rule involving only border sequents. They are of the form

$$\frac{\vdash \cdot \Uparrow \Theta_1 \quad \dots \quad \vdash \cdot \Uparrow \Theta_n}{\vdash \cdot \Uparrow \Theta}$$

which is *justified* by a derivation of the form

$$\begin{array}{c} \vdash \cdot \Uparrow \Theta_1 \quad \dots \quad \vdash \cdot \Uparrow \Theta_n \\ \Pi \\ \vdash \cdot \Uparrow \Theta \end{array}$$

Here, $n \geq 0$, and the derivation Π contains exactly one occurrence of the *decide* rule and that occurrence is the last inference rule (having the conclusion $\vdash \cdot \Uparrow \Theta$). If that *decide* rule selects as its focus the polarized formula $B \in \Theta$, we say that this derivation is a *synthetic inference rule for B*.

Consider again using the formula (from Section 2.4)

$$\forall x \forall y \forall z.(path(x, y) \supset path(y, z) \supset path(x, z))$$

as an assumption in a given fixed theory. In the one-sided sequent setting of LKF, consider instead the negation of this assumption, namely,

$$\exists x \exists y \exists z.(path(x, y) \wedge^{+} path(y, z) \wedge^{+} \neg path(x, z)).$$

Assuming that this positive polarized formula is a member of Θ, then consider the following derivation.

$$\cfrac{\cfrac{\cfrac{\begin{array}{c}\Xi_1\\ \vdash path(r,s) \Downarrow \Theta\end{array} \quad \begin{array}{c}\Xi_2\\ \vdash path(s,t) \Downarrow \Theta\end{array} \quad \begin{array}{c}\Xi_3\\ \vdash \neg path(r,t) \Downarrow \Theta\end{array}}{\vdash path(r,s) \wedge^+ path(s,t) \wedge^+ \neg path(r,t) \Downarrow \Theta}\ \wedge^+ \times 2}{\vdash \exists x \exists y \exists z.(path(x,y) \wedge^+ path(y,z) \wedge^+ \neg path(x,z)) \Downarrow \Theta}\ \exists \times 3}{\vdash \cdot \Uparrow \Theta}\ decide$$

In order to determine the shape of the proofs Ξ_1, Ξ_2, and Ξ_3, we must declare the polarization given to atoms with the *path* predicate. If all such atoms have a negative polarity assigned to them, then both Ξ_1 and Ξ_2 end with the *release* and *store* rules while the proof Ξ_3 must be trivial (just containing the *init* rule) and $path(r,t)$ must be a member of Θ. We can write the synthetic rule justified by the above derivation as

$$\frac{\vdash \cdot \Uparrow path(r,s), \Theta \qquad \vdash \cdot \Uparrow path(s,t), \Theta}{\vdash \cdot \Uparrow path(r,t), \Theta}$$

However, if all *path*-atoms have a positive polarity assigned to them, then Ξ_3 ends with the *release* and *store* rules while the proof Ξ_1 and Ξ_2 must be trivial and both $\neg path(r,s)$ and $\neg path(s,t)$ must be members of Θ. We can write the synthetic rule justified by the above derivation as

$$\frac{\vdash \cdot \Uparrow \neg path(r,s), \neg path(s,t), \neg path(r,t), \Theta}{\vdash \cdot \Uparrow \neg path(r,s), \neg path(s,t), \Theta}\ .$$

Note that these synthetic inference rules are the one-sided version of the back-chaining and forward-chaining synthetic inference rules for *path* displayed in Section 2.4.

The paper Marin, Miller, Pmentel, and Volpe (2020) develops the proof theory of synthetic inferences for both classical and intuitionistic logic by using the focused proof systems LKF and LJF. That paper also shows that cut and the general initial rule are both admissible in the LK and LJ proof systems augmented with such synthetic inference rules based on *geometric formulas*.

9.3 Herbrand's theorem

The completeness of LKF proofs yields a surprisingly simple proof of Herbrand's theorem, particularly the variant of Herbrand's theorem based on formulas with only existential quantifiers in prefix position. A richer connection between a more general form of Herbrand's theorem, based on expansion trees (Miller, 1987), and LKF proofs can be found in Chaudhuri, Hetzl, and Miller (2016).

Theorem 9.3 (Herbrand's theorem) *Let B be an (unpolarized) quantifier-free formula of first-order classical logic, $n \geq 1$, and $x_1, \ldots, x_n$ be a list of first-order variables containing all the free variable of B. The formula $\exists x_1 \ldots \exists x_n.B$ is provable in* LK *if and only if there is an $m \geq 1$ and substitutions $\theta_1, \ldots, \theta_m$ for the variables $x_1, \ldots, x_n$ such that $B\theta_1 \vee \cdots \vee B\theta_m$ is provable in* LK.

Proof Let $\hat{B}$ be a polarized version of B in which all logical connectives and units in B are polarized negatively. (For convenience, we abbreviate $\exists x_1 \ldots \exists x_n$ with $\exists \bar{x}$.) Since $\exists \bar{x}.B$ is provable in LK, the sequent $\vdash \exists \bar{x}.\hat{B} \Uparrow \cdot$ must have an LKF proof, say Ξ. Clearly, the last inference rule of Ξ is the *store* rule with premise $\vdash \cdot \Uparrow \exists \bar{x}.\hat{B}$. Given our choice of polarization, it is easy to show that every border sequent in Ξ is of the form $\vdash \cdot \Uparrow \exists \bar{x}.\hat{B}, \mathcal{L}$, where $\mathcal{L}$ is a set of literals. Thus, there are only two different ways that the *decide* rule is applied in Ξ. If the *decide* rule is used with a positive literal, the premise is immediately proved using the *init* rule. Otherwise, the *decide* rule starts the synchronous phase with the choice of $\exists \bar{x}.\hat{B}$ and the subproof determined by that occurrence of the *decide* rule ends with the following inference rules.

$$\dfrac{\dfrac{\vdash \hat{B}\theta \Uparrow \exists \bar{x}.\hat{B}, \mathcal{L}}{\vdash \hat{B}\theta \Downarrow \exists \bar{x}.\hat{B}, \mathcal{L}}\ \textit{release}}{\vdash \exists \bar{x}.\hat{B} \Downarrow \exists \bar{x}.\hat{B}, \mathcal{L}}\ \exists \times n$$

That is, every non-trivial synchronous phase encodes a substitution. Let $m \geq 1$ be the number of such non-trivial synchronous phases and let $\theta_1, \ldots, \theta_m$ be the substitutions that those phases encode.

Now let C be the polarized formula C equal to $\hat{B}\theta_1 \vee^+ \ldots \vee^+ \hat{B}\theta_m$ and consider building an LKF proof of $\vdash C \Uparrow \cdot$. In order to ensure that C is polarized positively, if $m = 1$, we take C to be $C \vee^+ f^+$ (essentially encoding a unary version of the binary $\vee^+$). It is now a simple matter to convert the proof Ξ of $\vdash \exists \bar{x}.\hat{B} \Uparrow \cdot$ into a proof of $\vdash \hat{B}\theta_1 \vee^+ \ldots \vee^+ \hat{B}\theta_m \Uparrow \cdot$ by copying the asynchronous phases directly and by replacing all the non-trivial synchronous phase in Ξ as follows.

$$\dfrac{\dfrac{\vdash \hat{B}\theta_i \Uparrow \exists \bar{x}.\hat{B}, \mathcal{L}}{\vdash \hat{B}\theta_i \Downarrow \exists \bar{x}.\hat{B}, \mathcal{L}}\ \textit{release}}{\vdash \exists \bar{x}.\hat{B} \Downarrow \exists \bar{x}.\hat{B}, \mathcal{L}}\ \exists \times n \quad \Longrightarrow \quad \dfrac{\dfrac{\vdash \hat{B}\theta_i \Uparrow C, \mathcal{L}}{\vdash \hat{B}\theta_i \Downarrow C, \mathcal{L}}\ \textit{release}}{\vdash \hat{B}\theta_1 \vee^+ \ldots \vee^+ \hat{B}\theta_m \Downarrow C, \mathcal{L}}\ \vee^+$$

In this way, the phase-by-phase structure of Ξ can be used to build an LKF proof for $\vdash \hat{B}\theta_1 \vee^+ \ldots \vee^+ \hat{B}\theta_m \Uparrow \cdot$. □

9.4 Hosting other focused proof systems

Proof systems with focusing-like behaviors can sometimes be hosted inside LKF. Such hosting is usually done by translating unpolarized classical logic formulas into polarized formulas in which *delays* have been inserted. These delays are written as $\partial_-(B)$ and $\partial_+(B)$ and are such that they are both logically equivalent to the polarized formula B and are such that $\partial_-(B)$ is negative and $\partial_+(B)$ is positive. The expression $\partial_-(B)$ can be defined to be either $f^- \vee^- B$, $t^- \wedge^- B$, or $\forall x B$ (where x is not free in B). Similarly, the expression $\partial_+(B)$ can be defined to be either $f^+ \vee^+ B$, $t^+ \wedge^+ B$, or $\exists x B$ (where x is not free in B).

The LKQ and LKT proof systems of Danos, Joinet, and Schellinx (1995) can be seen

	Atoms are negative		*Atoms are positive*
LKT	$(A)^l = \neg A$	LKQ	$(A)^l = \neg A$
	$(A)^r = A$		$(A)^r = A$
	$(B \supset C)^l = (B)^r \wedge^+ (C)^l$		$(B \supset C)^l = (B)^r \wedge^+ \partial_-((C)^l)$
	$(B \supset C)^r = (B)^l \vee^- \partial_+((C)^r)$		$(B \supset C)^r = \partial_+((B)^l \vee^- (C)^r)$

Fig. 6 Two different ways to translate classical logic formulas into polarized formulas.

as LKF proofs in which the following polarization functions are used. Figure 6 defines the left and right translations of unpolarized formulas containing only implications and atoms to polarized formulas. In that figure, A ranges over atomic formulas. It is the case that (cut-free) proofs in LKT of an unpolarized formula B using only implications correspond to LKF proofs of $(B)^r$ (using the LKT definition) and (cut-free) proofs in LKQ of an unpolarized formula B using only implications correspond to LKF proofs of $(B)^r$ (using the LKQ definition). LKT focuses only on the left and LKQ only on the right of two-sided sequents. These systems are also examples of "less aggressive" focused systems that designate a *"stoup"* formula: these systems impose fewer restrictions than the formula under focus in LKF. The delays emulate the one-sided focusing character of these systems as well as adopt the stoup to a strongly focused system.

10 Other variations for focusing in classical logic

There have been several variations on focusing systems studied in the literature. In fact, the general phenomena of focusing for classical logic can be seen as arising from Girard's uses of linear logic exponentials to encode classical logic (Girard, 1987) and from Andreoli's discovery of polarity (Andreoli, 1992).

The LKF proof system we have given here can be called a *strongly focused* system: the *decide* rule can only be invoked after *every* negative non-atomic polarized formula has been removed from the sequent. If we do not insist that all negative polarized formulas have been removed in this way, the resulting variant is called a *weakly focused* proof system following Laurent (2004) and Simmons and Pfenning (2011). Girard's LC proof system is an early example of a weakly focused proof system for classical logic (Girard, 1991). A variant on strong focusing is a system where one chooses a predetermined *suspension criterion* and then allows explicitly suspected negative polarized formulas to remain in the conclusion of the (suitably modified) *decide* rule: suspensions of this kind have proved useful in a setting where logic contains fixed point expressions (Gérard and Miller, 2017).

Let LKFm be the proof system that results from replacing the inference rules for LKF with the extended version of the synchronous introduction rules and the *release* and *decide* rules given in Figure 7. If the ‡ proviso on the *decide* rule requires that the multiset Δ contains exactly one positive polarized formula, then LKFm is the same as

SYNCHRONOUS INTRODUCTION RULES

$$\frac{}{\vdash t^+ \Downarrow \Gamma} \qquad \frac{\vdash B_1, \Theta_1 \Downarrow \Gamma \quad \vdash B_2, \Theta_2 \Downarrow \Gamma}{\vdash B_1 \wedge^+ B_2, \Theta_1, \Theta_2 \Downarrow \Gamma} \qquad \frac{\vdash B_i, \Theta \Downarrow \Gamma}{\vdash B_1 \vee^+ B_2, \Theta \Downarrow \Gamma}\ i \in \{1,2\} \qquad \frac{\vdash [s/x]B, \Theta \Downarrow \Gamma}{\vdash \exists x.B, \Theta \Downarrow \Gamma}$$

RELEASE AND DECIDE RULES

$$\frac{\vdash \Delta \Uparrow \Gamma}{\vdash \Delta \Downarrow \Gamma}\ release^{\dagger} \qquad \frac{\vdash \Delta \Downarrow \bar{\Delta}, \Gamma}{\vdash \cdot \Uparrow \bar{\Delta}, \Gamma}\ decide^{\ddagger}$$

The † proviso requires that Δ consists of only negative polarized formulas. In the *decide* rule, Δ is a non-empty multiset of positive polarized formulas and $\bar{\Delta}$ is its underlying set of polarized formulas. The ‡ proviso is discussed in the text.

Fig. 7 Variations in some of the LKF inference rules.

LKF. It is for this reason that we say that LKF is *single focused*: in such proofs, the zone to the left of the $\Downarrow$ always contains exactly one polarized formula (the focus of that sequent). If the ‡ proviso restricts Δ to be just a non-empty set of positive polarized formulas, then the resulting proof system is *multifocused* and that proof system contains more proofs than the single conclusion system. Multifocused proofs were first considered in Delande and Miller (2008) and Delande, Miller, and Saurin (2010) (in the context of linear logic) and the notion of *maximal multifocused* proofs has been used to describe canonical proof system in linear logic (Chaudhuri, Miller, and Saurin, 2008) and classical logic (Chaudhuri, Hetzl, and Miller, 2016) and to relate sequent calculus proofs to natural deduction proofs (Pimentel, Nigam, and Neto, 2016).

Note that the version of the $\wedge^+$ introduction rule in LKFm is not necessarily invertible, while the version of that introduction rule in LKF is invertible: it appears that the true status of $\wedge^+$ introduction as belonging to the synchronous phase only becomes apparent in the multifocused setting. Note also that it is immediate to prove the completeness of LKFm given the completeness of LKF.

Two simple changes to the LKF proof system yield a focused proof system for *multiplicative additive linear logic* MALL (Girard, 1987). First, the set of formulas to the right of the double arrows must be changed to multisets. Second, the following four inference rules must replace the corresponding inference rules in LKF (Figure 3).

$$\frac{A \text{ atomic}}{\vdash A \Downarrow \neg A}\ init \qquad \frac{\vdash P \Downarrow \Gamma}{\vdash \cdot \Uparrow P, \Gamma}\ decide \qquad \frac{}{\vdash t^+ \Downarrow \cdot}\ t^+ \qquad \frac{\vdash A \Downarrow \Theta_1 \quad \vdash B \Downarrow \Theta_2}{\vdash A \wedge^+ B \Downarrow \Theta_1, \Theta_2}\ \wedge^+$$

Here, the *init* and t^+ rules do not do an implicit weakening, the *decide* rule does not do an implicit contraction, and the side formulas of $\wedge^+$ are treated multiplicatively. The resulting proof system, called MALLF in Liang and Miller (2011), is a focused proof system for MALL. Of course, the usual presentation of MALL results from replacing the logical connectives t^-, t^+, f^-, f^+, $\wedge^-$, $\wedge^+$, $\vee^+$, and $\vee^-$ with $\top$, $\mathbf{1}$, $\bot$, $\mathbf{0}$, &, $\otimes$, ⅋, and $\oplus$, respectively. The fact that this proof system is sound and complete for MALL immediately follows from the results about focusing in full linear logic given by Andreoli (1992).

Another variation on focused proof systems uses a list, not a multiset, of formulas to the left of the $\Uparrow$: that is, the order by which the asynchronous inference rules are attempted is proscribed in a fixed fashion. This variation was used by Andreoli (1992) in his first focused proof system for linear logic.

The LKF proof system was designed to support automated proof checking and proof search (Chihani, Miller, and Renaud, 2017) as well as to provide new means for proving meta-theoretic results for first-order classical logic (see Section 9). Other researchers, concentrating on the Curry-Howard correspondence (proofs-as-programs) perspective, have designed other variants of focusing for classical logic. In particular, see the LC proof system (Girard, 1991), the $\mathsf{LK}^{\eta}_{\rho}$ (Danos, Joinet, and Schellinx, 1995; Danos, Joinet, and Schellinx, 1997), and the proof system used to define the $\bar{\lambda}\mu\tilde{\mu}$-calculus (Curien and Herbelin, 2000).

11 Conclusion

We have presented the proof system LKF and have proved that it is sound and complete for LK and that the cut rule and the initial rule are admissible. The proofs of these theorems were all done directly using permutation arguments. We have illustrated the utility of LKF by applying it to some standard topics that arise in the proof theory of classical logic. We hope that while the metatheory of LKF was established by tedious permutation arguments, many other properties of proofs in classical logic can be proved by applying LKF directly and without the need for such permutation arguments.

Acknowledgements We thank Beniamino Accattoli, Marianna Girlando, and the anonymous reviewers for their comments on an earlier version of this paper.

References

Andreoli, J. M. (1992). Logic programming with focusing proofs in linear logic. *Journal of Logic and Computation* 2, 297–347. DOI: 10.1093/logcom/2.3.297.

Chaudhuri, K., S. Hetzl, and D. Miller (2016). A multi-focused proof system isomorphic to expansion proofs. *Journal of Logic and Computation* 26, 577–603. DOI: 10.1093/logcom/exu030.

Chaudhuri, K., D. Miller, and A. Saurin (2008). Canonical sequent proofs via multi-focusing. In: *Fifth International Conference on Theoretical Computer Science*. Ed. by G. Ausiello, J. Karhumäki, G. Mauri, and L. Ong. Vol. 273. IFIP. Springer, 383–396. DOI: 10.1007/978-0-387-09680-3_26.

Chaudhuri, K., F. Pfenning, and G. Price (2008). A logical characterization of forward and backward chaining in the inverse method. *Journal of Automated Reasoning* 40, 133–177. DOI: 10.1007/s10817-007-9091-0.

Chihani, Z., D. Miller, and F. Renaud (2017). A semantic framework for proof evidence. *Journal of Automated Reasoning* 59, 297–330. DOI: 10.1007/s10817-016-9380-6.
Curien, P.-L. and H. Herbelin (2000). The duality of computation. In: *ICFP '00: Proceedings of the fifth ACM SIGPLAN international conference on Functional programming*. New York: ACM Press, 233–243. DOI: 10.1145/357766.351262.
Danos, V., J. B. Joinet, and H. Schellinx (1995). LKT and LKQ: sequent calculi for second order logic based upon dual linear decompositions of classical implication. In: *Advances in Linear Logic*. Ed. by J.-Y. Girard, Y. Lafont, and L. Regnier. Vol. 22. London Mathematical Society Lecture Note Series. Cambridge University Press, 211–224.
Danos, V., J.-B. Joinet, and H. Schellinx (1997). A new deconstructive logic: linear logic. *Journal of Symbolic Logic* 62, 755–807. DOI: 10.2307/2275572.
Delande, O. and D. Miller (2008). A neutral approach to proof and refutation in MALL. In: *23th Symposium on Logic in Computer Science*. Ed. by F. Pfenning. IEEE Computer Society Press, 498–508. DOI: 10.1016/j.apal.2009.07.017.
Delande, O., D. Miller, and A. Saurin (2010). Proof and refutation in MALL as a game. *Annals of Pure and Applied Logic* 161, 654–672. DOI: 10.1016/j.apal.2009.07.017.
Gentzen, G. (1935). Investigations into logical deduction. In: *The Collected Papers of Gerhard Gentzen*. Ed. by M. E. Szabo. Translation of articles that appeared in 1934-35. Collected papers appeared in 1969. Amsterdam: North-Holland, 68–131. DOI: 10.1007/BF01201353.
Gérard, U. and D. Miller (2017). Separating functional computation from relations. In: *26th EACSL Annual Conference on Computer Science Logic (CSL 2017)*. Ed. by V. Goranko and M. Dam. Vol. 82. LIPIcs, 23:1–23:17. DOI: 10.4230/LIPIcs.CSL.2017.23.
Girard, J. Y., P. Taylor, and Y. Lafont (1989). *Proofs and Types*. Cambridge University Press.
Girard, J.-Y. (1987). Linear logic. *Theoretical Computer Science* 50, 1–102. DOI: 10.1016/0304-3975(87)90045-4.
– (1991). A new constructive logic: classical logic. *Mathematical Structures in Computer Science* 1, 255–296. DOI: 10.1017/S0960129500001328.
Guglielmi, A. and T. Gunderson (2008). Normalisation control in deep inference via atomic flows. *Logical Methods in Computer Science* 1.
Hallnäs, L. and P. Schroeder-Heister (1990). A proof-theoretic approach to logic programming. I. Clauses as rules. *Journal of Logic and Computation* 1, 261–283. DOI: 10.1093/logcom/1.2.261.
Heath, Q. and D. Miller (2019). A proof theory for model checking. *Journal of Automated Reasoning* 63, 857–775. DOI: 10.1007/s10817-018-9475-3.
Herbelin, H. (1995). A lambda-calculus structure isomorphic to Gentzen-style sequent calculus structure. In: *Computer Science Logic, 8th International Workshop, CSL '94*. Vol. 933. Lecture Notes in Computer Science. Springer, 61–75. DOI: 10.1007/BFb0022247.
Kleene, S. C. (1952). Permutability of inferences in Gentzen's calculi LK and LJ. *Memoirs of the American Mathematical Society* 10, 1–26.

Laurent, O. (2004). A proof of the focalization property of linear logic. Unpublished note.

– (2011). Intuitionistic dual-intuitionistic nets. *Journal of Logic and Computation* 21, 561–587. DOI: 10.1093/logcom/exp044.

Liang, C. and C. Miller (2011). A focused approach to combining logics. *Annals of Pure and Applied Logic* 10, 679–697. DOI: 10.1016/j.apal.2011.01.012.

Liang, C. and D. Miller (2009). Focusing and polarization in linear, intuitionistic, and classical logics. *Theoretical Computer Science* 410, 4747–4768. DOI: 10.1016/j.tcs.2009.07.041.

Miller, D. (1987). A compact representation of proofs. *Studia Logica* 46, 347–370. DOI: 10.1007/BF00370646.

– (1989). A logical analysis of modules in logic programming. *Journal of Logic Programming* 6, 79–108. DOI: 10.1016/0743-1066(89)90031-9.

Miller, D., G. Nadathur, F. Pfenning, and A. Scedrov (1991). Uniform proofs as a foundation for logic programming. *Annals of Pure and Applied Logic* 51, 125–157. DOI: 10.1016/0168-0072(91)90068-W.

Negri, S. and J. von Plato (1998). Cut elimination in the presence of axioms. *Bulletin of Symbolic Logic* 4, 418–435. DOI: 10.2307/420956.

Pimentel, E., V. Nigam, and J. Neto (2016). Multi-focused proofs with different polarity assignments. In: *Proceedings of the Tenth Workshop on Logical and Semantic Frameworks, with Applications (LSFA 2015)*. Vol. 323. ENTCS, 163–179. DOI: 10.1016/j.entcs.2016.06.011.

Scherer, G. (2016). Which types have a unique inhabitant? Focusing on pure program equivalence. PhD thesis. Université Paris-Diderot.

Simmons, R. J. (2014). Structural focalization. *ACM Transaction on Computational Logic* 15. DOI: 10.1145/2629678.

Simmons, R. J. and F. Pfenning (2011). *Weak focusing for ordered linear logic*. Technical Report. CMU-CS-10-147, Carnegie Mellon University.

von Plato, J. (2012). Gentzen's proof systems: byproducts in a work of genius. *Bulletin of Symbolic Logic* 18, 313–367. DOI: 10.2178/bsl/1344861886.

Intensional Harmony as Isomorphism

Paolo Pistone and Luca Tranchini

Abstract In the present paper we discuss a recent suggestion of Schroeder-Heister concerning the possibility of defining an intensional notion of harmony using isomorphism in second-order propositional logic. The latter is not an absolute notion, but its definition is relative to the choice of criteria for identity of proofs. In the paper, it is argued that in order to attain a satisfactory account of harmony, one has to consider a notion of identity stronger than the usual one (based on β- and η-conversions) that the authors have investigated in recent work.

Key words: identity of proofs, equivalence of derivations, second-order logic, parametricity, System F, quantification

1 Introduction

The inferentialist thesis that the meaning of a logical constant is determined by its inferences rules has been famously challenged by Prior (1960) who put forward the following pair of introduction and elimination rules for the binary connective `tonk`:

$$\frac{A}{A \,\mathtt{tonk}\, B}\,\mathtt{tonkI} \qquad\qquad \frac{A \,\mathtt{tonk}\, B}{B}\,\mathtt{tonkE}$$

The strong intuition that `tonk` is semantically deficient has been taken to require a qualification of the inferentialist thesis: Not any arbitrary collection of inference rules

Paolo Pistone
Department of Computer Science and Engineering, University of Bologna, Italy,
e-mail: paolo.pistone@uniroma3.it

Luca Tranchini
Department of Computer Science, University of Tübingen, Germany,
e-mail: luca.tranchini@gmail.com

T. Piecha and K. F. Wehmeier (eds.), *Peter Schroeder-Heister on Proof-Theoretic Semantics*, Outstanding Contributions to Logic 29,
https://doi.org/10.1007/978-3-031-50981-0_10

can determine the meaning of a logical constant but only those satisfying a certain requirement, that (following Dummett, 1991) is commonly referred to as *harmony*.

The exact significance of harmony is open to interpretation. In particular, there is no agreement as to whether harmony should be regarded as a descriptive or normative criterion; nor as to whether harmony should be regarded as a criterion of "meaningfulness" or merely of "logicality" (i.e., whether expressions governed by rules which are not in harmony should be regarded as meaningless; or as meaningful, but not belonging to the logical vocabulary), or possibly something else. Moreover, different characterizations of the notion of harmony have been proposed in the literature, some of them being fully formal, some of them being less precise.

Harmony via second-order translations. In the context of natural deduction (Gentzen, 1935; Prawitz, 1965a), harmony is usually described by reference to the "perfect balance" between the introduction and elimination rules of the connectives of intuitionistic propositional logic NI (see Table 1).

Table 1 The natural deduction system NI.

$$\frac{A \quad B}{A \wedge B}\wedge\mathrm{I} \qquad \frac{A \wedge B}{A}\wedge\mathrm{E}_1 \quad \frac{A \wedge B}{B}\wedge\mathrm{E}_2$$

$$\frac{\begin{array}{c}[A]\\ B\end{array}}{A \supset B}\supset\mathrm{I} \qquad \frac{A \supset B \quad A}{B}\supset\mathrm{E}$$

$$\frac{A}{A \vee B}\vee\mathrm{I}_1 \quad \frac{B}{A \vee B}\vee\mathrm{I}_2 \qquad \frac{A \vee B \quad \begin{array}{c}[A]\\ C\end{array} \quad \begin{array}{c}[B]\\ C\end{array}}{C}\vee\mathrm{E}$$

$$\frac{}{\top}\top\mathrm{I} \qquad \frac{\bot}{C}\bot\mathrm{E}$$

But what does this "perfect balance" consist in, exactly? In spite of some attempts to answer the question in a precise way (see, e.g., Belnap, 1962; Tennant, 1978; Read, 2010) a first fully formal definition of harmony has been proposed only recently by Peter Schroeder-Heister (2014a; 2014b).

In a nutshell, Schroeder-Heister's proposal is that of characterizing collections of introduction rules and collections of elimination rules with a formula of quantified propositional intuitionistic logic NI^2, the extension of NI with universal quantification over propositions, governed by the following rules:

$$\frac{A}{\forall X.A}\forall\mathrm{I} \qquad \frac{\forall X.A}{A[C/X]}\forall\mathrm{E}$$

Schroeder-Heister's proposal is that two collections of introduction and elimination rules for a connective are in harmony if and only if their characteristic formulas are *interderivable* in NI^2.

Towards intensional harmony. Although this proposal represents a long-awaited step forward in the understanding of harmony, there are reasons of dissatisfaction with it. These hinge upon the fact that collections of rules that are interderivable are characterized by interderivable formulas. For instance, given $\wedge E_1$, the rule $\wedge E_2$ is obviously interderivable with the following rule:

$$\frac{A \wedge B \quad A}{B} \wedge E_2'$$

and the formulas $X \wedge Y$ and $X \wedge (X \supset Y)$ — characterizing the collections of elimination rules for $\wedge$ consisting in the pairs of $\wedge E_1$, $\wedge E_2$ and of $\wedge E_1$, $\wedge E_2'$ respectively — are obviously interderivable. Thus both pairs of rules qualify as in harmony with $\wedge I$ on Schroeder-Heister's criterion. However, one has a clear intuition that the pair $\wedge E_1$, $\wedge E_2'$ is "less" in harmony with $\wedge I$ than the pair $\wedge E_1$, $\wedge E_2$.

This has prompted the second author (see Tranchini, 2016a) to regard Schroeder-Heister's criterion as a characterization of a "weak" notion of harmony, and to call for a strengthening of it capable of capturing a notion of harmony on which only the pair $\wedge E_1$, $\wedge E_2$ (and not the pair $\wedge E_1$, $\wedge E_2'$) qualify as in harmony with $\wedge I$. What distinguishes such a would-be stronger notion of harmony — not just from Schroeder-Heister's weak harmony, but also from other proposals, such as those of Belnap (1962), and of Tennant (1978) — is its (hyper-)*intensional* nature, that is, its being capable of discriminating among collection of rules which are indistinguishable in terms of derivability.

Intensional inferentialism. The idea that a semantic framework should be able to draw (hyper-)intensional distinctions has a long tradition, going back at least to Carnap (1956), who notably regarded logical equivalence as too coarse a criterion for synonymy, and proposed instead to characterize synonymy using the notion of intensional isomorphism.

In the context of inferential accounts of meaning, especially in the proof-theoretic semantics tradition of Dummett and Prawitz, (hyper-)intensional aspects have been largely ignored. One of the reasons for this is that these theories of meaning have been shaped in analogy with traditional ones, by replacing the notion of truth conditions with the one of assertibility conditions. Consider for instance a language containing two distinct binary connectives $\sharp$ and $\flat$ whose inferential behaviour is described as follows: a proof of $A \sharp B$ is a triple consisting of a proof of A, a method to transform proofs of A into proofs of B and a method to transform proofs of B into proofs of A; a proof of $A \flat B$ differs from a proof $A \sharp B$ in that its first member is a proof of B. Clearly, $A \sharp B$ and $A \flat B$ are assertible under the same conditions: Whenever one has a proof of $A \sharp B$, one knows that a proof of $A \flat B$ could be constructed, and vice versa. However, to be in possession of a proof of $A \sharp B$ is clearly a different epistemic state from being in possession of a proof of $A \flat B$. If we take meaning to consist not only of what Frege called *Bedeutung* (the portion of reality referred to by an expression) but also of *Sinn* (the epistemic content of an expression), then an inferentialist account of meaning should be able to distinguish between the meanings of $A \sharp B$ and $A \flat B$.

As a matter of fact, proof theory offers a wide range of formal tools to analyze such issues, and in the present paper we will discuss the prospects of using some of these tools to deliver an intensional account of the notion of harmony.

Harmony via isomorphism. An obvious way of attaining a notion of harmony stronger than Schroeder-Heister's would be to require the characteristic formula of the collection of introduction rules to be *the same as* that of the collection of eliminations. But this would be too much. It is true that on such a strengthening the pair $\wedge E_1$, $\wedge E_2'$ would not count as in harmony with $\wedge I$. However, neither would $\vee E$ count as in harmony with the pair $\vee I_1$, $\vee I_2$: The characteristic formula (see below for precise definition) of the former is $\forall X((Y \supset X) \wedge (Z \supset X)) \supset X$ while that of the latter is $Y \vee Z$, i.e., two distinct, though interderivable NI^2-formulas. In other words, by adopting this notion of harmony (we dub it *strict harmony*) one would be led to deny that the rules of disjunction are harmonious.

The notion of formula isomorphism coming from categorial proof-theory and the study of typed lambda-calculi provides a middle ground between interderivability and identity. Inspired by the work of Došen (see, e.g., Došen, 2003), the second author (see again Tranchini, 2016a) used the notion of isomorphism to clarify the exact sense in which merely weakly harmonious rules are harmful (see also below Section 3) therefore pointing to the relevance of isomorphism for a characterization of harmony.

Different options as to defining harmony using isomorphism have been tentatively put forward by Schroeder-Heister (2016). Among the different options proposed, there is that of defining strong harmony by replacing interderivability (resp. identity) in the definition of weak (resp. strict) harmony with that of isomorphism. However, this proposal is discarded as inappropriate, due to the fact that — at least *prima facie* — $\forall Z((X \supset Z) \wedge (Y \supset Z)) \supset Z$ and $X \vee Y$ are not isomorphic in NI^2.

Main contribution. The isomorphism of two formulas in a given system is not an absolute notion, but it is relative to the choice of a notion of identity of proofs (that is, of an equational theory on the derivations of the system). Building on well-established results in the categorial semantics of second-order logic, in recent work the authors have introduced an equational theory stronger than the usual one using a class of equations referred to as ε-equations (see Tranchini, Pistone, and Petrolo, 2019; Pistone, Tranchini, and Petrolo, 2021; Pistone and Tranchini, 2021). The class of isomorphisms relative to ε-equivalence is rich enough to overcome the problem mentioned above (in particular $\forall X((Y \supset X) \wedge (Z \supset X)) \supset X$ and $Y \vee Z$ are ε-isomorphic formulas in NI^2). Moreover, although the equational theory induced by ε-equations (together with the standard conversions for NI^2) is not maximal on the whole of NI^2, it is the maximum equational theory of certain weak fragments of NI^2. Among these fragments there is the one whose formulas correspond to the "encodings" of collections of introduction and elimination rules for propositional connectives. In this paper, we present in an informal way the notion of identity of proofs captured by the ε-equations and show how the results obtained about them provide a firm footing for an intensional account of harmony between weak and strict harmony.

2 From reductions and expansions to isomorphism

The "perfect balance" between introduction and elimination rules that the notion of harmony aims at capturing has been described as obtaining when

> what can be inferred from a logically complex sentence by means of the *elimination rules* for its main connective is *no more* and *no less* than what has to be established in order to infer that very logically complex sentence using the *introduction rules* for its main connective.

When the rules for a connective are in harmony, two kinds of deductive patterns can be exhibited.

Patterns of the first kind are those giving rise to *maximal formulas occurrences*, that is formula occurrences which are the major premise of an application of an elimination rule (i.e., the premise whose main connective is the one to be eliminated) and that are the conclusion of an application of one of the introduction rules. Prawitz (1965b) defined certain operations on derivations called *reductions*. Reductions allow rewriting a derivation into another one thereby getting rid of a single maximal formula occurrence (though new ones may be generated in the process): In the case of conjunction, we have the following two reductions:

$$\dfrac{\dfrac{\begin{matrix}\mathscr{D}_1 \\ A\end{matrix} \quad \begin{matrix}\mathscr{D}_2 \\ B\end{matrix}}{A \wedge B}\,{\wedge}\mathrm{I}}{A}\,{\wedge}\mathrm{E}_1 \quad \text{reduces to} \quad \begin{matrix}\mathscr{D}_1 \\ A\end{matrix} \qquad\qquad \dfrac{\dfrac{\begin{matrix}\mathscr{D}_1 \\ A\end{matrix} \quad \begin{matrix}\mathscr{D}_2 \\ B\end{matrix}}{A \wedge B}\,{\wedge}\mathrm{I}}{B}\,{\wedge}\mathrm{E}_2 \quad \text{reduces to} \quad \begin{matrix}\mathscr{D}_2 \\ B\end{matrix}$$

Prawitz showed how — by successively applying reductions in a certain order — any given NI-derivation can be to transformed into one in *normal* form, that is one with no maximal formula occurrences.

The other kinds of patterns are those in which the premises of applications of introduction rules have been obtained by applying the corresponding elimination rules. Prawitz (1971) defined operations that are, in a sense, the dual of reductions, called immediate expansions. In the case of conjunction, the expansion looks as follows:

$$\begin{matrix}\mathscr{D} \\ A \wedge B\end{matrix} \quad \text{expands to} \quad \dfrac{\dfrac{\begin{matrix}\mathscr{D} \\ A \wedge B\end{matrix}}{A}\,{\wedge}\mathrm{E}_1 \qquad \dfrac{\begin{matrix}\mathscr{D} \\ A \wedge B\end{matrix}}{B}\,{\wedge}\mathrm{E}_2}{A \wedge B}\,{\wedge}\mathrm{I}$$

Prawitz showed that by successively applying expansions it is possible to transform any given normal derivation in NI into one in *long normal* form, i.e., into a derivation in which all minimal formula occurrences (those that are the conclusion of an elimination and the premise of an introduction rule) are atomic.

The reduction and expansion associated to the rules of implication are the following:

$$\dfrac{(u)\,\dfrac{\begin{matrix}\overset{u}{[A]}\\ \mathscr{D}\\ B\end{matrix}}{A \supset B}\supset\mathrm{I} \qquad \begin{matrix}\mathscr{D}'\\ A\end{matrix}}{B} \quad \text{reduces to} \quad \begin{matrix}\mathscr{D}'\\ [A]\\ \mathscr{D}\\ B\end{matrix}$$

$$\begin{matrix}\mathscr{D}\\ A \supset B\end{matrix} \quad \text{expands to} \quad (u)\,\dfrac{\dfrac{\begin{matrix}\mathscr{D}\\ A \supset B\end{matrix} \qquad \overset{u}{A}}{B}\supset\mathrm{E}}{A \supset B}\supset\mathrm{I}$$

Via the Curry-Howard correspondence between derivations in the implicational fragment of NI and terms of the simply typed λ-calculus, these rewriting operations on derivations correspond (respectively) to those of β-reduction and η-expansion on λ-terms:

$$(\lambda x.t)s \overset{\beta}{\rightsquigarrow} t[s/x] \qquad t \overset{\eta}{\rightsquigarrow} (\lambda x.t)x$$

Like in λ-calculus, reductions and expansions can be used to define an equivalence relation on natural deduction derivations. Two derivations $\mathscr{D}$ and $\mathscr{D}'$ are equivalent if and only if one can be obtained from the other by applying a finite number of times (β-)reduction, (η-)expansion and their inverse operations to $\mathscr{D}$ and $\mathscr{D}'$ or their sub-derivations (we indicate that $\mathscr{D}$ and $\mathscr{D}'$ are $\beta\eta$-equivalent as $\mathscr{D} \overset{\beta\eta}{\equiv} \mathscr{D}'$, and more in general given an equivalence relation E, we indicate E-equivalence as $\mathscr{D} \overset{E}{\equiv} \mathscr{D}'$).[1] As equivalent λ-terms can be seen as different ways of representing the same function, Prawitz (1971) observed — following a suggestion by Martin-Löf — that equivalent derivations can be seen as different linguistic representations of the same proof (where proofs are understood as abstract entities informally characterized by the so-called BHK-clauses; see Tranchini 2012; 2016b; 2019).

Given an equivalence relation on derivations, it is possible to use it to define an equivalence relation on formulas that is, in general, stricter than interderivability and that it is commonly referred to as *isomorphism*. Let E be an equivalence relation on derivations of a natural deduction system S. Two formulas are E-isomorphic (notation $A \overset{E}{\simeq} B$) iff

1. there exist two S-derivations $\mathscr{D}_1$ and $\mathscr{D}_2$ of A from B and of B from A respectively (i.e., A and B are interderivable in S);
2. such that their compositions are E-equivalent to the derivations consisting only of the assumptions of A and of B respectively:

$$B \overset{E}{\equiv} \begin{matrix}[B]\\ \mathscr{D}_1\\ [A]\\ \mathscr{D}_2\\ B\end{matrix} \qquad\qquad \begin{matrix}[A]\\ \mathscr{D}_2\\ [B]\\ \mathscr{D}_1\\ A\end{matrix} \overset{E}{\equiv} A$$

[1] We will always implicitly identify derivations up to renaming of discharge indexes, which corresponds to α-equivalence on λ-terms.

The derivation consisting of the assumption of a formula A can be viewed as representing the identity function on the set of proofs of A. Hence, the second condition of the definition of isomorphism can be expressed by saying that the two derivations $\mathscr{D}_1$ and $\mathscr{D}_2$ represent two functions from proofs of A to proofs of B and vice versa which are the inverse of each other. This in turn means that the set of proofs of A and of B are in bijection.

Typical examples of $\beta\eta$-isomorphic formulas in NI are pairs of formulas of the form $(A \wedge B) \wedge C$ and $A \wedge (B \wedge C)$, or $(A \wedge B) \supset C$ and $A \supset (B \supset C)$, whereas typical examples of interderivable but non-$\beta\eta$-isomorphic formulas are pairs of formulas of the form A and $A \wedge A$, or $A \wedge B$ and $A \wedge (A \supset B)$.

The notion of isomorphism has been proposed (notably by Došen, 2003) as a formal counterpart of the informal notion of synonymy, i.e., identity of meaning. Intuitively, interderivability is only a necessary, but not sufficient condition for synonymy. From an inferential perspective, isomorphic formulas can be regarded as synonymous in the sense that:

> "they behave exactly in the same manner in proofs: by composing, we can always extend proofs involving one of them, either as assumption or as conclusion, to proofs involving the other, so that nothing is lost, nor gained. There is always a way back. By composing further with the inverses, we return to the original proofs." (Došen, 2003, p. 498)

Clearly, a necessary condition for some notion of E-isomorphism not to collapse on that of interderivability is that the equivalence relation E used in the definition is non-trivial (i.e., there must be at least one formula A and two derivations of A belonging to distinct equivalence classes). In particular, if any two derivations $\mathscr{D}_1$ and $\mathscr{D}_2$ of any formula A from itself were E-equivalent, the second condition of the definition of E-isomorphism would be vacuously satisfied.

The notion of $\beta\eta$-equivalence (and consequently that of $\beta\eta$-isomorphism) plays a distinguished role in NI, since $\beta\eta$-equivalence is the maximum non-trivial equivalence relation definable on NI-derivations. As Došen (2003) and Widebäck (2001) argued, the maximality of an equivalence relation E on the derivation of a system S can be taken as supporting the claim that it is the *correct* way of analyzing the notion of identity of proofs underlying S.

For the $\{\supset, \wedge, \top\}$-fragment of NI, $\beta\eta$-equivalence and $\beta\eta$-isomorphism are well-understood: the decidability of $\beta\eta$-equivalence is an immediate consequence of normalization and confluence of $\beta\eta$-reduction in the $\{\supset, \wedge\}$-fragment, and its maximality was established by Statman (1983), Došen and Petrić (2001), and Widebäck (2001). Moreover, $\beta\eta$-isomorphism in this fragment is decidable and it has been fully axiomatized by Solov'ev (1983).

The extension of these results to richer language fragments has proven a difficult task. In presence of disjunction the decidability and maximality of $\beta\eta$-equivalence have been established only recently by Scherer (2017); the decidability of $\beta\eta$-isomorphism was established by Ilik (2014). In this case the difficulty were due to the form of the η-expansion for disjunction:

$$\begin{array}{c}\mathscr{D}'\\ [A\vee B]\\ \mathscr{D}''\\ C\end{array} \quad\text{expands to}\quad \dfrac{\begin{array}{c}\mathscr{D}'\\ A\vee B\end{array} \quad \begin{array}{c}\dfrac{\overset{n}{A}}{[A\vee B]}\vee\mathrm{I}\\ \mathscr{D}''\\ C\end{array}\quad \begin{array}{c}\dfrac{\overset{m}{B}}{[A\vee B]}\vee\mathrm{I}\\ \mathscr{D}''\\ C\end{array}}{C}\,(n,m)$$

(with n and m fresh for $\mathscr{D}'$)

which can be seen as the composition of the simpler form of expansion proposed by Prawitz

$$\begin{array}{c}\mathscr{D}\\ A\vee B\end{array} \quad\text{expands to}\quad \dfrac{\begin{array}{c}\mathscr{D}\\ A\vee B\end{array} \quad \dfrac{\overset{n}{A}}{A\vee B}\vee\mathrm{I}_1 \quad \dfrac{\overset{m}{B}}{A\vee B}\vee\mathrm{I}_2}{A\vee B}\,\vee\mathrm{E}\,(n,m)$$

(with n and m *fresh*[2] for $\mathscr{D}$)

and a generalization of the permutative conversions used in establishing the subformula property of normal derivations in NI:

$$\begin{array}{c}\dfrac{\begin{array}{c}\mathscr{D}\\ A\vee B\end{array}\quad\begin{array}{c}\overset{n}{[A]}\\ \mathscr{D}_1\\ C\end{array}\quad\begin{array}{c}\overset{m}{[B]}\\ \mathscr{D}_2\\ C\end{array}}{[C]}\,\vee\mathrm{E}\,(n,m)\\ \mathscr{D}_3\\ D\end{array} \quad\rightsquigarrow_{\gamma^+}\quad \dfrac{\begin{array}{c}\mathscr{D}\\ A\vee B\end{array}\quad\begin{array}{c}\overset{n}{[A]}\\ \mathscr{D}_1\\ [C]\\ \mathscr{D}_3\\ D\end{array}\quad\begin{array}{c}\overset{m}{[B]}\\ \mathscr{D}_2\\ [C]\\ \mathscr{D}_3\\ D\end{array}}{D}\,\vee\mathrm{E}\,(n,m)$$

(see, for a discussion, Tranchini 2016a; 2018).[3]

Whereas the maximality and decidability of $\beta\eta$-equality also hold in presence of $\bot$ (hence for the full language of NI), the decidability of $\beta\eta$-isomorphism in presence of $\bot$ is still an open problem.

3 Weak harmony and its limits

In order to define harmony formally, a useful preliminary move is that of identifying rules — that are usually taken to be meta-linguistic schemata — with expressions belonging to an object language of the appropriate kind. Slightly reformulating

[3] In this work, the first author proposed the existence of reduction and expansions of this more general form as an informal notion of harmony. Although informal, the requirement is enough to rule out "quantum disjunction" (i.e., the connective whose rules are almost the same as those of disjunction in NI, the only difference being that the elimination rule can be applied only when its minor premises depend on no assumption other than those discharged by the rule application; see Dummett, 1991) as disharmonious, since the restriction on the elimination rule blocks the possibility of defining an expansion of the more general form. See also concluding remarks in Section 5.

insights of Schroeder-Heister (2014a; 2014b), we propose to identify the *rules* of an arbitrary natural deduction system with a particular class of *formulas* of the second-order language extending that of the system NI with

1. universal quantification over propositions
2. for each n, denumerably many variables for n-ary connectives (to be indicated with $\dagger^n$, possibly with subscripts, and other *ad hoc* symbols), so that if $A_1, \dots, A_n$ are formulas and $\dagger^n$ is an n-ary connective variable, then $\dagger^n(A_1, \dots, A_n)$ is also a formula (the superscript n indicating the arity of $\dagger$ will be mostly omitted).

More precisely, given denumerably many propositional variables (to be indicated with X, Y, Z possibly with subscripts) and denumerably many connective variables as above, we define the set of formulas $\mathcal{L}$, to be indicated with A, B, C, possibly with subscripts, by the following grammar (we use A^n for a sequence of n comma-separated formulas):[4]

$$A ::= X \mid \top \mid \bot \mid A \wedge A \mid A \vee A \mid A \supset A \mid \dagger^n(A^n) \mid \forall X.A$$

(we will indicate with $\mathcal{L}^2$ the fragment of $\mathcal{L}$ lacking variables for connectives, and with $\mathcal{L}^{2\supset}$ fragment of $\mathcal{L}^2$ lacking all connectives apart from $\supset$).

The idea of identifying a rule with a formula may appear odd at first, but it is actually very natural. For instance, the rules $\vee I_1$, $\vee I_2$ and $\vee E$ of NI can be identified with the following $\mathcal{L}$-formulas:

$$(\vee i_1) \qquad \forall XY.X \supset \dagger(X,Y)$$
$$(\vee i_2) \qquad \forall XY.Y \supset \dagger(X,Y)$$
$$(\vee e) \qquad \forall XYZ.(\dagger(Y,Z) \wedge (Y \supset X) \wedge (Z \supset X)) \supset X$$

Observe that the standard propositional connectives and the universal quantifier are used to "encode" the different "structural features" implicit in natural deduction rules (i.e., conjunction "encodes" multiplicity of premises, implication "encodes" the passage from premises to conclusions, and universal quantification "encodes" the schematic nature of rules). As we are using $\mathcal{L}$ as a meta-language to investigate the notion of rule, disregarding the fact that rules can be associated to a specific piece of vocabulary, we replaced accordingly the disjunction of the rules of NI with the (binary) connective variable $\dagger$. In this way, the conjunction of the three formulas $(\vee i_1)$, $(\vee i_2)$ and $(\vee e)$ above is an $\mathcal{L}$-formula with a free connective variable that we can take to express the predicate "being a disjunction". Similarly, we can identify the rule $\supset$E with the formula $\forall XY.(\dagger(X,Y) \wedge X) \supset Y$ and $\wedge$I with the formula $\forall XY.(X \wedge Y) \supset \dagger(X,Y)$.

More generally, we call *structural formulas* (to be indicated with S, $S_1, \dots$) those formulas constructed using only propositional variables and connective variables.

[4] In other words, we are actually working in the fragment of the third-order language of the system F_1 of Girard (1986), in which variables for connectives occur only free. Thus the natural deduction system over $\mathcal{L}$ consists of the rules of the second-order natural deduction system NI^2, the extension of Girard's System F with primitive rules for $\top$, $\bot$, $\wedge$ and $\vee$.

More precisely, the set of structural formulas is the subset $\mathcal{L}^s$ of $\mathcal{L}$ defined as follows (we use S^n for a list of n comma-separated structural formulas):

$$S ::= X \mid \dagger^n(S^n)$$

and we call *rule formulas* (or simply rules, to be indicated with $R, R_1, \ldots$) the $\mathcal{L}$-formulas constructed according to the following grammar (we indicate the set of rule formulas as $\mathcal{L}^r$):

$$R ::= S \mid \forall\overrightarrow{X}(\textstyle\bigwedge_{i=1}^n R_i \supset S)$$

where $\forall\overrightarrow{X} = \forall X_1 \ldots \forall X_m$ if $m > 0$ or it is empty otherwise, and $\bigwedge_{i=1}^n A_i = A_1 \wedge (A_2 \wedge (\ldots \wedge A_n)\ldots)$ if $n > 0$ or $\bigwedge_1^n A_i = \top$ otherwise. The *level of a rule* R, indicated with $\ell(R)$ is the maximum number of nested implications in R, so that $\ell(S) = 0$ and $\ell(\forall\overrightarrow{X}(\bigwedge_{i=1}^n R_i \supset S)) = \mathtt{max}(\ell(R_i)) + 1$.

By an *introduction rule* for an n-ary connective $\dagger$ we understand a rule of the form

(INTRO) $$\forall\overrightarrow{X}\left(\textstyle\bigwedge_{i=1}^n R_i \supset \dagger(\overrightarrow{X})\right)$$

satisfying the following two conditions:

1. in each R_i neither the universal quantifier nor connective variables occurs at all;
2. no propositional variable occurs free (i.e., the only propositional variables occurring in each R_i are among those in $\overrightarrow{X}$).

If R is an introduction rule for $\dagger$ of the above form, we define the *content of R* (notation $\mathscr{C}(R)$) to be the $\mathcal{L}^2$-formula $\bigwedge_{i=1}^n R_i$. If $\mathscr{I}\dagger = \langle R_{I1}, \ldots, R_{Im_I}\rangle$ is a list of introduction rules for $\dagger$, we define the *content of* $\mathscr{I}\dagger$ (notation $\mathscr{C}(\mathscr{I}\dagger)$) to be the $\mathcal{L}^2$-formula $\bigvee_{j=1}^{m_I} \mathscr{C}(R_{Ii})$, where $\bigvee_{i=1}^n A_i = A_1 \vee (A_2 \vee (\ldots \vee A_n)\ldots)$ if $n > 0$ or $\bigvee_{i=1}^n A_i = \bot$ otherwise.

By an *elimination rule* for an n-ary connective $\dagger$ we understand a rule of the form

(ELIM) $$\forall X \forall\overrightarrow{Y}\forall\overrightarrow{X}\left(\left(\dagger(\overrightarrow{X}) \wedge \textstyle\bigwedge_1^n R_i\right) \supset X\right)$$

satisfying the following three conditions:

1. in each R_i neither the universal quantifier nor connective variables occurs at all;
2. no propositional variable occurs free (i.e., $\overrightarrow{Y}$ contains all propositional variables occurring in any of the R_i other than X and those in $\overrightarrow{X}$);
3. if X occurs in any of the R_i it occurs in rightmost position.[5]

If R is an elimination rule for $\dagger$ of the above form, its *content* $\mathscr{C}(R)$ is the $\mathcal{L}^2$-formula $\forall X\forall\overrightarrow{Y}\left(\bigwedge_{i=1}^n R_i \supset X\right)$. If $\mathscr{E}\dagger = \langle R_{E1}, \ldots, R_{Em_E}\rangle$ is a list of elimination rules for $\dagger$, we define the *content of* $\mathscr{E}\dagger$ (notation $\mathscr{C}(\mathscr{E}\dagger)$) to be the $\mathcal{L}^2$-formula $\bigwedge_{k=1}^{m_E} \mathscr{C}(R_{Ek})$.

[5] With the exception of this last conditions, the definitions of introduction and elimination rules follow those given in Schroeder-Heister (2014a). The reason for considering this final restriction will be discussed at the end of Section 4.

Suppose now we are given a list $\mathscr{I}\dagger$ of introduction rules and a list $\mathscr{E}\dagger$ of elimination rules for an n-ary connective $\dagger$. We say that the two collections of rules are in *weak harmony* if and only if the following holds in NI^2:[6]

$$\mathscr{C}(\mathscr{I}\dagger) \dashv\vdash \mathscr{C}(\mathscr{E}\dagger)$$

For example, the rule $\wedge\mathrm{I}$, $\wedge\mathrm{E}_1$ and $\wedge\mathrm{E}_2'$ discussed in the introduction are the formulas (we use the connective variable $\dagger$ for conjunction) $\forall XY.(X \wedge Y) \supset \dagger(X,Y)$, $\forall XY.\dagger(X,Y) \supset X$ and $\forall XY.(\dagger(X,Y) \wedge X) \supset Y$ respectively. Thus, the content of the collection of introduction rules for conjunction consisting of $\wedge\mathrm{I}$ and of the collection of elimination rules for conjunction consisting of $\wedge\mathrm{E}_1$ and $\wedge\mathrm{E}_2'$ are $X \wedge Y$ and $X \wedge (X \supset Y)$ respectively. As these two formulas are intederivable in NI^2, the two collections of rules are in weak harmony. As the reader can check, the collections of introduction and elimination rules consisting of the rules of NI are in weak harmony as well.

Prawitz (1979) and Schroeder-Heister (1981; 1984; 2014b) proposed a simple method to construct a weakly harmonious collection of elimination rules by "inverting" a collection of introduction rules for a given connective $\dagger$. In particular, let $\mathscr{I}\dagger$ be a sequence of m distinct introduction rules of the above form, i.e., $\mathscr{I}\dagger = \langle R_1, \ldots, R_m \rangle$, with $R_j = \forall\overrightarrow{X}\left(\bigwedge_{i=1}^{n_j} R_{ji} \supset \dagger(\overrightarrow{X})\right)$ for all $1 \leq j \leq m$. The Prawitz-Schroeder-Heister collection of canonical elimination rule associated to $\mathscr{I}\dagger$, to be indicated as $\mathtt{PSH}(\mathscr{I}\dagger)$ is the list containing only one element, namely the elimination rule $\forall X\forall\overrightarrow{X}.\left(\dagger(\overrightarrow{X}) \wedge \bigwedge_{j=1}^{m}\left(\bigwedge_{i=1}^{n_j} R_{ji} \supset X\right)\right) \supset X$ that we indicate with $\dagger\mathrm{E}_{\mathtt{PSH}(\mathscr{I}\dagger)}$. The collections of introduction rules $\mathscr{I}\dagger$ and that of elimination rules $\mathtt{PSH}(\mathscr{I}\dagger)$ are in weak harmony since in NI^2 the content of $\mathscr{I}\dagger$ is interderivable with that of $\mathtt{PSH}(\mathscr{I}\dagger)$ (i.e., with the content of $\dagger\mathrm{E}_{\mathtt{PSH}(\mathscr{I}\dagger)}$):

$$\bigvee_{j=1}^{m} \bigwedge_{i=1}^{n_j} R_{ji} \dashv\vdash \forall X.\left(\bigwedge_{j=1}^{m}\left(\bigwedge_{i=1}^{n_j} R_{ji} \supset X\right)\right) \supset X$$

As Schroeder-Heister (2014a) shows, the left-to-right direction of harmony (called "conservativity" criterion) warrants that the addition of the rules for $\dagger$ (understood as meta-linguistic schemata) to a given natural deduction system N yields a conservative extension of N; and that the right-to-left direction warrants the uniqueness of $\dagger$ (where conservativity and uniqueness are understood in the sense of Belnap, 1962).

As shown in Schroeder-Heister (1981) one can define reduction procedures to get rid of consecutive applications of an introduction rule for a connective followed immediately by the Prawitz-Schroeder-Heister elimination rule, and the same is true for expansions (see Tranchini, 2016a). Actually, reduction and expansions are available not only when the elimination rules follow the pattern of Prawitz and Schroeder-Heister: If two collections of introduction and elimination rules are in weak harmony, it is possible to equip them with expansions and reductions as well. A formalization of this claim hinges on a formal characterization of what sort of

[6] Observe that the two formulas contain no occurrence of connective variables, that is they belong to the proper second-order system NI^2. Cf. note 4 above.

operations can qualify as reductions and expansions. Here we limit to an informal sketch of how reduction and expansions for harmonious rules can be obtained from those associated to the "canonical pair" consisting of a collection of introduction rule and the Prawitz-Schroeder-Heister collection of elimination rules, and to discuss some examples.

Observe first that, if a collection of introduction rules $\mathscr{I}\dagger$ and a collection of elimination rules $\mathscr{E}\dagger = \langle R_1, \ldots, R_m \rangle$ are in weak harmony, then the content $\mathscr{E}$ is interderivable with the content of the Prawitz-Schroeder-Heister elimination rule and hence the content of each elimination rule in $\mathscr{E}\dagger$ is derivable from that of $\dagger\mathrm{E}_{\mathrm{PSH}(\mathscr{I}\dagger)}$. From each possible way of deriving the content of any of the rules R_j in $\mathscr{E}\dagger$ from that of $\dagger\mathrm{E}_{\mathrm{PSH}(\mathscr{I}\dagger)}$ one can "extract" a reduction procedure to get rid of consecutive applications of any introduction rule in $\mathscr{I}\dagger$ followed immediately by R_j. Moreover, from the derivation of the content of $\dagger\mathrm{E}_{\mathrm{PSH}(\mathscr{I}\dagger)}$ from the content of $\mathscr{E}\dagger$ we can "extract" an expansion for the collections of rules $\mathscr{I}\dagger$ and $\mathscr{E}\dagger$.

For example, let us reconsider the collection of introduction rules simply consisting of $\wedge\mathrm{I}$ and the "deviant" collection of elimination rules consisting of $\wedge\mathrm{E}_1$ and $\wedge\mathrm{E}'_2$ that we discussed in the introduction. We can obviously define the following reductions and expansion:

$$\dfrac{\dfrac{\begin{matrix}\mathscr{D}_1\\ A\end{matrix}\quad \begin{matrix}\mathscr{D}_2\\ B\end{matrix}}{A\wedge B}\wedge\mathrm{I}}{A}\wedge\mathrm{E}_1 \quad \overset{\beta-\wedge_1}{\rhd} \quad \begin{matrix}\mathscr{D}_1\\ A\end{matrix} \qquad\qquad \dfrac{\dfrac{\begin{matrix}\mathscr{D}_1\\ A\end{matrix}\quad \begin{matrix}\mathscr{D}_2\\ B\end{matrix}}{A\wedge B}\wedge\mathrm{I} \qquad \begin{matrix}\mathscr{D}'\\ A\end{matrix}}{B}\wedge\mathrm{E}'_2 \quad \overset{\beta-\wedge'_2}{\rhd} \quad \begin{matrix}\mathscr{D}_2\\ B\end{matrix}$$

$$\begin{matrix}\mathscr{D}\\ A\wedge B\end{matrix} \quad \text{expands to} \quad \dfrac{\dfrac{\begin{matrix}\mathscr{D}\\ A\wedge B\end{matrix}}{A}\wedge\mathrm{E}_1 \qquad \dfrac{\begin{matrix}\mathscr{D}\\ A\wedge B\end{matrix} \quad \dfrac{\begin{matrix}\mathscr{D}\\ A\wedge B\end{matrix}}{A}\wedge\mathrm{E}_1}{B}\wedge\mathrm{E}'_2}{A\wedge B}\wedge\mathrm{I}$$

Using these transformation one can prove a normalization theorem and the atomization of minimal formulas in normal derivation for the natural deduction system NI′ obtained by replacing $\wedge\mathrm{E}_2$ with the deviant $\wedge\mathrm{E}'_2$ by opportunely modifying the standard proofs of Prawitz (1965b; 1971) (from which conservativity and uniqueness results for the conjunction governed by these rules follow).

As we remarked in the introduction, however, there are reasons for regarding this as a characterization of a weak notion of harmony, and for looking for a stricter criterion capturing a notion of strong harmony.

The problem with weak harmony is that although one can equip weakly harmonious rules with reductions and expansions, the resulting notion of equivalence on derivation might collapse the notion of isomorphism on that of interderivability. Tranchini (2016a) considers the following collection of weakly harmonious rules:

$$
\dfrac{\begin{matrix}[A] & [B] & \\ B & A & A\end{matrix}}{A \natural B}\,\natural\mathrm{I}
\qquad
\begin{matrix}
\dfrac{A \natural B \quad A}{B}\,\natural\mathrm{E}_1 \\[2ex]
\dfrac{A \natural B \quad B}{A}\,\natural\mathrm{E}_2 \\[2ex]
\dfrac{A \natural B}{B}\,\natural\mathrm{E}_3
\end{matrix}
$$

In spite of the mismatch between the third premise of the introduction and the conclusion of the third elimination, it is not difficult to devise three reductions and an expansion for these rules as well. However, Tranchini shows that any two derivations of A from itself can be shown to belong to the same equivalence class induced by these operations. From this, it immediately follows that any two closed derivations of the same formula are equivalent. Hence, although the rules are weakly harmonious and thus conservative in Belnap's sense (i.e., with respect to derivability), they are non-conservative with respect to identity of proofs. Let S' be the result of extending a system S (including the rules for $\supset$) with the rules for $\natural$, and let E' be the smallest equivalence relation on derivations of S' extending an equivalence relation E on derivations on S (that is closed under the reduction for $\supset$) and closed under the reductions and expansions for $\natural$.[7] For any pair of closed derivations $\mathscr{D}_1$ and $\mathscr{D}_2$ of S we have that $\mathscr{D}_1 \overset{E'}{\equiv} \mathscr{D}_2$. Thus if E is non-trivial E' is a non-conservative extension of E, in fact a trivial non-conservative extension in which any two derivations (of the same formula) are equated. Similar consideration apply to isomorphism. The notion of isomorphism relative to the equational theory E' is trivial in the sense that any two interderivable formula qualify as E'-isomorphic. Even if the notion of E-isomorphism is non-trivial (i.e even if there are interderivable formulas which are not E-isomorphic), E'-isomorphism is trivial. Thus, we can say that weakly harmonious rules are unacceptable since their addition to a system has the consequence of blurring meaning distinctions.

4 Strong harmony via isomorphism

As recalled in the introduction, a tighter connection between introduction and elimination rules could be achieved by replacing interderivability with syntactic identity in Schroeder-Heister's definition of weak harmony. The resulting notion could be dubbed *strict harmony* since not only the standard introduction rule for conjunction and the deviant collection of elimination rules are not in strict harmony, but not even the two standard introductions rules for disjunction and its standard elimination rule are in strict harmony. One may expect that a middle ground between weak and strict harmony–a notion of *strong harmony* — could be obtained by using isomorphism instead of derivability in the definition of harmony. At first, it might

[7] The existence of E' is warranted by the fact that the class of equivalence relations with these properties are closed under infinite intersection.

seem that this move does not bring us very far. If strong harmony is defined using $\beta\eta$-isomorphism, not only the rules for ♮ would fail to qualify as harmonious, but also those for disjunction, since $Y \vee Z \overset{\beta\eta}{\not\simeq} \forall X.((Y \supset X) \wedge (Z \supset X)) \supset X$.

However, this counter-example does not rule out a definition of strong harmony based on isomorphism *per se*, but only one based on $\beta\eta$-isomorphism. Actually, there are independent reasons for adopting a different notion of equivalence and of isomorphism when working in NI^2. Both in NI and NI^2 the maximum non-trivial notion of equivalence can be defined as *contextual equivalence* (notation $\overset{C\mathcal{E}}{\equiv}$). In the setting of natural deduction this notion can be defined as follows: two derivations $\mathscr{D}_1$ and $\mathscr{D}_2$ of A are contextually equivalent iff for every derivation $\mathscr{D}$ of $\top \vee \top$ such that

1. the assumption A occurs exactly once in $\mathscr{D}$;
2. the result of replacing the assumption A in $\mathscr{D}$ with $\mathscr{D}_1$ and $\mathscr{D}_2$ (respectively) are closed derivations;

the following holds:[8]

$$\begin{array}{c} \mathscr{D}_1 \\ [A] \\ \mathscr{D} \\ \top \vee \top \end{array} \quad \overset{\beta}{\equiv} \quad \begin{array}{c} \mathscr{D}_2 \\ [A] \\ \mathscr{D} \\ \top \vee \top \end{array}$$

In contrast to what happens in NI, where $\beta\eta$-equivalence and contextual equivalence coincide (Scherer, 2017), in NI^2 $\beta\eta$-equivalence is much weaker than contextual equivalence (so much that the former is decidable, while contextual equivalence is undecidable; this is well-known, for a detailed proof of the latter claim see, e.g., Pistone and Tranchini, 2021). Similarly, the notions of isomorphism arising by contextual equivalence is strictly richer than $\beta\eta$-isomorphisms. In particular $Y \vee Z$ does qualify as $C\mathcal{E}$-isomorphic to $\forall X.((Y \supset X) \wedge (Z \supset X)) \supset X$.

From these considerations, a natural proposal would be that of defining strong harmony by replacing derivability with $C\mathcal{E}$-isomorphism in Schroeder-Heister definition of weak harmony. The resulting notion would allow to overcome the problems of weak harmony. However, also this notion is not entirely satisfactory. In general, it is very hard to decide whether two derivations of a formula A are contextually equivalent, since one has to consider *all* derivations of $\top \vee \top$ from A. Moreover, due to the undecidability of contextual equivalence there are few hopes for the decidability of the notion of $C\mathcal{E}$-isomorphism in NI^2 and hence for that of a notion of strong harmony defined in its terms.

For this reason Tranchini, Pistone, and Petrolo (2019) investigated notions of equivalence lying between $\beta\eta$-equivalence and contextual equivalence, in the hope of finding a notion more manageable than $C\mathcal{E}$-equivalence but still suitable for the definition of the notion of strong harmony. In particular, the authors focused on the

[8] The derivation $\mathscr{D}$ can be seen as a context, and thus $\mathscr{D}_1 \overset{C\mathcal{E}}{\equiv} \mathscr{D}_2$ means that $\mathscr{D}_1$ and $\mathscr{D}_2$ are (β-)equivalent in any context of type $\top \vee \top$. Note that $\top \vee \top$ is a proposition with exactly two distinct proofs, since $\top$ is the proposition with a unique (trivial) proof, and a proof of a disjunction is a proof of either of the disjuncts (together with a bit of information telling which of the two disjuncts is proven). Thus contextual *inequivalence* means that two proofs can be distinguished in some context in which the possible results of evaluating them are only two.

one arising from the functorial interpretation of second-order formulas (Bainbridge, Freyd, Scedrov, and Scott, 1990) and established some results suggesting that this notion fits the needs of the study of generalized intuitionistic connectives in the second-order setting (Pistone and Tranchini, 2021).

The key idea underlying this extension of $\beta\eta$-equivalence can be informally introduced starting from the statement of the proof-conditions of a formula of the form $\forall X.A$ in the style of the BHK-clauses:

> A proof of $\forall X.A$ (henceforth a *universal proof*) is a function that applied to a proposition B yields a proof $A[B/X]$.

As in the case of a proof of $A \supset B$ (taken to be a function from proofs of A to proofs of B) a universal proof cannot be merely understood as an infinite list of ordered pairs (each consisting of a proposition B and of a proof of $A[B/X]$) on pain of making the notion of proof epistemically unsurveyable, and thereby contradicting the assumptions that intuitionistic proofs are the result of an activity of mental construction performed by a knowing subject. Rather, these functions have to be understood as given in such a way as to make it possible for a knowing subject to grasp them. A way of meeting this demand is that of assuming that a universal proof of $\forall X.A$ is a function f that associates a proof of $A[B/X]$ to each proposition B "in a uniform way".[9]

To ask that the values of a universal proof f are assigned to its arguments in a uniform way amounts to requiring that the values of f themselves must be defined in a uniform manner. We take this to mean that in defining each value of f (i.e., the proofs of $A[B/X]$) no knowledge about the argument of f (i.e., the propositions B) should be assumed (for a more thorough discussion, see Pistone, 2018).

Consider for example the proposition $\forall X.X \supset X$. A proof of this proposition is a function f mapping each proposition B onto a proof fB of $B \supset B$, which in turn is a function from the set of proofs of B onto itself.[10] It is true that there may be different ways of mapping the set of proofs of a certain proposition onto itself. However, there does not seem to be many options for defining such a map without assuming any knowledge about the proposition and hence, about its sets of proofs. In fact it seems that the only function one can come up with is the identity function. In other words, if we assume the proofs of $\forall X.X \supset X$ to be uniform functions, there seems to be only one such proof, namely the one associating to each proposition B the identity function $\mathtt{id}_B$ on the set of proofs of B. Although this informal argument is non-conclusive, being based on considerations of a heuristic nature, it turns out that the strengthening of $\beta\eta$-equivalence considered by the authors captures exactly these intuitions.

That the assumption of uniformity has consequences for identity of proofs is not as surprising as it may appear at first. Consider proofs of propositions of the form

[9] The notion of uniformity has been widely investigated in theoretical computer science under the name "parametricity" (Strachey, 1967; Reynolds, 1983; Hermida, Reddy, and Robinson, 2014) and it is in direct line of descent from the "schematic" (as opposed to the "numerical") interpretation of second-order quantification (see, e.g., Carnap, 1931).

[10] Following the common notation in λ-calculus, we indicate the application of a function f to its argument a with (fa), where outermost parentheses will be dropped and application is assumed to be left associative, so that fgh is short for $((fg)h)$.

$(\forall X.A) \supset B$, i.e., functions from proofs of $\forall X.A$ to proofs of B. To assume that universal proofs are uniform functions means that one is restricting the domain of the proofs of $(\forall X.A) \supset B$. If two such proofs assign the same value when taking an arbitrary, but uniform, universal proof as argument, then they denote the same proof under the assumption that all universal proofs are uniform. Yet, it might still be possible that these two proofs of $\forall X.A \supset B$ assign a different value to some (non-uniform) proof of $\forall X.A$, so that they would no more denote the same function without the assumption.[11]

To see how the assumption of uniformity can be used to justify new equations between proofs, consider for example the following two derivations:

$$\dfrac{B \supset C \qquad \dfrac{\dfrac{\forall X.X \supset X}{B \supset B} \qquad B}{B}}{C} \qquad\qquad \dfrac{\dfrac{\forall X.X \supset X}{C \supset C} \qquad \dfrac{B \supset C \qquad B}{C}}{C}$$

On the assumption that the proofs of $\forall X.X \supset X$ are uniform, the two derivations should denote the same proof. This is best appreciated when the derivations are decorated with proofs terms:

$$\dfrac{h : B \supset C \qquad \dfrac{\dfrac{f : \forall X.X \supset X}{fB : B \supset B} \qquad b : B}{fBb : B}}{h(fBb) : C} \qquad\qquad \dfrac{\dfrac{f : \forall X.X \supset X}{fC : C \supset C} \qquad \dfrac{h : B \supset C \quad b : B}{hb : C}}{fC(hb) : C}$$

Uniformity warrants that fB and fC are the identity functions $\mathtt{id}_B$ and $\mathtt{id}_C$ on the sets of proofs of B and C respectively, and thus the two derivation encode (for any f, h, b, B and C) the same proof of C:

$$h(fBb) = h(\mathtt{id}_B b) = hb = \mathtt{id}_C(hb) = fC(hb)$$

It easy to see that $\beta\eta$-equivalence fails to capture the consequences of uniformity. The two derivations above are $\beta\eta$-normal and thus (as a consequence of the Church-Rosser theorem for $\beta\eta$-reduction in NI^2) they belong to two distinct $\beta\eta$-equivalence classes. Hence in order to capture the uniformity of the proofs of $\forall X.X \supset X$ we need to strengthen the equivalence relation on derivations by requiring it to be closed under the following scheme, the instances of which will be referred to as ε-equations (observe that the left-to-right orientation of these equations can be seen as an operation that permutes-up the derivation $\mathscr{D}'$ across the application of $\forall$E):

[11] It may also be worth observing that the restriction to uniform proofs does not require to modify in any way the rules for the second order quantifier in NI^2: in fact, the variable condition on the rule $\forall$I (that ensures that when inferring $\forall X.A$ from A no assumption is made on X) can be seen as a syntactic counterpart of the uniformity requirement informally described above. In other words, all NI^2-derivations of formulas of the form $\forall X.A$ actually denote uniform universal proofs. Moreover, it is well-known that extensions of the syntax of NI^2 with non-uniform constructors, although possible, might lead to inconsistencies (see, e.g., Harper and Mitchell, 1999).

$$\dfrac{\dfrac{\begin{array}{c}\mathscr{D}\\ \forall X.X\supset X\end{array}}{B\supset B}\qquad B}{B}\begin{array}{c}\\ \mathscr{D}'\\ C\end{array} \quad \overset{\varepsilon}{\equiv} \quad \dfrac{\dfrac{\begin{array}{c}\mathscr{D}\\ \forall X.X\supset X\end{array}}{C\supset C}\qquad \begin{array}{c}B\\ \mathscr{D}'\\ C\end{array}}{C}$$

Hence, the ε-equations are justified by the assumption that the only proof of $\forall X.X \supset X$ is the function associating to each proposition the identity function on its sets of proofs. Conversely, in every categorial model of NI^2 in which the ε-equations are satisfied, $\forall X.X \supset X$ has exactly one proof.

Analogous informal considerations show that the set of uniform proofs of $\forall X.(A \supset X) \supset X$ must be in bijection with the set of proofs of A itself (provided X does not occur free in A). In particular, a proof of $\forall X.(A \supset X) \supset X$ associates to each proposition B a function from proofs of $(A \supset B)$ (which in turn are functions from proofs of A into proofs of B) to proofs of B. But the only way to define such a proof in a uniform manner consists in taking a proof a of A (if any is available) and associate to each proposition B the function that maps each proof f of $A \supset B$ onto fa.

Syntactically, uniformity can be again expressed as the possibility of permuting-up a derivation across an application of $\forall$E with premise $\forall X.(A \supset X) \supset X$ using the ε-equations obtained from the scheme below. In this case observe that the derivation $\mathscr{D}'$ cannot be permuted as it stands, on pain of changing the open assumptions of the derivation, and due to the mismatch between the conclusion of $\mathscr{D}'$ and the minor premise required to apply $\supset$E. The mismatch can however be resolved by "surrounding" $\mathscr{D}'$ (whose conclusion is D and whose undischarged assumptions are C and possibly further assumptions Δ) with some applications of elimination and introduction rules yielding a derivation, that we indicate with $(A \supset X)\{\mathscr{D}'\}$, of $(A \supset X)[D/X]$ from $(A \supset X)[C/X], \Delta$:

$$\dfrac{\dfrac{\begin{array}{c}\mathscr{D}\\ \forall X.(A\supset X)\supset X\end{array}}{((A\supset X)\supset X)[C/X]}\quad (A\supset X)[C/X]}{C}\begin{array}{c}\\ \mathscr{D}'\\ D\end{array} \quad \overset{\varepsilon}{\equiv} \quad \dfrac{\dfrac{\begin{array}{c}\mathscr{D}\\ \forall X.(A\supset X)\supset X\end{array}}{((A\supset X)\supset X)[D/X]}\quad \begin{array}{c}(A\supset X)[C/X]\\ \boxed{(A\supset X)\{\mathscr{D}'\}}\\ (A\supset X)[D/X]\end{array}}{D}$$

where

$$(A\supset X)\{\mathscr{D}'\} = \quad (n)\,\dfrac{\begin{array}{c}\dfrac{(A\supset X)[C/X] \qquad \overset{n}{A}}{C}\\ \mathscr{D}\\ D\end{array}}{(A\supset X)[D/X]}$$

Since the uniform proofs of $\forall X.(A \supset X) \supset X$ are in bijection with those of A, it should be possible, syntactically, to show that $\forall X.(A \supset X) \supset X$ and A are E-isomorphic on any notion of equivalence E that is strong enough to encode the uniformity of universal proofs. This is actually the case: taken the two derivations

$$\mathscr{D}_1 \; = \; \dfrac{\dfrac{\dfrac{\overset{n}{A \supset X} \quad A}{X}}{(A \supset X) \supset X}\,(n)}{\forall X.(A \supset X) \supset X} \qquad\qquad \dfrac{\dfrac{\forall X.(A \supset X) \supset X}{(A \supset A) \supset A} \qquad \dfrac{\overset{n}{A}}{A \supset A}\,(n)}{A} \; = \; \mathscr{D}_2$$

it is easy to show that the composition $\begin{array}{c}\mathscr{D}_1\\ \forall X.(A \supset X) \supset X\\ \mathscr{D}_2\end{array}$ β-reduces to the derivation consisting only of the assumption of A and that the composition $\begin{array}{c}\mathscr{D}_2\\ A\\ \mathscr{D}_1\end{array}$, after the application of an ε-permutation, η-reduces to the derivation consisting only of the assumption of $\forall X.(A \supset X) \supset X$.

In a similar way, we can define ε-permutations for the formula $\forall X.(A \supset X) \supset ((B \supset X) \supset X)$ (provided X does not occur free in A and B) encoding the uniformity of the proofs of this proposition. On the one hand, this formula is $\beta\eta$-isomorphic to $\forall X.((A \supset X) \wedge (B \supset X)) \supset X$, and on the other hand, using the ε-permutation we can show that $\forall X.(A \supset X) \supset (B \supset X) \supset X \overset{\varepsilon}{\simeq} A \vee B$. Hence, we have that $A \vee B \overset{\varepsilon}{\simeq} \forall X.((A \supset X) \wedge (B \supset X)) \supset X$, that is, that by defining strong harmony using ε-isomorphism, the standard rules for $\vee$ qualify as strongly harmonious.

More in general, we can establish that:

$$\bigvee_{j=1}^{m} \bigwedge_{i=1}^{n_j} R_{ji} \overset{\varepsilon}{\simeq} \forall X. \left(\bigwedge_{j=1}^{m} \left(\bigwedge_{i=1}^{n_j} R_{ji} \supset X \right) \right) \supset X$$

and hence that any collection of introduction rules and its Prawitz–Schroeder-Heister collection of elimination rules are in strong harmony, by defining ε-permutations for all quantified formulas $\forall X.A$ in which A has a distinctively simple form that we call *nested sp-X* (see, for details, Tranchini, Pistone, and Petrolo, 2019). An $\mathcal{L}^{2\supset}$-formula A is *strictly positive in* X iff X does not occur to the left of $\supset$ in A, and a nested sp-X formula is a formula of the form $A_1 \supset (\ldots (A_n \supset X) \ldots)$ where each A_i is sp-X for all $1 \leq i \leq n$. The above isomorphism is established by showing that the right-hand side formula is $\beta\eta$-isomorphic to the nested sp-X formula

$$\forall X.\Big(R_{11} \supset (\ldots (R_{1n_1} \supset X) \ldots)\Big) \supset \Big(\ldots \Big(\Big(R_{m1} \supset (\ldots (R_{mn_m} \supset X) \ldots)\Big) \supset X\Big) \ldots \Big)$$

which in turn is ε-isomorphic to the left-hand side formula.

Let $\mathcal{L}^{2\supset}_{sp}$ be the fragment of $\mathcal{L}^{2\supset}$ obtained by allowing to prefix a formula A with $\forall X$ only if A is nested sp-X. The content of any introduction and elimination rule of the form we considered above are ε-isomorphic to formulas in $\mathcal{L}^{2\supset}_{sp}$, and so are the content of collections of introduction and the elimination rules.

Let $\mathsf{NI}^{2\supset}_{sp}$ be the restriction of NI^2 to the language $\mathcal{L}^{2\supset}_{sp}$. As shown by the authors (Pistone and Tranchini, 2021), the ε-equational theory has characteristically strong properties in this fragment, namely it is decidable and the maximum equivalence extending β-equivalence.[12] Thus, taking the stance of Došen and Widebäck, ε-equivalence (resp. ε-isomorphism) can be considered as the canonical notion of equivalence (resp. isomorphism) in $\mathsf{NI}^{2\supset}_{sp}$.

These decidability and maximality results are based on the fact that the derivations of $\mathsf{NI}^{2\supset}_{sp}$ modulo ε-equivalence form a category equivalent to that of the derivations of NI modulo $\beta\eta$-equivalence. This in turns implies that the question of the decidability of isomorphism in $\mathsf{NI}^{2\supset}_{sp}$ is equivalent to that of the decidability of isomorphism in NI, which is (as remarked in the previous section) still open.

The foregoing results speak in favor of defining strong harmony using ε-isomorphism rather than $\beta\eta$- or $C\mathcal{E}$-isomorphism, at least when the form of introduction and elimination rules follows the schemata given above.

Whenever confronted with two collections of introduction and elimination rules for $\dagger$, we are not in general capable of telling whether they are in harmony (since $\beta\eta$-isomorphism in NI, and hence ε-isomorphism in $\mathsf{NI}^{2\supset}_{sp}$ is — of today — not known to be decidable), but we can decide whether certain derivations do or do not testify their isomorphism.

5 Concluding remarks

The account of strong harmony using ε-isomorphism delivers a satisfactory sharpening of the notion of weak harmony developed by Schroeder-Heister for introduction and elimination rules of the form discussed above.

It is worth stressing, however, that Schroeder-Heister (2014b) considers introduction and elimination rules of a more general form, namely the following:

(INTRO*) $$\forall\overrightarrow{X}\forall\overrightarrow{Y}\left(\bigwedge_{i=1}^{n} R_i \supset \dagger(\overrightarrow{X})\right)$$

(ELIM*) $$\forall X\forall\overrightarrow{Y}\forall\overrightarrow{X}\left(\left(\dagger(\overrightarrow{X}) \wedge \bigwedge_{i=1}^{n} R_i\right) \supset X\right)$$

satisfying the following two conditions:

1. in each R_i no connective variable occurs at all;
2. no propositional variable occurs free (i.e., $\overrightarrow{Y}$ contains all propositional variables occurring free in any of the R_i other than X and those in $\overrightarrow{X}$);

By dropping the third condition on elimination rules (see footnote 5 above) and allowing nested quantification inside introduction rules and elimination rules, these less restricted forms of introduction and elimination rules significantly enrich the class of connectives amenable of a characterization in terms of "pure" introduction and elimination rules (i.e., introduction rules in which no connective occurs apart from the one being "defined"). For instance, Schroeder-Heister (2014b) observes that

[12] Moreover, it allows to show that any derivation is equivalent to one in which applications of $\forall$E are have an atomic witness (Pistone, Tranchini, and Petrolo, 2021).

it is possible to formulate an introduction rule for negation which does not mention $\bot$, namely $\forall X.(\forall Y.X \supset Y) \supset \dagger(X)$.

This more general form of introduction rules is however much more expressive than that, as it allows for instance to formulate an introduction rule for a zero-place connective:

$$(\forall Y.(Y \supset Y) \wedge Y \supset Y) \supset \dagger$$

which is essentially the impredicative encoding of the natural number predicate in NI^2.

In contrast to what we observed in the case of introduction and elimination of the more restricted form we considered throughout the paper, the content of introduction and elimination rules of this more general form are formulas which cannot be shown to be ε-isomorphic to formulas in the fragment $\mathcal{L}^{2\supset}_{sp}$.

It is true that the ε-equations can be formulated for any formula of the form $\forall X.A$. However, in contrast to what happens in the restricted fragment so far considered, as soon as one allows for encodings of inductive types the ε-equational theory is not decidable (Pistone and Tranchini, 2021) and might not even be maximal (this is suggested by the fact that the equivalence between the ε-equational theory for the fragment of $\mathsf{NI}^{2\supset}$ containing the encoding of the natural number predicate is related to a (non-maximal) equational theory for Gödel's System T investigated by Okada and Scott, 1999).[13]

Thus, on this more general understanding of introduction and elimination rules, it may be more appropriate to define the notion of strong harmony using an equational theory stronger than ε. Moreover, since the ε-equational theory is undecidable outside the $\mathcal{L}^{2\supset}_{sp}$ language fragment, there are few hopes for a decidable notion of strong harmony when introduction and elimination rules of this more general form are taken into consideration.

We conclude by observing that the notions of weak and strong harmony as defined in the work of Schroeder-Heister and in the present paper are not directly applicable to some prominent examples discussed in the literature, such as the rules for quantum-disjunction[14] and rules whose formulation requires first-order structure (as, e.g., those

[13] Equational theories stronger than ε for the whole of $\mathsf{NI}^{2\supset}$ have been studied, among others, by (Longo, Milsted, and Soloviev, 1993).

[14] As to "quantum disjunction" (see footnote 3 above), due to the restriction on its elimination rule, it does not seem that its elimination content is expressible using an NI^2 formula, and hence its rules fall outside the scope of weak (and hence strong) harmony. One could however consider extensions of NI^2 capable of expressing rules with restrictions of the kind displayed by the quantum disjunction elimination rule. The most natural possibility would be that of extending NI^2 with an implication of the kind described by Dummett (1991) (see also Tranchini, 2018), i.e., in which the introduction rule is restricted so that it can be applied only if its premise depends on no other assumption than those to be discharged by the rule. Using $\overline{\supset}$ for this connective, the content of the collection of elimination rule for quantum disjunction would be expressible as $\forall X.((Y \overline{\supset} X) \wedge (Z \overline{\supset} X)) \supset X$. Perhaps unsurprisingly, this formula is interderivable with $A \vee B$ in the envisaged extension of NI^2 and thus the rules of quantum disjunction would qualify as weakly harmonious. The question of strong harmony is harder to address, since it would require the definition of an appropriate equivalence relation on derivations for the system considered (and this might not be obvious, since no expansion

of the identity predicate). The extension of the notions of weak and strong harmony to a first-order setting is an interesting topic for further research.

References

Bainbridge, E. S., P. J. Freyd, A. Scedrov, and P. J. Scott (1990). Functorial polymorphism. *Theoretical Computer Science* 70, 35–64.

Belnap, N. D. (1962). Tonk, plonk and plink. *Analysis* 22, 130–134.

Carnap, R. (1931). Die logizistische grundlegung der mathematik. *Erkenntnis* 2, 91–105.

– (1956). *Meaning and Necessity*. Chicago Univerisity Press.

Došen, K. (2003). Identity of proofs based on normalization and generality. *Bulletin of Symbolic Logic* 9, 477–503.

Došen, K. and Z. Petrić (2001). The typed Böhm theorem. *Electronic Notes in Theoretical Computer Science* 50. BOTH 2001, Böhm's theorem: applications to Computer Science Theory (Satellite Workshop of ICALP 2001), 117–129.

Dummett, M. (1991). *The Logical Basis of Metaphysics*. London: Duckworth.

Gentzen, G. (1935). Untersuchungen über das logische Schließen. *Mathematische Zeitschrift* 39. eng. transl. 'Investigations into logical deduction', in M. E. Szabo (ed.), *The collected papers of Gerhard Gentzen*, North-Holland, 1969, pp. 68–131.

Girard, J.-Y. (1986). The System F of variable types, fifteen years later. *Theoretical Computer Science* 45, 159–192.

Harper, R. and J. C. Mitchell (1999). Parametricity and variants of Girard's J operator. *Information Processing Letters* 70 (1), 1–5.

Hermida, C., U. S. Reddy, and E. P. Robinson (2014). Logical relations and parametricity – A Reynolds programme for category theory and programming languages. In: *Proceedings of the Workshop on Algebra, Coalgebra and Topology (WACT 2013)*. Vol. 303. Electronic Notes in Theoretical Computer Science. Elsevier, 149–180.

Ilik, D. (2014). Axioms and decidability for type isomorphism in the presence of sums. In: *Proceedings of the Joint Meeting of the Twenty-Third EACSL Annual Conference on Computer Science Logic (CSL) and the Twenty-Ninth Annual ACM/IEEE Symposium on Logic in Computer Science (LICS)*. CSL-LICS '14. New York, NY, USA: Association for Computing Machinery. DOI: 10.1145/2603088.2603115. URL: https://doi.org/10.1145/2603088.2603115.

Longo, G., K. Milsted, and S. Soloviev (1993). The genericity theorem and the notion of parametricity in the polymorphic λ-calculus. *Theoretical Computer Science* 121, 323–349.

Okada, M. and P. J. Scott (1999). A note on rewriting theory for uniqueness of iteration. *Theory and Applications of Categories* 6, 47–64.

Pistone, P. (2018). Polymorphism and the obstinate circularity of second order logic, a victims' tale. *The Bulletin of Symbolic Logic* 24, 1–52.

seems to be available for the implication with restricted introduction rule; see Tranchini, 2018). This is however not the only option, and the issue requires further investigation.

Pistone, P. and L. Tranchini (2021). The Yoneda Reduction of Polymorphic Types. In: *29th EACSL Annual Conference on Computer Science Logic (CSL 2021)*. Ed. by C. Baier and J. Goubault-Larrecq. Vol. 183. Leibniz International Proceedings in Informatics (LIPIcs). Dagstuhl, Germany: Schloss Dagstuhl–Leibniz-Zentrum für Informatik, 35:1–35:22.

Pistone, P., L. Tranchini, and M. Petrolo (2021). *The naturality of natural deduction (II). Some remarks on atomic polymorphism.* To appear *Studia Logica*. Pre-print available at https://arxiv.org/abs/1908.11353.

Prawitz, D. (1965a). *Natural deduction, a proof-theoretical study*. Almqvist & Wiskell.

– (1971). Ideas and results in proof theory. In: *Proceedings of the Second Scandinavian Logic Symposium*. Ed. by J. E. Fenstad. Vol. 63. Studies in Logic and the Foundations of Mathematics. Elsevier, 235–307.

– (1979). Proofs and the meaning and completeness of the logical constants. In: *Essays on Mathematical and Philosophical Logic: Proceedings of the Fourth Scandinavian Logic Symposium and the First Soviet-Finnish Logic Conference, Jyväskylä, Finland, June 29–July 6, 1976*. Ed. by J. Hintikka, I. Niiniluoto, and E. Saarinen. Dordrecht: Kluwer, 25–40.

Prior, A. N. (1960). The runabout inference-ticket. *Analysis* 21, 38–39.

Reynolds, J. C. (1983). Types, abstraction and parametric polymorphism. In: *Information Processing '83*. Ed. by R. E. A. Mason. North-Holland, 513–523.

Scherer, G. (2017). Deciding equivalence with sums and the empty type. In: *Proceedings of the 44th ACM SIGPLAN Symposium on Principles of Programming Languages, POPL 2017, Paris, France, January 18-20, 2017*. Ed. by G. Castagna and A. D. Gordon. ACM, 374–386. URL: http://dl.acm.org/citation.cfm?id=3009901.

Schroeder-Heister, P. (1981). Untersuchungen zur regellogischen Deutung von Aussagenverknüpfungen. PhD thesis. Bonn University.

– (Dec. 1984). A natural extension of natural deduction. *Journal of Symbolic Logic* 49, 1284–1300.

– (2014a). Harmony in proof-theoretic semantics: A reductive analysis. In: *Dag Prawitz on Proof and Meaning*. Ed. by H. Wansing. Vol. 7. Outstanding Contributions to Logic. Springer.

– (2014b). The calculus of higher-level rules, propositional quantification, and the foundational approach to proof-theoretic harmony. English. *Studia Logica* 102, 1185–1216. DOI: 10.1007/s11225-014-9562-3.

– (2016). Open problems in proof-theoretic semantics. In: *Advances in Proof-Theoretic Semantics*. Ed. by T. Piecha and P. Schroeder-Heister. Springer International Publishing, 253–283. DOI: 10.1007/978-3-319-22686-6_16. URL: https://doi.org/10.1007/978-3-319-22686-6_16.

Solov'ev, S. V. (1983). The category of finite sets and cartesian closed categories. *Journal of Soviet Mathematics* 22, 1387–1400.

Statman, R. (1983). λ-definable functionals and $\beta\eta$-conversion. *Archiv für mathematische Logik und Grundlagenforschung* 23, 21–26.

Strachey, C. (1967). Fundamental concepts in programming languages. *Higher Order and Symbolic Computation* 13, 11–49.

Tennant, N. (1978). *Natural Logic*. Edimburgh: Edimburgh University Press.

Tranchini, L. (2012). Truth from a proof-theoretic perspective. *Topoi* 31, 47–57.
– (2016a). Proof-theoretic harmony: towards an intensional account. *Synthese*. DOI: 10.1007/s11229-016-1200-3.
– (2016b). Proof-theoretic semantics, paradoxes, and the distinction between sense and denotation. *Journal of Logic and Computation* 26, 495–512.
– (2018). Stabilizing quantum disjunction. *Journal of Philosophical Logic* 47, 1029–1047.
– (2019). Proof, meaning and paradox: some remarks. *Topoi* 38, 591–603.
Tranchini, L., P. Pistone, and M. Petrolo (2019). The naturality of natural deduction. *Studia Logica* 107, 195–231.
Widebäck, F. (2001). *Identity of proofs*. Stockholm: Almqvist & Wiksell.

A Note on Synonymy in Proof-Theoretic Semantics

Heinrich Wansing

Abstract The topic of identity of proofs was put on the agenda of general (or structural) proof theory at an early stage. The relevant question is: When are the differences between two distinct proofs (understood as linguistic entities, proof figures) of one and the same formula so inessential that it is justified to identify the two proofs? The paper addresses another question: When are the differences between two distinct formulas so inessential that these formulas admit of identical proofs? The question appears to be especially natural if the idea of working with more than one kind of derivations is taken seriously. If a distinction is drawn between proofs and disproofs (or refutations) as primitive entities, it is quite conceivable that a proof of one formula amounts to a disproof of another formula, and vice versa. A notion of inherited identity of derivations is introduced for derivations in a cut-free sequent system for Almukdad and Nelson's constructive paraconsistent logic N4 with strong negation. The notion is obtained by identifying sequent rules the application of which has no effect on the identity of derivations. Then the notion of inherited identity is used to define a bilateralist notion of synonymy between formulas, which is a relation drawing more fine-grained distinctions between formulas than strong equivalence.

Key words: proof-theoretic semantics, synonymy, identity of derivations, constructive logic N4, BHK interpretation

1 Introduction

Proof-theoretic semantics has largely been focusing on the central semantical notion of validity; cf. Francez (2015) and Schroeder-Heister (2018). But there are other important semantical notions as well, and one of them is the notion of synonymy. In linguistics, a distinction is drawn between total and partial (or relative) synonymy

Heinrich Wansing
Department of Philosophy I, Ruhr University Bochum, Germany, e-mail: Heinrich.Wansing@rub.de

T. Piecha and K. F. Wehmeier (eds.), *Peter Schroeder-Heister on Proof-Theoretic Semantics*, Outstanding Contributions to Logic 29,
https://doi.org/10.1007/978-3-031-50981-0_11

of expressions, where the former is synonymy in the very strict sense of sameness of linguistic meaning in all its aspects. The expressions "square" and "equilateral rectangle," for example, are often regarded as being strictly synonymous, whereas the expressions "discover" and "find" are only partially synonymous, and so are "new" and "novel" as well as "buy" and "purchase." Whilst some linguists doubt that there exist strict synonyms, it is generally agreed in denotational semantics that synonymy implies sameness of denotation.

It seems that the reasons for assuming only partial instead of strict synonymy in natural languages, i.e., reasons such as different connotations associated with only partially synonymous expressions or their belonging to different dialects, do not apply to formal languages. The present paper pursues the idea of explicating synonymy between formulas from a certain formal propositional language in terms of identity between derivations in a given sequent-style proof system. Therefore, no distinction will be drawn between strict and partial synonymy. Also, there is no given, independently motivated notion of synonymy that the paper aims to capture.[1]

2 Identity of derivation trees and synonymy

There is a research program that bases synonymy of formulas on a notion of identity of proofs. The topic of identity of proofs (or identity of derivations, if proofs are seen as entities represented by derivations)[2] was put on the agenda of general (or structural) proof theory by Dag Prawitz already in his seminal paper Prawitz (1971). On p. 237 Prawitz writes:

> In the same way as one asks when two formulas define the same set or two sentences express the same proposition, one asks when two derivations represent the same proof; in other words, one asks for identity criteria for proofs or for a "synonymity" (or equivalence) relation between derivations.

Moreover, Prawitz (1971, p. 257) formulates a conjecture, which he attributes to Per Martin-Löf (on p. 261) and for which he mentions an influence by ideas of William Tait: "Two derivations represent the same proof if and only if they are equivalent,"[3] where equivalence is the reflexive, symmetric, and transitive closure of a certain reducibility relation between two natural deduction derivations with the same set of assumptions and the same conclusion. The relevant question then is: When are the

[1] I dedicate this paper to Peter Schroeder-Heister, whom I appreciate for his path-breaking work in proof-theoretic semantics, who some thirty years ago pointed out to me Nuel Belnap's display logic and Franz von Kutschera's work on functional completeness, and who made it possible for me to attend the now legendary conference on logics without structural rules in Tübingen 1989 and various conferences on proof-theoretic semantics.

[2] There is a convention of referring to proofs as semantical counterparts of derivations, but this terminological convention in not always followed. In Harper, Honsell, and Plotkin (1993), for instance, proofs are explicitly listed together with terms and formulas as syntactic entities.

[3] A detailed study and critical discussion of this conjecture understood as saying that derivations represent the same proofs if they are $\beta\eta$-equal can be found in Rezende de Castro Alves (2019).

differences between two distinct proofs (understood as linguistic entities, proof figures or prooftrees) of one and the same formula so inessential that it is justified to identify the two proofs? Troelstra and Schwichtenberg (2000, p. 27), for example, raise the question "When are two prooftrees to be regarded as identical?" and continue by explaining that "[t]aking the formulas-as-types isomorphism as our guideline, we can say that two prooftrees are the same, if the corresponding terms of simple type theory are the same (modulo renaming bound variables)." The $\beta\eta$-equal terms of simply typed λ-calculi à la Church are terms of one and the same type,[4] and it seems that a certain question that is in a sense dual to the one concerning identifying syntactically distinct derivations of one and the same formula has received much less attention:

(Q) When are the differences between two distinct formulas so inessential that these formulas admit of identical proofs?

If there are distinct formulas with identical proofs, i.e., with essentially the same proof, these formulas could be declared to be synonymous.[5] One reason why question (Q) has, apparently, not been asked could be that one may expect *structurally different* formulas to have distinct proofs that reflect the differing syntactical structure of the formulas under consideration. Kosta Došen and Zoran Petrić (2012) point out that Rudolf Carnap envisaged an understanding of synonymy that focuses on the internal structure of formulas or sentences. Carnap (1947, p. 56) explains that

> If two sentences are built *in the same way* out of corresponding designators with the same intensions, then we shall say that they have the same intensional structure. We might perhaps also use for this relation the term '*synonymous*', because it is used in a similar sense by other authors (emphasis added).

In order to give an account of the synonymy of syntactic contexts *in which formulas occur*, one would then seem to need a notion of synonymy of formulas. While Carnap considered the internal grammatical structure of formulas, there is also an established theory of isomorphism of propositions in proof theory, isomorphism of types in functional programming, and isomorphism of objects in category theory that focuses on outer structures, such as derivations, in which formulas occur. As highlighted by Roberto di Cosmo (1995, p. 178), the problem of characterizing isomorphic propositions, types, or objects "is really just one problem, as the solution to one of these is a solution to all the others." This problem has been investigated by a number of category theorists, computer scientists working on typed lambda calculi, and proof theorists, see the survey Cosmo (2005), references and further information given there, Danos, Joinet, and Schellinx (2003), where the term "computational isomorphism" is used, and Došen (1997; 2006), Došen and Petrić (2012), Restall (2019), and Tranchini (2021).[6] In Došen (1997, p. 306) Kosta Došen introduces the

[4] In Curry-style simply typed λ-calculi, in general η-reduction is not type-preserving; a counter-example can be found in Sørensen and Urzyczyn (2006, p. 384).

[5] This thought calls into doubt, of course, the view that syntactically distinct proofs can be essentially the same proofs only if they are proofs of the very same formula.

[6] An approach to synonymy in a calculus of typed λ-terms that can be applied to a formal rendering of large fragments of English has been presented by Yiannis Moschovakis (2006), who defines a notion of referential synonymy between λ-terms as a relation of identity between meanings understood as abstract algorithms.

notion of isomorphic formulas in a syntactical category-theoretic framework, where the objects are formulas and the arrows are deductions, as follows:

> A formula A is isomorphic to a formula B if and only if there is a deduction f from A to B and a deduction f^{-1} from B to A such that f composed with f^{-1} reduces via normalization to the identity deduction $\mathbf{1}_A$ from A to A and f^{-1} composed with f reduces via normalization to the identity deduction $\mathbf{1}_B$ from B to B (these reductions are represented in categories by equations between arrows). That two formulae are isomorphic is equivalent to the assertion that the deductions involving one of them, either as premise or as conclusion, can be extended to deductions where this formula is replaced by the other, the deductions involving the first formula being in one-one correspondence with the deductions involving the second. Roughly speaking, whatever you can do in deductions with one of these formulae you can do as well with the other.

Došen then notes that isomorphism between formulas is an equivalence relation stronger than mutual derivability as for formula-to-formula derivations in intuitionistic logic, for example, the mutually derivable formulas $(C \wedge C)$ and C fail to be isomorphic, and explains that it

> seems reasonable to suppose that isomorphism analyzes propositional identity, i.e. identity of meaning for propositions: A and B stand for the same proposition, i.e., A means the same as B, if and only if A is isomorphic to B. This way we would base propositional identity upon identity of deductions, which is codified by equality between arrows.

More recently, Francesca Poggiolesi (2020) has taken up this idea to define a notion of hyper-isomorphic formulas as formulas that have the same meaning. The main move is to replace in the definition of isomorphic formulas the deducibility relation by another inferential relation, namely logical grounding, defined by single-conclusion grounding rules that identify the premises as the reasons why the conclusion is true.

Within the tradition of structural proof theory and proof-theoretic semantics, synonymy has also been discussed by Cesare Cozzo (1994), who outlines a translation-based account of synonymy, and by Tiago Rezende de Castro Alves (2019).

There thus exist proof-theoretic accounts of propositional identity, i.e., *synonymy* of formulas.[7] Another proof-theoretic road to addressing question (Q) emerges if the idea of working with more than one kind of derivations is taken seriously. If a distinction is drawn between proofs and disproofs (or refutations) as primitive entities, it is quite conceivable that a proof of one formula amounts to a disproof of another formula, and vice versa. In particular, if a proof of a formula A is identified with (amounts to) a disproof of $\sim A$, the negation of A, and a disproof of a formula A amounts to a proof of $\sim A$, then, obviously, for any formula A, a proof (disproof) of A is a proof (disproof) of $\sim\sim A$.

Once the possibility of identifying proofs and disproofs of certain distinct formulas is recognized, it may be used to define a notion of synonymy between formulas: Two formulas A and B are synonymous just in case the following two conditions are satisfied: (i) there is a proof $\mathcal{D}$ of B from the assumption that A is true and a proof $\mathcal{D}'$ of A from the assumption that B is true, such that $\mathcal{D}$ and $\mathcal{D}'$ are identical, and

[7] Note that Göran Sundholm (1999) distinguishes between three notions of identity, namely absolute, criterial, and *propositional identity* (my emphasis), where propositional identity is the propositional function expressed by the binary identity predicate symbol.

(ii) there is a disproof $\mathcal{D}$ of B from the assumption that A is false and a disproof $\mathcal{D}'$ of A from the assumption that B is false such that $\mathcal{D}$ and $\mathcal{D}'$ are identical. The present paper develops this idea, which abstains from an appeal to the composition of deductions, by considering sequent rules that have no effect on the identity of derivations and by defining a notion of inherited identity between derivations. This approach is in the spirit of what, following Rumfitt (2000), is called *proof-theoretic bilateralism*, "an approach to meaning taking *denial* as a primitive attitude, on par with *assertion*" Francez (2014, p. 239), see also Drobyshevich (2021) and Kürbis (2019). The idea is much older, however, and can be seen as evolving from work by David Nelson (1949) and, independently, Franz von Kutschera (1969). A treatment of proofs and disproofs on an equal footing has been suggested by Edgar López-Escobar (1972), and we shall turn to it in the next section.

3 The Brouwer-Heyting-Kolmogorov interpretation

We start our considerations with a discussion of the Brouwer-Heyting-Kolmogorov (BHK) interpretation of the connectives of intuitionistic propositional logic in terms of (canonical) proofs of compound formulas. There exist several versions of the BHK interpretation in the literature. Troelstra (1999, p. 235) presents the BHK interpretation of $\vee$ (disjunction), $\rightarrow$ (implication) and $\neg$ (negation) as follows:

1. A proof of $A \vee B$ is given by exhibiting either a proof of A or a proof of B.
2. A proof of $A \rightarrow B$ is a construction f transforming any proof d of A into a proof $f(d)$ of B.
3. A proof of $\neg A$ is a construction which transforms any possible proof of A into a contradiction, e.g. $0 = 1$.

Moreover, he notes that "[o]ne can take the notion of a falsehood or contradiction ($\bot$) as primitive; then $\neg A$ is the same as $A \rightarrow \bot$. A falsehood is simply a statement which cannot have a proof,"[8] and he adds that "[t]hese explanations are actually rather vague, since they make use of unexplained primitive notions such as "(constructive) proof" and "construction"." What also might call for further consideration is the notion of a possible proof.

Troelstra and van Dalen (1988, p. 9) present the BHK interpretation differently:

(H1) A proof of $A \wedge B$ is given by presenting a proof of A and a proof of B.
(H2) A proof of $A \vee B$ is given by presenting either a proof of A or a proof of B (plus the stipulation that we want to regard the proof presented as evidence for $A \vee B$).
(H3) A proof of $A \rightarrow B$ is a construction which permits us to transform any proof of A into a proof of B.
(H4) Absurdity $\bot$ (contradiction) has no proof; a proof of $\neg A$ is a construction

[8] Note that if $\neg A$ stands for $A \rightarrow \bot$, then a proof of $\neg A$ is a construction f which transforms any proof d of A not into a contradiction, e.g. $0 = 1$, but into a proof $f(d)$ of $\bot$.

which transforms any hypothetical proof of A into a proof of a contradiction.

According to this version of the BHK interpretation, a proof of $(A \wedge B)$ is given if and only if (iff) a proof of $(B \wedge A)$ is given, and a proof of $(A \vee B)$ is given iff a proof of $(B \vee A)$ is given. This explication identifies proofs of the distinct formulas $(A \wedge B)$ and $(B \wedge A)$ and proofs of the distinct formulas $(A \vee B)$ and $(B \vee A)$. Another version of the BHK interpretation as presented, for example, by Girard (1989, p. 5f.) is different in that respect:

2. A proof of $A \wedge B$ is a pair (p, q) consisting of a proof p of A and a proof q of B.
3. A proof of $A \vee B$ is a pair (i, p) with:
   a. $i = 0$, and p is a proof of A, or
   b. $i = 1$, and p is a proof of B.
4. A proof of $A \rightarrow B$ is a function f, which maps each proof p of A to a proof $f(p)$ of B.[9]
5. In general, the negation $\neg A$ is treated as $A \rightarrow \bot$ where $\bot$ is a sentence with no possible proof.

As per this version of the BHK interpretation, a proof of $(A \wedge B)$ differs from a proof of $(B \wedge A)$ and a proof of $(A \vee B)$ is distinct from a proof of $(B \vee A)$. However, one and the same construction can be a proof of formulas $(A \vee B)$ and $(A \vee C)$. If p is a proof of A, then $(0, p)$ is a proof of both $(A \vee B)$ and $(A \vee C)$, for any formulas B and C.

Troelstra and van Dalen (1988, p. 24) regard the BHK interpretation as a "natural semantics" for intuitionistic logic, and according to Girard (1989, p. 71) "Heyting's *semantics of proofs*" even is "[o]ne of the greatest ideas in logic." The Brouwer-Heyting-Kolmogorov interpretation can be and has been criticized for its treatment of intuitionistic negation, $\neg$, see, for example, Wansing (1993, p. 22 f.).[10] By clause (H4), a proof of the intuitionistically valid $\neg(A \wedge \neg A)$ is a construction c such that if d is a proof of $(A \wedge \neg A)$, then $c(d)$ is a proof of $\bot$, which does not exist. However, since $(A \wedge \neg A)$ has no proof, any construction whatsoever proves $\neg(A \wedge \neg A)$, but this conception of any construction converting a non-existent object into a non-existent object may be criticized as being non-constructive, which might have led Troelstra and van Dalen to considering hypothetical proofs.

López-Escobar (1972) suggested to supplement the BHK interpretation of positive intuitionistic logic with the primitive notion of *refutation* (or *disproof*) to give an interpretation for negation. As a result, one obtains a semantics for the four-valued paraconsistent constructive logic with strong negation introduced by David Nelson and his often not mentioned co-author, Ahmad Almukdad, and now known as the system N4, see Almukdad and Nelson (1984) and Odintsov (2008). López-Escobar

[9] Girard uses "$\Rightarrow$" instead of "$\rightarrow$".

[10] It has been criticized for other reasons as well. Georg Kreisel (1962) suggested to expand the clause for implication by postulating that the construction or function referred to in that clause indeed satisfies the desired property. This *second clause* is highly controversial and later has been excluded from presentations of the BHK interpretation; cf. Dean and Kurokawa (2016). Wagner de Campos Sanz and Thomas Piecha (2014) argue that intuitionistic implicational logic is incomplete with respect to the BHK interpretation.

gives the following disproof interpretation of the intuitionistic connectives $\wedge$, $\vee$, $\rightarrow$, and the strong negation $\sim$ (notation adjusted):

i.) the construction c refutes $A \wedge B$ iff c is of the form $\langle i, d \rangle$ with i either 0 or 1 and if $i = 0$, then d refutes A and if $i = 1$ then d refutes B,

ii.) the construction c refutes $A \vee B$ iff c is of the form $\langle d, e \rangle$ and d refutes A and e refutes B,

iii.) the construction c refutes $A \rightarrow B$ iff c is of the form $\langle d, e \rangle$ and d proves A and e refutes B, ...

viii.) [t]he construction c refutes $\sim A$ iff c proves A.[11]

Whilst the first three disproof clauses specify the form of (canonical) refutations of conjunctions, disjunctions, and implications, the clause for strong negation is different. It specifies that a construction c is a refutation of $\sim A$ just in case the very same construction c is a proof of A. Moreover, López-Escobar requires that a construction c proves $\sim A$ iff c itself refutes A (and not iff c is a proof of $A \rightarrow \bot$ and thus has a specific form).

A fundamental assumption made by López-Escobar is that for no formula A there exists a construction that both proves and disproves A. If it is assumed that for no formula A there exist proofs of both A and $\sim A$, then *ex contradictione quodlibet* expressed as a formula, $(A \wedge \sim A) \rightarrow B$, becomes provable. It is well known that the BHK interpretation in its various versions is sound for intuitionistic propositional logic in the following sense: If a formula A is provable in intuitionistic propositional logic, then there is a construction that proves A. Similarly, N4 is sound with respect to López-Escobar's proof/disproof interpretation: If a formula A is provable in N4, then there is a construction that proves A. Moreover, however, if a formula $\sim A$ is provable in N4, then there is a construction that disproves A.[12]

Remark 3.1 Usually the BHK-interpretation is stated in terms of proofs from the empty set of assumptions. López-Escobar (1972, p. 367) considers sequents (derivability statements) $\{A_1, \ldots, A_n\} \Rightarrow A$ (notation adjusted). A sequent $\{A_1, \ldots, A_n\} \Rightarrow A$ is valid iff there is a construction c such that $c(c_1, \ldots, c_n)$ is a proof of A whenever $c_1, \ldots, c_n$ are constructions proving $A_1, \ldots, A_n$, respectively. A sequent $\varnothing \Rightarrow A$ is valid iff there exists a construction that is a proof of A.

If derivations from sets or multisets of assumptions (hypotheses) are considered, the following weakening clause for derivations is sound:

> if A is derivable from $\{A_1, \ldots, A_n\}$, then A is derivable from $\{A_1, \ldots, A_n, B\}$ for any formula B.

A derivation d of A from $\{A_1, \ldots, A_n\}$ is a construction that transforms any proofs $d_1, \ldots, d_n$ of $A_1, \ldots, A_n$, respectively, into a proof $d(d_1, \ldots, d_n)$ of A. From such a construction d one can define the $(n+1)$-place function d' that for any formula

[11] The clauses for proofs are as one may expect, with the addition that a construction c proves $\sim A$ iff c refutes A.

[12] Note that the notion of refutation in the proof/disproof interpretation is to be distinguished from the notion of refutation in the theory of refutation calculi; cf. Goranko, Pulcini, and Skura (2020). The latter theory is concerned with the axiomatization of the non-theorems of given logics.

B maps any proof of B and any proofs $d_1, \ldots, d_n$ of $A_1, \ldots, A_n$, respectively, to $d(d_1, \ldots, d_n)$. Therefore, one and the same construction can be a proof of formulas $(A \rightarrow (B \rightarrow A))$ and $(A \rightarrow (C \rightarrow A))$.

4 Sequent calculi for N4

The language of Nelson's constructive propositional logic N4 makes use of the connectives $\rightarrow$ (implication), $\wedge$ (conjunction), $\vee$ (disjunction), and $\sim$ (strong negation). In what follows, lower-case letters $p, q, r, \ldots$ are used to denote propositional variables, capital letters $A, B, C, \ldots$ are used to denote formulas, and Greek capital letters $\Gamma, \Delta, \ldots$ are used to represent finite (possibly empty) multisets of formulas. For a singleton multiset $\{A\}$ we usually write just A, and A, Γ as well as Γ, A (Δ, Γ as well as Γ, Δ) designates the union of the multisets Γ and $\{A\}$ (Δ and Γ). We use $A \leftrightarrow B$ as an abbreviation of $(A \rightarrow B) \wedge (B \rightarrow A)$ and $A \Leftrightarrow B$ as an abbreviation of $(A \leftrightarrow B) \wedge ({\sim}A \leftrightarrow {\sim}B)$, and we refer to $\Leftrightarrow$ as the connective of strong equivalence.

A *sequent* is an expression of the form $\Gamma \Rightarrow C$. A sequent calculus is a non-empty set containing some axiomatic, initial sequents and rules of the form

$$\frac{s_1 \ \ldots \ s_n}{s}$$

where s and all s_i $(1 \leq i \leq n)$ are sequents. Derivations in a sequent calculus are inductively defined a usual. Every instance of an initial sequent is a derivation. Applications of sequent rules to instances of their schematic premise sequents as conclusions of derivations result in a derivation. If there is a derivation of a sequent s in a sequent calculus L, we say that s is provable in L and denote this as $L \vdash s$ (or just as $\vdash s$ if the sequent calculus in question is clear). If $\mathcal{D}$ is a derivation of a sequent s we often write $\overset{\mathcal{D}}{s}$, so that $\mathcal{D}$ and $\overset{\mathcal{D}}{s}$ stand for the same derivation.

A rule of inference R is *admissible* in a sequent calculus L if for any instance

$$\frac{s_1 \ \ldots \ s_n}{s}$$

of R, if $L \vdash s_i$ for all i, then $L \vdash s$.

There is a kind of standard sequent calculus for N4 (cf. Kamide and Wansing, 2015; López-Escobar, 1972; Pearce, 1992; Wansing, 1993), which we shall refer to as GN4.

Definition 4.1 The initial sequents of GN4 are of the form

$$p \Rightarrow p \qquad\qquad {\sim}p \Rightarrow {\sim}p$$

for any propositional variable p.

The structural inference rules of GN4 are of the form:

$$\frac{\Gamma \Rightarrow A \quad A, \Sigma \Rightarrow C}{\Gamma, \Sigma \Rightarrow C}\ (\text{cut}) \quad \frac{\Gamma \Rightarrow C}{A, \Gamma \Rightarrow C}\ (\text{we}) \quad \frac{A, A, \Gamma \Rightarrow C}{A, \Gamma \Rightarrow C}\ (\text{co}).$$

The logical inference rules of GN4 are of the form:

$$\frac{\Gamma \Rightarrow A \quad B, \Delta \Rightarrow C}{A \to B, \Gamma, \Delta \Rightarrow C}\ (\to \text{l}) \quad \frac{A, \Gamma \Rightarrow B}{\Gamma \Rightarrow A \to B}\ (\to \text{r})$$

$$\frac{A, B, \Gamma \Rightarrow C}{A \wedge B, \Gamma \Rightarrow C}\ (\wedge\text{l}) \quad \frac{\Gamma \Rightarrow A \quad \Gamma \Rightarrow B}{\Gamma \Rightarrow A \wedge B}\ (\wedge\text{r})$$

$$\frac{A, \Gamma \Rightarrow C \quad B, \Gamma \Rightarrow C}{A \vee B, \Gamma \Rightarrow C}\ (\vee\text{l}) \quad \frac{\Gamma \Rightarrow A}{\Gamma \Rightarrow A \vee B}\ (\vee\text{r1}) \quad \frac{\Gamma \Rightarrow B}{\Gamma \Rightarrow A \vee B}\ (\vee\text{r2})$$

$$\frac{A, \Gamma \Rightarrow C}{{\sim}{\sim}A, \Gamma \Rightarrow C}\ ({\sim}{\sim}\text{l}) \quad \frac{\Gamma \Rightarrow A}{\Gamma \Rightarrow {\sim}{\sim}A}\ ({\sim}{\sim}\text{r})$$

$$\frac{A, {\sim}B, \Gamma \Rightarrow C}{{\sim}(A \to B), \Gamma \Rightarrow C}\ ({\sim} \to \text{l}) \quad \frac{\Gamma \Rightarrow A \quad \Gamma \Rightarrow {\sim}B}{\Gamma \Rightarrow {\sim}(A \to B)}\ ({\sim} \to \text{r})$$

$$\frac{{\sim}A, \Gamma \Rightarrow C \quad {\sim}B, \Gamma \Rightarrow C}{{\sim}(A \wedge B), \Gamma \Rightarrow C}\ ({\sim} \wedge \text{l})$$

$$\frac{\Gamma \Rightarrow {\sim}A}{\Gamma \Rightarrow {\sim}(A \wedge B)}\ ({\sim} \wedge \text{r1}) \quad \frac{\Gamma \Rightarrow {\sim}B}{\Gamma \Rightarrow {\sim}(A \wedge B)}\ ({\sim} \wedge \text{r2})$$

$$\frac{{\sim}A, {\sim}B, \Gamma \Rightarrow C}{{\sim}(A \vee B), \Gamma \Rightarrow C}\ ({\sim} \vee \text{l}) \quad \frac{\Gamma \Rightarrow {\sim}A \quad \Gamma \Rightarrow {\sim}B}{\Gamma \Rightarrow {\sim}(A \vee B)}\ ({\sim} \vee \text{r}).$$

Remark 4.2 A sequent calculus GLJ for positive intuitionistic logic is obtained from GN4 by deleting all initial sequents of the form ${\sim}p \Rightarrow {\sim}p$ for any propositional variable p and deleting all the logical inference rules displaying ${\sim}$.

Remark 4.3 In Kamide and Wansing (2012) and Kamide and Wansing (2015) a syntactical embedding of GN4 into GLJ is used to prove that the rule (cut) is admissible in cut-free GN4. As corollaries to that result, it is observed that N4 is decidable, that N4 satisfies the constructible falsity property, namely, that in GN4:

$$\text{if } \vdash \varnothing \Rightarrow {\sim}(A \wedge B), \text{ then } \vdash \varnothing \Rightarrow {\sim}A \text{ or } \vdash \varnothing \Rightarrow {\sim}B$$

and that N4 is paraconsistent with respect to strong negation, namely, that there exist formulas A and B such that it is not the case that in GN4, $\vdash A, {\sim}A \Rightarrow B$.

The cut-free version of the above sequent calculus for N4 has two features which may be seen as disadvantages: the subformula property fails and the distinction between proofs and disproofs as primitive entities is hidden by being built into the distinction between left and right introduction rules for non-negated and for negated compound formulas. In Kamide and Wansing (2012) and Kamide and Wansing (2015), one can find a "subformula sequent calculus" that avoids the first problem, and a "dual sequent calculus" that highlights the distinction between proofs and disproofs.

For our purposes it is best to combine both systems into a sequent calculus, SN4, that avoids sequent rules for strongly negated compound formulas and makes use of two kinds of sequent arrows, one for proofs, $\Rightarrow^+$, and one for disproofs, $\Rightarrow^-$. Sequents of SN4 are of the form $\Gamma : \Delta \Rightarrow^+ A$ or $\Gamma : \Delta \Rightarrow^- A$, where A is a formula, and Γ and Δ are finite multisets of formulas. If a sequent $\Gamma : \Delta \Rightarrow^+ A$ or $\Gamma : \Delta \Rightarrow^- A$ is provable in a sequent calculus L, then A is said to be provable, respectively disprovable, in L from $\Gamma : \Delta$. If a formula A is (dis)provable in L from $\varnothing : \varnothing$, then A is said to be (dis)provable in L.

The sequent system SN4 is merely a notational variant of the subformula sequent calculus for N4. Moreover, sequents of the form:

$$A_1, \ldots, A_m : B_1, \ldots, B_n \Rightarrow^+ C \quad \text{respectively} \quad A_1, \ldots, A_m : B_1, \ldots, B_n \Rightarrow^- C$$

in SN4 can be understood as sequents of the form:

$$\sim A_1, \ldots, \sim A_m, B_1, \ldots, B_n \Rightarrow C \text{ respectively } \sim A_1, \ldots, \sim A_m, B_1, \ldots, B_n \Rightarrow \sim C$$

in GN4.

Definition 4.4 Let $* \in \{+, -\}$. The initial sequents of SN4 are of the form:

$$\varnothing : p \Rightarrow^+ p \qquad p : \varnothing \Rightarrow^- p$$

for any propositional variable p.

The interaction rules of SN4 are of the form:

$$\frac{\Gamma : \Delta \Rightarrow^- A}{\Gamma : \Delta \Rightarrow^+ \sim A}\ (\sim\mathrm{r}+) \qquad \frac{\Gamma : \Delta \Rightarrow^+ A}{\Gamma : \Delta \Rightarrow^- \sim A}\ (\sim\mathrm{r}-)$$

$$\frac{A, \Gamma : \Delta \Rightarrow^* C}{\Gamma : \sim A, \Delta \Rightarrow^* C}\ (\sim\mathrm{l}+) \qquad \frac{\Gamma : A, \Delta \Rightarrow^* C}{\sim A, \Gamma : \Delta \Rightarrow^* C}\ (\sim\mathrm{l}-)$$

$$\frac{\Gamma : \Delta \Rightarrow^+ \sim A}{\Gamma : \Delta \Rightarrow^- A}\ (\sim\mathrm{r} + i) \qquad \frac{\Gamma : \Delta \Rightarrow^- \sim A}{\Gamma : \Delta \Rightarrow^+ A}\ (\sim\mathrm{r} - i)$$

$$\frac{\Gamma : \sim A, \Delta \Rightarrow^* C}{A, \Gamma : \Delta \Rightarrow^* C}\ (\sim\mathrm{l} + i) \qquad \frac{\sim A, \Gamma : \Delta \Rightarrow^* C}{\Gamma : A, \Delta \Rightarrow^* C}\ (\sim\mathrm{l} - i).$$

The structural rules of SN4 are of the form:

$$\frac{\Gamma_1 : \Delta_1 \Rightarrow^- A \quad A, \Gamma_2 : \Delta_2 \Rightarrow^* C}{\Gamma_1, \Gamma_2 : \Delta_1, \Delta_2 \Rightarrow^* C}\ (\mathrm{cut}-)$$

$$\frac{\Gamma_1 : \Delta_1 \Rightarrow^+ A \quad \Gamma_2 : A, \Delta_2 \Rightarrow^* C}{\Gamma_1, \Gamma_2 : \Delta_1, \Delta_2 \Rightarrow^* C}\ (\mathrm{cut}+)$$

$$\frac{A, A, \Gamma : \Delta \Rightarrow^* C}{A, \Gamma : \Delta \Rightarrow^* C}\ (\mathrm{co}-) \qquad \frac{\Gamma : A, A, \Delta \Rightarrow^* C}{\Gamma : A, \Delta \Rightarrow^* C}\ (\mathrm{co}+)$$

$$\frac{\Gamma : \Delta \Rightarrow^* C}{A, \Gamma : \Delta \Rightarrow^* C}\ (\text{we}-) \quad \frac{\Gamma : \Delta \Rightarrow^* C}{\Gamma : A, \Delta \Rightarrow^* C}\ (\text{we}+).$$

The positive inference rules of SN4 are of the form:[13]

$$\frac{\Gamma_1 : \Delta_1 \Rightarrow^+ A \quad \Gamma_2 : B, \Delta_2 \Rightarrow^* C}{\Gamma_1, \Gamma_2 : A \to B, \Delta_1, \Delta_2 \Rightarrow^* C}\ (\to \text{l}+) \quad \frac{\Gamma : A, \Delta \Rightarrow^+ B}{\Gamma : \Delta \Rightarrow^+ A \to B}\ (\to \text{r}+)$$

$$\frac{\Gamma : A, \Delta \Rightarrow^* C}{\Gamma : A \wedge B, \Delta \Rightarrow^* C}\ (\wedge\text{l1}+) \quad \frac{\Gamma : B, \Delta \Rightarrow^* C}{\Gamma : A \wedge B, \Delta \Rightarrow^* C}\ (\wedge\text{l2}+)$$

$$\frac{\Gamma : \Delta \Rightarrow^+ A \quad \Gamma : \Delta \Rightarrow^+ B}{\Gamma : \Delta \Rightarrow^+ A \wedge B}\ (\wedge\text{r}+) \quad \frac{\Gamma : A, \Delta \Rightarrow^* C \quad \Gamma : B, \Delta \Rightarrow^* C}{\Gamma : A \vee B, \Delta \Rightarrow^* C}\ (\vee\text{l}+)$$

$$\frac{\Gamma : \Delta \Rightarrow^+ A}{\Gamma : \Delta \Rightarrow^+ A \vee B}\ (\vee\text{r1}+) \quad \frac{\Gamma : \Delta \Rightarrow^+ B}{\Gamma : \Delta \Rightarrow^+ A \vee B}\ (\vee\text{r2}+).$$

The negative inference rules of SN4 are of the form:

$$\frac{B, \Gamma : A, \Delta \Rightarrow^* C}{A \to B, \Gamma : \Delta \Rightarrow^* C}\ (\to \text{l}-) \quad \frac{\Gamma_1 : \Delta_1 \Rightarrow^- B \quad \Gamma_2 : \Delta_2 \Rightarrow^+ A}{\Gamma_1, \Gamma_2 : \Delta_1, \Delta_2 \Rightarrow^- A \to B}\ (\to \text{r}-)$$

$$\frac{A, \Gamma : \Delta \Rightarrow^* C \quad B, \Gamma : \Delta \Rightarrow^* C}{A \wedge B, \Gamma : \Delta \Rightarrow^* C}\ (\wedge\text{l}-)$$

$$\frac{\Gamma : \Delta \Rightarrow^- A}{\Gamma : \Delta \Rightarrow^- A \wedge B}\ (\wedge\text{r1}-) \quad \frac{\Gamma : \Delta \Rightarrow^- B}{\Gamma : \Delta \Rightarrow^- A \wedge B}\ (\wedge\text{r2}-)$$

$$\frac{A, \Gamma : \Delta \Rightarrow^* C}{A \vee B, \Gamma : \Delta \Rightarrow^* C}\ (\vee\text{l1}-) \quad \frac{B, \Gamma : \Delta \Rightarrow^* C}{A \vee B, \Gamma : \Delta \Rightarrow^* C}\ (\vee\text{l2}-)$$

$$\frac{\Gamma : \Delta \Rightarrow^- A \quad \Gamma : \Delta \Rightarrow^- B}{\Gamma : \Delta \Rightarrow^- A \vee B}\ (\vee\text{r}-).$$

The single-premise interaction rules are sequent rules that are taken to be rules the application of which has no effect on the identity of derivations.

Proposition 4.5 *In* SN4, *for any formula* A, $\vdash \varnothing : A \Rightarrow^+ A$ *and* $\vdash A : \varnothing \Rightarrow^- A$.

Derivations ending in a sequent of the form $\Gamma : \Delta \Rightarrow^+ C$ are said to be proofs, and derivations ending in a sequent of the form $\Gamma : \Delta \Rightarrow^- C$ are said to be disproofs. The following theorem is proved by simultaneous induction for the cases (a) and (b), i.e., for the first claim by simultaneous induction on the height of proofs and disproofs in SN4, and for the second claim by induction on the height of proofs GN4$-\{(\text{cut})\}$, and justifies the reading of sequents in SN4 mentioned above. We let $\sim\Gamma$ stand for the multiset of all formulas $\sim A$ with A in Γ if Γ is not empty, otherwise $\sim\Gamma$ is the empty multiset.

[13] The rules $(\wedge\text{l1}+)$ and $(\wedge\text{l2}+)$ do not follow the pattern of the single left introduction rule of $\wedge$ in GN4 simply for uniformity with the presentation in Kamide and Wansing (2015). Similarly for the rules $(\vee\text{l1}-)$ and $(\vee\text{l2}-)$ and the left introduction rule for negated disjunctions in GN4.

Theorem 4.6 *Let* Γ *and* Δ *be finite multisets of formulas, and* C *be a formula.*

1. *(a) If* $\mathsf{SN4} \vdash \Gamma : \Delta \Rightarrow^{+} C$*, then* $\mathsf{GN4} \vdash {\sim}\Gamma, \Delta \Rightarrow C$ *and*
   (b) if $\mathsf{SN4} \vdash \Gamma : \Delta \Rightarrow^{-} C$*, then* $\mathsf{GN4} \vdash {\sim}\Gamma, \Delta \Rightarrow {\sim}C$*.*
2. *(a) If* $\mathsf{GN4}{-}\{(\mathrm{cut})\} \vdash {\sim}\Gamma, \Delta \Rightarrow C$*, then* $\mathsf{SN4}{-}\{(\mathrm{cut}-), (\mathrm{cut}+)\} \vdash \Gamma : \Delta \Rightarrow^{+} C$ *and*
   (b) if $\mathsf{GN4}{-}\{(\mathrm{cut})\} \vdash {\sim}\Gamma, \Delta \Rightarrow {\sim}C$*, then* $\mathsf{SN4}{-}\{(\mathrm{cut}-), (\mathrm{cut}+)\} \vdash \Gamma : \Delta \Rightarrow^{-} C$*.*

Proof To illustrate the need for a simultaneous induction, we consider one case for item 2. (b). Suppose the last step in the derivation of ${\sim}\Gamma, \Delta \Rightarrow C$ is an application of $({\sim} \rightarrow \mathrm{r})$, and C has the form ${\sim}(A \rightarrow B)$:

$$\frac{\Gamma \Rightarrow A \quad \Gamma \Rightarrow {\sim}B}{\Gamma \Rightarrow {\sim}(A \rightarrow B).}$$

Then, by the induction hypotheses for both (a) and (b) we obtain the following derivation, where the dots indicate applications of (co+):

$$\dfrac{\dfrac{\varnothing : \Gamma \Rightarrow^{+} A \quad \varnothing : \Gamma \Rightarrow^{-} B}{\varnothing : \Gamma, \Gamma \Rightarrow^{-} (A \rightarrow B)}}{\vdots} \qquad \frac{}{\varnothing : \Gamma \Rightarrow^{-} (A \rightarrow B).}$$

□

Theorem 4.7 *The rules* (cut+) *and* (cut−) *are admissible in cut-free* SN4 *(i.e., in* $\mathsf{SN4}\ {-}\{(\mathrm{cut}-), (\mathrm{cut}+)\}$*).*

Proof Suppose that a sequent $\Gamma : \Delta \Rightarrow^{+} C$, respectively $\Gamma : \Delta \Rightarrow^{-} C$, is provable in SN4. Then ${\sim}\Gamma, \Delta \Rightarrow C$, respectively ${\sim}\Gamma, \Delta \Rightarrow {\sim}C$, is provable in GN4 by Theorem 4.6(1), and hence the sequent ${\sim}\Gamma, \Delta \Rightarrow C$, respectively ${\sim}\Gamma, \Delta \Rightarrow {\sim}C$, is provable in cut-free GN4 by the cut-elimination theorem for GN4. Therefore $\Gamma : \Delta \Rightarrow^{+} C$, respectively $\Gamma : \Delta \Rightarrow^{-} C$, is provable in cut-free SN4 by Theorem 4.6 (2). □

Theorem 4.8 *Let* Γ *and* Δ *be finite multisets of formulas, and* C *be a formula, and let* $\mathsf{SN4}^{-}$ *stand for* $\mathsf{SN4}{-}\{(\mathrm{cut}-), (\mathrm{cut}+), ({\sim}\mathrm{r} + i), ({\sim}\mathrm{r} - i), ({\sim}\mathrm{l} + i), ({\sim}\mathrm{l} - i)\}$.

(a) If $\mathsf{GN4}{-}\{(\mathrm{cut})\} \vdash {\sim}\Gamma, \Delta \Rightarrow C$*, then* $\mathsf{SN4}^{-} \vdash \Gamma : \Delta \Rightarrow^{+} C$ *and*
(b) if $\mathsf{GN4}{-}\{(\mathrm{cut})\} \vdash {\sim}\Gamma, \Delta \Rightarrow {\sim}C$*, then* $\mathsf{SN4}^{-} \vdash \Gamma : \Delta \Rightarrow^{-} C$*.*

Proof By simultaneous induction on the height of derivations in GN4−{(cut)} for (a) and (b). □

Theorem 4.9 *The subformula spoiling rules* $({\sim}\mathrm{r} + i)$*,* $({\sim}\mathrm{r} - i)$*,* $({\sim}\mathrm{l} + i)$*, and* $({\sim}\mathrm{l} - i)$ *are admissible in* $\mathsf{SN4}^{-}$*.*

Proof Suppose that a sequent $\Gamma : \Delta \Rightarrow^{+} C$, respectively $\Gamma : \Delta \Rightarrow^{-} C$, is provable in cut-free SN4. Then ${\sim}\Gamma, \Delta \Rightarrow C$, respectively ${\sim}\Gamma, \Delta \Rightarrow {\sim}C$, is provable in GN4 by Theorem 4.6(1), and hence the sequent ${\sim}\Gamma, \Delta \Rightarrow C$, respectively ${\sim}\Gamma, \Delta \Rightarrow {\sim}C$, is provable in cut-free GN4 by the cut-elimination theorem for GN4. Therefore $\Gamma : \Delta \Rightarrow^{+} C$, respectively $\Gamma : \Delta \Rightarrow^{-} C$, is provable in $\mathsf{SN4}^{-}$ by Theorem 4.8. □

Corollary 4.10 SN4 has the subformula property, i.e., if a sequent s is provable (disprovable) in SN4, then there is a proof (disproof) of s such that all formulas appearing in that proof (disproof) are subformulas of some formula in s.

5 Synonymy in cut-free SN4

In view of Theorem 4.7, we focus on cut-free SN4 and define a notion of identity between derivations in cut-free SN4. We want to have a notion of identity of derivations such that, for example, there is a proof $\mathcal{D}$ of $(\sim A \wedge \sim B)$ from $\sim(A \vee B)$ and a proof $\mathcal{D}'$ of $\sim(A \vee B)$ from $(\sim A \wedge \sim B)$ such that $\mathcal{D}$ and $\mathcal{D}'$ are identical because according to López-Escobar's proof/disproof interpretation, a proof of $(\sim A \wedge \sim B)$ is an ordered pair consisting of a proof of $\sim A$ as the first component and a proof of $\sim B$ as the second component. This pair is a pair consisting of a disproof of A and a disproof of B, which then is a disproof of $(A \vee B)$, and therefore it is also a proof of $\sim(A \vee B)$.

If $\mathcal{D}$ and $\mathcal{D}'$ are derivations in cut-free SN4, we shall write $\mathcal{D} \equiv \mathcal{D}'$ to express that $\mathcal{D}$ and $\mathcal{D}'$ are syntactically (that is, as strings of symbols) identical.

Definition 5.1 The relation $\approx$ of inherited identity (in-identity) between derivations $\mathcal{D}_1$ and $\mathcal{D}_2$ in cut-free SN4 is defined inductively. It is the smallest binary relation on the set of derivations in cut-free SN4 such that:

1. $\mathcal{D}_1 \approx \mathcal{D}_2$ if $\mathcal{D}_1 \equiv \mathcal{D}_2$.
2. $\mathcal{D}_1 \approx \mathcal{D}_2$ if either $\mathcal{D}_1 \approx \mathcal{D}$ and $\mathcal{D}_2 \equiv \dfrac{\mathcal{D}}{s}$ or $\mathcal{D}_2 \approx \mathcal{D}$ and $\mathcal{D}_1 \equiv \dfrac{\mathcal{D}}{s}$, where s is obtained from $\mathcal{D}$ by an application of an (instance of an) interaction rule.
3. $\mathcal{D}_1 \approx \mathcal{D}_2$ if $\mathcal{D}_1 \equiv \dfrac{\mathcal{D}_1^1 \dots \mathcal{D}_n^1}{s_1}$, $\mathcal{D}_2 \equiv \dfrac{\mathcal{D}_1^2 \dots \mathcal{D}_n^2}{s_2}$, and $\mathcal{D}_i^1 \approx \mathcal{D}_i^2$ $(1 \leq i \leq n \leq 2)$.

Proposition 5.2 *The relation $\approx$ is an equivalence relation.*

Proof Reflexivity of $\approx$ follows by Definition 5.1(1). Symmetry: If $\mathcal{D}_1 \approx \mathcal{D}_2$ by Definition 5.1(1), then $\mathcal{D}_2 \approx \mathcal{D}_1$ follows by Definition 5.1(1). If $\mathcal{D}_1 \approx \mathcal{D}_2$ by Definition 5.1(2), then $\mathcal{D}_2 \approx \mathcal{D}_1$ follows by Definition 5.1(2) since by the induction hypothesis, either $\mathcal{D} \approx \mathcal{D}_1$ or $\mathcal{D} \approx \mathcal{D}_2$. If $\mathcal{D}_1 \approx \mathcal{D}_2$ by Definition 5.1(3), then $\mathcal{D}_2 \approx \mathcal{D}_1$ follows by Definition 5.1(3) because by the induction hypothesis, $\mathcal{D}_i^1 \approx \mathcal{D}_i^2$ implies $\mathcal{D}_i^2 \approx \mathcal{D}_i^1$. Transitivity: There are nine cases, viz.:

[1] $\mathcal{D}_1 \overset{5.1(1)}{\approx} \mathcal{D}_2 \overset{5.1(1)}{\approx} \mathcal{D}_3$

[2] $\mathcal{D}_1 \overset{5.1(1)}{\approx} \mathcal{D}_2 \overset{5.1(2)}{\approx} \mathcal{D}_3$

[3] $\mathcal{D}_1 \overset{5.1(1)}{\approx} \mathcal{D}_2 \overset{5.1(3)}{\approx} \mathcal{D}_3$

[4] $\mathcal{D}_1 \overset{5.1(2)}{\approx} \mathcal{D}_2 \overset{5.1(1)}{\approx} \mathcal{D}_3$

[5] $\mathcal{D}_1 \overset{5.1(2)}{\approx} \mathcal{D}_2 \overset{5.1(2)}{\approx} \mathcal{D}_3$

[6] $\mathcal{D}_1 \overset{5.1(2)}{\approx} \mathcal{D}_2 \overset{5.1(3)}{\approx} \mathcal{D}_3$

[7] $\mathcal{D}_1 \overset{5.1(3)}{\approx} \mathcal{D}_2 \overset{5.1(1)}{\approx} \mathcal{D}_3$

[8] $\mathcal{D}_1 \overset{5.1(3)}{\approx} \mathcal{D}_2 \overset{5.1(2)}{\approx} \mathcal{D}_3$

[9] $\mathcal{D}_1 \overset{5.1(3)}{\approx} \mathcal{D}_2 \overset{5.1(3)}{\approx} \mathcal{D}_3$

In cases [1], [2], [3], [4], and [7], $\mathcal{D}_1 \approx \mathcal{D}_3$ holds by clause 5.1(1). In cases [5] and [9], $\mathcal{D}_1 \approx \mathcal{D}_3$ holds by the induction hypothesis and clause 5.1(2), respectively 5.1(3). Case [6]: $\mathcal{D}^2$ and $\mathcal{D}^3$ have the form

$$\frac{\mathcal{D}_1^2 \ldots \mathcal{D}_n^2}{s_2} \quad \text{respectively} \quad \frac{\mathcal{D}_1^3 \ldots \mathcal{D}_n^3}{s_3}\ .$$

Since $\mathcal{D}_1 \overset{5.1(2)}{\approx} \mathcal{D}_2$, $\mathcal{D}^1$, $\mathcal{D}^2$, $\mathcal{D}^3$ have the form

$$\frac{\dfrac{\mathcal{D}'}{s'}}{s_1} \quad \frac{\dfrac{\mathcal{D}''}{s''}}{s_2} \quad \frac{\dfrac{\mathcal{D}'''}{s'''}}{s_3}$$

respectively, and $\dfrac{\mathcal{D}'}{s'} \approx \dfrac{\mathcal{D}''}{s''} \approx \dfrac{\mathcal{D}''}{s''}$. By the induction hypothesis and clause (3), $\mathcal{D}_1 \approx \mathcal{D}_3$. Case [8] is analogous. □

Remark 5.3 We are working with a variant of the standard sequent calculus for N4 with (co+), (co-), (we+), and (we-) as primitive rules and not with a variant of a G3-style version for which not only cut but also weakening and contraction are admissible.[14] Different notions of in-identity can be obtained by excluding some structural rules from being applied in the last derivation step of $\mathcal{D}_1$ and $\mathcal{D}_2$ in the third clause of Definition 5.1. Moreover, the third clause of Definition 5.1 allows one to identify derivations such as

$$\frac{\begin{matrix}\mathcal{D}\\ \Gamma : \Delta \Rightarrow^{+} A\end{matrix}}{\Gamma : \Delta \Rightarrow^{+} (A \vee B)} \quad \text{and} \quad \frac{\begin{matrix}\mathcal{D}\\ \Gamma : \Delta \Rightarrow^{+} A\end{matrix}}{\Gamma : \Delta \Rightarrow^{+} (A \vee C)}$$

which is in accordance with the BHK interpretation allowing for identical proofs of $(A \vee B)$ and $(A \vee C)$.

Remark 5.4 Note that not any two cut-free derivations $\mathcal{D}_1$ and $\mathcal{D}_2$ of a formula A are in-identical. There are, e.g., syntactically distinct cut-free derivations of the sequent $\varnothing : \varnothing \Rightarrow^{+} (p \wedge q) \rightarrow (p \vee q)$ that are not in-identical.

Example 5.5 Let $\begin{matrix}\mathcal{D}_1\\ \Gamma : \Delta \Rightarrow^{+} {\sim}A\end{matrix}$ and $\begin{matrix}\mathcal{D}_2\\ \Gamma : \Delta \Rightarrow^{+} {\sim}B\end{matrix}$ be derivations in cut-free SN4. Then

$$\begin{matrix}\mathcal{D}_1\\ \Gamma : \Delta \Rightarrow^{+} {\sim}A\end{matrix} \approx \frac{\begin{matrix}\mathcal{D}_1\\ \Gamma : \Delta \Rightarrow^{+} {\sim}A\end{matrix}}{\Gamma : \Delta \Rightarrow^{-} A} \quad \text{and} \quad \begin{matrix}\mathcal{D}_2\\ \Gamma : \Delta \Rightarrow^{+} {\sim}B\end{matrix} \approx \frac{\begin{matrix}\mathcal{D}_2\\ \Gamma : \Delta \Rightarrow^{+} {\sim}B\end{matrix}}{\Gamma : \Delta \Rightarrow^{-} B}$$

by Definition 5.1(2). Hence, by Definition 5.1(3):

$$\frac{\begin{matrix}\mathcal{D}_1\\ \Gamma : \Delta \Rightarrow^{+} {\sim}A\end{matrix} \quad \begin{matrix}\mathcal{D}_2\\ \Gamma : \Delta \Rightarrow^{+} {\sim}B\end{matrix}}{\Gamma : \Delta \Rightarrow^{+} ({\sim}A \wedge {\sim}B)} \approx \frac{\dfrac{\begin{matrix}\mathcal{D}_1\\ \Gamma : \Delta \Rightarrow^{+} {\sim}A\end{matrix}}{\Gamma : \Delta \Rightarrow^{-} A} \quad \dfrac{\begin{matrix}\mathcal{D}_2\\ \Gamma : \Delta \Rightarrow^{+} {\sim}B\end{matrix}}{\Gamma : \Delta \Rightarrow^{-} B}}{\Gamma : \Delta \Rightarrow^{-} (A \vee B)}\ .$$

[14] A contraction-free Hudelmaier-Dyckhoff-style sequent calculus for N4 can be found in Kamide and Wansing (2012) and Kamide and Wansing (2015).

By Definition 5.1(2):

$$\dfrac{\dfrac{\overset{\mathcal{D}_1}{\Gamma:\Delta\Rightarrow^+ {\sim}A}}{\Gamma:\Delta\Rightarrow^- A} \quad \dfrac{\overset{\mathcal{D}_2}{\Gamma:\Delta\Rightarrow^+ {\sim}B}}{\Gamma:\Delta\Rightarrow^- B}}{\Gamma:\Delta\Rightarrow^- (A\vee B)} \approx \dfrac{\dfrac{\dfrac{\overset{\mathcal{D}_1}{\Gamma:\Delta\Rightarrow^+ {\sim}A}}{\Gamma:\Delta\Rightarrow^- A} \quad \dfrac{\overset{\mathcal{D}_2}{\Gamma:\Delta\Rightarrow^+ {\sim}B}}{\Gamma:\Delta\Rightarrow^- B}}{\Gamma:\Delta\Rightarrow^- (A\vee B)}}{\Gamma:\Delta\Rightarrow^+ {\sim}(A\vee B)}\,.$$

By transitivity of $\approx$:

$$\dfrac{\overset{\mathcal{D}_1}{\Gamma:\Delta\Rightarrow^+ {\sim}A} \quad \overset{\mathcal{D}_2}{\Gamma:\Delta\Rightarrow^+ {\sim}B}}{\Gamma:\Delta\Rightarrow^+ ({\sim}A\wedge{\sim}B)} \approx \dfrac{\dfrac{\dfrac{\overset{\mathcal{D}_1}{\Gamma:\Delta\Rightarrow^+ {\sim}A}}{\Gamma:\Delta\Rightarrow^- A} \quad \dfrac{\overset{\mathcal{D}_2}{\Gamma:\Delta\Rightarrow^+ {\sim}B}}{\Gamma:\Delta\Rightarrow^- B}}{\Gamma:\Delta\Rightarrow^- (A\vee B)}}{\Gamma:\Delta\Rightarrow^+ {\sim}(A\vee B)}\,.$$

Remark 5.6 Note that in-identity is not a congruence relation. If $\mathcal{D}_1 \approx \mathcal{D}_2$, then, in general, substituting $\mathcal{D}_1$ for $\mathcal{D}_2$ in a derivation tree does not result in a derivation. In the previous example, we have

$$\dfrac{\overset{\mathcal{D}_1}{\Gamma:\Delta\Rightarrow^+ {\sim}A} \quad \overset{\mathcal{D}_2}{\Gamma:\Delta\Rightarrow^+ {\sim}B}}{\Gamma:\Delta\Rightarrow^+ ({\sim}A\wedge{\sim}B)} \approx \dfrac{\dfrac{\overset{\mathcal{D}_1}{\Gamma:\Delta\Rightarrow^+ {\sim}A}}{\Gamma:\Delta\Rightarrow^- A} \quad \dfrac{\overset{\mathcal{D}_2}{\Gamma:\Delta\Rightarrow^+ {\sim}B}}{\Gamma:\Delta\Rightarrow^- B}}{\Gamma:\Delta\Rightarrow^- (A\vee B)}$$

but whereas

$$\dfrac{\dfrac{\dfrac{\overset{\mathcal{D}_1}{\Gamma:\Delta\Rightarrow^+ {\sim}A}}{\Gamma:\Delta\Rightarrow^- A} \quad \dfrac{\overset{\mathcal{D}_2}{\Gamma:\Delta\Rightarrow^+ {\sim}B}}{\Gamma:\Delta\Rightarrow^- B}}{\Gamma:\Delta\Rightarrow^- (A\vee B)}}{\Gamma:\Delta\Rightarrow^+ {\sim}(A\vee B)}$$

is a derivation in cut-free SN4,

$$\dfrac{\dfrac{\overset{\mathcal{D}_1}{\Gamma:\Delta\Rightarrow^+ {\sim}A} \quad \overset{\mathcal{D}_2}{\Gamma:\Delta\Rightarrow^+ {\sim}B}}{\Gamma:\Delta\Rightarrow^+ ({\sim}A\wedge{\sim}B)}}{\Gamma:\Delta\Rightarrow^+ {\sim}(A\vee B)}$$

is not.

Example 5.7 This example is very similar to the previous one but maybe it is nevertheless instructive. Let now $\overset{\mathcal{D}_1}{\Gamma:\Delta\Rightarrow^- A}$ and $\overset{\mathcal{D}_2}{\Gamma:\Delta\Rightarrow^- B}$ be derivations in cut-free SN4. Then

$$\overset{\mathcal{D}_1}{\Gamma:\Delta\Rightarrow^- A} \approx \dfrac{\overset{\mathcal{D}_1}{\Gamma:\Delta\Rightarrow^- A}}{\Gamma:\Delta\Rightarrow^+ {\sim}A} \text{ and } \overset{\mathcal{D}_2}{\Gamma:\Delta\Rightarrow^- B} \approx \dfrac{\overset{\mathcal{D}_2}{\Gamma:\Delta\Rightarrow^- B}}{\Gamma:\Delta\Rightarrow^+ {\sim}B}$$

by Definition 5.1(2). Also by Definition 5.1(2):

$$\dfrac{\dfrac{\begin{matrix}\mathcal{D}_1 & \mathcal{D}_2\\ \Gamma : \Delta \Rightarrow^- A & \Gamma : \Delta \Rightarrow^- B\end{matrix}}{\Gamma : \Delta \Rightarrow^- (A \vee B)}}{\Gamma : \Delta \Rightarrow^+ {\sim}(A \vee B)} \approx \dfrac{\begin{matrix}\mathcal{D}_1 & \mathcal{D}_2\\ \Gamma : \Delta \Rightarrow^- A & \Gamma : \Delta \Rightarrow^- B\end{matrix}}{\Gamma : \Delta \Rightarrow^- (A \vee B)}$$

By Definition 5.1(3):

$$\dfrac{\begin{matrix}\mathcal{D}_1 & \mathcal{D}_2\\ \Gamma : \Delta \Rightarrow^- A & \Gamma : \Delta \Rightarrow^- B\end{matrix}}{\Gamma : \Delta \Rightarrow^- (A \vee B)} \approx \dfrac{\dfrac{\begin{matrix}\mathcal{D}_1\\ \Gamma : \Delta \Rightarrow^- A\end{matrix}}{\Gamma : \Delta \Rightarrow^+ {\sim}A} \quad \dfrac{\begin{matrix}\mathcal{D}_2\\ \Gamma : \Delta \Rightarrow^- B\end{matrix}}{\Gamma : \Delta \Rightarrow^+ {\sim}B}}{\Gamma : \Delta \Rightarrow^+ ({\sim}A \wedge {\sim}B)} .$$

By transitivity of $\approx$:

$$\dfrac{\dfrac{\begin{matrix}\mathcal{D}_1 & \mathcal{D}_2\\ \Gamma : \Delta \Rightarrow^- A & \Gamma : \Delta \Rightarrow^- B\end{matrix}}{\Gamma : \Delta \Rightarrow^- (A \vee B)}}{\Gamma : \Delta \Rightarrow^+ {\sim}(A \vee B)} \approx \dfrac{\dfrac{\begin{matrix}\mathcal{D}_1\\ \Gamma : \Delta \Rightarrow^- A\end{matrix}}{\Gamma : \Delta \Rightarrow^+ {\sim}A} \quad \dfrac{\begin{matrix}\mathcal{D}_2\\ \Gamma : \Delta \Rightarrow^- B\end{matrix}}{\Gamma : \Delta \Rightarrow^+ {\sim}B}}{\Gamma : \Delta \Rightarrow^+ ({\sim}A \wedge {\sim}B)} .$$

Definition 5.8 Two formulas A and B are said to be synonymous with respect to cut-free SN4 ($A \rightleftarrows B$ in cut-free SN4) iff

1. (positive condition) there exists a derivation $\mathcal{D}$ of $\varnothing : A \Rightarrow^+ B$ and a derivation $\mathcal{D}'$ of $\varnothing : B \Rightarrow^+ A$ with $\mathcal{D} \approx \mathcal{D}'$,
2. (negative condition) there exists a derivation $\mathcal{D}$ of $A : \varnothing \Rightarrow^- B$ and a derivation $\mathcal{D}'$ of $B : \varnothing \Rightarrow^- A$ with $\mathcal{D} \approx \mathcal{D}'$.

Proposition 5.9 *The following pairs of formulas are synonymous with respect to cut-free* SN4*:*

1. *p and ${\sim}{\sim}p$,*
2. *$(p \wedge q)$ and ${\sim}({\sim}p \vee {\sim}q)$,*
3. *$(p \vee q)$ and ${\sim}({\sim}p \wedge {\sim}q)$.*

Proof 1: We have

$$\dfrac{\dfrac{\varnothing : p \Rightarrow^+ p}{\varnothing : p \Rightarrow^- {\sim}p}}{\varnothing : p \Rightarrow^+ {\sim}{\sim}p} \quad \dfrac{\dfrac{\varnothing : p \Rightarrow^+ p}{{\sim}p : \varnothing \Rightarrow^+ p}}{\varnothing : {\sim}{\sim}p \Rightarrow^+ p} \quad \dfrac{\dfrac{p : \varnothing :\Rightarrow^- p}{p : \varnothing \Rightarrow^+ {\sim}p}}{p : \varnothing \Rightarrow^- {\sim}{\sim}p} \quad \dfrac{\dfrac{p : \varnothing :\Rightarrow^- p}{\varnothing : {\sim}p \Rightarrow^- p}}{{\sim}{\sim}p : \varnothing \Rightarrow^- p}$$

2., positive condition: By clauses (1) and (2) of Definition 5.1, the derivations $\varnothing : p \Rightarrow^+ p$ and $\dfrac{\varnothing : p \Rightarrow^+ p}{{\sim}p : \varnothing \Rightarrow^+ p}$ are in-identical, and so are $\varnothing : q \Rightarrow^+ q$ and $\dfrac{\varnothing : q \Rightarrow^+ q}{{\sim}q : \varnothing \Rightarrow^+ q}$. Let $\mathcal{D}_1$ and $\mathcal{D}_2$ be the derivations

$$\dfrac{\dfrac{\varnothing : p \Rightarrow^+ p}{\varnothing : (p \wedge q) \Rightarrow^+ p}}{\varnothing : (p \wedge q) \Rightarrow^- {\sim}p} \qquad \text{respectively} \qquad \dfrac{\dfrac{\varnothing : q \Rightarrow^+ q}{\varnothing : (p \wedge q) \Rightarrow^+ q}}{\varnothing : (p \wedge q) \Rightarrow^- {\sim}q}$$

and let $\mathcal{D}_3$ and $\mathcal{D}_4$ be the derivations

$$\dfrac{\dfrac{\dfrac{\varnothing : p \Rightarrow^+ p}{\sim p : \varnothing \Rightarrow^+ p}}{(\sim p \vee \sim q) : \varnothing \Rightarrow^+ p}}{\varnothing : \sim(\sim p \vee \sim q) \Rightarrow^+ p} \quad \text{respectively} \quad \dfrac{\dfrac{\dfrac{\varnothing : q \Rightarrow^+ q}{\sim q : \varnothing \Rightarrow^+ q}}{(\sim p \vee \sim q) : \varnothing \Rightarrow^+ q}}{\varnothing : \sim(\sim p \vee \sim q) \Rightarrow^+ q} .$$

By clauses (3) and (2) of Definition 5.1, $\mathcal{D}_1 \approx \mathcal{D}_3$ and $\mathcal{D}_2 \approx \mathcal{D}_4$. Let $\mathcal{D}_5$ and $\mathcal{D}_6$ be the derivations

$$\dfrac{\mathcal{D}_1 \qquad \mathcal{D}_2}{\varnothing : (p \wedge q) \Rightarrow^- (\sim p \vee \sim q)} \quad \text{respectively} \quad \dfrac{\mathcal{D}_3 \qquad \mathcal{D}_4}{\varnothing : \sim(\sim p \vee \sim q) \Rightarrow^+ (p \wedge q)} .$$

By clause (3) of Definition 5.1, $\mathcal{D}_5 \approx \mathcal{D}_6$, and by clause (2) of Definition 5.1, $\mathcal{D}_7 \approx \mathcal{D}_6$, where $\mathcal{D}_7 \equiv \dfrac{\mathcal{D}_5}{\varnothing : (p \wedge q) \Rightarrow^+ \sim(\sim p \vee \sim q)}$.

2., negative condition: By clauses (1) and (2) of Definition 5.1,

$$\dfrac{p : \varnothing \Rightarrow^- p}{p : \varnothing \Rightarrow^+ \sim p} \approx \dfrac{p : \varnothing \Rightarrow^- p}{\varnothing : \sim p \Rightarrow^- p} \text{ and } \dfrac{q : \varnothing \Rightarrow^- q}{q : \varnothing \Rightarrow^+ \sim q} \approx \dfrac{q : \varnothing \Rightarrow^- q}{\varnothing : \sim q \Rightarrow^- q} .$$

Call these derivations $\mathcal{D}_1, \mathcal{D}_2, \mathcal{D}_3$, and $\mathcal{D}_4$, respectively. By clause (3) of Definition 5.1, therefore

$$\dfrac{\mathcal{D}_1}{p : \varnothing \Rightarrow^+ (\sim p \vee \sim q)} \approx \dfrac{\mathcal{D}_2}{\varnothing : \sim p \Rightarrow^- (p \wedge q)} \quad \text{and}$$

$$\dfrac{\mathcal{D}_3}{q : \varnothing \Rightarrow^+ (\sim p \vee \sim q)} \approx \dfrac{\mathcal{D}_4}{\varnothing : \sim q \Rightarrow^- (p \wedge q)} .$$

Let us call the latter derivations $\mathcal{D}_{11}$, $\mathcal{D}_{21}$, $\mathcal{D}_{31}$, and $\mathcal{D}_{41}$, respectively. Let

$$\mathcal{D}_6 \equiv \dfrac{\mathcal{D}_{11} \qquad \mathcal{D}_{31}}{(p \wedge q) : \varnothing \Rightarrow^+ (\sim p \vee \sim q)} \text{ and } \mathcal{D}_7 \equiv \dfrac{\mathcal{D}_{21} \qquad \mathcal{D}_{41}}{\varnothing : (\sim p \vee \sim q) \Rightarrow^- (p \wedge q)} .$$

By clause (3) of Definition 5.1, $\mathcal{D}_6 \approx \mathcal{D}_7$, and by clause (2) of Definition 5.1,

$$\dfrac{\mathcal{D}_6}{(p \wedge q) : \varnothing \Rightarrow^- \sim(\sim p \vee \sim q)} \approx \dfrac{\mathcal{D}_7}{\sim(\sim p \vee \sim q) : \varnothing \Rightarrow^- (p \wedge q)} .$$

3: This case is similar to the previous case. □

Definition 5.10 Two formulas A and B are strongly equivalent in cut-free SN4 iff $\vdash \varnothing; \varnothing \Rightarrow^+ A \Leftrightarrow B$.

It can easily be seen that two formulas A and B are strongly equivalent in SN4, and hence also in cut-free SN4, just in case $\vdash \varnothing; A \Rightarrow^+ B$, $\vdash \varnothing; B \Rightarrow^+ A$, $\vdash A : \varnothing \Rightarrow^- B$, and $\vdash B : \varnothing \Rightarrow^- A$.

Proposition 5.11 *If $A \rightleftarrows B$ in cut-free* SN4, *then A and B are strongly equivalent in cut-free* SN4.

Proof Assume that A and B are synonymous. By Proposition 4.5, $\vdash A : \varnothing \Rightarrow^- A$, and by synonymy of A and B, $\vdash A : \varnothing \Rightarrow^- B$. Analogously, we obtain $\vdash \varnothing; A \Rightarrow^+ B$, $\vdash \varnothing; B \Rightarrow^+ A$, and $\vdash B : \varnothing \Rightarrow^- A$. □

Corollary 5.12 *The following pairs are pairs of formulas that are not synonymous with respect to cut-free* SN4 *because they are not strongly equivalent in cut-free* SN4*:*

1. $\sim(p \to q)$ *and* $(p \wedge \sim q)$,
2. $((p \to p) \to (q \to q))$ *and* $((q \to q) \to (p \to p))$.

By Proposition 5.11, synonymy implies strong equivalence, and one may wonder whether two formulas are synonymous in cut-free SN4 if and only if they are strongly equivalent. Strong equivalence in SN4 does, however, not imply synonymy in cut-free SN4.

Proposition 5.13 *The converse of Proposition 5.11 does not hold.*

Proof The following pairs, for instance, are pairs of strongly equivalent formulas that are not in-identical:

$$p \text{ and } (p \vee p), \quad p \text{ and } (p \wedge p), \quad \sim p \text{ and } \sim(p \wedge p), \quad \sim p \text{ and } \sim(p \vee p).$$

If $\mathcal{D}$ is a derivation in cut-free SN4 with no branching and $\mathcal{D}'$ is a syntactically distinct derivation in cut-free SN4 with a branching, then $\mathcal{D}$ and $\mathcal{D}'$ cannot be in-identical. Every cut-free proof of $\varnothing : (p \vee p) \Rightarrow^{+} p$ is branching, whereas no cut-free proof of $\varnothing : p \Rightarrow^{+} (p \vee p)$ is branching. Something analogous holds for the other pairs. □

6 Conclusion

We have introduced a notion of equivalence between derivations, inherited identity, in a cut-free sequent system for Almukdad and Nelson's constructive paraconsistent logic N4 with strong negation. This notion was obtained by considering sequent rules the application of which has no effect on the identity of derivations. Then the notion of in-identity has been used to define a bilateralist notion of synonymy between formulas, which is a relation drawing more fine-grained distinctions between formulas than strong equivalence.

Note that this approach can also be applied to the connexive variant C of N4. A sequent calculus for C can be obtained from GN4 by replacing the rules for negated implications by the following ones that allow one to prove the sequents $\varnothing \Rightarrow \sim(A \to B) \to (A \to \sim B)$ and $\varnothing \Rightarrow (A \to \sim B) \to \sim(A \to B)$; cf. Wansing (2005), Omori and Wansing (2020):

$$\frac{\Gamma \Rightarrow A \quad \sim B, \Delta \Rightarrow C}{\sim(A \to B), \Gamma, \Delta \Rightarrow C}\ (\sim \to \mathrm{l})' \qquad \frac{A, \Gamma \Rightarrow \sim B}{\Gamma \Rightarrow \sim(A \to B)}\ (\sim \to \mathrm{r})'.$$

I conclude this note by giving a rather speculative glimpse into some further directions for studying the inherited identity of derivations that emerges from a distinction between proofs and refutations and the notion of synonymy based on in-identity.

1. It is conspicuous that there is a certain gap in the system of rules of SN4. There is no connective that stands to disjunction with respect to disproofs as conjunction stands to implication with respect to proofs. Closing this gap calls for adding to the language of N4 a binary connective that is in a certain sense dual to implication, namely the co-implication, $\prec$, from the bi-intuitionistic logic 2Int introduced in Wansing (2016a; 2016b; 2017).[15] Concretely, we can add the following rules for $\prec$ to GN4:

$$\frac{\Gamma \Rightarrow {\sim}A \quad {\sim}B, \Delta \Rightarrow C}{{\sim}(B \prec A), \Gamma, \Delta \Rightarrow C} \qquad \frac{{\sim}A, \Gamma \Rightarrow {\sim}B}{\Gamma \Rightarrow {\sim}(B \prec A)}$$

$$\frac{{\sim}A, B, \Gamma \Rightarrow C}{(B \prec A), \Gamma \Rightarrow C} \qquad \frac{\Gamma \Rightarrow {\sim}A \quad \Gamma \Rightarrow B}{\Gamma \Rightarrow (B \prec A)}$$

and the following rules for $\prec$ to SN4:

$$\frac{\Gamma_1 : \Delta_1 \Rightarrow^- A \quad B, \Gamma_2 : \Delta_2 \Rightarrow^* C}{\Gamma_1, \Gamma_2, (B \prec A) : \Delta_1, \Delta_2 \Rightarrow^* C}\ (\prec\text{l}-) \qquad \frac{A, \Gamma : \Delta \Rightarrow^- B}{\Gamma : \Delta \Rightarrow^- (B \prec A)}\ (\prec\text{r}-)$$

$$\frac{A, \Gamma : B, \Delta \Rightarrow^* C}{\Gamma : (B \prec A), \Delta \Rightarrow^* C}\ (\prec\text{l}+) \qquad \frac{\Gamma_1 : \Delta_1 \Rightarrow^- A \quad \Gamma_2 : \Delta_2 \Rightarrow^+ B}{\Gamma_1, \Gamma_2 : \Delta_1, \Delta_2 \Rightarrow^+ (B \prec A)}\ (\prec\text{r}+).$$

Consider the addition of $\prec$ to the language of N4. Let us refer to the addition of the former set of sequent rules for $\prec$ to GN4 as $\mathsf{GN4}^{\prec}$, and let us refer to the result of adding the latter sequent rules for $\prec$ and the following rules for the constants $\top$ and $\bot$:

$$\overline{\top, \Gamma : \Delta \Rightarrow^* A} \quad \overline{\Gamma : \bot, \Delta \Rightarrow^* A} \quad \overline{\Gamma : \Delta \Rightarrow^+ \top} \quad \overline{\Gamma : \Delta \Rightarrow^- \bot}$$

to SN4 as $\mathsf{SN4}^{\prec,\top,\bot}$. Moreover, let us refer to the result of deleting the rules displaying strong negation, i.e., the interaction rules of SN4, from $\mathsf{SN4}^{\prec,\top,\bot}$ as S2Int.

The language of S2Int is the language of the bi-intuitionistic logic 2Int from Wansing (2016a; 2016b; 2017).[16] With $\top$ and $\bot$ as primitive, one can define two negations, the intuitionistic negation, $\neg$, and its counterpart with respect to co-implication, namely co-negation, $-$, as follows:

$$\neg A := (A \to \bot) \qquad -A := (\top \prec A).$$

One natural question is whether in the absence of $\sim$, there are interaction rules that allow for an identification of proofs and disproofs of distinct formulas from the language of 2Int.

2. If we think of derivation trees from the perspective of model-theoretic semantics, it is natural to wonder about their Fregean denotation and sense. What do they

[15] For the notion of co-implication in Heyting-Brouwer logic see Goré (2000), Goré and Shillito (2020), Rauszer (1980), and Schroeder-Heister (2011) and references given there.

[16] Note that in the notation $(\Gamma; \Delta \vdash A)$ and $(\Gamma; \Delta \vdash^d A)$ from Wansing (2016a; 2016b; 2017), Γ stands for formulas assumed to be true, and Δ stands for formulas assumed to be false.

refer to and what do they express (what is their "mode of presentation" (Art des Gegebenseins))? Whatever the denotation and the sense of a derivation is, it seems to be clear that identical derivations ought to have the same sense and hence also the same denotation. There is not much work on the sense of derivations; I am aware of Ayhan (2021) and Tranchini (2016).

In order to make sure that two derivation trees are co-referential if they are encoded by co-referential, uniquely typed λ-terms, a two-sorted typed λ-calculus is needed. For a connexive variant of the bi-intuitionistic logic 2Int, a two-sorted typed λ-calculus has been presented in Wansing (2016b), see also Wansing (2010).

We are then working with two sorts of typed variables, proof variables $x^+, y^+, z^+, x_1^+, x_2^+, x_3^+, \ldots$, and disproof variables $x^-, y^-, z^-, x_1^-, x_2^-, x_3^-, \ldots$, together with the usual term-forming operations and new type-forming operations, $^+$ and $^-$.

The interaction rules of SN4 give rise to the following term assignment rules:

$$\frac{\Gamma : \Delta \Rightarrow^- M^- :: A}{\Gamma : \Delta \Rightarrow^+ M^{-+} :: {\sim}A} \qquad \frac{\Gamma : \Delta \Rightarrow^+ M^+ :: A}{\Gamma : \Delta \Rightarrow^- M^{+-} :: {\sim}A}$$

$$\frac{x^- :: A, \Gamma : \Delta \Rightarrow^* M}{\Gamma : x^{-+} :: {\sim}A, \Delta \Rightarrow^* M} \qquad \frac{\Gamma : x^+ :: A, \Delta \Rightarrow^* M}{x^{+-} :: {\sim}A, \Gamma : \Delta \Rightarrow^* M}$$

$$\frac{\Gamma : \Delta \Rightarrow^+ M^{-+} :: {\sim}A}{\Gamma : \Delta \Rightarrow^- M^- :: A} \qquad \frac{\Gamma : \Delta \Rightarrow^- M^{+-} :: {\sim}A}{\Gamma : \Delta \Rightarrow^+ M^+ :: A}$$

$$\frac{\Gamma : x^{-+} :: {\sim}A, \Delta \Rightarrow^* M}{x^- :: A, \Gamma : \Delta \Rightarrow^* M} \qquad \frac{x^{+-} :: {\sim}A, \Gamma : \Delta \Rightarrow^* M}{\Gamma : x^+ :: A, \Delta \Rightarrow^* M}$$

Example 6.1 (Example 5.7 continued) We noted that

$$\dfrac{\dfrac{\begin{matrix}\mathcal{D}_1\\ \Gamma : \Delta \Rightarrow^- A\end{matrix} \quad \begin{matrix}\mathcal{D}_2\\ \Gamma : \Delta \Rightarrow^- B\end{matrix}}{\Gamma : \Delta \Rightarrow^- (A \vee B)}}{\Gamma : \Delta \Rightarrow^+ {\sim}(A \vee B)} \approx \dfrac{\dfrac{\begin{matrix}\mathcal{D}_1\\ \Gamma : \Delta \Rightarrow^- A\end{matrix}}{\Gamma : \Delta \Rightarrow^+ {\sim}A} \quad \dfrac{\begin{matrix}\mathcal{D}_2\\ \Gamma : \Delta \Rightarrow^- B\end{matrix}}{\Gamma : \Delta \Rightarrow^+ {\sim}B}}{\Gamma : \Delta \Rightarrow^+ ({\sim}A \wedge {\sim}B)} .$$

If we encode the two in-identical derivations by typed λ-terms, we obtain:

$$\dfrac{\dfrac{\begin{matrix}\mathcal{D}_1\\ \Gamma : \Delta \Rightarrow^- M^- :: A\end{matrix} \quad \begin{matrix}\mathcal{D}_2\\ \Gamma : \Delta \Rightarrow^- N^- :: B\end{matrix}}{\Gamma : \Delta \Rightarrow^- \langle M^-, N^- \rangle^- :: (A \vee B)}}{\Gamma : \Delta \Rightarrow^+ \langle M^-, N^- \rangle^{-+} :: {\sim}(A \vee B)} \quad \dfrac{\dfrac{\begin{matrix}\mathcal{D}_1\\ \Gamma : \Delta \Rightarrow^- M^- :: A\end{matrix}}{\Gamma : \Delta \Rightarrow^+ M^{-+} :: {\sim}A} \quad \dfrac{\begin{matrix}\mathcal{D}_2\\ \Gamma : \Delta \Rightarrow^- N^- :: B\end{matrix}}{\Gamma : \Delta \Rightarrow^+ N^{-+} :: {\sim}B}}{\Gamma : \Delta \Rightarrow^+ \langle M^{-+}, N^{-+} \rangle^+ :: ({\sim}A \wedge {\sim}B)} .$$

We thus need a semantics for the two-sorted typed λ-calculus such that the equation

$$\langle M^- :: A, N^- :: B \rangle^{-+} :: {\sim}(A \vee B) = \langle M^{-+} :: {\sim}A, N^{-+} :: {\sim}B \rangle^+ :: ({\sim}A \wedge {\sim}B)$$

is valid.

I conjecture that in order to obtain a sound and complete semantics for such a two-sorted typed λ-calculus, for every formula (type) A, one will have to use of a pair of two separate domains D^{+A} and D^{-A}.

Acknowledgements A preliminary and substantially different work-in-progress version of this paper was presented at the *Third Tübingen Conference on Proof-Theoretic Semantics*, 27–30 March 2019, and the *Eleventh Smirnov Readings in Logic* at Lomonosov Moscow State University, 19–21 June 2019. I would like to thank Peter Schroeder-Heister and Thomas Piecha for inviting me to PTS III, Greg Restall and Lutz Straßburger for discussions at that conference, and Vladimir Markin for inviting me to the Smirnov Readings 2019. A version of the paper much closer to the above version was presented at *TULIPS – The Utrecht Logic in Progress Series*, October 6, 2020 and at the *Logic and Metaphysics Workshop*, City University of New York, November 2, 2020. I would like to thank Colin Caret and Rosalie Iemhoff for inviting me to the *TULIPS* workshop, Graham Priest for inviting me to the *Logic and Metaphysics Workshop*, and Sergei Artemov, Jan Broersen, Melvin Fitting, Rosalie Iemhoff, Valeria de Paiva, Graham Priest, and David Ripley for their questions and remarks. Moreover, I am grateful to Sara Ayhan, Tiago Rezende de Castro Alves, Göran Sundholm, and an anonymous referee for their helpful comments.

References

Almukdad, A. and D. Nelson (1984). Constructible falsity and inexact predicates. *Journal of Symbolic Logic* 49, 231–233.

Ayhan, S. (2021). What is the meaning of proofs? A Fregean distinction in proof-theoretic semantics. *Journal of Philosophical Logic* 50. Published online October 22, 571–59.

Carnap, R. (1947). *Meaning and Necessity. A Study in Semantics and Modal Logic*. Chicago: University of Chicago Press.

Cosmo, R. di (1995). *Isomorphisms of Types: from λ-Calculus to Information Retrieval and Language Design*. Basel: Birkhäuser.

– (2005). A short survey of isomorphisms of types. *Mathematical Structures in Computer Science* 15.

Cozzo, C. (1994). *Meaning and Argument: A Theory of Meaning Centred on Immediate Argumental Role*. Stockholm: Almqvist & Wiksel.

Danos, V., J.-B. Joinet, and H. Schellinx (2003). Computational isomorphisms in classical logic. *Theoretical Computer Science* 294, 353–378.

de Campos Sanz, W. and T. Piecha (2014). A critical remark on the BHK interpretation of implication. *Philosophia Scientiae* 18, 13–22.

Dean, W. and H. Kurokawa (2016). Kreisel's theory of constructions, the Kreisel-Goodman paradox, and the second clause. In: *Advances in Proof-Theoretic Semantics*. Ed. by T. Piecha and P. Schroeder-Heister. Trends in Logic Vol. 43, Cham: Springer, 27–63.

Došen, K. (1997). Logical consequence: a turn in style. In: *Logic and Scientific Methods. Volume One of the Tenth International Congress of Logic, Methodology and Philosophy of Science*. Ed. by M. L. D. Chiara, K. Doets, D. Mundici, and J. van Benthem. Florence, August 1995, Synthese Library Vol. 259. Dordrecht: Kluwer, 289–311.

– (2006). Models of deduction. *Synthese* 148, 639–657.

Došen, K. and Z. Petrić (2012). Isomorphic formulae in classical propositional logic. *Mathematical Logic Quarterly* 58, 5–17.

Drobyshevich, S. (2021). Tarskian consequence relations bilaterally: some familiar notions. *Synthese* 198. Published online May 30, 2019, 5213–5240.

Francez, N. (2014). Bilateralism in proof-theoretic semantics. *Journal of Philosophical Logic* 43, 239–259.

– (2015). *Proof-theoretic Semantics*. London: College Publications.

Girard, J.-Y. (1989). *Proofs and Types*. Translated and with appendices by P. Taylor and Y. Lafont. Cambridge: Cambridge University Press.

Goranko, V., G. Pulcini, and T. Skura (2020). Refutation systems: an overview and some applications to philosophical logics. In: *Knowledge, Proof and Dynamics*. Ed. by F. Liu, H. Ono, and J. Yu. Springer: Cham, 173–197.

Goré, R. (2000). Dual intuitionistic logic revisited. In: *Automated Reasoning with Analytic Tableaux and Related Methods*. Ed. by R. Dyckhoff. Springer Lecture Notes in Artificial Intelligence, Vol. 1847. Berlin: Springer, 252–267.

Goré, R. and I. Shillito (2020). Bi-intuitionistic logics: a new instance of an old problem. In: *Advances in Modal Logic. Volume 13*. Ed. by N. Olivetti, R. Verbrugge, S. Negri, and G. Sandu. London: College Publications, 296–288.

Harper, R., F. Honsell, and G. Plotkin (1993). A framework for defining logics. *Journal of the ACM* 40, 143–184.

Kamide, N. and H. Wansing (2012). Proof theory of Nelson's paraconsistent logic: A uniform perspective. *Theoretical Computer Science* 415, 1–38.

– (2015). *Proof Theory of N4-related Paraconsistent Logics*. London: College Publications.

Kreisel, G. (1962). Foundations of intuitionistic logic. In: *Logic Methodology and Philosophy of Science, Proceedings of the 1960 International Congress*. Ed. by E. Nagel, P. Suppes, and A. T. A. Tarski. Stanford: Stanford UP, 198–210.

Kürbis, N. (2019). *Proof and Falsity: A Logical Investigation*. Cambridge: Cambridge UP.

López-Escobar, E. G. K. (1972). Refutability and elementary number theory. *Indagationes. Mathematicae* 43, 362–374.

Moschovakis, Y. (2006). A logical calculus of meaning and synonymy. *Linguistics and Philosophy* 29, 27–89.

Nelson, D. (1949). Constructible falsity. *Journal of Symbolic Logic* 14, 16–26.

Odintsov, S. P. (2008). *Constructive Negations and Paraconsistency*. Dordrecht: Springer.

Omori, H. and H. Wansing (2020). An extension of connexive logic **C**. In: *Advances in Modal Logic*. Ed. by N. Olivetti, R. Verbrugge, S. Negri, and G. Sandu. Vol. 13. London: College Publications, 503–522.

Pearce, D. (1992). Reasoning with negative information, ii: hard negation, strong negation and logic programs. In: *Nonclassical Logics and Information Processing*. Ed. by D. Pearce and H. Wansing. Lecture Notes in AI 619. Berlin: Springer, Berlin, 63–79.

Poggiolesi, F. (2020). Grounding rules and (hyper-)isomorphic formulas. *Australasian Journal of Logic* 1, 70–80.

Prawitz, D. (1971). Ideas and results in proof theory. In: *Proceedings of the Second Scandinavian Logic Symposium*. Ed. by J. E. Fenstad. Amsterdam: North-Holland, 235–307.

Rauszer, C. (1980). An algebraic and Kripke-style approach to a certain extension of intuitionistic logic. *Dissertationes Mathematicae* 167.

Restall, G. (2019). *Isomorphisms in a category of proofs*. Tech. rep. Working Paper, 1–20.

Rezende de Castro Alves, T. (2019). Synonymy and Identity of Proofs: A Philosophical Essay. PhD thesis. University of Tübingen.

Rumfitt, I. (2000). "Yes" and "no". *Mind* 109, 781–823.

Schroeder-Heister, P. (2011). Schluß und Umkehrschluß. Ein Beitrag zur Definitionstheorie. In: *Lebenswelt und Wissenschaft. Akten des XXI. Deutschen Kongresses für Philosophie 2009*. Ed. by C. F. Gethmann. Vol. 3. Deutsches Jahrbuch Philosophie, 1065–1092.

– (2018). Proof-theoretic semantics. In: *The Stanford Encyclopedia of Philosophy*. Spring 2018 Edition. E. N. Zalta. URL: https://plato.stanford.edu/archives/spr2018/entries/proof-theoretic-semantics/.

Sørensen, M. and P. Urzyczyn (2006). *Lectures on the Curry-Howard Isomorphism*. Amsterdam: Elsevier.

Sundholm, G. (1999). Identity: propositional, criterial, absolute. In: *The Logca Yearbook 1998*. Ed. by T. Childers. Prague: Filosofia Publishers, 20–26.

Tranchini, L. (2016). Proof-theoretic semantics, paradoxes and the distinction between sense and denotation. *Journal of Logic and Computation*. Published online June 02, 2014, 495–512.

– (2021). Proof-theoretic harmony: towards an intensional account. *Synthese* 198. Published online September 21, 2016, 1145–1176.

Troelstra, A. (1999). From constructivism to computer science. *Theoretical Computer Science* 211, 233–252.

Troelstra, A. and D. van Dalen (1988). *Constructivism in Mathematics*. Vol. 1. Amsterdam: North-Holland.

Troelstra, A. and H. Schwichtenberg (2000). *Basic Proof Theory*. 2nd ed. Cambridge: Cambridge UP.

von Kutschera, F. (1969). Ein verallgemeinerter Widerlegungsbegriff für Gentzenkalküle. *Archiv für mathematische Logik* 12, 104–118.

Wansing, H. (1993). *The Logic of Information Structures*. Vol. 681. Springer Lecture Notes in AI. Berlin: Springer.

– (2005). Connexive modal logic. In: *Advances in Modal Logic*. Ed. by R. Schmidt, I. Pratt-Hartmann, M. Reynolds, and H. Wansing. Vol. 5. London: King's College Publications, 367–383. URL: http://www.aiml.net/volumes/volume5/.

– (2010). Proofs, disproofs, and their duals. In: *Advances in Modal Logic*. Ed. by L. Beklemishev, V. Goranko, and V. Shehtman. Vol. 8. London: College Publications, 483–505.

– (2016a). Falsification, natural deduction, and bi-intuitionistic logic. *Journal of Logic and Computation* 26. Published online July 17, 2013, 425–450.

Wansing, H. (2016b). On split negation, strong negation, information, falsification, and verification. In: *J. Michael Dunn on Information Based Logics*. Ed. by K. Bimbó. Cham: Springer, 161–189.

– (2017). A more general general proof theory. *Journal of Applied Logic* 25, 23–46.

Note added in proof. A sequel to the present paper has been accepted for publication in the *Bulletin of the Section of Logic*, 2023, in a special issue on Bilateralism and Proof-Theoretic Semantics, namely Sara Ayhan and Heinrich Wansing, On synonymy in proof-theoretic semantics. The case of 2Int. This paper contains a proof of cut-elimination for a bilateral sequent calculus for 2Int. Note that there in antecedent position of sequents the assumptions taken to be true come first and the premises taken to be false second; cf. also footnote 16 above.

A type-theory for 2Int with Curry-style typing is presented in chapter 6 of Sara Ayhan, *Meaning and Identity of Proofs in (Bilateralist) Proof-theoretic Semantics*, 2023, submitted as a PhD thesis at Ruhr University Bochum. Moreover, in the paper by Sergei Odintsov, Sergey Drobyshevich and Heinrich Wansing, Moisil's modal logic and related systems, in: K. Bimbó (ed.), *Relevance Logics and other Tools for Reasoning. Essays in Honour of Michael Dunn*, College Publications, London, 2022, 150–177, it is observed that the bi-intuitionistic logic also known as Heyting-Brouwer logic, referred to in footnote 15 above, was already studied by Grigore Moisil in a publication from 1942, *Logique modale*.

Paradoxes, Intuitionism, and Proof-Theoretic Semantics

Reinhard Kahle and Paulo Guilherme Santos

> Wenn wir einen mathematischen Beweis erst am Resultate auf seine Zulässigkeit prüfen können, so brauchen wir überhaupt keinen Beweis.
>
> David Hilbert[1]

Abstract In this note, we review paradoxes like Russell's, the Liar, and Curry's in the context of intuitionistic logic. One may observe that one cannot blame the underlying logic for the paradoxes, but has to take into account the particular concept formations. For proof-theoretic semantics, however, this comes with the challenge to block some forms of direct axiomatizations of the Liar. A proper answer to this challenge might be given by Schroeder-Heister's *definitional freedom*.

Key words: paradoxes, liar, intuitionism, proof-theoretic semantics, definitional freedom

1 Weyl on the Grelling–Nelson paradox

Kurt Grelling presented in 1908, in a paper published jointly with Leonard Nelson (Grelling and Nelsen, 1908), the now well-known paradox concerning whether or not the adjective "heterologic" is heterologic.

Reinhard Kahle
Carl Friedrich von Weizsäcker Center, University of Tübingen, Germany, and CMA, FCT, Universidade Nova de Lisboa, Portugal, e-mail: reinhard.kahle@uni-tuebingen.de

Paulo Guilherme Santos
Centro de Matemática e Aplicações, NOVA School of Science and Technology, Universidade Nova de Lisboa, Portugal, e-mail: pgd.santos@campus.fct.unl.pt

[1] Hilbert (1917, p. 135). English translation: "If we can verify the admissibility of a mathematical proof only at the result, we do not need any proof at all."

T. Piecha and K. F. Wehmeier (eds.), *Peter Schroeder-Heister on Proof-Theoretic Semantics*, Outstanding Contributions to Logic 29,
https://doi.org/10.1007/978-3-031-50981-0_12

Hermann Weyl discusses this paradox in some length in the opening section of his "predicative manifesto", *Das Kontinuum* (Weyl, 1918):[2]

> But anyone who forgets that a proposition with such a structure can be meaningless is in danger of becoming trapped in absurdity — as a famous "paradox," essentially due to Russell, shows. Let a word which signifies a property be called *autological* if this word itself possesses the property which it signifies; if it does not possess that property, let it be called *heterological*. For example, the German word "kurz" (meaning "short") is itself *kurz* (i.e., is itself short — for a word in the German language which consists of only four letters will without question have to be described as a short one); hence "kurz" is autological. The word "long," on the other hand, is not itself long and, so, is heterological. Now what about the word "heterological" itself? If it is autological, then it has the property which it expresses and, so, is heterological. If, on the other hand, it is heterological, then it does not have this property and, so, is autological. Formalism regards this as an insoluble contradiction; but in reality this is a matter of scholasticism of the worst sort: for the slightest consideration shows that absolutely no sense can be attached to the question of whether the word "heterological" is itself auto- or heterological.

Weyl's warning, that propositions can be meaningless, can be taken as an indication that one would have to renounce the *tertium non datur* for such propositions. In other words, a logic which admits the formulation of such propositions cannot be classical; one will have to allow "truth values" — if they still can be called this way — beyond or between "true" and "false".

Historically, the natural choice for such a logic appears to be the intuitionistic one, which is distinguished for leaving out the *tertium non datur* from classical logic.[3] In the following we see, however, that intuitionistic logic is not much help regarding the paradoxes.

2 Russell's paradox in an intuitionistic setting

Russell's paradox had a profound impact on the development of modern logic. On the one hand, it forced set theory to reconsider its formal base, resulting eventually in Zermelo's axiomatization which can be taken as the standard set-theoretical foundation today. On the other hand, it questioned the mathematical concept formation, including the very notion of mathematical proof, prompting Hilbert to conceive proof theory as a tool to investigate the foundations of mathematics.

Interestingly, a simple inspection of Russell's paradox shows that it does not depend on the *tertium non datur*; in other words, the proof that the allowance to define the "Russell set" $\{X \mid X \notin X\}$ leads to a contradiction is carried out by logical reasoning which is intuitionistically valid.[4] As far as we know, it was first put on

[2] The English citation is from Weyl (1987, p. 6f).

[3] Weyl joined Brouwer's intuitionism only shortly after the publication of *Das Kontinuum*. But his slight shift from predicativism to intuitionism does not affect the criticism stated in the paragraph above; rather to the contrary.

[4] See, for instance, Irvine and Deutsch (2016, §4). The first author learned this observation from Robert Lubarsky.

record by Neil Tennant (1982), see also footnote 6. He refers to Prawitz, who gives a proof which works in minimal logic (Prawitz, 1965, p. 95). At this point, however, Prawitz did not comment on the (weak) logical framework he is using.[5]

In consequence, this paradox (and, as we will see, others) cannot be resolved by just replacing the underlying classical logic by intuitionistic logic.[6]

3 The Liar and Curry's paradox

The Liar paradox — "This sentence is false." — is surely the oldest paradox in scientific history, and it appears to be obvious that it just contradicts the law of bivalance. In this case, the standard argument uses indeed classical logic, arguing by a case distinction on assuming that it is true or that it is false, both leading to contradictions.

Curry (1942) presented a paradox, now named after him, to simplify the *Kleene–Rosser paradox* which showed the inconsistency of a "λ-calculus logic" of Church. Curry points out that Kleene and Rosser had used the *Richard paradox*, while his argument is based on Russell or the Liar. In essence, Curry's paradox is based in the definability of sentences saying "This sentence implies φ" for any sentence φ. Only requiring some very simple rules for implication one can obtain, from the defined sentence, φ. As, in this way, every formula of the system is derivable, the system is inconsistent.

Apparently, this paradox does not even involve negation — but, we will argue in a minute that this is only apparent —, and therefore, it was taken to question our very intuition about implication. In particular, the reasoning used by Curry is clearly intuitionistically valid.

One can replace φ in Curry's sentence by the *falsum*, $\bot$, a propositional constant for a false sentence. The single instances follow then with the intuitionistically valid principle *ex falso quodlibet*. But taking into account that intuitionistic[7] negation $\neg\varphi$

[5] A more recent discussion of Prawitz and Tennant's treatment of Russell's paradox can be found in Schroeder-Heister and Tranchini (2017).

[6] See Tennant (1982, p. 268f.): "I shall always try to present my proof-theoretic considerations within *intuitionistic* logic. This should at least allay the suspicion that bivalence or excluded middle or some jaundiced relative is the source of contagion."

[7] See Troelstra and Schwichtenberg (2000, p. 3). Of course, in principle, also classical negation can be defined this way; but while, in intuitionistic logic, negation is directly *reduced* to $\bot$, in classical logic, we would have merely a syntactic variation, still requiring axioms or rules involving negated formulas as such — see, for instance, the difference of the absurdity rules $\bot_i$ and $\bot_c$ in Troelstra and Schwichtenberg (2000, p. 37).

can be defined as $\varphi \rightarrow \bot$, one observes that Curry's paradox (using $\bot$)[8] is essentially the Liar in intuitionistic terms.

"At the referee's suggestion" Löb (1955, Fn. 4) remarks that Curry's "paradox is derived without the word 'not' " (Löb, 1955, p. 117). Löb was carefully enough to speak only about the *word* "not". Beall and Murzi (2013, p. 144) rephrase it by saying that Curry's paradox "arises even in negation-free languages". But, as one can see from the definition of negation in an intuitionistic language without a primitive symbol for negation, the absence of the word does not imply necessarily the absence of the concept of negation. Even if Curry's paradox is still applicable in languages which would not even allow a definition by use of *falsum*, an implicit representation of negation will be present in any case.[9]

At the end — and as for Russell — revoking the bivalence is not enough to ban Liar-like paradoxes; rather, it appears that the paradoxes are independent of the underlying logic.

Concerning the given reading of intuitionistic negation we like to recall that a similar perspective was already given by Bernays (1979, p. 4) as a distinctive feature in comparison with classical logic:[10]

> As one knows, the use of the "tertium non datur" in relation to infinite sets, in particular in Arithmetic, was disputed by L. E. J. Brouwer, namely in the form or an opposition of the traditional logical principle of the excluded middle. Against this opposition is to say that it is just based on a reinterpretation of the negation. Brouwer avoids the usual negation non-A, and takes instead "A is absurd". It is then obvious that the general alternative "Every sentence A is true or absurd" is not justified.

4 Intuitionism

The fact that changing from classical to intuitionistic logic does not resolve the paradoxes, neither Russell's nor the Liar, leads to the conclusion that one cannot hold

---

[8] The original version of Curry, allowing arbitrary formulas in the consequence is, in some sense, more general, as it involves as a particular case "Löb's Theorem" (which, however, is not really a paradox any longer, at least not in the sense that it leads to a contradiction). The relation between Curry's paradox and Löb's Theorem is an interesting issue in itself; see, for instance, Ruitenburg (1991).

[9] Therefore, it seems to be misleading to say, as Benthem (1978, p. 49), that "Curry's paradox shows that negation is not essential in this connection"; it is just the negation *symbol* which appears not to be essential.

[10] German original: "Wie man weiß, ist die Verwendung des ‚tertium non datur' in bezug auf unendliche Gesamtheiten, insbesondere schon in der Arithmetik, von L. E. J. Brouwer angefochten worden, und zwar in der Form einer Opposition gegen das traditionelle logische Prinzip vom ausgeschlossenen Dritten. Gegenüber dieser Opposition ist zu bemerken, daß sie ja auf einer Umdeutung der Negation beruht. Brouwer vermeidet die übliche Negation nicht-A, und nimmt stattdessen ‚A ist absurd'. Es ist dann klar, daß eine allgemeine Alternative ‚Jede Aussage A ist wahr oder ist absurd' nicht berechtigt ist."

the logic responsible for them.[11] Thus, one should look for the concept formations involved in the paradoxes when searching for solutions.

In standard semantics, one takes care of concept formations by careful choices of interpretations. This involves either a staunch platonistic insight in the interpretation or, at least, a firm confidence in set-theoretic constructions for them.[12]

Zermelo (1908) gave the classical example of *concept formation* when he axiomatized set theory with the explicit aim to ban the paradoxes:

> Under these circumstances there is this point nothing left for us to do but [. . .] to seek out the principles requried for establishing the foundations of this mathematical discipline. In solving the problem we must, on the one hand, restrict these principles sufficiently to exclude all contradictions and, on the other, take them sufficiently wide to retain all that is valuable in this theory.[13]

The principles of set theory, of course, serve as (implicit) definitions of the set-theoretical concepts. In practice, Zermelo's set theory fully satisfies the needs of the mathematicians. But Poincaré was not convinced (cited according to Gray, 2013, p. 540): "But even though he has closed his sheepfold carefully, I am not sure that he has not set the wolf to mind the sheep." Thus, without a consistency proof for the axiomatized set theory, the situation remains unsatisfactory from a philosophical point of view.

Interestingly, also Brouwer can cope with the problem, insofar one puts the concept formation ahead of the logic. This is in line with his idea that mathematics goes ahead of logic.[14] For Russell's paradox, one may note that Brouwer clearly rejected Cantorian set theory as such and abstract set formation principles are plainly anti-intuitionistic. In the same way, formalizations of the Liar and Curry's paradox depend on self-referential features of formal languages — to be implemented by some kind of Gödelization. But such formal languages are not the subject of intuitionism. In this perspective, the paradoxes may even support Brouwer's anti-logical convictions.

This perspective also vindicates Weyl (1987, p. 5), who was using his criticism of (the scholasticism around) the Grelling–Nelson paradox, not to advocate a many-valued logic, but rather to demand a careful delimitation of the "categories" to which a meaningful proposition is affiliated.

Here is not the place to evaluate the success of intuitionism to provide convincing techniques for concept formation. Brouwer coined the name with reference to the

---

[11] Here, we are not going into the attempts to mutilate further the logical framework (as, for instance, by questioning *modus ponens*); nor do we discuss informal notions of provability (Weaver, 2012) or validity versions of the Liar (Beall and Murzi, 2013) which are sometimes used to clarify the situation.

[12] See, for instance, Feferman (2000, p. 72).

[13] Zermelo (1967, p. 200). German orginial (Zermelo, 1908, p. 261): "Unter diesen Umständen bleibt gegenwärtig nichts anderes übrig, als [. . .] die Prinzipien aufzusuchen, welche zur Begründung dieser mathematischen Disziplin erforderlich sind. Diese Aufgabe muß in der Weise gelöst werden, daß man die Prinzipien einmal eng genug einschränkt, um alle Widersprüche auszuschließen, gleichzeitig aber auch weit genug ausdehnt, um alles Wertvolle diese Lehre beizubehalten."

[14] See the third chapter of Brouwer's dissertation, reprinted in English translation in L. E. J. Brouwer (1975), which contains the theses: *Mathematics is independent of logic* and *Logic depends upon mathematics*.

intuition of mathematicians. The resulting risk of subjectivity was criticised by Lorenzen with respect to the rejection of the *tertium non datur*:[15] "Unfortunately, the explanation which Brouwer himself offers for this phenomenon is an esoteric issue: only one who listened the Master himself understands him." Anyhow, it is the supposed intuition which should save the intuitionist from contradictions, not the underlying logic.

To avoid misunderstandings, we have to stress that Brouwer's conception of intuitionism was by no means motivated by the paradoxes — quite contrary to Hilbert's motivations for his foundational research.[16] In fact, Brouwer did not comment on the paradoxes at all, except for a plain rejection of Cantorian or axiomatic set theory in his dissertation and in a paper of 1912; cf. L. E. J. Brouwer (1975, pp. 80ff. and 130ff.).

5 Proof-theoretic semantics

Proof-theoretic semantics "is based on the fundamental assumption that the central notion in terms of which meanings are assigned to certain expressions of our language, in particular to logical constants, is that of *proof* rather than *truth*. In this sense proof-theoretic semantics is *semantics in terms of proofs*" (Schroeder-Heister, 2016b). "Proof-theoretic semantics is intuitionistically biased" (Schroeder-Heister, 2016b, §3.5)[17]. As such, it is confronted with the paradoxes in the very same way as intuitionism itself; but, as we will see, there is an additional challenge.

In a first step, proof-theoretic semantics may follow the "solution" we attributed to Brouwer: turning to the particular concept formations. In the case of Russell's paradox, this means that one would have to provide a proof-theoretic semantics for the set formation principles (expecting that such a semantics blocks the possibility to introduce the "Russell set"). Such an approach was, in fact, already initiated by Hallnäs (2016). With respect to the Liar in its usual form, one needs a framework axiomatizing truth and providing some form of term representation of formulas (as *Gödelization*). There are plenty of truth theories around and giving them a proof-theoretic semantics can be subsumed under the "open problem" of *Proof-Theoretic Semantics Beyond Logic* addressed by Schroeder-Heister (2016a, §4).

However, there is another form of treating the liar in a formal theory, which constitute a genuine challenge to proof-theoretic semantics. One may axiomatize a self-contradicting atom R with $R \leftrightarrow \neg R$. Schroeder-Heister (2012a) introduced such an R in a sequent calculus by the following two rules:

[15] German original in Lorenzen (1960): "Unglücklicherweise ist die Erklärung, die Brouwer selbst für dieses Phänomen anbietet, eine esoterische Angelegenheit: nur, wer den Meister selber hörte, versteht ihn."

[16] For Hilbert's motivation, see Kahle (2006).

[17] "Most forms of proof-theoretic semantics are intuitionistic in spirit, which means in particular that principles of classical logic such as the law of excluded middle or the double negation law are rejected or at least considered problematic." (Schroeder-Heister, 2016b, §1.2)

Paradoxical rules

$$\frac{\Gamma, \neg R \vdash C}{\Gamma, R \vdash C}\,(R \vdash) \qquad\qquad \frac{\Gamma \vdash \neg R}{\Gamma \vdash R}\,(\vdash R)$$

In a fine analysis of the contradiction, which one can derive from these rules (together with the usual structural rules and rules for negation), Schroeder-Heister shows that the contradiction can be blocked by imposing some restriction on each of the structures rules of Identity, Contraction, or Cut — restrictions which would not harm " 'ordinary' mathematical reasoning" (Schroeder-Heister 2012a; 2016c). From the point of proof-theoretic semantics, however, we don't see that such restrictions — in fact, any switch to sub-structural logics — is justifiable, as one would not like to dismiss the usual structural rules in other contexts.

Substructural logics (Došen and Schroeder-Heister, 1993) do not intended to mutilate logics (see footnote 11), but they rather serve as tactical modifications of standard logic to obtain a fine-grained analysis of the interplay of different strutural operations. Some of the substructural logics turned out to be of interest in special applications, as the Lambek calculus (Lambek, 1958) in linguistics and Linear Logic (Girard, 1987) for resource aware reasoning. The "sub"structural character of the latter is manifest in the possibility to reestablish classical reasoning by use of the bang operator. Intuitionistic logic, however, when considered as a substructural logic in the form of mono-succedent sequent calculus has further claims. Brouwer and Weyl were aiming, indeed, to replace classical reasoning in Mathematics. And they invoked sofisticated philosophical arguments — although these arguments were dismissed (or ignored) by the mathematical community at large. But they neither addressed the paradoxes nor were concerned with technical properties of calculi.

If, thus, classical reasoning should not be dismissed in general, proof-theoretic semantics should provide an argument to invalidate directly the mentioned *paradoxical rules*. We are facing here a "*tonk*-like phenomenon",[18] and as such it is discussed in Tranchini (2016). As for *tonk*, the sheer definition of $(R \vdash)$ and $(\vdash R)$ would spoil our calculus. To deal with *tonk*, *proof-theoretic harmony* was conceived as a possible solution.[19] But the two paradoxical rules appear to be in perfect harmony; and this was already observed by Read (2010, §7).

Next to harmony — as the "first principle" of proof-theoretic semantics —, we can consider a "second principle": *normalizability of proofs*. In fact, Tennant (1982) used

[18] The "tonk" connective was introduced by Prior (1960) by the following two rules:

$$\frac{A}{A \text{ tonk } B}\,(\text{tonk-I}) \qquad\qquad \frac{A \text{ tonk } B}{B}\,(\text{tonk-E})$$

It is widely discussed in the literature and, for our context, we may refer to Read (2010) or Tranchini (2016) for further information.

[19] This concept is due to Dummett (1981; 1991); for a recent discussion see, for instance, Tranchini (2021).

normalizability[20] exactly to block the paradoxical rules under discussion.[21] This is today one of the main directions proof-theoretic semantics is following and Tranchini (2015; 2016) proposed, quite convincingly, a combined treatment of *tonk* and the paradoxical rules in this vein by combining harmony and normalizability.

The problem with the "second principle" is that it is *global* and not any longer *local*; i.e., we cannot assign a proof-theoretic meaning to the connectives by solely inspect the given rules, but we have to prove properties of derivability in general.[22] In the last consequence, we are confronted with Hilbert's concern as expressed in the citation at the beginning of the paper: the admissibility of a proof can only be verified *a posteriori*.[23]

But even using normalizability as a proof-theoretic principle, we have no superordinate philosophical argument that this blocks *all* potential contradictions; we just verified empirically that it will block the liar.[24] There is an important lesson to learn from Martin-Löf's first type theory (Martin-Löf, 1971). It was conceived in a way that it was not subject to a direct liar-like contradiction; only a much more subtle reasoning, expressed in *Girard's paradox* showed its inconsistency (Coquand, 1986; Hurkens, 1995).[25] Thus, global proof-theoretic conditions which block "one or another" paradox might be far from being sufficient to convince one from the consistency of a system as a whole.

With reference to Hallnäs (1991; 2006), Schroeder-Heister proposes a possible solution: *Definitional Freedom* (Schroeder-Heister 2012b, §2 and 2016a, p. 276). Under this freedom, one does not forbid any rules, but has to single out the "well-behaved" ones by (*a posteriori*) mathematical arguments. Qualitatively, this was already done by Gödel, when he formalized the sentence "I am not provable in T" in an arithmetical (consistent) theory T.[26] The definition is perfectly fine, but the

---

[20] Ekman (2016, p. 212) observed that non-normalizability can be related to some form of "overloading" of (the use of) propositions: "A *self-contradictory argument* is, informally, an argument [. . .] in which there is a proposition which is used in two or more ways such that not all of the ways of using the proposition are compatible."

[21] For critical evaluations of Tennant's approach see, for instance, Schroeder-Heister and Tranchini (2017) and Petrolo and Pistone (2019).

[22] See also *Local and Global Proof-Theoretic Semantics* (Schroeder-Heister, 2016a, §2.4).

[23] In principle, normalizability could be proven, for a given set of axioms, before performing the single proofs. But there are two problems: first, normalizability will be, in general, an undecidable property; second, proof-theoretic semantics would depend on such a (meta-)proof of normalizability, which is rather delicate with respect to the philosophical claim of proof-theoretic semantics.

[24] More exactly: the liar and some other known paradoxes; see Tennant (1982). Tennant is well aware of the limitations of his approach: "I fully realise how inadequate any supposedly final word on this matter [the paradoxes] would be." (Tennant, 1982, p. 278). Thus, we are still in need for a *philosophical* argument that normalizability is more than an "ad hoc reply" to the known paradoxes.

[25] Admittedly, Girard's paradox will not be detected by a "slightest consideration"; thus, the situation is far more complex than Weyl might have judged it in 1917.

[26] Gödel explicitely refers to the paradoxes as heuristic motivation: "The analogy between this result and Richard's antinomy leaps to the eye; there is also a close relationship with the 'liar' antinomy"; in a footnote he continues: "Every epistemological antinomy can likewise be used for a similar undecidability proof" Gödel (1931, p. 175, translated); see also Lethen (2021).

provability predicate of T turns out to be incomplete, i.e., not every sentence is either provable or refutable in terms of such a formal predicate.[27]

The situation is also exemplified in recursion theory, where one does not like to forbid the definition of partial functions, but rather likes to single out, *a posteriori*, the functions which are total (or the domain of a partial function).[28] For recursion theory, there are adequate formal systems to incorporate the reasoning about partiality, namely *free logics* (Bencivenga, 2002) or the *logic of partial terms* (Beeson, 1985). For logical calculi, there exist a largely forgotten attempt by Behmann (1959)[29], but a modern worked out formalism is still a desideratum. Incorporating the reasoning about the "well-behaviour" of definitions would, in fact, vindicate both, Weyl and Hilbert: for Weyl, the slightest (or not so slight) consideration about the sense of a definition would turn explicit; for Hilbert, the admissibility of (the concepts used in) a proof would be checked not only at the result, but — if not globally, which we cannot expect any longer in view of undecidability phenomena — at least locally, for every proof in advance.

In fact, it is one of the features of the paradoxes that they work with *locally correct reasoning*.[30] According to our analysis, the paradoxes are not phenomena of the underlying logic, but of the concept formations; therefore, a proper treatment has to take into account the reasoning about these concepts. For proof-theoretic semantics such a reasoning should be a part of the game, and we agree with Schroeder-Heister (2012b, p. 78) to allow such reasoning within the formal frameworks:

> We strongly propose definitional freedom in the sense that there should be one or several formats for definitions, but within this format one should be free. Whether a certain definition is well-behaved is a matter of (mathematical) 'observation', and not something to be guaranteed from the very beginning.

References

Beall, J. C. and J. Murzi (2013). Two flavors of Curry's paradox. *Journal of Philosophy* 110, 143–165.

Beeson, M. (1985). *Foundations of Constructive Mathematics*. Ergebnisse der Mathematik und ihrer Grenzgebiete; 3. Folge, Band 6. Springer.

[27] Apparently, it was Russell, who didn't get the point here, when he argues as if Gödel had shown an inconsistency in mathematics in a letter to Leon Henkin of 1 April 1963, cited in Dawson, Jr. (1988, p. 89ff).

[28] The common underlying reason for partiality in recursion theory and incompleteness in arithmetic is, of course, *diagonalization*. It is the source of the very most paradoxes, and *definitional freedom* requires to deal positively with it, rather than forbitting it. Discussions of such positive handling in Mathematics can be found, for instance, in Sommaruga-Rosolemos (1991) and Santos (2020).

[29] See also Thiel (2019).

[30] Schroeder-Heister (2016a, §4.2) remarked this in the context of clausal definitions: "This connects the proof theory of clausal definitions with theories of paradoxes, which conceive paradoxes as based on locally correct reasoning."

Behmann, H. (1959). Der Prädikatenkalkül mit limitierten Variablen. Grundlegung einer natürlichen exakten Logik. *Journal of Symbolic Logic* 24, 112–140.

Bencivenga, E. (2002). Free logics. In: *Handbook of Philosophical Logic*. Ed. by D. M. Gabbay and F. Guenthner. Vol. 5. Dordrecht: Springer, 147–196.

Benthem, J. F. A. K. van (1978). Four paradoxes. *Journal of Philosophical Logic* 7, 49–72.

Bernays, P. (1979). Bemerkungen zu LORENZEN's Stellungnahme in der Philosophie der Mathematik. In: *Konstruktionen versus Positionen, Vol. 1*. Ed. by K. Lorenz. Berlin: de Gruyter, 3–16.

Coquand, T. (1986). An analysis of Girard's paradox. In: *Proceedings Symposium on Logic in Computer Science*. IEEE Computer Society Press, 227–236.

Curry, H. (1942). The inconsistency of certain formal logics. *Journal of Symbolic Logic* 7, 115–117.

Dawson, Jr., J. W. (1988). The reception of Gödel's incompleteness theorems. In: *Gödel's Theorem in focus*. Ed. by S. G. Shanker. Routledge, 74–95.

Došen, K. and P. Schroeder-Heister (1993). *Substructural Logics*. Vol. 2. Studies in Logic and Computation. Oxford University Press.

Dummett, M. (1981). *Frege. Philosophy language*. 2nd ed. Duckworth.

– (1991). *The Logical Basis of Metaphysics*. Duckworth.

Ekman, J. (2016). Self-contradictory reasoning. In: *Advances in Proof-Theoretic Semantics*. Ed. by T. Piecha and P. Schroeder-Heister. Springer, 211–229.

Feferman, S. (2000). Does reductive proof theory have a viable rationale? *Erkenntnis* 53, 63–96.

Girard, J.-Y. (1987). Linear logic. *Theoretical Computer Science* 50, 1–102.

Gödel, K. (1931). Über formal unentscheidbare Sätze der Principia Mathematica und verwandter Systeme I. *Monatshefte für Mathematik und Physik* 38, 173–198.

Gray, J. (2013). *Henri Poincaré*. Princeton University Press.

Grelling, K. and L. Nelsen (1908). Bemerkungen zu den Paradoxien von Russell und Burali-Forti. *Abhandlungen der Fries'schen Schule II*. reprinted in: Nelson, Leonard. Gesammelte Schriften III. Die kritische Methode in ihrer Bedeutung für die Wissenschaften. Hamburg: Felix Meiner Verlag, 1974, pp. 95–127, 301–334.

Hallnäs, L. (1991). Partial inductive definitions. *Theoretical Computer Science* 87, 115–142.

– (2006). On the proof-theoretic foundation of general definition theory. *Synthese* 148, 589–602.

– (2016). On the proof-theoretic foundations of set theory. In: *Advances in Proof-Theoretic Semantics*. Ed. by T. Piecha and P. Schroeder-Heister. Springer, 161–171.

Hilbert, D. (1917). *Mengenlehre*. Vorlesung Sommersemester 1917, Ausarbeitung (Bibliothek des Mathematischen Instituts der Universität Göttingen).

Hurkens, A. J. C. (1995). A simplification of Girard's paradox. In: *Typed Lambda Calculi and Applications: Second International Conference on Typed Lambda Calculi and Applications, TLCA '95 Edinburgh, United Kingdom, April 10–12, 1995 Proceedings*. Ed. by M. Dezani-Ciancaglini and G. Plotkin. Springer, 266–278.

Irvine, A. D. and H. Deutsch (2016). Russell's paradox. In: *The Stanford Encyclopedia of Philosophy*. Ed. by E. N. Zalta. Winter 2016 edition. Metaphysics Research Lab, Stanford University.

Kahle, R. (2006). *David Hilbert über Paradoxien*. Pré-Publicações. Preprint Number 06-17. Departamento de Matemática, Universidade de Coimbra.

L. E. J. Brouwer (1975). *Collected Works*. Vol. 1: Philosophy and Foundations of Mathematics. Edited by A. Heyting. North-Holland.

Lambek, J. (1958). The mathematics of sentence structure. *American Mathematical Monthly* 65, 154–169.

Lethen, T. (2021). Kurt Gödel on logical, theological, and physical antinomies. *The Bulletin of Symbolic Logic* 27, 267–297.

Löb, M. (1955). Solution of a problem of Leon Henkin. *Journal of Symbolic Logic* 20, 115–118.

Martin-Löf, P. (1971). *A Theory of Types*. Tech. rep. Department of Mathematics, University of Stockholm.

Petrolo, M. and P. Pistone (2019). On Paradoxes in Normal Form. *Topoi* 38, 605–617.

Prawitz, D. (1965). *Natural Deduction*. Vol. 3. Acta Universitatis Stockholmiensis, Stockholm Studies in Philosophy. Almqvist & Wiksell.

Prior, A. N. (1960). The runabout inference-ticket. *Analysis* 21, 38–39.

Read, S. (2010). General-elimination harmony and the meaning of the logical constants. *Journal of Philosophical Logic* 39, 557–576.

Ruitenburg, W. (1991). Constructive logic and the paradoxes. *Modern Logic* 1, 271–301.

Santos, P. G. (2020). *Diagonalization in Formal Mathematics*. BestMasters. Springer.

Schroeder-Heister, P. (2012a). Paradoxes and structural rules. In: *Insolubles and Consequences: Essays in honour of Stephen Read*. Ed. by C. D. Novaes and O. T. Hjortland. College Publications, 203–211.

– (2012b). Proof-theoretic semantics, self-contradiction, and the format of deductive reasoning. *Topoi* 31, 77–85.

– (2016a). Open problems in proof-theoretic semantics. In: *Advances in Proof-Theoretic Semantics*. Ed. by T. Piecha and P. Schroeder-Heister. Springer, 253–283.

– (2016b). Proof-theoretic semantics. In: *The Stanford Encyclopedia of Philosophy*. Ed. by E. N. Zalta. Winter 2016 edition. Metaphysics Research Lab, Stanford University.

– (2016c). Restricting initial sequents: the trade-offs between identity, contraction and cut. In: *Advances in Proof Theory*. Ed. by R. Kahle, T. Strahm, and T. Studer. Springer, 339–351.

Sommaruga-Rosolemos, G. (1991). *Fixed point constructions in various theories of Mathematical Logic*. Bibliopolis.

Tennant, N. (1982). Proof and paradox. *Dialectica* 36, 265–296.

Thiel, C. (2019). Heinrich Behmanns Beitrag zur Grundlagendebatte. In: *Siegener Beiträge zur Geschichte und Philosophie der Mathematik*. Ed. by D. Koenig, G. Nickel, S. Shokrani, and R. Krömer. Vol. 11. Universitätsverlag Siegen, 191–202.

Tranchini, L. (2015). Harmonising harmony. *The Review of Symbolic Logic* 8, 41–423.

Tranchini, L. (2016). Proof-theoretic semantics, paradoxes and the distinction between sense and denotation. *Journal of Logic and Computation* 26, 495–512.
– (2021). Proof-theoretic harmony: towards an intensional account. *Synthese* 198 (Suppl 5), 1145–1176.
Troelstra, A. S. and H. Schwichtenberg (2000). *Basic Proof Theory*. 2nd ed. Cambridge University Press.
Weaver, N. (2012). Intuitionism and the liar paradox. *Annals of Pure and Applied Logic* 163, 1437–1445.
Weyl, H. (1918). *Das Kontinuum*. Veit.
– (1987). *The Continuum: A Critical Examination of the Foundation of Analysis*. Thomas Jefferson University Press. Corrected re-publication, Dover 1994. English translation of Weyl (1918).
Zermelo, E. (1908). Untersuchungen über die Grundlagen der Mathematik. I. *Mathematische Annalen*. English translation in Zermelo (1967), 261–281.
– (1967). Inverstigations in the foundations of set theory I. In: *From Frege to Gödel*. Ed. by J. van Heijenoort. Harvard University Press, 199–215.

On the Structure of Proofs

Lars Hallnäs

Abstract The initial premise of this paper is that the structure of a proof is inherent in the definition of the proof. Side conditions to deal with the discharging of assumptions means that this does not hold for systems of natural deduction, where proofs are given by monotone inductive definitions. We discuss the idea of using higher order definitions and the notion of a functional closure as a foundation to avoid these problems. In order to focus on structural issues we introduce a more abstract perspective, where a structural proof theory becomes part of a more general theory of functional closures. A notion of proof equations is discussed as a structural classifier and we compare the Russell and Ekman paradoxes to illustrate this.

1 Introduction

If we consider proofs represented in a formal calculi such as a system of natural deduction or a sequent calculus it is a tree structure of some sort. From the elementary *monotone* inductive definition of a proof in a sequent calculus we can derive an explicit tree structure, but it is an indirect representation as it is a calculus of derivability. In natural deduction the situation is complicated by the discharging of assumptions, a straightforward monotone inductive definition has to be complemented by side conditions. Such a calculus does not in a direct manner provide a mathematically satisfying foundation for a structure theory of proofs. The basic problem in systems of natural deduction is that we try to represent a *functional* structure in terms of a function closure which force us to add side conditions that makes structural issues difficult to handle and unclear with respect to classifications.

While rules of conjunction introduction and implication elimination, i.e., modus ponens, relates to the application of an elementary function that builds the proof from given proofs of the premises, the implication introduction rule is different as it builds

Lars Hallnäs
The Swedish School of Textiles, University of Borås, Sweden, e-mail: lars.hallnas@hb.se

T. Piecha and K. F. Wehmeier (eds.), *Peter Schroeder-Heister on Proof-Theoretic Semantics*, Outstanding Contributions to Logic 29,
https://doi.org/10.1007/978-3-031-50981-0_13

on *conditional* reasoning. It is *functional* in nature, it is not naturally element of a function closure, but of a *functional closure*.

The inferences we make as we build a proof can be local in nature, i.e., given a series of premises we draw a direct conclusion. Given a proposition stating that a is an element of a set M we look up the definition of the set and draw the conclusion that a satisfy these definitional conditions. As a rule of inference this is at the level of functions on a certain domain of objects. The proof we obtain so far is simply built by applying a function to the proofs of the given premises.

Inferences may also be more non local in nature as they depend not only on the proofs of the immediate given premises, but further more on a conditional argument. We may assume a certain proposition P to be true and using this assumption proving absurdity, thus concluding that $\neg P$ must be true. Here, the conclusion that $\neg P$ must be true depends on a function that given a proof of P produces a proof of absurdity. The rule of inference we apply here is at the level of functionals operating on functions over a certain domain of proofs.

In this paper we will discuss a general and more abstract approach to a structure theory of proofs, where proofs simply will be viewed as a special class of functional closures. These ideas have been discussed briefly in earlier work (Hallnäs, 2006). Here we will discuss this approach in more detail and introduce some foundational notions that can be used to discuss the structural complexity and structural similarities of proofs.

As we discuss structural properties of proofs in terms of functional closures in this paper we disregard, or bracket, provability more or less completely, i.e., we bracket what it is a proof proves. It could be tempting to use some sort of propositions-as-type system in order to relate functional closure *terms* with the propositions that a proof proves. But that would be rather strange since the whole idea here is to search for a mathematically simpler abstract structure theory of proofs and introducing a type system would be just to complicate matters without any real gain.

2 Definitions

To model functional closures we will use the framework of partial inductive definitions (Hallnäs, 1991; Hallnäs and Schroeder-Heister, 1991; Schroeder-Heister, 1993), which extends the monotone inductive definitions with higher order, possible non-monotone, definitions.

A *definition* D is a function

$$D : Dom(D) \rightarrow \mathcal{P}(CoDom(D)).$$

Intuitively we may think of D as a collection of, possible non deterministic, equations

$$a = A$$

saying that the *atom* a is defined by the *condition* A, where $a \in Dom(D)$ for some given universe and where $A \in CoDom$ is a condition built up from objects in $Dom(D)$, $\top$ and $\bot$ using (possibly infinitary) conjunctions $\bigwedge_I$ and arrows $\to$. The logical interpretation of a definition D is given by a *local* notion of consequence $\Gamma \vdash_D A$, where $\Gamma \subset CoDom(D)$ and $A \in CoDom$:

$$\Gamma, a \vdash_D a$$

$$\Gamma \vdash_D \top \qquad\qquad \Gamma, \bot \vdash_D C$$

$$\frac{\Gamma \vdash_D A_i}{\Gamma \vdash_D \bigwedge_I A_i}\,(i \in I) \qquad\qquad \frac{\Gamma, A_i \vdash_D C}{\Gamma, \bigwedge_I A_i \vdash_D C}\,(i \in I)$$

$$\frac{\Gamma, A \vdash_D B}{\Gamma \vdash_D A \to B} \qquad\qquad \frac{\Gamma \vdash_D A \quad \Gamma, B \vdash_D C}{\Gamma, A \to B \vdash_D C}$$

$$\frac{\Gamma \vdash_D A}{\Gamma \vdash_D a}\,(A \in D(a)) \qquad\qquad \frac{\Gamma, A \vdash_D C}{\Gamma, a \vdash_D C}\,(A \in D(a))$$

The property over $Dom(D)$ defined by D is the collection of atoms true in the derived logic $Def(D) = \{a \mid \vdash_D a\}$ and $Cov(D) = \{A \mid \vdash_D A\}$ is the conditions that *covers* the *true* objects and *valid* connections.

3 The functional closure

Let $X \subset U$ and let $f_1, \ldots, f_n$ be functions with arities $k_1, \ldots, k_n$ over U. The *function closure* defined by $(X, f_1, \ldots, f_n)$ is the class obtained by starting with X and closing it under the given functions, i.e., the class given by the definition $D(X, f_1, \ldots, f_n)$:

(1) $$a = \top \quad (a \in X)$$

(2) $$f_i(x_1, \ldots, x_{k_i}) = (x_1, \ldots, x_{k_i}) \quad (i \leq n)$$

the smallest class containing X and closed under the functions $f_1, \ldots, f_n$.

The intuition behind the function closure is simply that we build a collection of objects by starting with some given *atoms* and the successively close our collection under the given functions—a monotone operation.

Now assume $X \subset U$, functions $f_1, \ldots, f_n$ with arities $k_1, \ldots, k_n$ over U, a functional $F : [U \to U] \to U$ and a class $\mathcal{K} \subset [U \to U]$, the *functional closure* with respect to $(X, f_1, \ldots, f_n, F, \mathcal{K})$ is given by the following definition $D(X, f_1, \ldots, f_n, F, \mathcal{K})$:

$$a = \top \quad (a \in X)$$
$$f_i(x_1, \ldots, x_{k_i}) = (x_1, \ldots, x_{k_i}) \quad (i \leq n)$$
$$F(f) = \bigwedge\nolimits_{\mathrm{dom}(f)} (x \to f(x)) \quad (f \in \mathcal{K})$$

The functional closure builds on this idea adding a further abstraction, the closure under given functionals. Intuitively this means that we build our collection of objects at the sam time as we build the function space of this very collection of objects—a possible non monotone operation.

4 Proof structures

The idea of using the functional closure, or more precisely a class of functional closures, as a foundation for a structure theory of proofs, is that this is a very elementary and general foundation that captures the distinction between local and non-local rules of proofs in a precise manner. Another way to say this is that it is a general abstraction of natural deduction based on inductive definitions without side conditions for non-local rules, i.e., that the structure of the proofs is internal to the inductive definition of them. The definition of a proof gives a local logic of the proof and the definition itself is closely related to the idea of validity (Prawitz 1971; 1973b; 1973a; Schroeder-Heister 2006; 2012) of a proof modulo rules of proof contractions.

As a certain kind of abstract proof theory this type of structure theory becomes part of the general study of the functional closure. Initially this calls for axioms that set up boundaries and frames such a theory.

We will identify a *proof structure* with a functional closure

$$DF_1, \ldots, F_m = D(X, f_1, \ldots, f_n, F_1, \ldots, F_m)$$

(we use $D\overline{F}$ for short). We assume that $D\overline{F}$ is deterministic.

A *proof* is then intuitively a focal point $a \in \mathit{Def}(D\overline{F})$, we refer to these focal points as *proof objects*. But what is of main interest here is the definition of the proof that presents the structure of the proof.

As an example consider the following proof structure $IE(U, a, b, ie, ki, F)$:

$$\begin{aligned} a &= \top \\ b &= \top \\ ie(x, y) &= (x, y) \\ ki(x, y) &= (x, y) \\ F(f) &= \bigwedge\nolimits_{\mathrm{dom}(f)} (x \to f(x)). \end{aligned}$$

In this example U is universe of terms, a and b constants, the functions ie and ki are *terms functions* and F a *term functional*, i.e., $ie(a, b), ki(a, b))$ and $F(f)$ are the term themselves. The *proof terms* , the notation, we use for proof objects builds on variables and constants for objects in a given domain, function terms $(f(t_1, \ldots, t_n)$ and functional terms $(F(f))$. We will refer to a proof structure that builds on a term universe and where the functions $f_1, \ldots, f_n$ and the functionals $F_1, \ldots, F_m$ are all term functions and term functionals as a *term proof structure*.

Consider the following proof object

$$ie(F(f), ie(a, b)), \text{ where } f(x) = ki(b, x).$$

Given that $a = p \longrightarrow q$ and $b = p$ this proof object is a representation of a proof of $p \wedge q$ in the following sense:

Assume p and assume q. By $\wedge$-introduction we get $p \wedge q$, this is $ki(b, x)$. It is clear that the definition is locally provable closed under the function $f(x) = ki(b, x)$. By $\longrightarrow$-introduction we get $q \longrightarrow p \wedge q$. This is $F(f)$. Now assume $p \longrightarrow q$ and p, by $\longrightarrow$-elimination we have q. This is $ie(a, b)$. Finally by $\longrightarrow$-elimination we have $p \wedge q$, i.e., $ie(F(f), ie(a, b))$.

The *normal form* of this proof is the proof object $ki(b, ie(a, b))$. In natural deduction for propositional logic this corresponds to:

$$\dfrac{\dfrac{\dfrac{p \quad [q]}{p \wedge q}}{q \to (p \wedge q)} \qquad \dfrac{p \to q \quad p}{q}}{p \wedge q} \quad \Rightarrow \quad \dfrac{p \qquad \dfrac{p \to q \quad p}{q}}{p \wedge q}$$

Note that $\longrightarrow$-introduction will be represented by functional objects $F(f)$ so a *cut* will have the form $ie(F(f), a)$, whereas the situation will be reversed for the $\vee$-rules where $\vee$-elimination is represented by functional objects $F(f, g, a)$ and the $\vee$-introduction-rules are represented by function objects $h_1(b)$, $h_2(c)$, which somehow introduce a reverse cut.

It is also possible to develop a certain *model theory* of proofs using the local logic of a proof structure, i.e., a certain proof semantics.

Given $M \subset Dom(D)$, let $D + M$ be the definition obtained by adding definitional clauses $a = \top$ for all $a \in M$ to the clauses of D. Now we define $M \models_D A$ to hold iff $A \in Cov(D + M)$. Using this we can define a notion of logical consequence in model theoretical terms:

$$A \models_D B \text{ iff } M \models_D A \text{ implies } M \models_D B.$$

5 Proof unfolders

The local logic of a proof structure places proof objects in a logical context. A more direct structural approach would be to *unfold* the proof structure at the given object, i.e., going backwards to trace the way in which the object is built up in the given structure. What this means with respect to the function closure is obvious, it is the local step of simply going one step back.

Given a proof structure $D\overline{F}$ and a proof object $a \in Def(D\overline{F})$, we define the notion of *subobjects*, $S(D\overline{F}, a)$ as follows:

$a \in S(D\overline{F}, a)$.

If $f(b_1, \ldots, b_n) \in S(D\overline{F}, a)$, then $b_1, \ldots, b_n \in S(D\overline{F}, a)$.

If $F(f) \in S(D\overline{F}, a)$, $b \in S(D\overline{F}, a)$ and $b \in \text{dom}(f)$, then $f(b) \in S(D\overline{F}, a)$.

Given a proof structure DF and a proof object $a \in Def(DF)$ we define two elementary operators, *proof unfolders* for objects in $S(DF, a)$:

(Δ) $\Delta(DF, f(b_1, \ldots, b_n), i) = b_i$

(Φ) $\Phi(DF, F(f), b) = f(b),$ this operation is defined whenever $f(b)$ is defined.

The Δ-unfolder unfolds the function closure, i.e., the structure build up by applying functions to given objects. In rule terms this correspond to looking at the premises in a rule application where no assumptions are discharged, note that there is no distinction here between introduction and elimination rules. We make for example no specific distinction between $\wedge$-introduction and $\wedge$-elimination and in a system of set theory between $\in$-introduction and $\in$-elimination. The reason for this is simply that there is no structural difference, at least locally, to consider. This opens up for questions about how to characterize directions and ways of reason in a proof.

The Φ-unfolder unfolds the functional closure, i.e., the structure built up by applying a functional to a function over a given domain. In rule terms this correspond to look at the function premise in a rule application where assumptions are discharged, and apply the function to a subproof for which the function is defined, note that this unfolder is not local in nature as it both works on a function and also use a subproof that does not need to be local to the rule application

In a natural way we may build more complex unfolders by composition:

$$\Delta\Phi(DF, F(f), b, i) = \Delta(DF, f(b), i)$$
$$\Phi\Delta(DF, F(f), g(b_1, \ldots, b_n), i) = \Phi(DF, F(f), b_i)$$
$$\Delta\Delta(DF, f(a, g(b, c)), 2, 2) = \Delta(DF, g(b, c), 2)$$

Alternations of unfolders, i.e., $\Delta\Phi\Delta\Phi \ldots$, involves increasing complexity. We will refer to the index $\bar{i}$ as the Δ-index.

These proof unfolders can be generalized to unfolders for a larger collections of functional closures in the following way:

$a \in S(D\overline{F}, a)$.

If $b \in S(D\overline{F}, a)$, then $D\overline{F}(b) \subset S(D\overline{F}, a)$.

If $\bigwedge_I A_i \in S(D\overline{F}, a)$, then $A_i \in S(D\overline{F}, a)$ for $i \in I$.

If $(b \to f(b)) \in S(D\overline{F}, a)$ and $b \in S(D\overline{F}, a)$, then $f(b) \in S(D\overline{F}, a)$.

For the unfolder operators we need to add a projection operator:

(Δ) $$\Delta(DF, b) = B, \text{ where } B \in DF(b),$$

(Φ) $$\Phi(D, F(f), b) = f(b),$$

(Θ) $$\Theta(DF, \bigwedge(B_i)_I, j) = B_j.$$

The Δ-unfolder unfolds the definition of the proof by giving us definiens for definiendum, which includes unfolding the function closure, i.e., the structure build up by applying functions to given objects. Note that in the general case when the given definition is non deterministic the Δ unfolder will also be non deterministic. The Θ-unfolder is the projection operator.

Initial composition of these unfolders looks like this:

$$\Delta\Phi(DF, F(f), b) = \Delta(DF, f(b))$$
$$\Theta\Delta(DF, b, j) = \Theta(DF, \bigwedge_I B_i, j) \text{ if } DF(b) = \{\bigwedge_I B\}.$$

We leave the proof structure term out in the unfolders when the context is clear, i.e., we simply write $\Phi(F(f), a)$ instead of $\Phi(D\overline{F}, F(f), a)$ when the proof structure context is given.

Le us consider the example above in Section 4 of a proof structure $IE(X, F)$ again, and look at the proof object $ie(F(f), ie(p \rightarrow q, p))$:

$$\Phi(F(f), ie(p \rightarrow q, p)) = ki(p, ie(p \rightarrow q, p)),$$

which is the *normal form* of the proof that $ie(F(f), ie(p \rightarrow q, p)$ represent.

$$\Delta(ki(p, ie(p \rightarrow q, p))1) = p,$$
$$\Delta\Delta(ki(p, ie(p \rightarrow q, p)), 2, 1) = p \rightarrow q.$$

6 Proof equations

Proof unfolders are, in general terms, *operators O* on proof structures. *Proof equations*

$$\overline{O}_1(\overline{T}_1, \bar{i}_1) = \overline{O}_2(\overline{T}_2, \bar{i}_2)$$

can be used to characterize structural properties of proofs as they, in a schematic way, display properties of unfolding sequences of proofs. $\overline{O}_i, \overline{T}_i, \bar{i}_i$ are unfolder operator terms, proof object terms and numerals or numeral variables for the Δ operator indexes, respectively.

A proof structure *solves* a given proof equation if there are suitable substitutions over the given proof structure validating the equation.

A typical simple proof equation could look like this:

$$\Delta^n(D, x, i) = \top$$

where a proof object validating the equation would have the form $f_1, \ldots, f_n(\top)$.

Another elementary example could be taken from Section 4:

$$\Phi\Delta(F(f), ie(F(f), ie(x, y)), i) = ki(y, ie(x, y)).$$

The proof structure $IE(U, F)$ has then a solution to this equation, i.e.,

$$\Phi\Delta(F(f), ie(F(f), ie(a,b),2)) = ki(b, ie(a,b)).$$

In the general case a simple example could look like:

$$\Delta^2\Theta\Delta\Phi(D, X, x, i) = X.$$

7 The paradoxes of Russell and Ekman

There are two fundamental questions here to dwell on:

(Q1) How could a systematic methodology for functional closure analysis of mathematical proofs look like, a methodology that opens up for an abstract structural analysis of actual proofs?
(Q2) How can we use the abstract functional closure theory to discuss structural issues in formal systems of natural deduction?

With respect to the question Q2, it is clear that we of course can use the ideas of structural unfolders and proof equations to discuss the classification of proofs in formal system of natural deduction. As a somewhat canonical example we may compare the Russell and Ekman paradoxes (Prawitz, 1965; Ekman 1994; 1998; Schroeder-Heister and Tranchini, 2017) across different formal systems of natural deduction. It is clear that these two proofs build on the same proof idea, i.e., diagonalization. How could we make this more precise? In the present context we can say that both proof structures are circular in terms of functional fixed points, they solve similar fixed-point proof equations. In these cases a functional closure analysis works as a tool to discuss formal proofs in systems of natural deduction, we assign a functional closure to each of these proofs and look for proof equations that we can use to compare the structure of the two proofs. In more general terms this, of course, leads us to a special case of question Q1.

Normalization as a structural procedure makes a distinct difference between the two proofs. The original Russell derivation is a derivation of *absurdity* in implication calculus extended with introduction and elimination rules for *set-membership*, corresponding to full set comprehension (Prawitz, 1965), while the Ekman derivation (Ekman, 1998) is a proof in propositional implication calculus of $\neg(p \leftrightarrow \neg p)$. Both proofs are intuitively circular with respect to normalization as a procedure of removing round-about reasoning, but the example of Ekman is for sure formally normalizable with respect to standard procedures. See Schroeder-Heister and Tranchini (2017) for an in depth discussion on this issue.

As a formal natural deduction derivation $\mathcal{R}$ the Russell paradox looks like this, where $t = \{x \mid \neg(x \in x)\}$ and where $\neg A$ is short for $A \rightarrow \bot$:

$$\dfrac{\mathcal{R}_1 \qquad \begin{matrix}\mathcal{R}_1\\ t \in t\end{matrix}}{\bot}$$

where

$$\mathcal{R}_1 = \dfrac{\dfrac{\dfrac{[t \in t]}{\neg(t \in t)} \qquad [t \in t]}{\bot}}{\neg(t \in t)} .$$

The last introduction of $\neg(t \in t)$ is a cut formula in the derivation and if we apply the standard rule of reduction we obtain the following derivation

$$\dfrac{\dfrac{\begin{matrix}\mathcal{R}_1 \\ t \in t\end{matrix}}{\neg(t \in t)} \qquad \begin{matrix}\mathcal{R}_1 \\ t \in t\end{matrix}}{\bot}$$

Here $t \in t$ becomes a cut formula. Applying the following rule of reduction

$$\dfrac{\dfrac{\neg(t \in t)}{t \in t}}{\neg(t \in t)} \quad \Rightarrow \quad \neg(t \in t)$$

we will come back to the derivation we started with.

Question Q1 is of a programmatic nature. It would, of course, be possible to perform such an analysis in two steps. First a formalization of proofs in some formal system and then an abstract analysis of these proofs in terms of functional closures. A more direct methods would be to make a *local* and direct analysis of a given proof. There is naturally nothing absolute about such an approach as an informal proof is not in itself a precise mathematical object. The analysis would give a, somewhat abstract, mathematical definition of the proof in structural terms. Central parts of any non trivial proof involves complex constructions, i.e., definitions, and it is in each case an interesting problem how to isolate the reasoning core of the proof. The Ekman paradox (Ekman 1994; 1998) is for instance the result of an analysis of the proof of Cantors theorem formalized in a natural deduction system for set theory. Consider the following derivation $\mathcal{E}$ of $\bot$ from $p \leftrightarrow \neg p$, where we shorten the proof by using rules for equivalence elimination:

$$\dfrac{\mathcal{E}_1 \qquad \dfrac{p \leftrightarrow \neg p \qquad \mathcal{E}_1}{p}}{\bot}$$

where

$$\mathcal{E}_1 = \dfrac{\dfrac{\dfrac{p \leftrightarrow \neg p \qquad [p]}{\neg p} \qquad [p]}{\bot}}{\neg p}$$

Let us consider a *Russell object*, in a term proof system R representing the structures of proofs in set theory with full comprehension, i.e., where R is a term proof structure including functions that represent the rules for implication and set membership.

Let, once again, $t = \{x \mid \neg(x \in x)\}$. Assume $t \in t$, then $\neg(t \in t)$ by $\in$-elimination. This is $r_2^{-1}(x)$. From this by $\rightarrow$-elimination we have $\bot$. This is $r_1(r_2^{-1}(x), x)$ and we have our function f. By $\rightarrow$-introduction we may conclude $\neg(t \in t)$. This is

$F(f)$. Using this we may conclude $\perp$ which is $r_1(F(f), r_2(F(f)))$, where $r_2(F(f))$ represent the $\in$-introduction of $t \in t$, i.e.,

$$r_1(F(f), r_2(F(f))), \text{ where } r_2(F(f)) \in \mathrm{dom}(f) \text{ and } f(x) = r_1(r_2^{-1}(x), x).$$

We then have:

$\Delta\Delta\Delta\Phi(R, F(f), r_2(F(f)), 1, 1, 1) = F(f)$,

(1) $\Phi(F(f), r_2(F(f))) = r_1(r_2^{-1}(r_2(F(f))), r_2(F(f)))$,

(2) $\Delta(r_1(r_2^{-1}(r_2(F(f))), r_2(F(f)), 1) = r_2^{-1}(r_2(F(f)))$,

(3) $\Delta(r_2^{-1}(r_2(F(f)), 1) = r_2(F(f))$,

(4) $\Delta r_2(F(f)), 1) = F(f)$.

So the pair $(F(f), r_2(F(f)))$ solves the following proof equation in R:

$$\Delta^3\Phi(R, X, Y, 1, 1, 1) = X.$$

An *Ekman object* in a term proof system E representing the structures of proofs in propositional logic could be defined in the following way.

Let $a = p \leftrightarrow \neg p$. Assume p, then we may conclude $\neg p$ by $\leftrightarrow$-elimination (2). This is $e_2'(a, x)$ and we may then further conclude $\perp$ by $\rightarrow$-elimination. This is $e_1(e_2'(a, x), x)$ and here we have f and we can conclude $\neg p$ by $\rightarrow$-introduction which is $F(f)$. Using $F(f)$ we may conclude p by $\leftrightarrow$-elimination (1), i.e., $e_2(a, F(f))$. By $\rightarrow$-elimination we then get $\perp$, which is $e_1(F(f), e_2(a, F(f)))$, that is,

$$e_1(F(f), e_2(a, F(f))), \text{ where } e_2(a, F(f)) \in \mathrm{dom}(f) \text{ and } f(x) = e_1(e_2'(a, x), x).$$

Now we have

$\Delta\Delta\Delta\Phi(E, F(f), e_2(a, F(f)), 1, 2, 2) = F(f)$,

(1) $\Phi(F(f), e_2(a, (F(f)))) = e_1(e_2'(a, e_2(a, F(f))), e_2(a, F(f)))$,

(2) $\Delta(e_1(e_2'(a, e_2(a, F(f))), e_2(a, F(f))), 1) = e_2'(a, e_2(a, F(f)))$,

(3) $\Delta(e_2'(a, e_2(a, F(f))), 2) = e_2(a, F(f))$,

(4) $\Delta(e_2(a, F(f)), 2) = F(f)$,

So the pair $(F(f), e_2(F(f), a))$ solves the same equation, modulo Δ-index, in E as the pair $(F(f), r_2(F(f)))$ does in R, i.e.

$$\Delta^3\Phi(E, X, Y, i_1, i_2, i_3) = X.$$

This is then one way to make precise what is intuitively clear, namely that these two proofs have, in principle, the same structure, that the proof idea is the same in the two proofs. The differences in Δ-index reflect the difference between explicit rules in the Russell paradox and implicit rules in the form of assumptions/axioms in the Ekman paradox. But still, both proof systems satisfy the same elementary fixed point equation.

A solution to the equation

$$\Delta^3\Phi(DF, X, Y, i_1, i_2, i_3) = X$$

must be of the form $(F(f), a)$ for some a, where $f(a)$ is defined and where $F(f) \in S(f(a))$.

As a function we think of an unfolder term $\overline{O}(D, \overline{F}, \overline{f}, \overline{n})$ as a *proof form*. Proof equations such as

$$\Delta^3\Phi(DF, X, Y, \bar{i}) = X$$

then display structural *proof expressions*.

Within the abstract framework of functional closures it is natural to say that a *cut* is a pair $(F(f), a)$, and if we consider a proof object $g(F(f), a)$ to say that $(F(f), a)$ is the *main cut* in this proof object. So with respect to the paradoxes of Russell and Ekman we have main cuts producing the same proof expression in relation to the proof form $\Delta^3\Phi(DF, X, Y, i_1, i_2, i_3)$. The proof equation

$$\Delta^3\Phi(DF, X, Y, i_1, i_2, i_3) = X$$

then expresses the central proof idea common to both proofs.

In this fixed point equation the fixed point $F(f)$ is *hidden* in $f(a)$ for some given a and *unlocked* by iterations of the Δ unfolder. The simplest example of such a proof object is the definition *Id*:

$$F(id) = \bigwedge\nolimits_{Dom(Id)}(x \to x),$$

where we of course need zero iterations to unlock $F(id)$.

8 General functional closure interpretations

Given a functional closure $L = (U, \Lambda(f), \phi(x, y), \psi(x))$ we may interpret natural deduction derivations in the implication calculus in terms of locally definable functions over this functional closure in the following way.

Let $\mathcal{D}(\overline{p})$ be a derivation in implication calculus with open assumptions $\overline{p}$. By recursion on $\mathcal{D}$ we

(i) define an interpretation $\pi\mathcal{D} : U^{\overline{p}} \to U$ and
(ii) prove that $\pi\mathcal{D}$ is locally definable in L, i.e., $\overline{a} \vdash_L \pi\mathcal{D}(\overline{a})$.

If $\mathcal{D} = p$, let $\pi\mathcal{D}(a) = \psi(a)$.

Clearly $a \vdash \pi\mathcal{D}(a)$, since $a \vdash \psi(a)$ holds by definition.

If $\mathcal{D}(\overline{p})$ ends with an implication elimination with premise derivations $\mathcal{D}_1(\overline{p}_1)$ and $\mathcal{D}_2(\overline{p}_2)$, let

$$\pi\mathcal{D}(\overline{a}) = \phi(\pi\mathcal{D}_1(\overline{a_1}), \pi\mathcal{D}_2(\overline{a_2})), \text{ where } \overline{a}_i = \overline{a} \mid \overline{p}_i \text{ for } i = 1, 2.$$

By IH $\overline{a}_i \vdash \pi\mathcal{D}_i(\overline{a}_i)$ for $i = 1, 2$. Clearly $\overline{a} \vdash \pi\mathcal{D}_i(\overline{a}_i)$ (for $i = 1, 2$), so $\overline{a} \vdash \phi(\pi\mathcal{D}_1(\overline{a}_1), \pi\mathcal{D}_2(\overline{a}_2))$. Thus $\overline{a} \vdash \pi\mathcal{D}(\overline{a})$.

If $\mathcal{D}(\overline{p})$ ends with an implication introduction with premise derivation $\mathcal{D}_1(\overline{p}_1)$, where $\overline{p}_1 = \overline{p} + p$ let

$$\pi\mathcal{D}(\overline{a}) = \Lambda(f_{(\overline{a})}) \text{ where } f_{(\overline{a})}(a) = \pi\mathcal{D}_1(\overline{a}_1) \text{ for } \overline{a} = \overline{a}_1 \mid \overline{p} \text{ and } \overline{a}_1(p) = a.$$

$\overline{a} \vdash \Lambda(f_{(\overline{a})})$ follows from $\overline{a}, a \vdash f_{(\overline{a})}(a)$ for $a \in U$. By definition this is $\overline{a}, a \vdash \pi\mathcal{D}_1(\overline{a}_1)$. This follows from $\overline{a}_1 \vdash \pi\mathcal{D}_1(\overline{a}_1)$, which follows from IH.

The informal interpretations of the paradoxes of Russell and Ekman corresponds to the obvious term interpretation τ, where we for the Russell example extend τ to include functions representing rules for $\in$.

9 Reduction characterizations

Given an interpretation π of derivations in natural deduction for implication calculus. An unfolding of proof objects $\pi\mathcal{D}$ can be seen as one way to both define and characterize reductions, this relates to the notion of simple operations on derivations discussed in Hallnäs (1988).

Take the example from Section 4.

$$\dfrac{\dfrac{\dfrac{p \quad [q]}{p \wedge q}}{q \to (p \wedge q)} \qquad \dfrac{p \to q \quad p}{q}}{p \wedge q} \quad \Rightarrow \quad \dfrac{p \qquad \dfrac{p \to q \quad p}{q}}{p \wedge q}$$

$$\tau(\mathcal{D}) = ie(F(f), ie(a, b)),$$
$$\tau(\text{red}\,\mathcal{D}) = ki(b, ie(a, b)) = \Phi(F(f), ie(a, b)).$$

For the reductions of the Russell derivation we have

$$\dfrac{\mathcal{R}_1 \qquad \dfrac{\mathcal{R}_1}{t \in t}}{\bot} \quad \Rightarrow \quad \dfrac{\dfrac{\dfrac{\mathcal{R}_1}{t \in t}}{\neg(t \in t)} \qquad \dfrac{\mathcal{R}_1}{t \in t}}{\bot} \quad \Rightarrow \quad \mathcal{R},$$

$$\tau(\mathcal{R}) = r_1(F(f), r_2(F(f))),$$

$$\begin{aligned}\tau(\text{red}(\mathcal{R})) &= r_1(r_2^{-1}(r_2(F(f))), r_2(F(f))) \\ &= \Phi(F(f), r_2(F(f))),\end{aligned}$$

$$\begin{aligned}\tau(\text{red}(\text{red}(\mathcal{R}))) &= r_1(F(f), r_2(F(f))) \\ &= r_1(\Delta^2(r_2^{-1}(r_2(F(f))), 1, 1), r_2(F(f))) \\ &= \tau\mathcal{R}.\end{aligned}$$

So Φ defines/characterizes the first reduction and Δ^2 the second one. In the form of a diagram:

$$\begin{array}{ccccc} \mathcal{R} & \xrightarrow{\text{red}} & \mathcal{R}' & \xrightarrow{\text{red}} & \mathcal{R}'' \\ \downarrow & & \downarrow & & \downarrow \\ \tau\mathcal{R} & \xrightarrow{\Phi} & \tau\mathcal{R}' & \xrightarrow{\Delta^2} & \tau\mathcal{R}'' \end{array}$$

This is a structural definition and wider than the standard one as we know from the Ekman paradox where the interesting reduction of course is the following reduction:

$$\dfrac{\dfrac{p \leftrightarrow \neg p \qquad \dfrac{p \leftrightarrow \neg p \qquad \overset{\mathcal{E}_1}{\neg p}}{p}}{\neg p} \qquad \dfrac{p \leftrightarrow \neg p \qquad \overset{\mathcal{E}_1}{\neg p}}{p}}{\bot}$$

$$\Rightarrow \quad \dfrac{\overset{\mathcal{E}_1}{\neg p} \qquad \dfrac{p \leftrightarrow \neg p \qquad \overset{\mathcal{E}_1}{\neg p}}{p}}{\bot}$$

$$\tau(\text{red}(\text{red}(\mathcal{E}))) = e_1(\Delta^2(e_2'(a, e_2(a, F(f))), 2, 2), e_2(a, F(f))) = \tau\mathcal{E}.$$

The expected objection here is that this definition/characterization does not take into account the distinction between introduction and elimination rules and thus does not respect the intended semantics of the derivation and reduction rules.

Structurally with respect to unfoldings

$$\dfrac{\dfrac{t \in t}{\neg(t \in t)}}{t \in t}$$

and

$$\dfrac{\dfrac{p \wedge (q \wedge r)}{q \wedge r}}{r}$$

are isomorphic, whereas there is a huge difference semantically. But this isomorphism is local in nature, the semantical difference will leave its traces on the level of global closure properties. So, yes there is really no big differences between these two derivations if we isolate them. The difference shows itself in the ways we can use them in a bigger context, i.e., as $\Phi\Delta^3(F(f), f(F(f)))$ for example.

So why does the Ekman reduction make sense? It is of course a matter of removing some clear roundabout reasoning, but we can also argue from a more structural point of view. The slogan will then be that any unfolding for which $\Gamma \vdash A$ is an invariant will define a valid reduction.

$$\begin{array}{ccc} \begin{array}{c}\Gamma \vdash A\\ \mathcal{D}_1\end{array} & \xrightarrow{\text{red}} & \begin{array}{c}\Gamma \vdash A\\ \mathcal{D}_2\end{array} \\ \downarrow & & \downarrow \\ \mathcal{T}\mathcal{D}_1 & \xrightarrow{o} & \mathcal{T}\mathcal{D}_2 \end{array}$$

Unfolding a proof object tell us about how the proof argument is built in terms of function and functional structure. A proof equation is then a *representation* of such a structure. A fixed point equation represent in that sense a circular argument, i.e., the unfolding sequence just circles around itself. The reason why this circular structure is not visible through standard reductions in propositional logic, the Ekman paradox, is of course that it is hidden in minor premises in contrast to the explicit rules of full comprehension in natural deduction for naive set theory. This is simply something that is not captured by the semantically motivated duality between introduction and elimination rules. It seems as if this is a structural phenomenon, but so is the idea of roundabout reasoning.

Is the Russell paradox $\mathcal{R}$ the same proof as the Ekman paradox $\mathcal{E}$? It is clearly a matter of the same proof idea and the differences between them are definitely less interesting than the similarities. Using the functional closure objects $\tau\mathcal{R}$ and $\tau\mathcal{E}$ we may say that the proof expression $(\Delta^3\Phi(F(f),Y) = F(f))$ *show* that they are the same proof modulo a difference δ in Δ index, i.e., a difference in the choices of premises of Δ unfoldings. So one way of answering the question is to say that we should look at the difference in the ways in which these two derivations solve the fixed point equation:

$$(\Delta^3\Phi(F(f),Y) = F(f))(\mathcal{R}) \setminus (\Delta^3\Phi(F(f),Y) = F(f))(\mathcal{E}).$$

References

Ekman, J. (1994). Normal proofs in set theory. PhD thesis. Gothenburg: Department of Computing Science, Chalmers University of Technology.

– (1998). Propositions in propositional logic provable only by indirect proofs. *Mathematical Logic Quarterly*.

Hallnäs, L. (1988). *On normalization of proofs in set theory*. Dissertationes Mathematicae. Warsaw: Institute of Mathematics, Polish Academy of Sciences.

– (1991). Partial inductive definitions. *Theoretical Computer Science*.

– (2006). On the proof-theoretic foundation of general definition theory. *Proof theoretic Semantics, Synthese* 148.

Hallnäs, L. and P. Schroeder-Heister (1991). A proof theoretic characterization of logic programming II: programs as definitions. *Journal of Logic and Computation* 1.

Prawitz, D. (1965). *Natural Deduction: A Proof-Theoretical Study*. Stockholm: Almqvist & Wiksell.

– (1971). Ideas and results in proof theory. In: *Proceedings of the Second Scandinavian Logic Symposium*. Ed. by J. E. Fenstad. Amsterdam: North Holland.
– (1973a). On the idea of a general proof theory. *Synthese*.
– (1973b). Towards a general proof theory. In: *Methodology and the Philosophy of Science IV*. Ed. by P. Suppes. Amsterdam: North Holland.
Schroeder-Heister, P. (1993). Rules of definitional reflection. In: *Proceedings of the 8th Annual IEEE Symposium on Logic in Computer Science, Montreal*. Montreal: IEEE Computer Society Press.
– (2006). Validity concepts in proof-theoretic semantics. *Synthese* 148.
– (2012). Proof-theoretic semantics, self-contradiction, and the format of deductive reasoning. *Topoi*.
Schroeder-Heister, P. and L. Tranchini (2017). Ekman's paradox. *Notre Dame Journal of Formal Logic*, 567–581.

Truth-Value Constants in Multi-Valued Logics

Nissim Francez and Michael Kaminski

Abstract In some presentations of classical and intuitionistic logics, the object-language is assumed to contain (two) *truth-value constants*: $\top$ (verum) and $\bot$ (falsum), that are, respectively, true and false under every bivalent valuation. We are interested to define and study analogical constants $\ddagger_i$, $1 \leq i \leq n$, that in an arbitrary multi-valued logic over truth-values $\mathcal{V} = \{v_1, \ldots, v_n\}$ have the truth-value v_i under every (multi-valued) valuation. As is well known, the absence or presence of such constants has a significant deductive impact on the logics studied. We define such constants proof-theoretically via their associated I/E-rules in a natural-deduction proof system. In particular, we propose a generalization of the notions of *contradiction* and *explosiveness* of a logic to the context of multi-valued logics.

1 Introduction

In some presentations of classical logic, and more often of intuitionistic logic, the object-language is assumed to contain (two) *truth-value constants*[1] (0-ary connectives): $\top$ (verum) and $\bot$ (falsum). These are, respectively, true under every valuation (truth-value assignment) and false under every valuation.

In classical and intuitionistic natural-deduction (ND) proof-systems, the respective introduction and elimination rules (I/E-rules) for those constants are (with φ ranging over object-language formulas)

Nissim Francez
Technion – Israel Institute of Technology, Haifa, Israel, e-mail: francez@cs.technion.ac.il

Michael Kaminski
Technion – Israel Institute of Technology, Haifa, Israel, e-mail: kaminski@cs.technion.ac.il

[1] Those constants should not be confused by other uses of those symbols as additional truth-values, as is done, for example, in bilattice logics.

T. Piecha and K. F. Wehmeier (eds.), *Peter Schroeder-Heister on Proof-Theoretic Semantics*, Outstanding Contributions to Logic 29,
https://doi.org/10.1007/978-3-031-50981-0_14

$$\frac{\bot}{\varphi}\ (\bot E) \qquad\qquad \frac{\varphi}{\top}\ (\top I)$$

and there are no $(\bot I)$ and $(\top E)$ rules.

Recall that $(\bot E)$ is the proof-theoretic source of the *explosion* property of logics admitting this rule.

We are interested to study analogical constants in arbitrary multi-valued logics (*mv*-logics) and their associated I/E-rules. As is well known, the absence or presence of such constants may have a significant deductive impact on such logics, in several respects[2]. For example:

1. In Avron (2014) it is shown that as long as the propositional constants **t** and **f** are not included in the object-language, any language-preserving extension of any important fragment of the relevance logics R and RMI can have only classical tautologies as theorems.
2. In Avron and Konikowska (2005), it is shown that if the object-language of a multi–valued logic contains truth-value constant for *every* truth-value, there is a significant simplification of sequent calculi for such a language. Those constants are used to facilitate a distinction between two kinds of derivations in proof-systems for such logics.
3. In Pelletier and Hazen (2019) it is observed that adding constants for the four truth-values of FDE (augmented with a classical-like conditional) renders the language functionally complete (a property not holding for the augmented FDE without such constants).

2 Located formulas and sequents

For $n \geq 2$, let $\mathcal{V} = \{v_1, \ldots, v_n\}$ be a collection of truth-values underlying a multi-valued logic $\mathcal{L}^n$ with a propositional object-language L_n with some unspecified connectives defined by truth-tables over $\mathcal{V}$. Furthermore, $\mathcal{L}^n$ is assumed to contain the n constants $\ddagger_i$, $1 \leq i \leq n$, where, by definition, for every valuation σ

$$\sigma[\![\ddagger_i]\!] = v_i.$$

Let $\hat{n} = \{1, \ldots, n\}$.

Definition 2.1 A *located formula* (l-formula) is a pair (φ, k), where φ is an object-language formula and $k \in \hat{n}$.

The intended interpretation of (φ, k) is that φ is associated with the truth-value $v_k \in \mathcal{V}$.

[2] Those examples are mentioned only for exemplifying the existence of an impact of the presence of the constants. The actual content of those examples is not directly related to what we have to say.

Definition 2.2 A *located sequent* (l-sequent) Π has the form $\Gamma : \Delta$, where Γ, Δ are (possibly empty) finite collections[3] of l-formulas.

We use $\mathbf{\Pi}$ for sets of l-sequents. Let σ range over valuations, mapping formulas to truth-values in $\mathcal{V}$; for atomic sentences the mapping is arbitrary, and it is extended to compound formulas so as to respect the truth-tables of the operators. Below, we define the central semantic notions as applicable to l-sequents.

Definition 2.3 *Satisfaction* is defined as follows:

$$\sigma \models \Gamma : \Delta \iff$$
$$\text{if } \sigma[\![\varphi]\!] = v_k \text{ for all } (\varphi, k) \in \Gamma, \text{ then } \sigma[\![\psi]\!] = v_j \text{ for some } (\psi, j) \in \Delta.$$

Call such a (ψ, j) a *witness* for $\Gamma : \Delta$'s satisfaction by σ.

Consequence is defined as follows:

$$\mathbf{\Pi} \models \Pi \;:\Longleftrightarrow\; \text{for every } \sigma\text{: } \sigma \models \Pi' \text{ for all } \Pi' \in \mathbf{\Pi} \text{ implies } \sigma \models \Pi.$$

We are interested in proof-systems sound and (strongly) complete for this consequence relation. In Francez and Kaminski (2019) and Kaminski and Francez (2021), we present natural-deduction proof-systems over l-sequents sound and (strongly) complete for the above consequence relation, constructed from the truth-tables in a uniform way. The multi-valued ND-systems $\mathcal{N}^n$ (over l-sequents) with their structural and logical rules for an arbitrary n-ary connective are presented in the appendix.

l-sequents have a natural way to represent the analogue of inconsistency (or non-satisfiability) in a multi-valued setting.

Definition 2.4 Consider any *mv*-logic $\mathcal{L}^n$. The $\mathcal{L}^n$ *mv-contradictions* are the $\mathcal{N}^n$-derivable l-sequents $\Gamma : \emptyset$, in which case Γ is *mv*-inconsistent.

Clearly, for every valuation σ and for any *mv*-consistent Γ, $\sigma \not\models \Gamma : \emptyset$.

Definition 2.5 $\mathcal{L}^n$ is *mv-explosive* iff for *every* Δ:

$$\Gamma : \emptyset \vdash_{\mathcal{N}^n} \Gamma : \Delta.$$

This notion of multi-valued explosion, the derivability of every l-sequent from an *mv*-contradiction, is the natural generalization of the usual notion of explosion.

Observation 2.6 *Due to the presence of the right-weakening structural rule (see the appendix),* every *mv-inconsistent* $\mathcal{L}^n$ *is explosive.*

[3] The exact nature of a collection, e.g., a set or a multi-set, depends on the specific logic being defined.

3 Bivalent l-sequents

As a preliminary step towards the introduction of general truth-value constants, we recast the traditional I/E-rules for $\bot$ and $\top$ in terms of l-sequents. For readability, we use the mnemonic $\mathcal{V} = \{t, f\}$ instead of $\{v_1, v_2\}$.

I-rules

From their definition, the located constants $(\bot, t)$ and $(\top, f)$ can never be non-trivially introduced, because there is no σ s.t. $\sigma[\![\bot]\!] = t$, and no σ s.t. $\sigma[\![\top]\!] = f$. The trivial I-rules

$$\frac{\Gamma : \Delta}{\Gamma : \Delta, (\top, f)}\ (\top I_f) \qquad \frac{\Gamma : \Delta}{\Gamma : \Delta, (\bot, t)}\ (\bot I_t)$$

are just instances of (right) weakening.

On the other hand, every σ is such that $\sigma[\![\bot]\!] = f$, and $\sigma[\![\top]\!] = t$. Thus, we have

$$\frac{}{\Gamma : \Delta, (\top, t)}\ (\top I_t) \qquad \frac{}{\Gamma : \Delta, (\bot, f)}\ (\bot I_f)$$

Note that the assignments of f to $\bot$ and t to $\top$ depend on no (sub)formulas of Γ or Δ, so the rules $(\bot I_f)$ and $(\top I_t)$ *have no premises*, and are, therefore, special cases of the general I/E-rules in the appendix.

E-rules

$$\frac{\Gamma : \Delta, (\bot, t)}{\Gamma : \Delta}\ (\bot E_t) \qquad \frac{\Gamma : \Delta, (\top, f)}{\Gamma : \Delta}\ (\top E_f)$$

Clearly, $(\bot, t)$ cannot be a witness for the premise, and there must be one in Δ itself, which is, therefore, also a witness for the conclusion of $(\bot E_t)$; similarly for $(\top E_f)$. It is quite remarkable that the $(\bot E_t)$ and $(\top E_f)$ rules lead to explosion (as shown below for the general case), naturally generalizing $(\bot E)$. Again, both rules can be seen as special cases of the general E-rules in the appendix.

4 The general case

We now apply our uniform construction, yielding the general I/E-rules in the appendix, to obtain $(\ddagger_k I)$ and $(\ddagger_k E)$ for every $1 \leq k \leq n$.

$(\ddagger I)$: Again, from their definition, the located constants $(\ddagger_k, k')$ can never be non-trivially introduced for $k \neq k'$, because there is no σ s.t. $\sigma[\![\ddagger_k]\!] = v_{k'}$. The trivial I-rules are again instances of (right) weakening.
On the other hand, every σ is such that $\sigma[\![\ddagger_k]\!] = v_k$. Thus, we have for every

$1 \le k \le n$:

$$\frac{}{\Gamma : \Delta, (\ddagger_k, k)} \ (\ddagger_k I_k)$$

Note again that the assignments of v_k to $\ddagger_k$ depend on no (sub)formulas, so the rules $(\ddagger_k I_k)$ *have no premises*, and are special cases of the general I-rules in the appendix.

$(\ddagger E)$: For every $1 \le k \le n$:

$$\frac{\Gamma : \Delta, (\ddagger_k, k')}{\Gamma : \Delta} \ (\ddagger_k E_{k'}), \quad k \neq k'.$$

$(\ddagger_k, k')$ cannot be a witness for the premise, and there must be one in Δ itself, which is, therefore, also a witness for the conclusion of $(\ddagger_k E_k)$. The $(\ddagger_k E_{k'})$ rules can be seen as special cases of the general E-rules in the appendix.

Remark 4.1 Note that the I/E-rules or the truth-value constant are *harmonious* (Dummett, 1993; Francez, 2015, Ch. 3), albeit *vacuously*: there are no maximal formulas consisting of located constants. This generalizes the vacuous harmony of $\top$ and $\bot$.

It is once again quite remarkable that the $(\ddagger_k E_{k'})$ rules lead to explosion (see Proposition 4.2), naturally generalizing $(\bot E_t)$ and $(\top E_f)$.

Proposition 4.2 *For any* Δ, $1 \le k,\ k' \le n,\ k \neq k'$*:*

$$\Gamma : (\ddagger_k, k') \vdash_{\mathcal{N}^n} \Gamma : \Delta.$$

Proof The derivation is

$$\frac{\dfrac{\Gamma : (\ddagger_k, k')}{\Gamma : \Delta, (\ddagger_k, k')} \ (WRs)}{\Gamma : \Delta} \ (\ddagger_k E_{k'}).$$

5 Conclusion

We have shown how to introduce truth-value constants into an arbitrary multi-valued logic over l-sequents. This is done by devising I/E-rules for those constants, generalizing the traditional I/E-rules for '$\top$' and '$\bot$' in classical logic.

We also generalized the classical notions of a contradiction and an explosive logic to our setting.

Interesting issues that might be investigated in continuation of this work include:

1. All our truth-value constants are the analogs of *additive* constants in logics of formulas. How can an analog of *multiplicative* constants be defined?
2. All our logics are explosive. How can *paraconsistent* multi-valued l-sequent logics be defined to include such constants?

Appendix: The proof-system $\mathcal{N}^n$

Initial l-sequents: For every $1 \leq i \leq n$: $\Gamma, (\varphi, i) : \Delta, (\varphi, i)$. Those initial l-sequents render the following *Weakening*-rules admissible:

$$\frac{\Gamma : \Delta}{\Gamma, (\varphi, i) : \Delta} (WL_i) \qquad \frac{\Gamma : \Delta}{\Gamma : \Delta, (\varphi, i)} (WR_i)$$

Shifting rules:

$$\frac{\Gamma, (\varphi, i) : \Delta}{\Gamma : \Delta, \varphi \times (\hat{n} \setminus \{i\})} (\overrightarrow{s}_i) \qquad \frac{\Gamma : \Delta, (\varphi, i)}{\Gamma, (\varphi, j) : \Delta} (\overleftarrow{s}_{i,j}), j \neq i$$

The rule says, roughly, that the n truth-values are *exhaustive*.

1. If φ does not have the truth-value v_i, it has the truth-value v_j for some $j \neq i$.
2. If φ has some truth-value v_i, it does not have any truth-value v_j, $j \neq i$.

Coordination:

$$\frac{\Gamma : \Delta, (\varphi, i) \quad \Gamma : \Delta, (\varphi, j)}{\Gamma : \Delta} (c_{i,j}), i \neq j$$

This rule is a generalization to multi-valued logic of a structural rule by the same name in Rumfitt (2000), in a bilateral setting.

The rule says, roughly, that different truth-values associated with the same formula are contradictory, which allows one to eliminate these formula-truth-value pairs, as in the rule of propositional resolution.

Operational rules: Those are not really used here, and are presented for completeness only. The guiding lines for the construction are the following, expressed in terms of a generic p-ary operator, say '$*$'. Notably, the operational rules for the constants can be seen as limit cases of those rules ($p = 0$).

($*I$): Such rules introduce a conclusion $\Gamma : \Delta, (*(\varphi_1, \ldots, \varphi_p), k)$. In general, if in the truth-table for '$*$' the values v_{i_j} for φ_j, $1 \leq j \leq p$, yield the value v_k for $*(\varphi_1, \ldots, \varphi_p)$, then there is a rule

$$\frac{\{\Gamma : \Delta, (\varphi_j, i_j) \mid 1 \leq j \leq p\}}{\Gamma : \Delta, (*(\varphi_1, \ldots, \varphi_p), k)} (*I_{i_1,\ldots,i_p,k})$$

The rule $(*I_{i_1,\ldots,i_p,k})$ has, thus, p premises.

($*E$): Such rules have a major premise $\Gamma : \Delta, (*(\varphi_1, \ldots, \varphi_p), k)$. The rule $(*E_k)$ has the form

$$\frac{\Gamma : \Delta, (*(\varphi_1, \ldots, \varphi_p), k) \quad \{\Gamma, (\varphi_1, k_1), \ldots, (\varphi_p, k_p) : \Delta \mid *(v_{k_1}, \ldots, v_{k_p}) = v_k\}}{\Gamma : \Delta}$$

for each $k = 1, \ldots, n$.

A detailed discussion of this system, presented in a different but equivalent notation, can be found in Francez and Kaminski (2019). Equivalent sequent calculi can be found in Kaminski and Francez (2021).

References

Avron, A. (2014). The classical constraint on relevance. *Logica Universalis* 8, 1–15.

Avron, A. and B. Konikowska (2005). Multi-valued calculi for logics based on non-determinism. *IGPL* 13, 365–387.

Dummett, M. (1993). *The Logical Basis of Metaphysics*. Cambridge, MA: Harvard University Press.

Francez, N. (2015). *Proof-theoretic Semantics*. London: College Publications.

Francez, N. and M. Kaminski (2019). On poly-logistic natural-deduction for finitely-valued propositional logics. *Journal of Applied Logic* 6 (2). Special issue for papers presented at ISRALOG17, Haifa, October 2017, 255–288.

Kaminski, M. and N. Francez (2021). Calculi for multi-valued logics. *Logica Universalis* 15.

Pelletier, F. J. and A. P. Hazen (2019). K3, L3, LP, RM3, A3, FDE, M: how to make many-valued logics work for you. In: *New Essays on Belnap–Dunn Logic*. Ed. by H. Omori and H. Wansing. Synthese Library (Studies in Epistemology, Logic, Methodology, and Philosophy of Science), vol 418. https://doi.org/10.1007/978-3-030-31136-0-11. Cham: Springer.

Rumfitt, I. (2000). 'Yes' and 'No'. *Mind* 169, 781–823.

Counterfactual Assumptions and Counterfactual Implications

Bartosz Więckowski

Abstract We define intuitionistic subatomic natural deduction systems for reasoning with elementary *would*-counterfactuals and causal *since*-subordinator sentences. The former kind of sentence is analysed in terms of counterfactual implication, the latter in terms of factual implication. Derivations in these modal proof systems make use of modes of assumptions which are sensitive to the factuality status of the formula that is to be assumed. This status is determined by means of the reference proof system on top of which a modal proof system is defined. The introduction and elimination rules for counterfactual (resp. factual) implication draw on this status. It is shown that derivations in the systems normalize and that normal derivations have the subexpression/subformula property. An intuitionistically acceptable proof-theoretic semantics is formulated in terms of canonical derivations. The systems are applied to so-called counterpossibles and to related constructions.

Key words: assumption, conditional logic, counterfactuals, counterpossibles, intuitionistic logic, natural deduction, proof-theoretic semantics

1 Introduction

The notion of assumption is essential to reasoning insofar as reasoning can be characterized as the activity of drawing conclusions from assumptions. It is the purpose of natural deduction systems (Gentzen, 1934; Jaśkowski, 1934) to depict this inferential activity as closely as possible and to lay it down formally in terms of inference rules. In his study of the notion of assumption in proof systems Schroeder-Heister (2004), Peter Schroeder-Heister stresses, using a tree-style format, the following asymmetry between assumptions and assertions (i.e., conclusions) in natural deduction:

Bartosz Więckowski
Institut für Philosophie, Goethe-Universität Frankfurt am Main, Germany,
e-mail: wieckowski@em.uni-frankfurt.de

T. Piecha and K. F. Wehmeier (eds.), *Peter Schroeder-Heister on Proof-Theoretic Semantics*, Outstanding Contributions to Logic 29,
https://doi.org/10.1007/978-3-031-50981-0_15

there is only an unspecific way of introducing assumptions, but there is both an unspecific and a specific way of introducing assertions. In order to introduce an *assumption* of formula A, one only has to state A as an assumption. Schroeder-Heister calls this way of introducing an assumption unspecific, because "the form of A is not specified" (Schroeder-Heister, 2004, p. 27). In order to introduce an assertion of A in an unspecific way, one has to proceed like in the case of introducing an assumption of A, thereby making A dependent on itself. To introduce an assertion of A in a specific way, one has to derive A as a conclusion using an inference rule (where A is an axiom, in case the inference rule considered has no premisses). Schroeder-Heister argues that this asymmetry is responsible for limitations of expressive power and considers proof systems which, like the sequent calculus in Gentzen (1934), on his preferred reading, allow for both an unspecific and a specific way of introducing assumptions.

In what follows, we take up Schroeder-Heister's call to contribute to the exploration of the notion of assumption and of its significance to philosophical logic (cf. Schroeder-Heister, 2004, p. 45). However, we shall remain within the confines of natural deduction and suggest a way to widen its conception of assumption so as to put natural deduction in a position to "capture and codify reasoning" (Schroeder-Heister, 2004, p. 28) with *would-counterfactuals*, i.e., constructions of the form

(1) If A were the case, B would be the case.

More precisely, the aim of this contribution is to outline an intuitionistically (or constructively; cf. Dalen, 2002) acceptable formal approach to counterfactual reasoning and to the semantics of *would*-counterfactuals in terms of *modal proof systems* which are motivated directly by the practice of counterfactual inference making. Specifically, the systems use different *modes of making assumptions*.

Modes of assumptions, as we shall understand them in what follows, are dependent on the *factuality status* of the formula that is to be assumed. In a modal proof system this status is determined by means of a *reference proof system* on top of which the modal system is defined. In a nutshell, we assign factual status to a formula A, in case A has been derived *canonically* (i.e., by means of an application of an introduction rule in the last inference step; cf. Dummett, 1991; Prawitz 2006; 2012) in the reference proof system S. We shall distinguish three modes of making assumptions in modal proof systems. Intuitively, in order to assume A in the *factual* mode, we need to make sure that a canonical derivation of A in S has been constructed. To assume A in the *counterfactual* mode, we need to make sure that a canonical derivation of A in S has not been constructed. Finally, to assume A in the *independent* mode, we just assume A (without any proviso).

Whereas sentences of the form (1) are most adequately used in case A does not count as a fact, *causal since-subordinator sentences* of the form

(2) Since A is the case, it is the case that B.

are most adequately used in case A does count as a fact (e.g., Dancygier and Sweetser, 2000, p. 126). Due to their sensitivity to the factuality status of the formula that is to be assumed, our modal natural deduction systems will be also in a position to deal with such *since*-constructions. Our analysis will focus exclusively on such intuitive uses. More specifically, constructions of the first form will be analysed as counterfactual implications, and those of the second form as factual implications. The meaning of the former will be explained by appeal to counterfactual assumptions, that of the latter by appeal to factual assumptions. In order to outline the main idea more clearly, the analysis will be confined to very elementary instances of (1) and (2).

The idea of using different ways of making assumptions for the purpose of a proof-theoretic analysis of counterfactual reasoning goes back at least to Thomason's (1970) Fitch-style natural deduction system FCS for Stalnaker's (1970) preferred conditional logic CS. However, not only the details but also the motivation of modal natural deduction systems differs from that underlying Thomason's FCS. We do not aim to formulate a structural proof system which is equivalent with a specific Hilbert-style axiom system for some specific conditional logic whose semantics is given non-inferentially. Rather, we shall develop our systems without any Hilbert-system in mind. One reason for this is that the axioms of such systems ultimately depend for their intelligibility on specific model-theoretic conditions. As a result, axioms may inherit undesirable features from these conditions. Recall, for example, that D. Lewis felt he should apologize for the "long and obscure" axiom $(\phi \mathrel{\Box\!\!\rightarrow} {\sim}\psi) \vee (((\phi \,\&\, \psi) \mathrel{\Box\!\!\rightarrow} \chi) \equiv (\phi \mathrel{\Box\!\!\rightarrow} (\psi \supset \chi)))$ of his simplest Hilbert-style system for VC (cf. Lewis, 2011, p. 133). It is for this reason that he preferred an equivalent axiomatization of VC in terms of his notion of "comparative possibility" of possible worlds (cf. Lewis, 2011, §2.5) rather than in terms of his *would*-counterfactual $\mathrel{\Box\!\!\rightarrow}$. Furthermore, Hilbert-systems are not indispensable, if our aim is to formulate inferentially intuitive natural deduction systems which have good structural properties (e.g., normalization, subformula property) and which admit a *proof-theoretic semantics* (see Francez, 2015; Kahle and Schroeder-Heister, 2006; Piecha and Schroeder-Heister, 2016; Schroeder-Heister, 2018; Wansing, 2001) that is acceptable from an intuitionistic point of view (see Dummett, 1991, Prawitz 2006; 2012; for model-theoretic semantical considerations on Hilbert-systems for intuitionistic conditional logic see, e.g., Ciardelli and Liu, 2020 and Weiss, 2018).

Specifically, the intended proof-theoretic semantics is to be *semantically autarkic*, i.e., not defined in terms of a structural proof system that is itself defined by appeal to a formal semantics of a different kind (cf. Więckowski, 2021a). Since our approach to counterfactual reasoning is intended to be acceptable from an *intuitionistic* perspective, it will support a verification oriented conception of truth (cf. Dummett, 1991, Prawitz 2006; 2012).

The way in which this contribution is organized reflects the architecture of a modal natural deduction system. Section 2 defines the kind of proof system that will be used as reference proof system of such a system. Modal natural deduction systems for reasoning with factual and counterfactual implications will be then defined in Section 3. This section also contains the main results of this contribution (i.e., normalization and the subexpression/subformula property for the intended modal systems) and presents a proof-theoretic semantics with the desired properties. In Section 4, the modal proof systems will be used in an analysis of so-called counterpossibles (see, e.g., Berto,

French, Priest, and Ripley, 2018 and Williamson, 2007) and related constructions. Section 5 makes some concluding remarks.

2 Reference proof systems

We shall now define the kind of reference proof system on top of which modal natural deduction systems will be defined in Section 3. We first specify the language and then formulate, in three steps, the intended kind of reference proof system. We choose subatomic natural deduction systems (cf. Więckowski 2011; 2016; 2021b; 2021a; 2023) as our reference proof systems, since we need, as mentioned above, a way to obtain canonical derivations of atomic sentences, if A in (1) (resp. (2)) is atomic and if we want to assume A in the counterfactual (factual) mode. Unlike standard natural deduction systems, subatomic natural deduction systems maintain introduction and elimination rules also for atomic sentences.

2.1 Subatomic systems

We first define the language $L0$ that we shall use in the formulation of reference proof systems.

Definition 2.1 $L0$ is a first-order language which is defined in the usual inductive way. C and $\mathcal{P}$ are the sets of individual (or nominal) constants (metavariables: α, α_i) and n-ary predicate constants (metavariables: φ^n, φ^n_i), respectively. $L0$-formulae are atomic formulae (form: $\varphi^n o_1 \dots o_n$), absurdity ($\bot$), conjunctions ($A \& B$), disjunctions ($A \vee B$), implications ($A \supset B$), universal quantifications ($\forall x A$), and existential quantifications ($\exists x A$). In addition to defined operators for negation and bi-implication, $L0$ contains also a special non-primitive identity predicate:

1. $\neg A =_{def} A \supset \bot$
2. $A \leftrightarrow B =_{def} (A \supset B) \;\&\; (B \supset A)$
3. Let φ^n be an n-ary predicate constant.

$$\begin{aligned} K^n_{\varphi^n}(o_1, o_2) =_{def} \forall z_1 \dots \forall z_{n-1} \forall z_n ((&\varphi^n o_1 z_2 \dots z_n \leftrightarrow \varphi^n o_2 z_2 \dots z_n) \\ &\& \; (\varphi^n z_1 o_1 \dots z_n \leftrightarrow \varphi^n z_1 o_2 \dots z_n) \\ &\& \dots \& \; (\varphi^n z_1 \dots z_{n-1} o_1 \leftrightarrow \varphi^n z_1 \dots z_{n-1} o_2)). \end{aligned}$$

Let $\varphi_1^{k_1}, \dots, \varphi_m^{k_m}$ be all the predicate constants in $\mathcal{P}$, where φ_i is k_i-ary.

$$o_1 \ddot{=} o_2 =_{def} K^{k_1}_{\varphi_1}(o_1, o_2) \;\&\; \dots \;\&\; K^{k_m}_{\varphi_m}(o_1, o_2).$$

Atm is the set of atomic sentences. $Atm(\alpha) =_{def} \{A \in Atm : A$ contains at least one occurrence of $\alpha \in C\}$ and $Atm(\varphi^n) =_{def} \{A \in Atm : A$ contains an occurrence of $\varphi^n \in \mathcal{P}\}$. Due to the presence of $\doteq$ in $L0$ we take $\mathcal{P}$ to be finite.

The first step of the definition of the intended kind of reference proof system consists in the definition of a subatomic system. In such systems we may introduce and eliminate atomic sentences using term assumptions for non-logical constants.

Definition 2.2 A *subatomic system* $\mathcal{S}$ is a pair $\langle \mathcal{I}, \mathcal{R} \rangle$, where $\mathcal{I}$ is a *subatomic base* and $\mathcal{R}$ is a set of *introduction and elimination rules for atomic sentences*. $\mathcal{I}$ is a 3-tuple $\langle C, \mathcal{P}, v \rangle$, where C and $\mathcal{P}$ are as above, and where v is such that:

1. For any $\alpha \in C$, $v : C \rightarrow \wp(Atm)$, where $v(\alpha) \subseteq Atm(\alpha)$.
2. For any $\varphi^n \in \mathcal{P}$, $v : \mathcal{P} \rightarrow \wp(Atm)$, where $v(\varphi^n) \subseteq Atm(\varphi^n)$.

We let $\tau\Gamma =_{def} v(\tau)$ for any $\tau \in C \cup \mathcal{P}$, and call $\tau\Gamma$ the set of *term assumptions* for τ. $\mathcal{R}$ contains I/E-rules of the following form:

$$\dfrac{\begin{matrix}\mathcal{D}_0 \\ \varphi_0^n\Gamma\end{matrix} \quad \begin{matrix}\mathcal{D}_1 \\ \alpha_1\Gamma\end{matrix} \quad \dots \quad \begin{matrix}\mathcal{D}_n \\ \alpha_n\Gamma\end{matrix}}{\varphi_0^n\alpha_1 \dots \alpha_n}\,(as\mathrm{I}) \qquad\qquad \dfrac{\begin{matrix}\mathcal{D}_1 \\ \varphi_0^n\alpha_1 \dots \alpha_n\end{matrix}}{\tau_i\Gamma}\,(as\mathrm{E}_i)$$

where $\varphi_0^n\alpha_1 \dots \alpha_n \in \varphi_0^n\Gamma \cap \alpha_1\Gamma \cap \dots \cap \alpha_n\Gamma$ | where $i \in \{0, \dots, n\}$ and $\tau_i \in \{\varphi_0^n, \alpha_1, \dots, \alpha_n\}$

Intuitively, a term assumption stores the elementary information which is associated with a non-logical constant and the asI-rule allows us to establish the truth of an atomic sentence on the basis of this information.

Definition 2.3 *Derivations in $\mathcal{S}$-systems.*

Basic step. Any term assumption $\tau\Gamma$ and any atomic sentence A (i.e., a derivation from the open assumption of A) is an $\mathcal{S}$-derivation.

Induction step. If $\mathcal{D}_i$, for $i \in \{0, \dots, n\}$, are $\mathcal{S}$-derivations, then an $\mathcal{S}$-derivation can be constructed by means of the asI/E-rules displayed above.

Example 2.4 Let the $\mathcal{S}$-system contain only two predicates (i.e., F, R) and two nominal constants (i.e., a, b), and let the term assumptions be as follows: $F\Gamma = \{Fa, Fb\}$, $R\Gamma = \{Rab, Rba\}$, $a\Gamma = \{Fa, Rab, Rba\}$, and $b\Gamma = \{Fb, Rab, Rba\}$.

$$(3) \qquad \dfrac{\dfrac{\dfrac{\dfrac{R\Gamma \quad a\Gamma \quad b\Gamma}{Rab}\,(as\mathrm{I})}{R\Gamma}\,(as\mathrm{E}_0) \qquad b\Gamma \qquad \dfrac{Fa}{a\Gamma}\,(as\mathrm{E}_1)}{Rba}\,(as\mathrm{I})}{a\Gamma}\,(as\mathrm{E}_2)$$

Derivation (3) contains two detours and is, therefore, not in normal form (or normal).

Definition 2.5 *Detour conversion for as:*

$$\dfrac{\dfrac{\begin{matrix}\mathcal{D}_0 \\ \varphi_0^n\Gamma\end{matrix} \quad \begin{matrix}\mathcal{D}_1 \\ \alpha_1\Gamma\end{matrix} \quad \dots \quad \begin{matrix}\mathcal{D}_n \\ \alpha_n\Gamma\end{matrix}}{\varphi_0^n\alpha_1 \dots \alpha_n}\,(as\mathrm{I})}{\tau_i\Gamma}\,(as\mathrm{E}_i) \quad \text{conv} \quad \begin{matrix}\mathcal{D}_i \\ \tau_i\Gamma\end{matrix}$$

Theorem 2.6 *Any derivation $\mathcal{D}$ in an $\mathcal{S}$-system can be transformed into a normal $\mathcal{S}$-derivation.*

Proof Immediate. □

Definition 2.7 Let $\mathcal{D}$ be a derivation in an $\mathcal{S}$-system.

1. An *$\mathcal{S}$-unit* in $\mathcal{D}$ is either an occurrence of (i) an atomic sentence or (ii) a term assumption $\tau\Gamma$ in $\mathcal{D}$. We use $U_{\mathcal{S}}, U'_{\mathcal{S}}$ (possibly with subscripts) for $\mathcal{S}$-units.
2. In case $U_{\mathcal{S}}$ is a term assumption $\tau\Gamma$ in $\mathcal{D}$, τ is *the expression in $U_{\mathcal{S}}$*.

Theorem 2.8 *If $\mathcal{D}$ is a normal $\mathcal{S}$-derivation of an $\mathcal{S}$-unit $U_{\mathcal{S}}$ from a set of $\mathcal{S}$-units Γ, then each $\mathcal{S}$-unit in $\mathcal{D}$ is a subexpression of an expression in $\Gamma \cup \{U_{\mathcal{S}}\}$.*

Proof Immediate. □

2.2 Subatomic identity systems

The next step of the definition of the intended kind of reference proof system consists in the extension of subatomic systems to subatomic identity systems by adding I/E-rules for non-primitive identity sentences. Roughly, two nominal constants are identical in this sense if they are indistinguishable with respect to the elementary information associated with them (cf. Definition 2.2).

Definition 2.9 Atomic sentences $\varphi(\alpha_1)$ and $\varphi(\alpha_2)$ are *mirror atomic sentences* if and only if they are exactly alike except that the former contains occurrences of α_1 at all the places at which the latter contains occurrences of α_2, and vice versa.

Definition 2.10 A *subatomic identity system* $\mathcal{S}^{\ddot{=}}$ is a 3-tuple $\langle \mathcal{I}, \mathcal{R}, \mathcal{R}^{\ddot{=}} \rangle$ which extends a subatomic system with a set $\mathcal{R}^{\ddot{=}}$ of *I/E-rules for $\ddot{=}$-sentences*:

$$\dfrac{\begin{matrix}[\varphi_1(\alpha_1)]^{(1_1)} & [\varphi_1(\alpha_2)]^{(1_2)} & & [\varphi_k(\alpha_1)]^{(k_1)} & [\varphi_k(\alpha_2)]^{(k_2)} \\ \mathcal{D}_{1_1} & \mathcal{D}_{1_2} & & \mathcal{D}_{k_1} & \mathcal{D}_{k_2} \\ \varphi_1(\alpha_2) & \varphi_1(\alpha_1) & \dots & \varphi_k(\alpha_2) & \varphi_k(\alpha_1)\end{matrix}}{\alpha_1 \ddot{=} \alpha_2}\ {\scriptstyle (\ddot{=}\mathrm{I}),\, 1_1,\, 1_2, \dots,\, k_1,\, k_2}$$

$$\dfrac{\begin{matrix}\mathcal{D}_1 & \mathcal{D}_{i_2} \\ \alpha_1 \ddot{=} \alpha_2 & \varphi_i(\alpha_1)\end{matrix}}{\varphi_i(\alpha_2)}\ (\ddot{=}\mathrm{E}_i 1) \qquad \dfrac{\begin{matrix}\mathcal{D}_1 & \mathcal{D}_{i_1} \\ \alpha_1 \ddot{=} \alpha_2 & \varphi_i(\alpha_2)\end{matrix}}{\varphi_i(\alpha_1)}\ (\ddot{=}\mathrm{E}_i 2)$$

where $i \in \{1, \dots, k\}$, and $\varphi_i(\alpha_1)$ and $\varphi_i(\alpha_2)$ are mirror atomic sentences.

Remark 2.11 In the $\ddot{=}$I/E-rules the operators figuring in the definiens of $\ddot{=}$ (Definition 2.1) have been absorbed, so to speak, into the metalanguage.

Definition 2.12 *Derivations in $\mathcal{S}^{\ddot{=}}$-systems.*

Basic step. Any derivation in an $\mathcal{S}$-system and any identity sentence $\alpha_i \doteqdot \alpha_j$ (i.e., a derivation from the open assumption of $\alpha_i \doteqdot \alpha_j$), where possibly $i = j$, is an $\mathcal{S}^{\doteqdot}$-derivation.

Induction step. If $\mathcal{D}_1, \mathcal{D}_{1_1}, \mathcal{D}_{1_2}, \ldots, \mathcal{D}_{k_1}, \mathcal{D}_{k_2}, \mathcal{D}_{i_1}$, and $\mathcal{D}_{i_2}$ are $\mathcal{S}$-derivations, then an $\mathcal{S}^{\doteqdot}$-derivation can be constructed using the $\doteqdot$I/E-rules listed above.

Example 2.13 For simplicity, let the $\mathcal{S}^{\doteqdot}$-system contain only one predicate (i.e., F) and two nominal constants (i.e., a and b). And let the term assumptions be as follows: $F\Gamma = \{Fa, Fb\}$, $a\Gamma = \{Fa\}$, and $b\Gamma = \{Fb\}$.

(4)
$$\dfrac{\dfrac{\dfrac{[Fa]^{(1_1)}}{F\Gamma}\,(asE_0) \qquad b\Gamma}{Fb}\,(asI) \qquad \dfrac{\dfrac{[Fb]^{(1_2)}}{F\Gamma}\,(asE_0) \qquad a\Gamma}{Fa}\,(asI)}{a \doteqdot b}\,(\doteqdot \mathrm{I}),\ 1_1,\ 1_2$$

(5)
$$\dfrac{\dfrac{\dfrac{[Fa]^{(1_1)}}{F\Gamma}\,(asE_0) \qquad \dfrac{[Fa]^{(1_1)}}{a\Gamma}\,(asE_1)}{Fa}\,(asI) \qquad \dfrac{\dfrac{[Fa]^{(1_2)}}{F\Gamma}\,(asE_0) \qquad \dfrac{[Fa]^{(1_2)}}{a\Gamma}\,(asE_1)}{Fa}\,(asI)}{a \doteqdot a}\,(\doteqdot \mathrm{I}),\ 1_1,\ 1_2$$

Example 2.14 Let $\varphi_1(\alpha), \ldots, \varphi_k(\alpha) \in Atm(\alpha)$ for any $\alpha \in C$. The following is a derivation in any $\mathcal{S}^{\doteqdot}$-system:

(6)
$$\dfrac{[\varphi_1(\alpha)]^{(1_1)} \quad [\varphi_1(\alpha)]^{(1_2)} \quad \ldots \quad [\varphi_k(\alpha)]^{(k_1)} \quad [\varphi_k(\alpha)]^{(k_2)}}{\alpha \doteqdot \alpha}\,(\doteqdot \mathrm{I}),\ 1_1, 1_2, \ldots, k_1, k_2$$

Remark 2.15 According to Example 2.14, $\alpha \doteqdot \alpha$ does not need to be postulated as an axiom. In particular, it is not declared, as it is usually the case, a conclusion of a zero premiss I-rule. Rather it is inferred, on a non-empty basis, by appeal to mirror formulae.

Definition 2.16 *Detour conversions for* $\doteqdot$:

$$\dfrac{\dfrac{\begin{matrix}[\varphi_1(\alpha_1)]^{(1_1)}\\ \mathcal{D}_{1_1}\\ \varphi_1(\alpha_2)\end{matrix} \quad \begin{matrix}[\varphi_1(\alpha_2)]^{(1_2)}\\ \mathcal{D}_{1_2}\\ \varphi_1(\alpha_1)\end{matrix} \quad \ldots \quad \begin{matrix}[\varphi_k(\alpha_1)]^{(k_1)}\\ \mathcal{D}_{k_1}\\ \varphi_k(\alpha_2)\end{matrix} \quad \begin{matrix}[\varphi_k(\alpha_2)]^{(k_2)}\\ \mathcal{D}_{k_2}\\ \varphi_k(\alpha_1)\end{matrix}}{\alpha_1 \doteqdot \alpha_2}\,(\doteqdot \mathrm{I}) \qquad \begin{matrix}\mathcal{D}_{i_2}\\ \varphi_i(\alpha_1)\end{matrix}}{\varphi_i(\alpha_2)}\,(\doteqdot \mathrm{E}_i 1) \qquad \text{conv} \qquad \begin{matrix}\mathcal{D}_{i_2}\\ [\varphi_i(\alpha_1)]\\ \mathcal{D}_{i_1}\\ \varphi_i(\alpha_2)\end{matrix}$$

$$\dfrac{\dfrac{\begin{matrix}[\varphi_1(\alpha_1)]^{(1_1)}\\ \mathcal{D}_{1_1}\\ \varphi_1(\alpha_2)\end{matrix} \quad \begin{matrix}[\varphi_1(\alpha_2)]^{(1_2)}\\ \mathcal{D}_{1_2}\\ \varphi_1(\alpha_1)\end{matrix} \quad \ldots \quad \begin{matrix}[\varphi_k(\alpha_1)]^{(k_1)}\\ \mathcal{D}_{k_1}\\ \varphi_k(\alpha_2)\end{matrix} \quad \begin{matrix}[\varphi_k(\alpha_2)]^{(k_2)}\\ \mathcal{D}_{k_2}\\ \varphi_k(\alpha_1)\end{matrix}}{\alpha_1 \doteqdot \alpha_2}\,(\doteqdot \mathrm{I}) \qquad \begin{matrix}\mathcal{D}_{i_1}\\ \varphi_i(\alpha_2)\end{matrix}}{\varphi_i(\alpha_1)}\,(\doteqdot \mathrm{E}_i 2) \qquad \text{conv} \qquad \begin{matrix}\mathcal{D}_{i_1}\\ [\varphi_i(\alpha_2)]\\ \mathcal{D}_{i_2}\\ \varphi_i(\alpha_1)\end{matrix}$$

Theorem 2.17 *Any derivation $\mathcal{D}$ in an $\mathcal{S}^{\doteqdot}$-system can be transformed into a normal $\mathcal{S}^{\doteqdot}$-derivation.*

Proof Cf. Więckowski (2016). □

Definition 2.18 Let $\mathcal{D}$ be a derivation in an $\mathcal{S}^{\doteqdot}$-system.

1. An $\mathcal{S}^{\doteqdot}$-*unit* in $\mathcal{D}$ is either an occurrence of (i) an atomic sentence, (ii) an identity sentence, or (iii) a term assumption $\tau\Gamma$ in $\mathcal{D}$. We use $U_{\mathcal{S}^{\doteqdot}}, U'_{\mathcal{S}^{\doteqdot}}$ (possibly with subscripts) for $\mathcal{S}^{\doteqdot}$-units.
2. In case $U_{\mathcal{S}^{\doteqdot}}$ is a term assumption $\tau\Gamma$ in $\mathcal{D}$, τ is *the expression in* $U_{\mathcal{S}^{\doteqdot}}$.

Theorem 2.19 *If $\mathcal{D}$ is a normal $\mathcal{S}^{\doteqdot}$-derivation of an $\mathcal{S}^{\doteqdot}$-unit $U_{\mathcal{S}^{\doteqdot}}$ from a set of $\mathcal{S}^{\doteqdot}$-units Γ, then each $\mathcal{S}^{\doteqdot}$-unit in $\mathcal{D}$ is a subexpression of an expression in $\Gamma \cup \{U_{\mathcal{S}^{\doteqdot}}\}$.*

Proof Cf. Więckowski (2016). □

2.3 Subatomic natural deduction systems

We now complete the definition of the intended kind of reference proof system. In order to reduce complexity and to focus on the main idea underlying modal proof systems, we define these reference systems, $\mathbf{I}(\mathcal{S}^{\doteqdot})$-systems, only for a fragment of $L0$.

Definition 2.20 *The language $L0'$.* $L0'$ is the fragment of $L0$ which comprises only $\bot$, atomic, $\doteqdot$-, and $\supset$-formulae.

Definition 2.21 *Derivations in $\mathbf{I}(\mathcal{S}^{\doteqdot})$-systems.*

Basic step. Any derivation in an $\mathcal{S}^{\doteqdot}$-system and any $L0'$-formula A (i.e., a derivation from the open assumption of A) is an $\mathbf{I}(\mathcal{S}^{\doteqdot})$-derivation.

Induction step. If $\mathcal{D}_1$ and $\mathcal{D}_2$ are $\mathbf{I}(\mathcal{S}^{\doteqdot})$-derivations, then an $\mathbf{I}(\mathcal{S}^{\doteqdot})$-derivation can be constructed by means of the following rules:

$$\dfrac{\begin{matrix}[A]^{(u)}\\ \mathcal{D}_1\\ B\end{matrix}}{A \supset B}\,(\supset\text{I}), u \qquad \dfrac{\begin{matrix}\mathcal{D}_1\\ A \supset B\end{matrix} \quad \begin{matrix}\mathcal{D}_2\\ A\end{matrix}}{B}\,(\supset\text{E}) \qquad \dfrac{\begin{matrix}\mathcal{D}_1\\ \bot\end{matrix}}{A}\,(\bot\text{i})$$

Definition 2.22 1. A derivation $\mathcal{D}$ of a formula A in an $\mathbf{I}(\mathcal{S}^{\doteqdot})$-system is a *canonical derivation* iff it derives A by means of an application of an I-rule in the last step of $\mathcal{D}$.

2. A canonical derivation $\mathcal{D}$ of A in an $\mathbf{I}(\mathcal{S}^{\doteqdot})$-system is a *canonical proof* of A in that system iff there are no applications of *as*-rules and no undischarged assumptions in $\mathcal{D}$.
3. The conclusions of canonical $\mathbf{I}(\mathcal{S}^{\doteqdot})$-derivations are $\mathbf{I}(\mathcal{S}^{\doteqdot})$-*theses* and the conclusions of $\mathbf{I}(\mathcal{S}^{\doteqdot})$-proofs are also $\mathbf{I}(\mathcal{S}^{\doteqdot})$-*theorems*.

Example 2.23 Let the $\mathbf{I}(\mathcal{S}^{\doteqdot})$-system maintain only one predicate (i.e., F) and two nominal constants (i.e., a, b). Let the term assumptions be like in Example 2.13: $F\Gamma = \{Fa, Fb\}$, $a\Gamma = \{Fa\}$, and $b\Gamma = \{Fb\}$.

(7)
$$\dfrac{\dfrac{\dfrac{\dfrac{\dfrac{[a\doteqdot a]^{(1)} \quad [Fa]^{(2_1)}}{Fa}\,(\doteqdot\mathrm{E})}{F\Gamma}\,(as\mathrm{E}_0) \qquad b\Gamma}{Fb}\,(as\mathrm{I}) \qquad \dfrac{\dfrac{[Fb]^{(2_2)}}{F\Gamma}\,(as\mathrm{E}_0) \qquad a\Gamma}{Fa}\,(as\mathrm{I})}{a\doteqdot b}\,(\doteqdot\mathrm{I}),\, 2_1,\, 2_2}{a\doteqdot a \supset a\doteqdot b}\,(\supset\mathrm{I}),\, 1$$

(8)
$$\dfrac{[a\doteqdot b]^{(1)}}{a\doteqdot b \supset a\doteqdot b}\,(\supset\mathrm{I}),\, 1$$

It can be readily verified, largely relying on standard methods (cf. Prawitz, 1965; Troelstra and Schwichtenberg, 2000), that derivations in $\mathbf{I}(\mathcal{S}^{\doteqdot})$-systems can be transformed into normal derivations and that normal derivations possess the subexpression/subformula property.

Definition 2.24 *Detour conversions in $\boldsymbol{I}(\mathcal{S}^{\doteqdot})$-systems.* The detour conversions for *as* and $\doteqdot$ are like those in Definitions 2.5 and 2.16. These are supplemented with the detour conversion for $\supset$:

$$\dfrac{\dfrac{\begin{matrix}[A]^{(u)}\\ \mathcal{D}_1\\ B\end{matrix}}{A \supset B}\,(\supset\mathrm{I}),\, u \qquad \begin{matrix}\mathcal{D}_2\\ A\end{matrix}}{B}\,(\supset\mathrm{E}) \quad \text{conv} \quad \begin{matrix}\mathcal{D}_2\\ [A]\\ \mathcal{D}_1\\ B\end{matrix}$$

Theorem 2.25 *Normalization for $\boldsymbol{I}(\mathcal{S}^{\doteqdot})$-systems: Any derivation $\mathcal{D}$ in an $\boldsymbol{I}(\mathcal{S}^{\doteqdot})$-system can be transformed into a normal $\boldsymbol{I}(\mathcal{S}^{\doteqdot})$-derivation.*

Proof A consequence of the normalization proof in Więckowski (2016). □

Remark 2.26 (3) is neither normal nor canonical, but it can be transformed into a normal $\mathbf{I}(\mathcal{S}^{\doteqdot})$-derivation (that is not canonical). (4)-(5) and (7)-(8) are normal canonical $\mathbf{I}(\mathcal{S}^{\doteqdot})$-derivations. (6) and (8) have the form of normal canonical $\mathbf{I}(\mathcal{S}^{\doteqdot})$-proofs.

Definition 2.27 Let $\mathcal{D}$ be a derivation in an $\mathbf{I}(\mathcal{S}^{\doteqdot})$-system.

1. A *unit* in $\mathcal{D}$ is either (i) a formula or (ii) the occurrence of an $\mathcal{S}^{\doteqdot}$-unit in $\mathcal{D}$. We use U, U' (possibly with subscripts) for units.
2. In case U is a term assumption $\tau\Gamma$ in $\mathcal{D}$, τ is *the expression in U*.

Theorem 2.28 *Subexpression property for $\boldsymbol{I}(\mathcal{S}^{\doteqdot})$-systems: If $\mathcal{D}$ is a normal $\boldsymbol{I}(\mathcal{S}^{\doteqdot})$-derivation of a unit U from a set of units Γ, then each unit in $\mathcal{D}$ is a subexpression of an expression in $\Gamma \cup \{U\}$.*

Proof A consequence of the corresponding proof in Więckowski (2016). □

Corollary 2.29 *Subformula property for $\boldsymbol{I}(\mathcal{S}^{\doteqdot})$-systems: If $\mathcal{D}$ is a normal $\boldsymbol{I}(\mathcal{S}^{\doteqdot})$-derivation of formula A from a set of formulae Γ, then each formula in $\mathcal{D}$ is a subformula of a formula in $\Gamma \cup \{A\}$.*

Remark 2.30 (4)-(8) possess the subexpression property, and so does (3) after its transformation into a normal derivation. Concerning the subformula property, we have to bear in mind that $\doteqdot$-formulae are abbreviations according to Definition 2.1.

3 Modal proof systems

Modal proof systems, as we shall understand them, are proof systems which, given a reference proof system that serves to determine what counts as a fact, distinguish between various modes of making assumptions. Section 3.1 defines modal proof systems for reasoning with elementary *would*-counterfactuals and causal *since*-subordinator sentences. Section 3.2 formulates a proof-theoretic semantics for such constructions.

3.1 IFC-systems

The natural deduction systems to be defined in this section maintain three modes of assumption. Derivations in these systems reflect, as it were, how the modes of assumptions are related to the moods of implications. The systems are defined for the language $L1$.

Definition 3.1 *The language* $L1$. The notion of a formula of $L1$ is inductively defined by the following clauses:

1. Any formula of $L0'$ is a formula of $L1$.
2. If A, B are formulae of $L1$, then $A \supset_f B$ (factual implication), $A \supset_c B$ (counterfactual implication), and $A \supset B$ (mode-sensitive implication) are formulae of $L1$.

Note that we do not use a different symbol for mode-sensitive implication. Let $\circ \in \{\supset_f, \supset_c, \supset\}$. Call the $\circ$-operators *implication-operators* and formulae with principal $\circ$ *implication-formulae*. *Defined operators of* $L1$: $\neg_f A =_{def} A \supset_f \bot$ (factual negation), $\neg_c A =_{def} A \supset_c \bot$ (counterfactual negation), $\neg A =_{def} A \supset \bot$ (mode-sensitive negation).

For instance, we symbolize sentences of the form (1) by $A \supset_c B$ and sentences of the form (2) by $A \supset_f B$.

Definition 3.2 Let Fml_0 be the set of formulae of $L0'$ and let Fml_1 be the set of formulae of $L1$. An $L1$-formula A is a *modal formula* in case $A \in Fml_1 \setminus Fml_0$.

We now define the intended modal proof systems.

Definition 3.3 An IFC-*system* is a modal natural deduction system for intuitionistic factual, counterfactual, and mode-sensitive implication which, given a reference proof system (Definition 3.4), distinguishes three modes of making assumptions (Definition 3.5): factual, counterfactual, and independent. Derivations in IFC-systems are defined on the basis of these modes (Definition 3.7).

Definition 3.4 *Reference proof system* S. Let S be an $\mathbf{I}(\mathcal{S}^{\doteqdot})$-system. Let an *established thesis of* S be an $L0'$-formula for which a canonical S-derivation (Definition 2.22(1)) has been constructed. And let Θ_S be the set of established theses (or facts).

Definition 3.5 *Modes of assumptions (IFC-systems).* There are three modes of making assumptions in IFC-systems:

1. $|A|$ indicates that A is assumed in the *factual mode*, given that $A \in Fml_0$ and $A \in \Theta_\mathsf{S}$.
2. $\wr A \wr$ indicates that A is assumed in the *counterfactual mode*, given that $A \in Fml_0$ and $A \in \Theta^c_\mathsf{S}$, where $\Theta^c_\mathsf{S} =_{def} Fml_0 \setminus \Theta_\mathsf{S}$.
3. A (no markers) indicates that A is assumed in the *independent mode*, where $A \in Fml_1$. Specifically, in case A is also an $L0'$-formula, A is assumed independently of whether it is contained in Θ_S or Θ^c_S.

We write $/A/$ to indicate that A is assumed in one of the three modes.

Remark 3.6 A consequence of Definition 3.5 is that modal $L1$-formulae can be assumed only in the independent mode.

Definition 3.7 *Derivations in* IFC*-systems.*

Basic step. Any derivation in the reference system S of an IFC-system, any $L0'$-formula A assumed in the factual (resp. counterfactual) mode $|A|$ ($\wr A \wr$), i.e., a derivation from the open factual (counterfactual) assumption of A, and any $L1$-formula A assumed in the independent mode, i.e., a derivation from the open independent assumption of A, is a derivation in that IFC-system.

Induction step. If $\mathcal{D}_1$ and $\mathcal{D}_2$ are IFC-derivations, then an IFC-derivation can be constructed by means of I/E-rules for *as* and $\doteqdot$, which now also take the modes of assumptions into account, and the following rules:

$$\dfrac{\begin{matrix}[|A|]^{(u)}\\ \mathcal{D}_1\\ B\end{matrix}}{A \supset_f B}\,(\supset_f\mathrm{I}), u \qquad \dfrac{\begin{matrix}\mathcal{D}_1\\ A \supset_f B\end{matrix} \quad \begin{matrix}\mathcal{D}_2\\ A\end{matrix}}{B}\,(\supset_f\mathrm{E}) \qquad \dfrac{\begin{matrix}[\wr A \wr]^{(u)}\\ \mathcal{D}_1\\ B\end{matrix}}{A \supset_c B}\,(\supset_c\mathrm{I}), u \qquad \dfrac{\begin{matrix}\mathcal{D}_1\\ A \supset_c B\end{matrix} \quad \begin{matrix}\mathcal{D}_2\\ A\end{matrix}}{B}\,(\supset_c\mathrm{E})$$

$$\dfrac{\begin{matrix}[/A/]^{(u)}\\ \mathcal{D}_1\\ B\end{matrix}}{A \supset B}\,(\supset\mathrm{I}), u \qquad \dfrac{\begin{matrix}\mathcal{D}_1\\ A \supset B\end{matrix} \quad \begin{matrix}\mathcal{D}_2\\ A\end{matrix}}{B}\,(\supset\mathrm{E}) \qquad \dfrac{\begin{matrix}\mathcal{D}_1\\ \bot\end{matrix}}{A}\,(\bot\mathrm{i})$$

1. *Side conditions*:

   SC1. $\supset_f$I: No empty discharge; and no empty discharge contained in $\mathcal{D}_1$.

   SC2. $\supset_f$E: The minor premiss A has *factual status*, i.e.: A depends on no counterfactual assumption in $\mathcal{D}_2$; and either A depends on at least one factual assumption in $\mathcal{D}_2$, or $\mathcal{D}_2$ contains at least one term assumption, or $\mathcal{D}_2$ is a derivation in S.

SC3. $\supset_c$I: Like SC1.

SC4. $\supset_c$E: (a) The minor premiss A has *counterfactual status*, i.e.: A depends on at least one counterfactual assumption in $\mathcal{D}_2$. (b) In case A and B are distinct formulae, the conclusion of $\supset_c$E must not be the minor premiss of another application of $\supset_c$E or of $\supset$E (*break formula*, for short).

2. *Terminology*: The minor premiss A of $\supset$E has *independent status* in case A depends on no factual or counterfactual assumption in $\mathcal{D}_2$ and has not been derived by means of a term assumption in $\mathcal{D}_2$.

3. *Assumption Principles*: The following principles are respected by derivations in IFC-systems:

AP1. No formula is assumed in more than one mode in $\mathcal{D}$.

AP2. The mode in which an antecedent A is assumed in $\circ$I-applications in $\mathcal{D}$ determines the *modal status* (factual, counterfactual, independent) of all antecedent A-nodes (i.e., minor premisses of $\circ$E-applications) in $\mathcal{D}$.

Remark 3.8 1. *Factual implication*: SC1 ensures that A is indeed assumed factually and that factual implication behaves inferentially in an intuitively required non-monotonic manner. SC2 ensures that the minor premiss A is rooted in facts and does not rest on the unestablished.

2. *Counterfactual implication*: SC3 ensures that A is indeed assumed counterfactually and that counterfactual implication behaves non-monotonically. SC4a ensures that the minor premiss A is neither based entirely on facts nor on assumptions made in the independent mode. SC4b excludes break formulae in order to block transitivity.

3. *(Mode-sensitive) implication*: There are no side conditions on the I/E-rules for $\supset$. The $\supset$-rules of S-systems are special cases of the $\supset$-rules of IFC-systems. We may regard these special cases, allowing only for independent assumptions, as governing $\supset$ in the usual (i.e., mode-less) sense of implication. As mentioned, if A is a modal $L1$-formula (Definition 3.2), $[/A/]^{(u)}$ in an $\supset$I-application can be only of the form $[A]^{(u)}$.

4. We obtain a minimal system (abbr. MFC-system) from an IFC-system, if we remove the $\bot$i-rule from the latter.

Definition 3.9 *Canonical derivation, canonical proof, thesis, and theorem (*IFC*-systems).* Analogous to Definition 2.22.

Example 3.10

$$\dfrac{\dfrac{\dfrac{\dfrac{\dfrac{\dfrac{[\wr Rab \wr]^{(1)}}{R\Gamma}(asE_0) \quad \dfrac{[|Fc|]^{(2)}}{c\Gamma}(asE_1) \quad d\Gamma}{Rcd}(asI)}{Fc \supset_f Rcd}(\supset_f I), 2}{Rab \supset_c (Fc \supset_f Rcd)}(\supset_c I), 1 \quad \wr Rab \wr}{Fc \supset_f Rcd}(\supset_c E) \quad \dfrac{F\Gamma \quad c\Gamma}{Fc}(asI)}{\dfrac{Rcd}{c\Gamma}(asE_1)}(\supset_f E) \tag{9}$$

(10) (a)
$$\dfrac{\dfrac{\dfrac{[\neg_c A]^{(2)} \qquad [\wr A \wr]^{(1)}}{\bot}\,(\supset_c \mathrm{E})}{\neg\neg_c A}\,(\supset \mathrm{I}),\,2}{A \supset_c \neg\neg_c A}\,(\supset_c \mathrm{I}),\,1$$

(b)
$$\dfrac{\dfrac{\dfrac{\dfrac{[\neg_f B]^{(2)} \qquad \dfrac{[A \supset_f B]^{(1)} \qquad [|A|]^{(3)}}{B}\,(\supset_f \mathrm{E})}{\bot}\,(\supset_f \mathrm{E})}{\neg_f A}\,(\supset_f \mathrm{I}),\,3}{\neg_f B \supset \neg_f A}\,(\supset \mathrm{I}),\,2}{(A \supset_f B) \supset (\neg_f B \supset \neg_f A)}\,(\supset \mathrm{I}),\,1$$

The side conditions imposed on the $\supset_c$-rules guarantee that IFC-systems do justice to traditional *counterfactual fallacies* (cf. Stalnaker, 1968, pp. 48–49; see also Lewis, 2011, §1.8). The failures of transitivity and contraposition are reflected by the illegal (11a) and (11b), respectively. In both derivations B is a break formula. (12a) and (12b) contain violations of weakening and are related to the fallacy of strengthening of the antecedent.

(11) (a)
$$\dfrac{\dfrac{B \supset_c C \qquad \dfrac{A \supset_c B \qquad [\wr A \wr]^{(1)}}{B}\,(\supset_c \mathrm{E})\ \textit{illeg.}}{C}\,(\supset_c \mathrm{E})}{A \supset_c C}\,(\supset_c \mathrm{I}),\,1$$

(b)
$$\dfrac{\dfrac{\dfrac{\dfrac{[\wr\neg B\wr]^{(2)} \qquad \dfrac{[A \supset_c B]^{(1)} \qquad [\wr A \wr]^{(3)}}{B}\,(\supset_c \mathrm{E})\ \textit{illeg.}}{\bot}\,(\supset \mathrm{E})}{\neg A}\,(\supset \mathrm{I}),\,3}{\neg B \supset_c \neg A}\,(\supset_c \mathrm{I}),\,2}{(A \supset_c B) \supset (\neg B \supset_c \neg A)}\,(\supset \mathrm{I}),\,1$$

(12) (a)
$$\dfrac{\dfrac{[\wr A \wr]^{(1)}}{B \supset_c A}\,(\supset_c \mathrm{I})\ \textit{illeg.}}{A \supset_c (B \supset_c A)}\,(\supset_c \mathrm{I}),\,1$$

(b)
$$\dfrac{\dfrac{\dfrac{\dfrac{[\neg(A \supset_c B)]^{(1)} \qquad \dfrac{[/B/]^{(2)}}{A \supset_c B}\,(\supset_c \mathrm{I})\ \textit{illeg.}}{\bot}}{\neg B}\,(\supset \mathrm{I}),\,2}{A \supset_c \neg B}\,(\supset_c \mathrm{I})\ \textit{illeg.}}{\neg(A \supset_c B) \supset (A \supset_c \neg B)}\,(\supset \mathrm{I}),\,1$$

Derivations in IFC-systems may contain detours (i.e., cut or maximum formulae). (9) is an example. Such derivations can be transformed into normal derivations.

Definition 3.11 1. The occurrence of a formula in a derivation $\mathcal{D}$ in and IFC-system is a *cut (or maximum) formula*, if it is the conclusion of an application of an

I-rule and the (major) premiss of an E-rule. A *maximal* cut formula in $\mathcal{D}$ is a cut formula with maximal rank r.

2. The *cut rank* of $\mathcal{D}$, $cr(\mathcal{D})$, is $\langle d, n\rangle$, where $d = max\{r(A) : A \text{ cut formula in } \mathcal{D}\}$, and where n is the number of maximal cut formulae in $\mathcal{D}$. A derivation is *normal (or in normal form)*, if it contains no cut formulae.

Definition 3.12 *Detour conversions in* IFC-*systems*. The detour conversions for as and $\doteq$ are like those in Definitions 2.5 and 2.16, except that assumptions of as- and $\doteq$-formulae now occur in $//$. These are supplemented with detour conversions for the $\circ$-operators:

$$\frac{\dfrac{\begin{matrix}[|A|]^{(u)}\\ \mathcal{D}_1\\ B\end{matrix}}{A \supset_f B}(\supset_f\mathrm{I}), u \qquad \begin{matrix}\mathcal{D}_2\\ A\end{matrix}}{B}(\supset_f\mathrm{E}) \qquad \text{conv} \qquad \begin{matrix}\mathcal{D}_2\\ [|A|]\\ \mathcal{D}_1\\ B\end{matrix}$$

$$\frac{\dfrac{\begin{matrix}[\wr A\wr]^{(u)}\\ \mathcal{D}_1\\ B\end{matrix}}{A \supset_c B}(\supset_c\mathrm{I}), u \qquad \begin{matrix}\mathcal{D}_2\\ A\end{matrix}}{B}(\supset_c\mathrm{E}) \qquad \text{conv} \qquad \begin{matrix}\mathcal{D}_2\\ [\wr A\wr]\\ \mathcal{D}_1\\ B\end{matrix}$$

$$\frac{\dfrac{\begin{matrix}[/A/]^{(u)}\\ \mathcal{D}_1\\ B\end{matrix}}{A \supset B}(\supset\mathrm{I}), u \qquad \begin{matrix}\mathcal{D}_2\\ A\end{matrix}}{B}(\supset\mathrm{E}) \qquad \text{conv} \qquad \begin{matrix}\mathcal{D}_2\\ [/A/]\\ \mathcal{D}_1\\ B\end{matrix}$$

Remark 3.13 1. The $\supset$-conversion of S-systems (Definition 3.12) is a special case of the $\supset$-conversion of IFC-systems.

2. Recall that, in virtue of AP1, a formula can be only assumed in exactly one mode in a derivation. Because of $[|A|]^{(u)}$, the minor premiss A of the $\supset_f$E-application in the $\supset_f$-conversion has, by AP2, factual status. Similarly, since we have $[\wr A\wr]^{(u)}$ on the left-hand side of the $\supset_c$-conversion, the minor premiss A of the $\supset_c$E-application in the $\supset_c$-conversion has counterfactual status. Finally, in the $\supset$-conversion, A is assumed in exactly one of the three modes in $[/A/]^{(u)}$. By AP2, this mode determines the modal status of the two A-nodes in the derivation on the left-hand side of the conversion.

The following considerations supplement Remark 3.13(2).

Remark 3.14 Let F be an implication-formula $A \circ B$. Let $\mathcal{D}^F$ be a derivation which derives F by means of an $\circ$I-application in its last step, and let $\mathcal{D}^B$ be the subderivation of $\mathcal{D}^F$ which derives the premiss B of that $\circ$I-application. Let $\mathcal{D}^A$ be a derivation which derives A. The tables below list the cases in which an $\circ$E-application can be used to construct a derivation $\mathcal{D}^*$ of B (i.e., a detour derivation with F being a detour formula) from derivations $\mathcal{D}^F$ and $\mathcal{D}^A$. (Since no mode-related side conditions are imposed on the I/E-rules for *as*- and $\doteqdot$-formulae, we consider only cases in which B and A have been obtained by means of $\circ$-rules.) The columns with the heading $\mathcal{D}^B$ (resp. $\mathcal{D}^A$) indicate the last rule applied in $\mathcal{D}^B$ ($\mathcal{D}^A$). '+' means that the construction of $\mathcal{D}^*$ is legal and that $\mathcal{D}^*$'s conversion is successful. '−' means that a conversion is precluded, since the construction is not legal. In case $\mathcal{D}^B$ ends with an $\circ$E-application there are two entries. The first [second] entry indicates the result for the case in which F is introduced discharging an assumption used to derive the major [minor] premiss of that $\circ$E-application.

T.1.	$\mathcal{D}^B$:	$\mathcal{D}^A$:	$\supset_f$-c	$\supset_c$-c	$\supset$-c
a.	$\supset_f$I	$\supset_f$I	−	−	+
b.	$\supset_f$I	$\supset_f$E	+	+	+
c.	$\supset_f$E	$\supset_f$I	$- \mid -$	$- \mid -$	$+ \mid +$
d.	$\supset_f$E	$\supset_f$E	$+ \mid +$	$+ \mid -$	$+ \mid +$

T.2.	$\mathcal{D}^B$:	$\mathcal{D}^A$:	$\supset_f$-c	$\supset_c$-c	$\supset$-c
a.	$\supset_f$I	$\supset_c$I	−	−	+
b.	$\supset_f$I	$\supset_c$E	−	−	−
c.	$\supset_f$E	$\supset_c$I	$- \mid -$	$- \mid -$	$+ \mid +$
d.	$\supset_f$E	$\supset_c$E	$- \mid -$	$- \mid -$	$- \mid -$

T.3.	$\mathcal{D}^B$:	$\mathcal{D}^A$:	$\supset_f$-c	$\supset_c$-c	$\supset$-c
a.	$\supset_f$I	$\supset$I	+	+	+
b.	$\supset_f$I	$\supset$E	+	+	+
c.	$\supset_f$E	$\supset$I	$+ \mid +$	$+ \mid -$	$+ \mid +$
d.	$\supset_f$E	$\supset$E	$+ \mid +$	$+ \mid -$	$+ \mid +$

T.4.	$\mathcal{D}^B$:	$\mathcal{D}^A$:	$\supset_f$-c	$\supset_c$-c	$\supset$-c
a.	$\supset_c$I	$\supset_f$I	−	−	+
b.	$\supset_c$I	$\supset_f$E	+	+	+
c.	$\supset_c$E	$\supset_f$I	$- \mid -$	$- \mid -$	$+ \mid +$
d.	$\supset_c$E	$\supset_f$E	$+ \mid +$	$+ \mid +$	$+ \mid +$

T.5.	$\mathcal{D}^B$:	$\mathcal{D}^A$:	$\supset_f$-c	$\supset_c$-c	$\supset$-c
a.	$\supset_c$I	$\supset_c$I	−	−	+
b.	$\supset_c$I	$\supset_c$E	−	−	−
c.	$\supset_c$E	$\supset_c$I	$- \mid -$	$- \mid -$	$+ \mid +$
d.	$\supset_c$E	$\supset_c$E	$- \mid -$	$- \mid -$	$- \mid -$

T.6.	$\mathcal{D}^B$:	$\mathcal{D}^A$:	$\supset_f$-c	$\supset_c$-c	$\supset$-c
a.	$\supset_c$I	$\supset$I	+	+	+
b.	$\supset_c$I	$\supset$E	+	+	+
c.	$\supset_c$E	$\supset$I	$+ \mid +$	$+ \mid +$	$+ \mid +$
d.	$\supset_c$E	$\supset$E	$+ \mid +$	$+ \mid +$	$+ \mid +$

T.7.	$\mathcal{D}^B$:	$\mathcal{D}^A$:	$\supset_f$-c	$\supset_c$-c	$\supset$-c
a.	$\supset$I	$\supset_f$I	−	−	+
b.	$\supset$I	$\supset_f$E	+	+	+
c.	$\supset$E	$\supset_f$I	$- \mid -$	$- \mid -$	$+ \mid +$
d.	$\supset$E	$\supset_f$E	$+ \mid +$	$+ \mid +$	$+ \mid +$

T.8.	$\mathcal{D}^B$:	$\mathcal{D}^A$:	$\supset_f$-c	$\supset_c$-c	$\supset$-c
a.	$\supset$I	$\supset_c$I	−	−	+
b.	$\supset$I	$\supset_c$E	−	−	−
c.	$\supset$E	$\supset_c$I	$- \mid -$	$- \mid -$	$+ \mid +$
d.	$\supset$E	$\supset_c$E	$- \mid -$	$- \mid -$	$- \mid -$

T.9.	$\mathcal{D}^B$:	$\mathcal{D}^A$:	$\supset_f$-c	$\supset_c$-c	$\supset$-c
a.	$\supset$I	$\supset$I	+	+	+
b.	$\supset$I	$\supset$E	+	+	+
c.	$\supset$E	$\supset$I	$+ \mid +$	$+ \mid +$	$+ \mid +$
d.	$\supset$E	$\supset$E	$+ \mid +$	$+ \mid +$	$+ \mid +$

Theorem 3.15 *Normalization for* IFC*-systems: Any derivation $\mathcal{D}$ in an* IFC*-system can be transformed into a normal* IFC*-derivation.*

Proof We proceed in the familiar way by applying detour conversions (Definition 3.12) to $\mathcal{D}$, in order to arrive at $cr(\mathcal{D}) = \langle 0, 0 \rangle$. □

(9), for instance, can be transformed into $c\Gamma$, a derivation in normal form. (10a) and (10b) are normal derivations.

Normal derivations have a simple structure. It can be shown, adapting standard methods (cf. Prawitz, 1965; Troelstra and Schwichtenberg, 2000), that they possess the subexpression property (of which the subformula property is a special case).

Definition 3.16 Let $\mathcal{D}$ be a derivation in an IFC-system.

1. A *unit* in $\mathcal{D}$ is either (i) a formula or (ii) the occurrence of an $\mathcal{S}^{\doteqdot}$-unit in $\mathcal{D}$. We use U, U' (possibly with subscripts) for units.
2. In case U is a term assumption $\tau\Gamma$ in $\mathcal{D}$, τ is *the expression in* U.

Definition 3.17 Let $\mathcal{D}$ be a normal derivation in an IFC-system, let $\circ \in \{\supset_f, \supset_c, \supset\}$. A sequence of unit occurrences $U_0, \ldots, U_n$ such that

1. U_0 is a top formula occurrence A_0 enclosed in // or a top occurrence of a term assumption $\tau\Gamma_0$ in $\mathcal{D}$, and
2. for $0 \leq i < n$, U_{i+1} is immediately below U_i, and
3. U_i is not the minor premiss of $\doteqdot$E or $\circ$E,

is a *track* of $\mathcal{D}$. A track of *order* 0 in $\mathcal{D}$ is a track ending in the conclusion of $\mathcal{D}$. A track of *order* $n + 1$ in $\mathcal{D}$ is a track ending in the minor premiss of an application of $\doteqdot$E or $\circ$E with the major premiss belonging to a track of order n.

Theorem 3.18 *Let $\mathcal{D}$ be a normal derivation in an* IFC-*system and let π be a track $U_0, \ldots, U_n$ in $\mathcal{D}$. Then there is a unit U_i in π, the minimum part of the track, which divides π into two (possibly empty) parts, an E-part $U_0, \ldots, U_{i-1}$ and an I-part $U_{i+1}, \ldots, U_n$. The E-part is constructed exclusively by E-rule applications. The I-part is constructed exclusively by I-rule applications. U_i is the conclusion of an E-rule, and in case $i < n$, a premiss of an I-rule or of* $\bot i$.

Proof By Theorem 3.15, a major premiss of an E-rule application cannot be a conclusion of an I-rule application. The result is a consequence of this insight. □

Theorem 3.19 *Subexpression property for* IFC-*systems: If $\mathcal{D}$ is a normal* IFC-*derivation of a unit U from a set of units Γ, then each unit in $\mathcal{D}$ is a subexpression of an expression in $\Gamma \cup \{U\}$.*

Proof Making use of Theorem 3.18, the result is established by induction of the order of tracks n. □

Corollary 3.20 *Subformula property for* IFC-*systems: If $\mathcal{D}$ is a normal* IFC-*derivation of formula A from a set of formulae Γ, then each formula in $\mathcal{D}$ is a subformula of a formula in $\Gamma \cup \{A\}$.*

As a consequence of Theorem 3.19, full analyticity can be claimed for the systems. We may use, relying on Corollary 3.20, the following method (cf. Więckowski, 2021a) in order to show that a formula of the language of IFC-systems cannot be derived as a theorem in these systems:

Definition 3.21 *Method of counter-derivations.* Construct a candidate for a normal canonical IFC-proof of formula A by proceeding bottom-up using the rules for the operators ignoring the side conditions on them. In case (i) the construction has been successful, check whether the candidate violates a side condition. If this is the case, (ia) we obtain a counter-derivation for A, otherwise (ib) we obtain a normal IFC-proof of A. In case (ii) the construction of a candidate has not been successful, we may conclude that A cannot be derived as a theorem. Consequently, we get a decision concerning the IFC-derivability of A as a theorem. It is derivable as a theorem in case (ib), and underivable in cases (ia) and (ii).

Remark 3.22 Some of the derivations in (11) and (12) can be seen as counter-derivations which show that their conclusions are not theorems.

Remark 3.23 1. As mentioned in Section 1, the idea of using different ways of making assumptions in the context of natural deduction for counterfactuals can be traced back at least to Thomason's (1970) FCS. Crucially, Thomason introduces the notion of a *strict derivation*. He takes it that in an ordinary derivation from an assumption A, we suppose that A is the case in the actual situation. By contrast, in a strict derivation, we may "hold in abeyance certain portions of our knowledge about our actual situation, and envisage another situation in which something is supposed to hold" (Thomason, 1970, p. 398). Since it may happen that in the envisaged alternative situations not all our knowledge about the actual situation is available, Thomason imposes restrictions on the availability of that knowledge in strict derivations. Formally, this is achieved by introducing special *reiteration rules* which govern reiteration into strict derivations. One may infer a *would*-counterfactual (Thomason's notation: $A > B$), by means of an introduction rule on the basis of a strict derivation of B from the, as it were, "strict" assumption of A. In particular, one may assume also known propositions in this counterfactual way. Thomason establishes the equivalence of FCS and CS. However, he does not discuss the proof-theoretic properties of FCS.

2. It seems possible to use factual implication for the formal analysis of those constructions of the form (2) in which "since" can be equivalently replaced by "because". For discussion of the relation between these two kinds of causal subordinator see, e.g., Dancygier and Sweetser (2000) and Guillaume (2013). For an outline of a formal system for reasoning with "because" see Schnieder (2011).

Remark 3.24 In defining modal natural deduction systems there are several decisions to be made. These may concern, for instance, the choice of the reference proof system, the conception of an established fact, the modes, in which modal formulae can be legally assumed, the shape of the rules, or the side conditions that are to be imposed on them. Thus, the complexity that pertains to formal accounts of counterfactual reasoning is not moved to conditions on external (e.g., model-theoretic) structures, but rather enters the systems via such proof-theoretic design options.

3.2 A proof-theoretic semantics

On the basis of Theorem 3.15, we may formulate a proof-theoretic semantics for the non-logical constants, the atomic sentences, the identity sentences, and for the formulae composed of the operators of IFC-systems.

Definition 3.25 *Meaning*: Let the modal proof system M be an IFC-system.

1. The meaning of a *non-logical constant* τ is given by the term assumptions $\tau\Gamma$ for τ which are determined by the subatomic base of the $\mathcal{S}^{\doteq}$-system of the reference proof system S (Definition 2.2) of M.
2. The meaning of an $L1$-*formula* A is given by the set of canonical derivations of A in M (Definition 3.9).

Remark 3.26 1. Definition 3.25 contains a proof-theoretic semantics for the non-logical constants and formulae of $L0'$, defined in terms of S-derivations, as a special case.

2. Since meaning is defined in terms of canonical derivations (cf. Dummett, 1991; Prawitz, 2006), the semantics specified above is acceptable from an *intuitionistic* point of view.

3. The proposed proof-theoretic semantics is *semantically autarkic*, since the modal natural deduction systems do not draw on a formal semantics of a different kind (e.g., a possible worlds similarity semantics; cf. Lewis, 2011; Nute and Cross, 2001; Stalnaker, 1968; Stalnaker and Thomason, 1970). For instance, labelled (e.g., Negri and Olivetti, 2015; Negri and Sbardolini, 2016; Poggiolesi, 2016) or internal (e.g., Lellmann and Pattinson, 2012; Olivetti and Pozzato, 2015) structural proof systems for standard counterfactual logics all of which are formulated in a classical context do not allow for an autarkic proof-theoretic semantics. Labelled proof systems incorporate model-theoretic structures in terms of which truth conditions are formulated into their rules by means of labels (for worlds) and labelled formulae (for similarity). A proof-theoretic semantics based on such a calculus (envisaged in Girlando, Negri, and Olivetti, 2018) would certainly not be autarkic. By contrast, internal proof systems for a given counterfactual logic can be characterized as not involving a syntax that cannot be defined in terms of the object language of that logic. As a result, the sequents of such a calculus do not wear their genesis on their sleeves. However, the internal systems mentioned above make use of structural operators and specific rules which directly imitate model-theoretic structures involved in the semantics. (Translations of internal into labelled systems and back are considered in Girlando, 2019; Girlando, Negri, and Olivetti, 2018.) From a foundational point of view—or seeing proof-theoretic semantics as an "alternative to truth-condition semantics" (Schroeder-Heister, 2018, p. 1)—neither an internalization of model-theoretic truth conditions nor an imitation of model-theoretic structures seems to be appealing.

Definition 3.27 A [subatomic] proof system is *meaning-integral*, if a proof-theoretic semantics is available for it that is based on [term assumptions and] canonical derivations.

Given a meaning-integral proof system, we may define a notion of *derivation-based intuitionistic truth* (cf. Więckowski, 2023).

Definition 3.28 Let I be an IFC-system and call $\{A : \Gamma \vdash_c^I A\}$, i.e., the set of formulae which have been canonically derived in I from a set of units Γ, the *canonical I-set*. The *I-truth* of A *with respect to* Γ is defined by: $\Gamma \Vdash^I A =_{def} A \in \{A : \Gamma \vdash_c^I A\}$. Special case: A is a *logical I-truth* (i.e., $\Vdash^I A$) in case A is the conclusion of a canonical I-proof (cf. Definition 3.9).

Remark 3.29 I-truth may serve to single out certain atomic sentences, identity sentences, and logical compounds. And it is logical I-truth, rather than plain I-truth, that can be used to single out certain I-true identity sentences (i.e., self-identities) and logical compounds further. In general, the canonical derivations on which I-truths are based can be seen as formal verifications (for verificationism in the context of proof-theoretic semantics see, e.g., Dummett, 1991, Prawitz 2006; 2012).

4 A philosophical application: counterpossibles

We shall now apply the modal proof systems developed in the previous section to the following constructions (cf. Williamson, 2007, p. 174):

(13) If Hesperus had not been Phosphorus, Hesperus would not have been Hesperus.

(14) If Hesperus had not been Phosphorus, Hesperus would not have been Phosphorus.

Given certain philosophical presuppositions, in particular, the doctrine of the *necessity of identity* (NI, for short; cf. Kripke, 1980; Marcus, 1961), conditional sentences of this kind are sometimes called "counterpossibles" (see, e.g., Berto, French, Priest, and Ripley, 2018; Williamson, 2007), since their antecedents turn out to be impossible: If Hesperus is Phosphorus, then, given NI, this is so of necessity. Moreover, given the interdefinability of the operators for necessity and possibility (cf. Williamson, 2007, p. 295), guaranteed by *classical* modal logic, "their" distinctness is, therefore, impossible.

A further common presupposition in the discussion of counterpossibles is "orthodoxy" (cf. Berto, French, Priest, and Ripley, 2018, p. 694), that is, the aforementioned similarity semantics for counterfactuals. A semantics of this kind makes use of truth conditions and explains the formal meaning of counterfactuals in terms of subset relations on possible worlds. Roughly, a sentence of the form (1) is true at world w exactly if B is true in all the possible worlds in which A is true that are most similar to w. On this account, there are no A-worlds, in case A is impossible. This means that an instance of (1) with impossible A is true, since, given orthodoxy, for any B, B is true (vacuously) at all the most similar A-worlds. So-called *vacuists* (e.g., Williamson, 2007) accept this consequence for both (13) and (14). By contrast, *non-vacuists* argue, typically admitting also impossible worlds into the orthodox picture (see, e.g., Berto, French, Priest, and Ripley, 2018 and the references therein), that the first counterpossible is false and that the second is true. Our discussion of (13) and (14) below will presuppose neither NI, nor classicality, nor orthodoxy.

We symbolize (13) and (14) as $\neg a \ddot{=} b \supset_c \neg a \ddot{=} a$ and $\neg a \ddot{=} b \supset_c \neg a \ddot{=} b$, respectively. For simplicity, let the reference proof system S of the IFC-system contain only a single predicate constant (i.e., F) and two nominal constants (i.e., a, b). Moreover, let the term assumptions be, again, like in Example 2.13: $F\Gamma = \{Fa, Fb\}$, $a\Gamma = \{Fa\}$, and $b\Gamma = \{Fb\}$. (15) is a canonical derivation for (13):

(15)
$$\dfrac{\dfrac{[\wr\neg a \ddot{=} b\wr]^{(1)} \qquad \dfrac{\dfrac{\dfrac{\dfrac{[a \ddot{=} a]^{(2)} \quad [Fa]^{(3_1)}}{Fa}\,(\ddot{=}\mathrm{E})}{F\Gamma}\,(as\mathrm{E}_0) \quad b\Gamma}{Fb}\,(as\mathrm{I}) \qquad \dfrac{\dfrac{[Fb]^{(3_2)}}{F\Gamma}\,(as\mathrm{E}_0) \quad a\Gamma}{Fa}\,(as\mathrm{I})}{a \ddot{=} b}\,(\ddot{=}\mathrm{I}), 3_1, 3_2}{\dfrac{\bot}{\neg a \ddot{=} a}\,(\supset\mathrm{I}), 2}\,(\supset\mathrm{E})}{\neg a \ddot{=} b \supset_c \neg a \ddot{=} a}\,(\supset_c\mathrm{I}), 1$$

The conclusion of (15) is only a *thesis* of the specific IFC-system. (An alternative derivation would be, e.g., one in which the subderivation of $a \ddot{=} b$ in (15) were replaced by $|a \ddot{=} b|$, or one in which it were replaced, e.g., by the S-derivation (4).) Note that we would not be in a position to assume $\neg a \ddot{=} b$ in the counterfactual mode, if we had established $\neg a \ddot{=} b$ as a fact. If we had done so, we would not be in a position to arrive at the intended conclusion. (16) is a canonical derivation for (14). Its conclusion is a *theorem* of the IFC-system:

(16)
$$\dfrac{[\wr\neg a \ddot{=} b\wr]^{(1)}}{\neg a \ddot{=} b \supset_c \neg a \ddot{=} b}\,(\supset_c\mathrm{I}), 1$$

Comments:

1. We may regard both (13) and (14) as true (cf. Definition 3.28). The latter can be taken to be also logically true, since its canonical derivation is a proof. Thus, the present assessment of these sentences as true seems to be closer to the vacuist one. Note, however, that no appeal to some notion of vacuity is being made in the explanation of their truth.

2. (15) shows how the *self-distinctness* of Hesperus can be inferred form the counterfactual assumption of the distinctness of Hesperus and Phosphorus. On the present semantics, (15) is one of those derivations which constitute the meaning of its conclusion.

3. By Definition 3.25, the meaning of $\neg a \ddot{=} b \supset_c \neg a \ddot{=} a$ does not coincide with that of $\neg a \ddot{=} b \supset_c \neg a \ddot{=} b$, nor do the meanings of any two theorems, or those of any two logically equivalent formulae. As a consequence, the present semantics is sensitive to *hyperintensional* distinctions (a recent collection on hyperintensionality is Duží and Jespersen, 2015; for discussion in the context of proof-theoretic semantics see, e.g., Pezlar, 2018).

4. Consider the structure of (4) and (6) which canonically derive $a \mathrel{\ddot{=}} b$ and $a \mathrel{\ddot{=}} a$ (an instance of $\alpha \mathrel{\ddot{=}} \alpha$), respectively. $a \mathrel{\ddot{=}} a$ can be derived as a theorem in any IFC-system, whereas $a \mathrel{\ddot{=}} b$ cannot be derived as a theorem in any such system. If we regard formulae that have been derived as theorems as *necessities*, and those that have been derived only as theses as *contingencies*, we may classify $a \mathrel{\ddot{=}} a$ as necessary and $a \mathrel{\ddot{=}} b$ as contingent.

5. In order to derive $a \mathrel{\ddot{=}} b$ canonically, we have to look to the term assumptions and to apply the asI-rule. However, in order to derive $a \mathrel{\ddot{=}} a$ canonically these steps are not required. If we regard, taking a derivation-oriented perspective, the conclusion of a canonical derivation as *a posteriori* in case the derivation requires an application of an as-rule, and if we regard the conclusion of a canonical derivation as *a priori* in case the derivation does not need to make use of such rules, we may classify $a \mathrel{\ddot{=}} b$ as *a posteriori* and $a \mathrel{\ddot{=}} a$ as *a priori*. Moreover, depending on whether the constants symbolize denoting or non-denoting names, we may also distinguish between a denotational (or referential) and a non-denotational kind of the *a posteriori*. Furthermore, taking this derivation-oriented perspective, we may also consider adding a distinction between an empirical and a non-empirical kind of the denotational *a posteriori*.

6. The above categorization of 'Hesperus is Hesperus' as necessary *a priori* and of 'Hesperus is Phosphorus' as contingent *a posteriori* is relatively old-fashioned in nature. It differs from Kripke's well-known proposal (cf. Kripke, 1980), according to which, given NI (and other prerequisites), 'Hesperus is Phosphorus' expresses an *a posteriori* necessity.

7. Analogous remarks apply to instances of (13) and (14) which feature *empty names* (e.g., let a symbolize 'Superman' and b 'Clark Kent'). For such instances, the idea that a proper name is a rigid designator (and so denotes the same object in every possible world) which lies at the hart of NI, does not seem to be appealing as, intuitively, there is nothing for such names to designate whether rigidly or not.

We shall next look at constructions related to (13) and (14), in order to obtain a sharper contrast. First, consider the following counterfactuals:

(17) If Hesperus were Phosphorus, Hesperus would be Hesperus.

(18) If Hesperus were Hesperus, Hesperus would be Phosphorus.

We symbolize (17) and (18) as $a \mathrel{\ddot{=}} b \supset_c a \mathrel{\ddot{=}} a$ and $a \mathrel{\ddot{=}} a \supset_c a \mathrel{\ddot{=}} b$, respectively. Let the IFC-system be like that in the analysis of (13) and (14), but with $F\Gamma = \{Fa\}$, $a\Gamma = \{Fa\}$, and $b\Gamma = \{Fb\}$.

(19)
$$\dfrac{\dfrac{\dfrac{\dfrac{\dfrac{[\wr a \mathrel{\ddot{=}} b \wr]^{(1)} \quad [Fa]^{(2_1)}}{Fb}\,(\ddot{=}\mathrm{E})}{F\Gamma}\,(as\mathrm{E}_0) \quad a\Gamma}{Fa}\,(as\mathrm{I}) \qquad \dfrac{\dfrac{[Fa]^{(2_2)}}{F\Gamma}\,(as\mathrm{E}_0) \quad a\Gamma}{Fa}\,(as\mathrm{I})}{a \mathrel{\ddot{=}} a}\,(\ddot{=}\mathrm{I}),\, 2_1,\, 2_2}{a \mathrel{\ddot{=}} b \supset_c a \mathrel{\ddot{=}} a}\,(\supset_c\mathrm{I}),\, 1$$

(20)

$$\dfrac{\dfrac{\dfrac{\dfrac{\dfrac{[\wr a\ddot{=}a\wr]^{(1)}\mathit{illeg.} \qquad [Fa]^{(2_1)}}{Fa}(\ddot{=}\mathrm{E})}{F\Gamma}(as\mathrm{E}_0) \qquad b\Gamma}{Fb}(as\mathrm{I})\ \mathit{illeg.} \qquad \dfrac{\dfrac{[Fb]^{(2_2)}}{F\Gamma}(as\mathrm{E}_0) \qquad a\Gamma}{Fa}(as\mathrm{I})}{a\ddot{=}b}(\ddot{=}\mathrm{I}),\ 2_1,\ 2_2}{a\ddot{=}a \supset_c a\ddot{=}b}(\supset_c\mathrm{I}),\ 1$$

In the present IFC-system, we can neither make a factual assumption of $a\ddot{=}b$ nor can we derive that formula by means of a canonical $\mathcal{S}^{\ddot{=}}$-derivation. In (20), $a\ddot{=}a$ cannot be assumed in the counterfactual mode given (6), and Fb cannot be derived by means of asI given that $Fb \notin F\Gamma$.

Now, consider the following two factuals:

(21) Since Hesperus is Hesperus, Hesperus is Phosphorus.

(22) Since Hesperus is Phosphorus, Hesperus is Hesperus.

We symbolize (21) as $a\ddot{=}a \supset_f a\ddot{=}b$ and (22) as $a\ddot{=}b \supset_f a\ddot{=}a$. Let the IFC-system be exactly like that in the discussion of (13) and (14).

(23)

$$\dfrac{\dfrac{\dfrac{\dfrac{\dfrac{[|a\ddot{=}a|]^{(1)} \qquad [Fa]^{(2_1)}}{Fa}(\ddot{=}\mathrm{E})}{F\Gamma}(as\mathrm{E}_0) \qquad b\Gamma}{Fb}(as\mathrm{I}) \qquad \dfrac{\dfrac{[Fb]^{(2_2)}}{F\Gamma}(as\mathrm{E}_0) \qquad a\Gamma}{Fa}(as\mathrm{I})}{a\ddot{=}b}(\ddot{=}\mathrm{I}),\ 2_1,\ 2_2}{a\ddot{=}a \supset_f a\ddot{=}b}(\supset_f\mathrm{I}),\ 1$$

(24)

$$\dfrac{\dfrac{\dfrac{\dfrac{\dfrac{[|a\ddot{=}b|]^{(1)} \qquad [Fa]^{(2_1)}}{Fb}(\ddot{=}\mathrm{E})}{F\Gamma}(as\mathrm{E}_0) \qquad a\Gamma}{Fa}(as\mathrm{I}) \qquad \dfrac{\dfrac{[Fa]^{(2_2)}}{F\Gamma}(as\mathrm{E}_0) \qquad a\Gamma}{Fa}(as\mathrm{I})}{a\ddot{=}a}(\ddot{=}\mathrm{I}),\ 2_1,\ 2_2}{a\ddot{=}b \supset_f a\ddot{=}a}(\supset_f\mathrm{I}),\ 1$$

None of the conclusions of (19), (20), (23), and (24) is a theorem.

5 Concluding remarks

We have defined rudimentary modal natural deduction systems for reasoning with relatively simple *would*-counterfactuals and causal *since*-subordinator sentences. The systems are motivated by inferential practice. They allow for different modes of making assumptions relative to their reference proof systems which serve to determine the factuality status of the formulae that are to be assumed.

Normalization and the subexpression/subformula property have been established for these systems along largely familiar lines. Due to the subexpression/subformula result, the systems are fully analytic. Due to the normalization result, the systems admit a proof-theoretic semantics. The proposed proof-theoretic semantics is acceptable from an intuitionistic point of view, since it is defined in terms of canonical derivations. Moreover, it is semantically autarkic, since the modal natural deduction systems do not draw on a formal semantics of a different kind (e.g., by internalizing a possible worlds similarity semantics).

Some aspects of the proposal are philosophically significant. Due to an absence of a semantic ontology (e.g., possible or impossible worlds), neither metaphysical nor epistemological considerations concerning such entities are triggered. Furthermore, an approach to formal epistemology is supported, according to which we arrive at knowledge of counterfactuals and factuals by means of constructive derivation and proof.

It is hoped that useful and less rudimentary modal proof systems can be obtained along the lines suggested in this outline.

Acknowledgements I would like to thank the reviewers of earlier versions of the material included in this contribution for valuable comments. I also thank the audiences at the Universities of Tübingen (Third Tübingen Conference on Proof-Theoretic Semantics, 2019) and Frankfurt am Main (LoSe colloquium, 2019) where versions of this content have been presented for useful feedback. Support by the DFG (Project: Proof-theoretic foundations of intensional semantics; Grants: WI 3456/4-1 and WI 3456/4-2) is gratefully acknowledged.

References

Berto, F., R. French, G. Priest, and D. Ripley (2018). Williamson on counterpossibles. *Journal of Philosophical Logic* 47, 693–713.

Ciardelli, I. and X. Liu (2020). Intuitionistic conditional logics. *Journal of Philosophical Logic* 49, 807–832. DOI: 10.1007/s10992-019-09538-4.

Dalen, D. van (2002). Intuitionistic logic. In: *Handbook of Philosophical Logic*. Ed. by D. M. Gabbay and F. Guenthner. 2nd ed. Dordrecht: Springer, 1–114.

Dancygier, B. and E. Sweetser (2000). Constructions with if, since, and because: causality, epistemic stance, and clause order. In: ed. by E. Couper-Kuhlen and B. Kortmann. Berlin: Mouton de Gruyter, 111–142.

Dummett, M. (1991). *The Logical Basis of Metaphysics*. Cambridge/MA: Harvard University Press.

Duží, M. and B. Jespersen (2015). Hyperintensionality. *Special section of Synthese* 192, 525–677.

Francez, N. (2015). *Proof-Theoretic Semantics*. London: College Publications.

Gentzen, G. (1934). Untersuchungen über das logische schließen i, ii. *Mathematische Zeitschrift* 39, 176–210, 405–431.

Girlando, M. (2019). On the Proof Theory of Conditional Logics. PhD thesis. University of Helsinki.

Girlando, M., S. Negri, and N. Olivetti (2018). Counterfactual logic: labelled and internal calculi, two sides of the same coin? In: *Advances in Modal Logic 2018*. Ed. by G. Bezhanishvili, G. D'Agostino, G. Metcalfe, and T. Studer. London: College Publications, 291–310.

Guillaume, B. (2013). Clause order in sentences containing a since-subordinate. *e-Rea*. DOI: 10.4000/erea.3476.

Jaśkowski, S. (1934). On the rules of suppositions in formal logic. *Studia Logica* 1, 5–32.

Kahle, R. and P. Schroeder-Heister (2006). Proof-Theoretic Semantics. *Special issue of Synthese* 148, 503–743.

Kripke, S. (1980). *Naming and Necessity*. Cambridge/MA: Harvard University Press.

Lellmann, B. and D. Pattinson (2012). Sequent systems for lewis' conditional logics. In: *Logics in Artificial Intelligence, 13th European Conference, JELIA 2012, Toulouse, France, September 26-28, 2012*. Ed. by L. Fariñas del Cerro, A. Herzig, and J. Mengin. Berlin: Springer, 320–332.

Lewis, D. (2011). *Counterfactuals*. First published in 1973. Blackwell Publishers.

Marcus, R. B. (1961). Modalities and intensional languages. *Synthese* 13, 303–322.

Negri, S. and N. Olivetti (2015). A sequent calculus for preferential conditional logic based on neighbourhood semantics. In: *International Conference on Automated Reasoning with Analytic Tableaux and Related Methods*. Ed. by H. D. Nivelle. Springer, 115–134.

Negri, S. and G. Sbardolini (2016). Proof analysis for lewis counterfactuals. *The Review of Symbolic Logic* 9, 44–75.

Nute, D. and C. B. Cross (2001). Conditional logic. In: *Handbook of Philosophical Logic, Vol. 4*. Ed. by D. M. Gabbay and F. Guenthner. 2nd ed. Kluwer Academic Publishers, 1–98.

Olivetti, N. and G. L. Pozzato (2015). A standard internal calculus for lewis' counterfactual logics. In: *TABLEAUX 2015, LNAI 9323*. Ed. by H. D. Nivelle. Springer, 266–282.

Pezlar, I. (2018). Proof-theoretic semantics and hyperintensionality. *Logique et Analyse* 242, 163–182.

Piecha, T. and P. Schroeder-Heister, eds. (2016). *Advances in Proof-Theoretic Semantics*. Cham: Springer.

Poggiolesi, F. (2016). Natural deduction calculi and sequent calculi for counterfactual logics. *Studia Logica* 104, 1003–1036.

Prawitz, D. (1965). Natural deduction: a proof-theoretical study. Reprint: Mineola/NY, Dover Publications, 2006.

– (2006). Meaning approached via proofs. *Synthese* 148, 507–524.

– (2012). Truth and proof in intuitionism. In: *Epistemology versus Ontology. Essays on the Philosophy and Foundations of Mathematics in Honour of Per Martin-Löf*. Ed. by P. Dybjer, S. Lindström, E. Palmgren, and G. Sundholm. Dordrecht: Springer, 45–67.

Schnieder, B. (2011). A logic for 'because'. *The Review of Symbolic Logic* 4, 445–465.

Schroeder-Heister, P. (2004). On the notion of assumption in logical systems. In: *Selected Papers Contributed to the Sections of GAP5 (Fifth International Congress*

of the Society for Analytical Philosophy, Bielefeld, 22-26 September 2003). Ed. by R. Bluhm and C. Nimtz. Paderborn: Mentis, 27–48.
– (2018). Proof-theoretic semantics. In: *The Stanford Encyclopedia of Philosophy*. Ed. by E. N. Zalta. Spring 2018 edition. URL: https://plato.stanford.edu/archives/spr2018/entries/proof-theoretic-semantics/.
Stalnaker, R. C. (1968). A theory of conditionals. In: *Studies in Logical Theory*. Ed. by N. Rescher. Oxford: Basil Blackwell, 98–112.
Stalnaker, R. C. and R. H. Thomason (1970). A semantic analysis of conditional logic. *Theoria* 36, 23–42.
Thomason, R. H. (1970). A fitch-style formulation of conditional logic. *Logique et Analyse* 13, 397–412.
Troelstra, A. S. and H. Schwichtenberg (2000). *Basic Proof Theory*. 2nd ed. Cambridge: Cambridge University Press.
Wansing, H. (2001). The idea of a proof-theoretic semantics and the meaning of the logical operations. *Studia Logica* 64, 3–20.
Weiss, Y. (2018). Basic intuitionistic conditional logic. *Journal of Philosophical Logic* 48, 447–469.
Więckowski, B. (2011). Rules for subatomic derivation. *The Review of Symbolic Logic* 4, 219–236.
– (2016). Subatomic natural deduction for a naturalistic first-order language with non-primitive identity. *Journal of Logic, Language and Information* 25, 215–268.
– (2021a). Intuitionistic multi-agent subatomic natural deduction for belief and knowledge. *Journal of Logic and Computation* 31 (3). Special issue on *External and Internal Calculi for Non-Classical Logics*, edited by A. Ciabattoni, D. Galmiche, N. Olivetti, and R. Ramanayake, 704–770. DOI: 10.1093/logcom/exab013.
– (2021b). Subatomic negation. *Journal of Logic, Language and Information* 30, 207–262.
– (2023). Negative predication and distinctness. *Logica Universalis* 17, 103–138.
Williamson, T. (2007). *The Philosophy of Philosophy*. Blackwell Publishing.

of the Society for Analytical Philosophy, Bielefeld, 22–26 September 2003). Ed. by R. Bluhm and C. Nimtz. Paderborn: Mentis, ...

Some Set-Theoretic Reduction Principles

Michael Bärtschi and Gerhard Jäger

Abstract In this article we study several reduction principles in the context of Simpson's set theory ATR_0^S and Kripke-Platek set theory KP (with infinity). Since ATR_0^S is the set-theoretic version of ATR_0 there is a direct link to second order arithmetic and the results for reductions over ATR_0^S are as expected and more or less straightforward. However, over KP we obtain several interesting new results and are lead to some open questions.

Dedicated to Peter Schroeder-Heister

1 Introduction

Peter Schroeder-Heister has been interested in the foundations of inference for many decades, most prominently in connection with a program that he baptized "proof-theoretic semantics". Though not related to this program in the strict sense, the work presented here is in direct connection to a talk given by the second author at the *Third Tübingen Conference on Proof-Theoretic Semantics* in 2019. Proof theory is the conceptual link between foundational questions considered under the heading of proof-theoretic semantics and our research on subsystems of second order arithmetic and set theory.

About terminology: Let $\mathcal{K}$ be a class of formulas of second order arithmetic. What Simpson calls the $\mathcal{K}$ separation principle in second order arithmetic is the collection of all

$$\neg\exists i(\varphi[i] \wedge \psi[i]) \rightarrow \exists Z \forall i(\varphi[i] \rightarrow i \in Z \rightarrow \neg\psi[i]),$$

Michael Bärtschi
Institute of Computer Science, University of Bern, Switzerland,
e-mail: michael.baertschi@inf.unibe.ch

Gerhard Jäger
Institute of Computer Science, University of Bern, Switzerland, e-mail: gerhard.jaeger@inf.unibe.ch

T. Piecha and K. F. Wehmeier (eds.), *Peter Schroeder-Heister on Proof-Theoretic Semantics*, Outstanding Contributions to Logic 29,
https://doi.org/10.1007/978-3-031-50981-0_16

where $\varphi[i]$ and $\psi[i]$ are formulas from $\mathcal{K}$. In Simpson (2009) various such separation principles have been studied. They play an interesting role in reverse mathematics and are equivalent — over a weak base theory — to certain comprehension principles. In particular, it is shown in Simpson (2009) that ACA_0 plus Σ^1_1 separation (Σ^1_1-Sep) is equivalent to the famous theory ATR_0 of arithmetical transfinite recursion.

However, this form of separation must not be confused with separation in set theory. There, $\mathcal{K}$ separation for a class $\mathcal{K}$ of formulas of set theory consists of all assertions

$$\forall x \exists y (y = \{z \in x : \varphi[z]\})$$

with $\varphi[z]$ ranging over $\mathcal{K}$. In order to avoid this conflict of notation we decided to call "reduction" what Simpson calls "separation"; see Definition 2.3. Thus we can use the same terminology in second order arithmetic and set theory.

In this article we consider several reduction principles, analog to Simpson's separation principles, though in the context of his set theory ATR^S_0 and in the context of Kripke-Platek set theory KP (with infinity). Since ATR^S_0 is the set-theoretic version of ATR_0 the results for reductions over ATR^S_0 are as expected and more or less straightforward. However, over KP we obtain several interesting results and are lead to some open questions.

This article begins with a review of separation principles in second order arithmetic — now, of course, under the new term "reduction principles" — and some important equivalences to comprehension principles. Then we have a section in which some basics about the theory ATR^S_0 and its extension ATR^S are presented before we address Kripke-Platek set theory KP and its relationship to ATR^S_0. Finally, we turn to Σ_1 reduction (Σ_1-Red) and Π_1 reduction (Π_1-Red). The respective strengths of (Σ_1-Red) and (Π_1-Red) over ATR^S_0 and ATR^S are then determined, in general by making use of the quantifier theorem and the fact that ATR^S_0 is equivalent to ATR_0. Afterwards, we change the environment and study (Σ_1-Red) and (Π_1-Red) in the context of KP. We end with some general comments and open problems. This paper is a mix of a survey article and new technical work.

2 Well-known reduction principles in second order arithmetic

Let $\mathcal{L}_2$ be a standard language of second order arithmetic with countably infinite supplies of two distinct sorts of variables; we also have the constant symbols 0 and 1 and function symbols for addition and multiplication plus relation symbols for the equality and less relation on the natural numbers. The first order variables are called *number variables* and supposed to range over natural numbers. The second order variables are known as *set variables* and intended to range over all sets of natural numbers. The number terms and formulas of $\mathcal{L}_2$ are built up as usual. See, for example, Simpson (2009).

We use the following categories of letters (possibly with subscripts) as metavariables:

1. i, j, k for first order variables;
2. X, Y, Z for second order variables;
3. $\eta, \theta, \varphi, \psi$ for formulas.

A formula of $\mathcal{L}_2$ without bound set variables is called *arithmetical*. For $1 \leq n \in \mathbb{N}$, a formula φ is said to be Σ^1_n or Π^1_n iff it is of the form

$$\exists X_1 \forall X_2 \ldots X_n \theta \quad \text{or} \quad \forall X_1 \exists X_2 \ldots X_n \theta,$$

respectively, where θ is arithmetical.

Throughout this paper we work in classical logic with equality for the first sort. Equality for sets in $\mathcal{L}_2$ is defined by saying that two sets are identical iff they contain the same elements.

ACA_0 is the system of second order arithmetic whose non-logical axioms comprise the defining axioms for all primitive recursive functions and relations, the axiom schema of *arithmetical comprehension*

$$\exists X \forall i (i \in X \leftrightarrow \varphi[i])$$

for all arithmetical formulas $\varphi[i]$, and the *induction axiom*

$$\forall X (0 \in X \land \forall i (i \in X \to i{+}1 \in X) \to \forall i (i \in X)).$$

ACA_0 is known to be a conservative extension of Peano arithmetic PA. The theory ACA is obtained from ACA_0 by adding the *schema of induction*

$$\varphi[0] \land \forall i (\varphi[i] \to \varphi[i{+}1]) \to \forall i \varphi[i]$$

for all $\mathcal{L}_2$ formulas $\varphi[i]$. Below we will make use of several further axiom schemas:

1. $(\Sigma^1_1\text{-}\mathsf{AC})$ is the schema

$$\forall i \exists X \varphi[i, X] \to \exists Y \forall i \varphi[i, (Y)_i]$$

for arbitrary Σ^1_1 formulas $\varphi[i, X]$;

2. $(\Delta^1_2\text{-}\mathsf{CA})$ is the schema

$$\forall i (\varphi[i] \leftrightarrow \psi[i]) \to \exists X \forall i (i \in X \leftrightarrow \varphi[i])$$

for all Σ^1_2 formulas $\varphi[i]$ and Π^1_2 formulas $\psi[i]$;

3. $(\Pi^1_2\text{-}\mathsf{CA})$ is the schema

$$\exists X \forall i (i \in X \leftrightarrow \varphi[i])$$

for all Π^1_2 formulas $\varphi[i]$.

To simplify the notation we write $\Delta^1_2\text{-}\mathsf{CA}_0$ for the theory $\mathsf{ACA}_0 + (\Delta^1_2\text{-}\mathsf{CA})$ and $\Pi^1_2\text{-}\mathsf{CA}_0$ for $\mathsf{ACA}_0 + (\Pi^1_2\text{-}\mathsf{CA})$.

Additional notation is necessary to formulate the principles of *arithmetical transfinite recursion* (ATR) and *bar induction* (BI), and to this end we follow Simpson (2009) as closely as possible.

Working in ACA_0, we code binary relations on the natural numbers $\mathbb{N}$ as subsets of $\mathbb{N}$ via the pairing function

$$(i, j) := (i + j)^2 + i.$$

A set X of natural numbers is said to be *reflexive* iff

$$\forall i, j((i, j) \in X \to ((i, i) \in X \land (j, j) \in X)).$$

If X is reflexive, then $\mathit{Field}[X]$ is defined to be the set $\{i : (i, i) \in X\}$, and we write

$$i \leq_X j := (i, j) \in X,$$

$$i <_X j := (i, j) \in X \land (j, i) \notin X.$$

Furthermore, if X is reflexive we say that X is *well-founded* iff every non-empty subset of $\mathit{Field}[X]$ has an X-minimal element.[1] We say that X is a *linear ordering* if it is a reflexive linear ordering of its field, i.e.,

$$\forall i, j, k(i \leq_X j \land j \leq_X k \to i \leq_X k),$$

$$\forall i, j(i \leq_X j \land j \leq_X i \to i = j),$$

$$(\forall i, j \in \mathit{Field}[X])(i \leq_X j \lor j \leq_X i).$$

We say that X is a *well-ordering* iff it is both well-founded and a linear ordering. Let $\mathit{WF}[X], \mathit{LO}[X]$, and $\mathit{WO}[X]$ be formulas saying that X is, respectively, well-founded, a linear ordering, and a well-ordering.

Definition 2.1 Given an $\mathcal{L}_2$ formula $\varphi[i]$, let $\mathit{TI}[X, \varphi]$ be the formula

$$\forall j((\forall i <_X j)\varphi[i] \to \varphi[j]) \to \forall j \varphi[j].$$

The schema (BI) of *bar induction* consists of all formulas

$$\forall X(\mathit{WF}[X] \to \mathit{TI}[X, \varphi]),$$

where φ ranges over all $\mathcal{L}_2$ formulas.

Now let $\varphi[i, Y]$ be any formula with distinguished free number variable i and distinguished free set variable Y. Define $\mathcal{H}_\varphi[X, Y]$ to be the formula

$$\mathit{LO}[X] \;\land\; Y = \{(i, j) : j \in \mathit{Field}[X] \land \varphi[i, Y^j]\},$$

where $Y^j := \{(i, k) \in Y : k <_X j\}$. Intuitively, $\mathcal{H}_\varphi[X, Y]$ says that X is a linear ordering and Y is the result of iterating φ along X.

---

[1] This is equivalent over ACA_0 to the definition given in Simpson (2009).

Definition 2.2 The schema (ATR) of *arithmetical transfinite recursion* comprises

$$\forall X(WO[X] \rightarrow \exists Y \mathcal{H}_{\varphi}[X,Y])$$

for all arithmetical formulas $\varphi[i,Y]$. Accordingly, we set

$$\mathsf{ATR}_0 := \mathsf{ACA}_0 + (\mathsf{ATR}) \quad \text{and} \quad \mathsf{ATR} := \mathsf{ACA} + (\mathsf{ATR}).$$

ACA_0 and ATR_0 belong to the "big five" in the Friedman-Simpson program of *reverse mathematics*:

$$\mathsf{RCA}_0 \subsetneq \mathsf{WKL}_0 \subsetneq \mathsf{ACA}_0 \subsetneq \mathsf{ATR}_0 \subsetneq \Pi^1_1\text{-}\mathsf{CA}_0.$$

For more about these theories and the program of reverse mathematics in general we refer to Simpson (2009).

It is also known that the proof-theoretic ordinals of ATR_0 and ATR are the ordinals Γ_0 and Γ_{ε_0}, respectively. For these results cf., for example, Friedman, McAloon and Simpson (1982) and Jäger (1980; 1984).

In the following the theory ATR_0 will play a major role. Arithmetical transfinite recursion is relevant here because of its remarkable equivalence to Π^1_1 reduction over ACA_0. This and related reduction principles are introduced now.

Definition 2.3 Let n be a natural number greater than 0.

1. Σ^1_n *reduction* (Σ^1_n-Red) is the schema consisting of all formulas

$$\forall i(\varphi[i] \rightarrow \psi[i]) \rightarrow \exists X \forall i(\varphi[i] \rightarrow i \in X \rightarrow \psi[i]),$$

   where $\varphi[i]$ is a Π^1_n and $\psi[i]$ a Σ^1_n formula.
2. Π^1_n *reduction* (Π^1_n-Red) is the schema consisting of all formulas

$$\forall i(\varphi[i] \rightarrow \psi[i]) \rightarrow \exists X \forall i(\varphi[i] \rightarrow i \in X \rightarrow \psi[i]),$$

   where $\varphi[i]$ is a Σ^1_n and $\psi[i]$ a Π^1_n formula.

As mentioned in the introduction, what we call Σ^1_n reduction [Π^1_n reduction] has been called Π^1_n separation [Σ^1_n separation] by Simpson. Modulo this renaming the following characterizations are known.

Theorem 2.4 (Buchholz-Schütte, Simpson)

(i) The theory ACA_0 + (Σ^1_1-AC) *proves* (Σ^1_1-Red).

(ii) ACA_0 + (Π^1_1-Red) *is equivalent to* ATR_0.

(iii) ACA_0 + (Σ^1_2-Red) *is equivalent to* Δ^1_2-CA_0.

(iv) ACA_0 + (Π^1_2-Red) *is equivalent to* Π^1_2-CA_0.

Assertion (i) is an easy observation; (ii) is a fairly complicated and technically demanding result presented in Simpson (2009); (iii) has been proved in Buchholz and

Schütte (1988); (iv) is mentioned in Simpson (2009) as an exercise. The following theorem is an immediate consequence of Simpson (2009), Corollary VII.2.19.

Theorem 2.5 (Simpson) *The theory* $\mathsf{ACA}_0 + (\mathsf{BI})$ *proves all instances of* (ATR).

3 Basic set theory BS_0 and Simpson's ATR_0^S

After these introductory remarks we now turn to the two set theories that are in the center of this work: Simpson's ATR_0^S and Kripke-Platek set theory KP (with infinity).

The set-theoretic language $\mathcal{L}_\in$ is a one-sorted first order language with two binary relation symbols $\in$ and $=$, countably many set variables, and the usual connectives and quantifiers of first order logic. Terms and formulas of $\mathcal{L}_\in$ are as usual.

We shall make use of the common set-theoretic terminology and employ the standard notational conventions. In addition, we use as metavariables (possibly with subscripts):

1. $a, b, c, f, i, j, k, r, u, v, w, x, y, z$ for set-theoretic variables,
2. $\eta, \theta, \varphi, \psi$ for formulas.

We also follow the general conventions in defining the $\Delta_0, \Sigma, \Pi, \Sigma_n$, and Π_n formulas of $\mathcal{L}_\in$ $(1 \leq n \in \mathbb{N})$.

Basic set theory BS_0 is a theory in the language $\mathcal{L}_\in$ — based on classical first order logic with equality — whose non-logical axioms are the universal closures of the following formulas:

1. Extensionality: $\forall x(x \in a \leftrightarrow x \in b) \rightarrow a = b$.
2. Regularity: $a \neq \varnothing \rightarrow (\exists x \in a)(\forall y \in x)(y \notin a)$.
3. Infinity: $\exists a(\varnothing \in a \;\wedge\; (\forall x, y \in a)(x \cup \{y\} \in a))$.
4. Rudimentary closure: We have axioms that formalize that the universe is closed under a series of rudimentary set-theoretic operations; cf. Simpson (2009), Definition VII.3.3. We also claim that every set is a subset of a transitive set.

It is set-theoretic folklore and mentioned, for example, in Simpson (2009) that BS_0 proves Δ_0 separation.

Lemma 3.1 BS_0 *proves for all* Δ_0 *formulas* $\varphi[x]$ *and all sets* a *that*

$$(\Delta_0\text{-}\mathsf{Sep}) \qquad \exists y \forall x(x \in y \;\leftrightarrow\; x \in a \wedge \varphi[x]).$$

The theory BS is the extension of BS_0 resulting from extending regularity from sets to arbitrary formulas $\varphi[x]$,

$$\exists x \varphi[x] \rightarrow \exists x(\varphi[x] \wedge (\forall y \in x)\neg\varphi[y]).$$

Thus $\in$-induction is available in BS for arbitray $\mathcal{L}_\in$ formulas.

Lemma 3.2 BS *proves, for any* $\mathcal{L}_\in$ *formula* $\varphi[x]$,

$$(\mathcal{L}_\in\text{-}\mathsf{I}_\in) \qquad \forall x((\forall y \in x)\varphi[y] \to \varphi[x]) \to \forall x\varphi[x].$$

Proof Aiming at the contrapositive, assume $\neg\varphi[a]$ for some a and let b be a transitive set such that $\{a\} \subseteq b$. Now set

$$\psi[x] := x \in b \wedge \neg\varphi[x].$$

Clearly, $\exists x\psi[x]$ and, therefore, the schema of regularity for formulas gives us an x satisfying

$$\psi[x] \ \wedge \ (\forall y \in x)\neg\psi[y],$$

i.e.,

$$x \in b \ \wedge \ \neg\varphi[x] \ \wedge \ (\forall y \in x)(y \notin b \vee \varphi[y]).$$

Since b is transitive, this can be simplified to

$$\neg\varphi[x] \ \wedge \ (\forall y \in x)\varphi[y],$$

and we have the desired statement. □

Moreover, we shall employ the common set-theoretic terminology and the standard notational conventions, for example:

$$Tran[a] := (\forall x \in a)(\forall y \in x)(y \in a),$$

$$Ord[a] := Tran[a] \wedge (\forall x \in a)Tran[x],$$

$$Succ[a] := Ord[a] \wedge (\exists x \in a)(a = x \cup \{x\}),$$

$$FinOrd[a] := \begin{cases} Ord[a] \wedge (a = \varnothing \vee Succ[a]) \wedge \\ (\forall x \in a)(x = \varnothing \vee Succ[x]). \end{cases}$$

In addition, we let ω be the collection of all finite ordinals and observe that it forms a set in BS_0.

There is a natural translation of $\mathcal{L}_2$ into $\mathcal{L}_\in$: The number variables of $\mathcal{L}_2$ are interpreted in $\mathcal{L}_\in$ as ranging over ω and the set variables of $\mathcal{L}_2$ are interpreted in $\mathcal{L}_\in$ as ranging over the subsets of ω. This means that

1. first order quantifiers $\exists i$ and $\forall i$ of $\mathcal{L}_2$ translate into $(\exists i \in \omega)$ and $(\forall i \in \omega)$,
2. second order quantifiers $\exists X$ and $\forall X$ of $\mathcal{L}_2$ translate into $(\exists x \subseteq \omega)$ and $(\forall x \subseteq \omega)$.

Then one has to verify that BS_0 proves the existence of set-theoretic functions on ω that correspond to the number-theoretic addition and multiplication. The number-theoretic less and equality relation go over into $<$ and $=$ on ω.

Clearly, each axiom of ACA_0 becomes a theorem of BS_0 under this translation. When working in $\mathcal{L}_\in$ we shall from now on identify $\mathcal{L}_2$ formulas with their translations into $\mathcal{L}_\in$.

Simpson's ATR_0^S is obtained from BS_0 by adding the axiom of countability (C) and the axiom (Beta). In the definition below we write $Inj[f, a, \omega]$ to state that f is an injective function from a to ω.

Definition 3.3 A set a is called *hereditarily countable* iff there exist a transitive superset x of a and an injection from x to ω,

$$HC[a] := \exists x, f(a \subseteq x \wedge Tran[x] \wedge Inj[f, x, \omega]).$$

The *axiom of countability* (C) claims that all sets are hereditarily countable,

$$(\mathsf{C}) := \forall x HC[x].$$

In the formulation of the axiom (Beta) we write $Dom[f, a]$ to express that f is a function with domain a. We write $\langle x, y \rangle$ for the ordered pair of x and y and $a \times b$ for the Cartesian product of a and b. Also, $r \subseteq a \times a$ is called *well-founded on a* iff every non-empty subset of a has an r-minimal element,

$$Wf[a, r] := (\forall b \subseteq a)(b \neq \varnothing \rightarrow (\exists x \in b)(\forall y \in b)(\langle y, x \rangle \notin r)).$$

Definition 3.4 The *axiom* (Beta) is the universal closure of the formula

$$Wf[a, r] \rightarrow \exists f(Dom[f, a] \wedge (\forall x \in a)(f(x) = \{f(y) : y \in a \wedge \langle y, x \rangle \in r\})).$$

This function f is said to be the *collapsing function* for r on a.

The axiom (Beta) has the effect of making the Π_1 predicate $Wf[a, r]$ a Δ_1 predicate since the existence of a collapsing function for r on a obviously implies the well-foundedness of r on a.

With these definitions the stage is set to introduce the theory ATR_0^S and its extension ATR^S:

$$\mathsf{ATR}_0^S := \mathsf{BS}_0 + (\mathsf{Beta}) + (\mathsf{C}) \quad \text{and} \quad \mathsf{ATR}^S := \mathsf{BS} + (\mathsf{Beta}) + (\mathsf{C}).$$

Below we write $\mathsf{ATR}_0^S \backslash (\mathsf{C})$ for the subsystem of ATR_0^S without the axiom of countability and $\mathsf{ATR}^S \backslash (\mathsf{C})$ for ATR^S without (C).

As we have mentioned above, there is a natural translation of $\mathcal{L}_2$ into $\mathcal{L}_\in$. The converse direction, i.e., the translation of $\mathcal{L}_\in$ into $\mathcal{L}_2$, is more complicated. In a nutshell: The sets of $\mathcal{L}_\in$ are represented by so-called *suitable trees*. Any suitable tree T is a well-founded subset of $\mathbb{N}^{<\mathbb{N}}$, and if $\langle n \rangle \in T$ then $T^{\langle n \rangle}$ is the subtree $\{\sigma : \langle n \rangle * \sigma \in T\}$ of T, with σ ranging over (the codes of) finite sequences. Elementhood is then coded by defining

$$S \in^* T \;:=\; \exists n(\langle n \rangle \in T \wedge S \simeq T^{\langle n \rangle}),$$

where $S \simeq T^{\langle n \rangle}$ says that there exists a specific tree isomorphism between S and $T^{\langle n \rangle}$. For all details concerning this representation we refer to Simpson (2009). There it is also described how to each formula φ of $\mathcal{L}_\in$ we associate a formula $|\varphi|$ of $\mathcal{L}_2$.

The following two results of Simpson (2009) make it clear that ATR_0^S is the set-theoretic variant of ATR_0.

Theorem 3.5 (Simpson)

(i) Every axiom of ATR_0 *is a theorem of* ATR_0^S.
(ii) If φ *is an axiom of* ATR_0^S, *then* $|\varphi|$ *is a theorem of* ATR_0.

This theorem can be easily extended to ATR^S. On the side of second order arithmetic, arithmetical transfinite recursion simply has to be replaced by bar induction.

Theorem 3.6

(i) Every axiom of $\mathsf{ACA}_0 + (\mathsf{BI})$ *is a theorem of* $\mathsf{ATR}^S \setminus (\mathsf{C})$.
(ii) If φ *is an axiom of* ATR^S, *then* $|\varphi|$ *is a theorem of* $\mathsf{ACA}_0 + (\mathsf{BI})$.

Proof It is a classic result that all instances of (BI) can be proved by means of (Beta) and $\in$-induction ($\mathcal{L}_\in$-$\mathsf{I}_\in$); see, for example, Jäger (1986). For (ii) we refer to Simpson (2009), Theorem VII.3.34 and Exercise VII.3.38. □

Corollary 3.7 *If we write* $|T|$ *for the proof-theoretic ordinal of the theory* T, *then we have:*

(i) $|\mathsf{ATR}_0| = |\mathsf{ATR}_0^S| = \Gamma_0$.
(ii) $|\mathsf{ATR}| = \Gamma_{\varepsilon_0}$.
(iii) $|\mathsf{ATR}^S| = \Psi(\varepsilon_{\Omega+1})$ *(Bachmann-Howard ordinal).*

We end this introductory section by stating the so-called *quantifier theorem* which relates formulas of $\mathcal{L}_2$ with formulas of $\mathcal{L}_\in$. The formulation below is from Simpson (2009). Its first part also follows from corresponding results in Jäger (1979; 1986).

Theorem 3.8 (Quantifier theorem; Jäger, Simpson) *Let n be any natural number.*

(i) Each Σ^1_{n+2} *formula of* $\mathcal{L}_2$ *is equivalent — provably in* $\mathsf{ATR}_0^S \setminus (\mathsf{C})$ *— to a* Σ_{n+1} *formula of* $\mathcal{L}_\in$.
(ii) If φ *is a* Σ_n *formula of* $\mathcal{L}_\in$, *then* $|\varphi|$ *is equivalent — provably in* ATR_0 *— to a* Σ^1_{n+1} *formula of* $\mathcal{L}_2$.

In many set theories — Kripke-Platek set theory is a typical example — we do not have to make a big difference between Δ_0 and Δ_1 formulas. That this is not the case for ATR_0^S is an immediate consequence of the quantifier theorem. Recall from Lemma 3.1 that ATR_0^S proves Δ_0 separation. However:

Corollary 3.9 *There are instances of* Δ_1 *separation that are not provable in* ATR^S.

Proof In view of the quantifier theorem, $\mathsf{ATR}^S + (\Delta_1\text{-}\mathsf{Sep})$ comprises the theory Δ^1_2-CA_0, whose proof-theoretic ordinal is much greater than the Bachmann-Howard ordinal. Therefore, $(\Delta^1_2\text{-}\mathsf{CA})$ cannot be provable in ATR^S. □

4 Kripke-Platek set theory KP and its relationship to ATR_0^S

Kripke-Platek set theory KP (with infinity) is one of the best studied subsystems of Zermelo-Fraenkel set theory ZF. The transitive models of KP are called *admissible sets*, and $L_{\omega_1^{CK}}$, where ω_1^{CK} denotes the first non-recursive ordinal, is the least standard model of KP. Kripke-Platek set theory and admissible sets play an important role in generalized recursion theory, definability theory and, of course, in proof theory.

Kripke-Platek set theory KP is obtained from BS by adding the schema of Δ_0 collection, i.e.,

$$(\Delta_0\text{-}\mathsf{Col}) \qquad (\forall x \in a)\exists y\varphi[x, y] \to \exists z(\forall x \in a)(\exists y \in z)\varphi[x, y]$$

for all Δ_0 formulas $\varphi[x, y]$;

$$\mathsf{KP} := \mathsf{BS} + (\Delta_0\text{-}\mathsf{Col}).^{2}$$

KP_0 is the subsystem of KP where regularity is restricted to sets (as in BS_0), i.e.,

$$\mathsf{KP}_0 := \mathsf{BS}_0 + (\Delta_0\text{-}\mathsf{Col}).$$

Thus, obviously, $\mathsf{KP}_0 + (\mathsf{Beta})$ and $\mathsf{KP} + (\mathsf{Beta})$ are the same theories as $\mathsf{ATR}_0^S\backslash(\mathsf{C}) + (\Delta_0\text{-}\mathsf{Col})$ and $\mathsf{ATR}^S\backslash(\mathsf{C}) + (\Delta_0\text{-}\mathsf{Col})$, respectively.

Below we list a series of known result that indicate that the relationship between ATR_0^S and KP is quite intricate. They are mostly taken from Simpson (1980; 2009) and Jäger (1982; 1986).

Theorem 4.1 (Overview)

(i) The proof-theoretic ordinal of KP *is the Bachmann-Howard ordinal* $\Psi(\varepsilon_{\Omega+1})$. KP *is proof-theoretically equivalent to* ATR^S *and to the theory of non-iterated inductive definitions* ID_1.
(ii) $\mathsf{KP}_0 + (\mathsf{Beta})$ *is proof-theoretically equivalent to* $\Delta_2^1\text{-}\mathsf{CA}_0$.
(iii) $\mathsf{KP} + (\mathsf{Beta})$ *is proof-theoretically equivalent to* $\Delta_2^1\text{-}\mathsf{CA}_0 + (\mathsf{BI})$.
(iv) Any well-founded model of ATR_0^S *of height* ω_1^{CK} *is not a model of* KP.
(v) Any well-founded model of KP *of height* ω_1^{CK} *is not a model of* ATR_0^S.

From that we immediately deduce that ATR_0^S and KP are not compatible in the sense that

$$\mathsf{KP} \not\subseteq \mathsf{ATR}_0^S \quad \text{and} \quad \mathsf{ATR}_0^S \not\subseteq \mathsf{KP}.$$

We even have $\mathsf{KP}_0 \not\subseteq \mathsf{ATR}^S$. Otherwise, ATR^S would comprise $\mathsf{KP} + (\mathsf{Beta})$, and the ordinal of this theory is greater than the Bachmann-Howard ordinal.

---

[2] Traditionally, KP is defined to be the set theory whose non-logical axioms consist of Extensionality, Pairing, Union, Infinity, $(\mathcal{L}_\in\text{-}\mathsf{I}_\in)$, $(\Delta_0\text{-}\mathsf{Sep})$, and $(\Delta_0\text{-}\mathsf{Col})$; but both definitions are equivalent.

5 Reduction axioms in set theory

The next step is to add reduction principles, similar to those of second order arithmetic in Definition 2.3, to our set theories.

Definition 5.1

1. Σ_1 *reduction* (Σ_1-Red) is the schema consisting of all formulas

$$(\forall x \in a)(\varphi[x] \to \psi[x]) \to (\exists y \subseteq a)(\forall x \in a)(\varphi[x] \to x \in y \to \psi[x]),$$

where $\varphi[x]$ is a Π_1 and $\psi[x]$ a Σ_1 formula.

2. Π_1 *reduction* (Π_1-Red) is the schema consisting of all formulas

$$(\forall x \in a)(\varphi[x] \to \psi[x]) \to (\exists y \subseteq a)(\forall x \in a)(\varphi[x] \to x \in y \to \psi[x]),$$

where $\varphi[x]$ is a Σ_1 and $\psi[x]$ a Π_1 formula.

Our aim is to analyze the strengths of (Σ_1-Red) and (Π_1-Red) in the context of Simpson's ATR_0^S and Kripke-Platek set theory. We begin with ATR_0^S and ATR^S where the situation is clear.

5.1 ATR_0^S and ATR^S plus (Σ_1-Red) and (Π_1-Red)

Theorem 3.5 and Theorem 3.6, in combination with the quantifier theorem, immediately give us

$$\begin{aligned}
\mathsf{ATR}_0 + (\Sigma^1_2\text{-Red}) &\subseteq \mathsf{ATR}_0^S + (\Sigma_1\text{-Red}),\\
\mathsf{ACA}_0 + (\mathsf{BI}) + (\Sigma^1_2\text{-Red}) &\subseteq \mathsf{ATR}^S + (\Sigma_1\text{-Red}),\\
\mathsf{ATR}_0 + (\Pi^1_2\text{-Red}) &\subseteq \mathsf{ATR}_0^S + (\Pi_1\text{-Red}),\\
\mathsf{ACA}_0 + (\mathsf{BI}) + (\Pi^1_2\text{-Red}) &\subseteq \mathsf{ATR}^S + (\Pi_1\text{-Red}).
\end{aligned}$$

Therefore, together with Theorem 2.4 we have the following lower bound results for (Σ_1-Red) and (Π_1-Red).

Theorem 5.2

(i) $\Delta^1_2\text{-}\mathsf{CA}_0 \subseteq \mathsf{ACA}_0 + (\Sigma^1_2\text{-Red}) \subseteq \mathsf{ATR}_0^S + (\Sigma_1\text{-Red})$.
(ii) $\Delta^1_2\text{-}\mathsf{CA}_0 + (\mathsf{BI}) \subseteq \mathsf{ACA}_0 + (\Sigma^1_2\text{-Red}) + (\mathsf{BI}) \subseteq \mathsf{ATR}^S + (\Sigma_1\text{-Red})$.
(iii) $\Pi^1_2\text{-}\mathsf{CA}_0 \subseteq \mathsf{ACA}_0 + (\Pi^1_2\text{-Red}) \subseteq \mathsf{ATR}_0^S + (\Pi_1\text{-Red})$.
(iv) $\Pi^1_2\text{-}\mathsf{CA}_0 + (\mathsf{BI}) \subseteq \mathsf{ACA}_0 + (\Pi^1_2\text{-Red}) + (\mathsf{BI}) \subseteq \mathsf{ATR}^S + (\Pi_1\text{-Red})$.

Turning to the converse directions, we first make use of the quantifier theorem again and observe that for every (closed) instance φ of (Σ_1-Red) and (Π_1-Red) the corresponding $\mathcal{L}_2$ formula $|\varphi|$ is derivable in $\mathsf{ATR}_0 + (\Sigma^1_2\text{-Red})$ and $\mathsf{ATR}_0 + (\Pi^1_2\text{-Red})$, respectively. Therefore, Theorem 3.5 and Theorem 3.6 yield for all sentences φ of $\mathcal{L}_\in$:

$$\mathsf{ATR}^S_0 + (\Sigma_1\text{-Red}) \vdash \varphi \implies \mathsf{ATR}_0 + (\Sigma^1_2\text{-Red}) \vdash |\varphi|,$$

$$\mathsf{ATR}^S + (\Sigma_1\text{-Red}) \vdash \varphi \implies \mathsf{ACA}_0 + (\Sigma^1_2\text{-Red}) + (\mathsf{BI}) \vdash |\varphi|,$$

$$\mathsf{ATR}^S_0 + (\Pi_1\text{-Red}) \vdash \varphi \implies \mathsf{ATR}_0 + (\Pi^1_2\text{-Red}) \vdash |\varphi|,$$

$$\mathsf{ATR}^S + (\Pi_1\text{-Red}) \vdash \varphi \implies \mathsf{ACA}_0 + (\Pi^1_2\text{-Red}) + (\mathsf{BI}) \vdash |\varphi|.$$

The following upper bounds for $(\Sigma_1\text{-Red})$ and $(\Pi_1\text{-Red})$ are straightforward consequences of Theorem 2.4.

Theorem 5.3 *We have for every sentence φ of $\mathcal{L}_\in$:*

(i) $\mathsf{ATR}^S_0 + (\Sigma_1\text{-Red}) \vdash \varphi \implies \Delta^1_2\text{-}\mathsf{CA}_0 \vdash |\varphi|$.
(ii) $\mathsf{ATR}^S + (\Sigma_1\text{-Red}) \vdash \varphi \implies \Delta^1_2\text{-}\mathsf{CA}_0 + (\mathsf{BI}) \vdash |\varphi|$.
(iii) $\mathsf{ATR}^S_0 + (\Pi_1\text{-Red}) \vdash \varphi \implies \Pi^1_2\text{-}\mathsf{CA}_0 \vdash |\varphi|$.
(iv) $\mathsf{ATR}^S + (\Pi_1\text{-Red}) \vdash \varphi \implies \Pi^1_2\text{-}\mathsf{CA}_0 + (\mathsf{BI}) \vdash |\varphi|$.

To sum up, we know that $(\Sigma_1\text{-Red})$ and $(\Pi_1\text{-Red})$ added to ATR^S_0 and ATR^S lead to the following proof-theoretic equivalences:

$$\mathsf{ATR}^S_0 + (\Sigma_1\text{-Red}) \equiv \Delta^1_2\text{-}\mathsf{CA}_0,$$

$$\mathsf{ATR}^S + (\Sigma_1\text{-Red}) \equiv \Delta^1_2\text{-}\mathsf{CA}_0 + (\mathsf{BI}),$$

$$\mathsf{ATR}^S_0 + (\Pi_1\text{-Red}) \equiv \Pi^1_2\text{-}\mathsf{CA}_0,$$

$$\mathsf{ATR}^S + (\Pi_1\text{-Red}) \equiv \Pi^1_2\text{-}\mathsf{CA}_0 + (\mathsf{BI}).$$

So for $(\Sigma_1\text{-Red})$ and $(\Pi_1\text{-Red})$ the situation is clear as long as we stay in the context of ATR^S_0 and its extension ATR^S. The picture is completely different when we move to Kripke-Platek set theory.

5.2 KP_0 and KP plus $(\Sigma_1\text{-Red})$ and $(\Pi_1\text{-Red})$

A first observation is that $(\Sigma_1\text{-Red})$ is irrelevant for Kripke-Platek set theory; it is provable there.

Lemma 5.4 *Every instance of* $(\Sigma_1\text{-Red})$ *is provable in* KP_0.

Proof Suppose that, for some Π_1 formula $\varphi[x]$ and Σ_1 formula $\psi[x]$,

$$(\forall x \in a)(\varphi[x] \to \psi[x]).$$

By Σ reflection there exists a set b such that

$$(\forall x \in a)(\varphi^b[x] \to \psi^b[x]).$$

Then $c := \{n \in a : \psi^b[n]\}$ is the set we need. □

A further obvious observation is that an upper bound for (Π_1-Red) is provided by Σ_1 separation, i.e., the schema

$$(\Sigma_1\text{-Sep}) \qquad \exists y \forall x(x \in y \leftrightarrow x \in a \wedge \varphi[x])$$

for all Σ_1 formulas $\varphi[x]$. To see why, take a Σ_1 formula $\varphi[x]$ and a Π_1 formula $\psi[x]$ such that

$$(\forall x \in a)(\varphi[x] \to \psi[x])$$

for some set a. By (Σ_1-Sep) we can define the set $b := \{x \in a : \varphi[x]\}$ which is an obvious witness for (Σ_1-Red). Thus we have the following upper bounds.

Theorem 5.5

(i) $\mathsf{KP}_0 + (\Pi_1\text{-Red}) \subseteq \mathsf{KP}_0 + (\Sigma_1\text{-Sep})$.
(ii) $\mathsf{KP} + (\Pi_1\text{-Red}) \subseteq \mathsf{KP} + (\Sigma_1\text{-Sep})$.

It is still an open question whether these bounds are sharp. However, we have some partial results.

Theorem 5.6 *We have for all formulas φ of $\mathcal{L}_2$:*

(i) $\Pi^1_2\text{-}\mathsf{CA}_0 \vdash \varphi \implies \mathsf{KP}_0 + (\mathsf{Beta}) + (\Pi_1\text{-Red}) \vdash \varphi$.
(ii) $\Pi^1_2\text{-}\mathsf{CA}_0 + (\mathsf{BI}) \vdash \varphi \implies \mathsf{KP} + (\mathsf{Beta}) + (\Pi_1\text{-Red}) \vdash \varphi$.

Proof We first recall from Theorem 2.4 that $\Pi^1_2\text{-}\mathsf{CA}_0$ is equivalent to $\mathsf{ACA}_0 + (\Pi^1_2\text{-Red})$. From the first part of the quantifier theorem we deduce that, within $\mathsf{KP}_0 + (\mathsf{Beta})$, every Σ^1_2 formula of $\mathcal{L}_2$ is equivalent to a Σ_1 formula and every Π^1_2 formula of $\mathcal{L}_2$ is equivalent to a Π_1 formula. Therefore, every instance of (Π^1_2-Red) is provable in $\mathsf{KP}_0 + (\mathsf{Beta}) + (\Pi_1\text{-Red})$. The rest follows from Theorem 3.6. □

Now we recall from, for example, Rathjen (1999) that $\mathsf{KP}_0 + (\Sigma_1\text{-Sep})$ and $\mathsf{KP} + (\Sigma_1\text{-Sep})$ prove the same $\mathcal{L}_2$ sentences as $\Pi^1_2\text{-}\mathsf{CA}_0$ and $\Pi^1_2\text{-}\mathsf{CA}_0 + (\mathsf{BI})$, respectively. Moreover, by following Barwise (1975) (with some small modifications), we can also show that $\mathsf{KP}_0 + (\Sigma_1\text{-Sep})$ proves (Beta). Thus the following assertions are direct consequences of the previous two theorems.

Corollary 5.7

(i) The theories $\Pi^1_2\text{-}\mathsf{CA}_0$, $\mathsf{KP}_0 + (\mathsf{Beta}) + (\Pi_1\text{-Red})$, *and* $\mathsf{KP}_0 + (\Sigma_1\text{-Sep})$ *prove the same $\mathcal{L}_2$ sentences.*
(ii) The theories $\Pi^1_2\text{-}\mathsf{CA}_0 + (\mathsf{BI})$, $\mathsf{KP} + (\mathsf{Beta}) + (\Pi_1\text{-Red})$, *and* $\mathsf{KP} + (\Sigma_1\text{-Sep})$ *prove the same $\mathcal{L}_2$ sentences.*

Shown schematically, we therefore have the following proof-theoretic equivalences:

$$\mathsf{KP}_0 + (\mathsf{Beta}) + (\Pi_1\text{-}\mathsf{Red}) \equiv \mathsf{KP}_0 + (\Sigma_1\text{-}\mathsf{Sep}) \equiv \Pi^1_2\text{-}\mathsf{CA}_0,$$

$$\mathsf{KP} + (\mathsf{Beta}) + (\Pi_1\text{-}\mathsf{Red}) \equiv \mathsf{KP} + (\Sigma_1\text{-}\mathsf{Sep}) \equiv \Pi^1_2\text{-}\mathsf{CA}_0 + (\mathsf{BI}).$$

The axiom of constructibility states that every set belongs to some level L_α of Gödel's hierarchy of constructible sets:

$$(V{=}L) \qquad \forall x \exists \alpha (x \in L_\alpha).$$

In addition, we write $(a <_L b)$ to express that a is smaller than b according to the well-order $<_L$ of the constructible universe. It is well-known that $(a \in L_\alpha)$ and $(a <_L b)$ are Δ over KP. Moreover, $(\exists x <_L a)$ and $(\forall x <_L a)$ may be treated as bounded quantifiers. $(x \in L)$ is short for $\exists \alpha (x \in L_\alpha)$. For more on the constructible universe see, e.g., Barwise (1975) or Kunen (1980).

In Jäger and Steila (2018) an interesting separation principle is introduced and studied. Call a quantifier *subset bounded* iff it ranges over the subsets of a given set. Then let $\exists^{\mathcal{P}}(\Delta_1)$ separation be the separation principle that, given any set a, allows the introduction of all subsets of a defined by a subset bounded Σ_1 formula over a Δ_1 matrix, i.e.,

$$(\exists^{\mathcal{P}}(\Delta_1)\text{-}\mathsf{Sep}) \qquad \begin{aligned} &(\forall x \in a)(\forall y \subseteq a)(\varphi[x,y] \leftrightarrow \psi[x,y]) \to \\ &\quad (\exists z \subseteq a)(\forall x \in a)(x \in z \leftrightarrow (\exists y \subseteq a)\varphi[x,y]), \end{aligned}$$

for all Σ_1 formulas $\varphi[x,y]$ and Π_1 formulas $\psi[x,y]$.

The relationship between $(\Sigma_1\text{-}\mathsf{Sep})$ and $(\exists^{\mathcal{P}}(\Delta_1)\text{-}\mathsf{Sep})$ is interesting. It is easy to see that all instances of $(\exists^{\mathcal{P}}(\Delta_1)\text{-}\mathsf{Sep})$ are provable in $\mathsf{KP} + (\Sigma_1\text{-}\mathsf{Sep})$. For the converse direction see the follwing theorem from Jäger and Steila (2018).

Theorem 5.8 (Jäger and Steila) *All instances of* $(\Sigma_1\text{-}\mathsf{Sep})$ *are provable in* $\mathsf{KP} + (V{=}L) + (\exists^{\mathcal{P}}(\Delta_1)\text{-}\mathsf{Sep})$.

Therefore, in order to show that the theory $\mathsf{KP} + (V{=}L) + (\Pi_1\text{-}\mathsf{Red})$ contains $\mathsf{KP} + (\Sigma_1\text{-}\mathsf{Sep})$, it is sufficient to prove that it contains $(\exists^{\mathcal{P}}(\Delta_1)\text{-}\mathsf{Sep})$.

Lemma 5.9 $\mathsf{KP} + (V{=}L) + (\Pi_1\text{-}\mathsf{Red})$ *proves all instances of* $(\exists^{\mathcal{P}}(\Delta_1)\text{-}\mathsf{Sep})$.

Proof Working in $\mathsf{KP} + (V{=}L) + (\Pi_1\text{-}\mathsf{Red})$, let us assume that

$$(\forall x \in a)(\forall y \subseteq a)(\varphi[x,y] \leftrightarrow \psi[x,y])$$

for a Σ_1 formula $\varphi[x,y]$ and a Π_1 formula $\psi[x,y]$. We define

$$\widetilde{\varphi}[x,y] := \varphi[x,y] \ \wedge \ (\forall z <_L y)(z \subseteq a \to \neg\psi[x,z]),$$

$$\widetilde{\psi}[x,y] := \psi[x,y] \ \wedge \ (\forall z <_L y)(z \subseteq a \to \neg\varphi[x,z]).$$

Thus $\widetilde{\varphi}[x,y]$ is a Σ formula and $\widetilde{\psi}[x,y]$ is a Π formula, and we have:

(1) $(\forall x \in a)(\forall y \subseteq a)(\widetilde{\varphi}[x,y] \leftrightarrow \widetilde{\psi}[x,y])$.

(2) $\widetilde{\varphi}[x, y] \to \varphi[x, y]$.
(3) $y \subseteq a \land \varphi[x, y] \to (\exists z \subseteq a)\widetilde{\varphi}[x, z]$.
(4) $(\forall y, z \subseteq a)(\widetilde{\varphi}[x, y] \land \widetilde{\varphi}[x, z] \to y = z)$.

We also define

$$\mathfrak{A}[u] := (\exists v, w \in a)(\exists y \subseteq a)(u = \langle v, w\rangle \land \widetilde{\varphi}[v, y] \land w \in y),$$

$$\mathfrak{B}[u] := (\forall v, w \in a)(\forall y \subseteq a)(u = \langle v, w\rangle \land \widetilde{\varphi}[v, y] \to w \in y).$$

Hence $\mathfrak{A}[u]$ is equivalent to a Σ_1 formula and $\mathfrak{B}[u]$ to a Π_1 formula. In addition, because of (4) we have

$$(\forall u \in a \times a)(\mathfrak{A}[u] \to \mathfrak{B}[u]).$$

So by (Π_1-Red) there is a $b \subseteq a \times a$ such that

$$(\forall u \in a \times a)(\mathfrak{A}[u] \to u \in b \to \mathfrak{B}[u]).$$

Claim: For all $v \in a$,

$$(*) \qquad (\exists y \subseteq a)\varphi[v, y] \leftrightarrow \widetilde{\varphi}[v, (b)_v],$$

where $(b)_v$ stands for the set $\{w \in a : \langle v, w\rangle \in b\}$.

Proof of the claim. The direction from right to left is obvious. To prove the converse, assume $(\exists y \subseteq a)\varphi[v, y]$. Then there is a $c \subseteq a$ with $\widetilde{\varphi}[v, c]$ according to (3), and for this c we have $c = (b)_v$. This can be seen as follows: For all $w \in a$,

$$w \in c \to \mathfrak{A}[\langle v, w\rangle] \to \langle v, w\rangle \in b \to w \in (b)_v,$$

$$w \in (b)_v \to \langle v, w\rangle \in b \to \mathfrak{B}[\langle v, w\rangle] \to w \in c.$$

Since $c = (b)_v$ and $\widetilde{\varphi}[v, c]$, we have $\widetilde{\varphi}[v, (b)_v]$ as desired, finishing the proof of the claim.

In view of (1) the set $\{v \in a : \widetilde{\varphi}[v, (b)_v]\}$ exists by Δ separation and is the set we need according to $(*)$. □

The following is an immediate consequence of Theorem 5.5, Theorem 5.8, and the previous lemma.

Corollary 5.10 *We have for all $\mathcal{L}_\in$ formulas φ that*

$$\mathsf{KP} + (V{=}L) + (\Sigma_1\text{-}\mathsf{Sep}) \vdash \varphi \iff \mathsf{KP} + (V{=}L) + (\Pi_1\text{-}\mathsf{Red}) \vdash \varphi.$$

In order to get rid of the axiom $(V{=}L)$ on the left-hand side of the previous implication, we show that L is an inner model of KP + (Σ_1-Sep).

Theorem 5.11 *If φ is the universal closure of an axiom of* KP + (Σ_1-Sep), *we have that*

$$\mathsf{KP} + (\Sigma_1\text{-}\mathsf{Sep}) \vdash \varphi^L.$$

Here φ^L is the result of restricting all unbounded quantifiers in φ to L.

Proof In view of Barwise (1975) we only have to deal with the instances of $(\Sigma_1\text{-}\mathsf{Sep})$. So let $\varphi[x, y, z]$ be a Δ_0 formula with all free variables indicated; we suppress mentioning additional parameters. Given elements $a, b \in L$ we have to show that

$$(\exists z \in L)(\forall x \in L)(x \in z \leftrightarrow x \in a \land \exists y(y \in L \land \varphi[x, y, b])).$$

By $(\Sigma_1\text{-}\mathsf{Sep})$ there exists the set

$$c = \{x \in a : \exists y(y \in L \land \varphi[x, y, b])\}$$

and thus we have

$$(\forall x \in c)\exists \xi(\exists y \in L_\xi)\varphi[x, y, b].$$

By Σ collection there is an α such that

$$(\forall x \in c)(\exists \xi < \alpha)(\exists y \in L_\xi)\varphi[x, y, b]$$

and so, by the properties of the L-hierarchy,

$$(\forall x \in c)(\exists y \in L_\alpha)\varphi[x, y, b].$$

This implies that $c = \{x \in a : (\exists y \in L_\alpha)\varphi[x, y, b]\}$. Hence c is the required witness in L. □

This theorem implies that $\mathsf{KP} + (\Sigma_1\text{-}\mathsf{Sep}) + (V{=}L)$ is conservative over $\mathsf{KP} + (\Sigma_1\text{-}\mathsf{Sep})$ for formulas which are absolute w.r.t. $\mathsf{KP} + (\Sigma_1\text{-}\mathsf{Sep})$, in particular for all arithmetical formulas.

Corollary 5.12 *We have for all arithmetical formulas φ that*

$$\mathsf{KP} + (\Sigma_1\text{-}\mathsf{Sep}) \vdash \varphi \iff \mathsf{KP} + (V{=}L) + (\Pi_1\text{-}\mathsf{Red}) \vdash \varphi.$$

If we summarize our results, we have the following proof-theoretic equivalences:

$$\mathsf{KP} + (V{=}L) + (\Pi_1\text{-}\mathsf{Red}) \equiv \mathsf{KP} + (\Sigma_1\text{-}\mathsf{Sep}) \equiv \Pi^1_2\text{-}\mathsf{CA}_0 + (\mathsf{BI}).$$

6 Comments and questions

This article did not discuss the theory $\mathsf{KP}_0 + (V{=}L) + (\Pi_1\text{-}\mathsf{Red})$. There is the question of how to deal with the constructible hierarchy in KP_0 and whether there is an analogue of Theorem 5.8 with KP replaced by KP_0. However, we are not sure whether this leads to something interesting.

Our real concern in the present context is the question of the strength of $\mathsf{KP} + (\Pi_1\text{-}\mathsf{Red})$. We know that $(\Pi_1\text{-}\mathsf{Red})$ is not provable in KP. This can be seen as follows:

(i) In KP+(Π_1-Red) we can prove (Π^1_1-Red) and therefore, according to Theorem 2.4, KP + (Π_1-Red) contains ATR_0.
(ii) In view of a result in Avigad (1996) we thus know that KP + (Π_1-Red) proves that every X-positive arithmetical formula $\varphi[x, X^+]$ has a fixed point (which is a set).
(iii) On the other hand, we also know by results due to Gregoriades (2019) and Probst (2005) that there are positive arithmetical formulas that do not have hyperarithmetical fixed points.
(iv) So we conclude that $L_{\omega_1^{CK}}$ is not a model of KP + (Π_1-Red), implying that KP does not prove (Π_1-Red).

But is the proof-theoretic strength of KP + (Π_1-Red) greater than that of KP? As a preparatory step for the analysis of the proof-theoretic strength of KP + (Π_1-Red) it could be useful to check whether KP + (Π_1-Red) proves (Π^1_1-CA) or (Δ^1_2-CA).

A different line of research is to look at reduction principles of the form as discussed above in theories of sets and classes. Some first results are known, but in general this field is wide open.

References

Avigad, J. (1996). On the relationship between ATR_0 and $\widehat{\mathsf{ID}}_{<\omega}$. *Journal of Symbolic Logic* 62, 768–779.

Barwise, J. (1975). *Admissible Sets and Structures*. Springer.

Buchholz, W. and K. Schütte (1988). Proof Theory of Impredicative Subsystems of Analysis. *Studies in Proof Theory. Monographs* 2.

Friedman, H., K. McAloon, and S. G. Simpson (1982). A finite combinatorial principle which is equivalent to the 1-consistency of predicative analysis. In: *Patras Logic Symposion*. Ed. by G. Metakides. Vol. 109. Studies in Logic and the Foundations of Mathematics. Elsevier, 197–230.

Gregoriades, V. (2019). On a question of Jaeger's. arXiv: 1905.09609.

Jäger, G. (1979). Die konstruktible Hierarchie als Hilfsmittel zur beweistheoretischen Untersuchung von Teilsystemen der Mengenlehre und Analysis. PhD thesis. University of Munich.

– (1980). Theories for iterated jumps. Notes. Oxford.

– (1982). Zur Beweistheorie der Kripke-Platek-Mengenlehre über den natürlichen Zahlen. *Archive for Mathematical Logic* 22, 121–139.

– (1984). The strength of admissibility without foundation. *Journal of Symbolic Logic* 49, 867–879.

– (1986). Theories for Admissible Sets. A Unifying Approach to Proof Theory. In: *Studies in Proof Theory. Lecture Notes*. Vol. 2. Bibliopolis.

Jäger, G. and S. Steila (2018). About some fixed point axioms and related principles in Kripke-Platek environments. *Journal of Symbolic Logic* 83, 642–668.

Kunen, K. (1980). *Set Theory. An Introduction to Independence Proofs*. Vol. 102. Studies in Logic and the Foundations of Mathematics. North-Holland.

Probst, D. (2005). Pseudo-Hierarchies in Admissible Set Theories without Foundation and Explicit Mathematics. PhD thesis. University of Bern.

Rathjen, M. (1999). Explicit mathematics with the monotone fixed point principle. II: models. *Journal of Symbolic Logic* 64.

Simpson, S. G. (1980). Set theoretic aspects of ATR_0. In: *Logic Colloquium '80*. Ed. by D. van Dalen, D. Lascar, and T. J. Smiley. Studies in Logic and the Foundations of Mathematics. Elsevier, 255–271.

– (2009). *Subsystems of second order arithmetic*. 2nd. Perspectives in Logic. Association for Symbolic Logic and Cambridge University Press.

Comments on the Contributions

Peter Schroeder-Heister

Abstract The contributions to this volume represent a broad range of aspects of proof-theoretic semantics. Some do so in the narrower, and some in the wider sense of the term. Some deal with issues I have been concerned with directly, and some tackle further problems. All of them open interesting new perspectives and develop the field in different directions. I will briefly comment on the significance of each contribution here.

Sundholm on Frege's anticipation of the deduction theorem. Two papers are historically oriented: Göran Sundholm's and to some extent that by Neil Tennant (which will be discussed later). Both deal with Frege. I welcome this very much, not only as Frege has been a main topic in my own historical studies, but also because Frege's work is of high systematic relevance to proof-theoretic semantics. He was the first to develop a precise notion of logical deduction, based on the idea that we proceed from judgements already seen as true to new true judgement in a gap-free way. Even the sentences of his concept script (*Begriffsschrift*) can be read as a two-dimensional notation of sequents, so that his proofs proceed in a sort of sequent calculus (Schroeder-Heister 1997; 2014). Sundholm draws attention to the much neglected logical content of Section 17 of Frege's *Foundations of Arithmetic* (1884). He shows that Frege in a sense anticipates the deduction theorem as well as its proof in Hilbert-Bernays (1934). In pointing this out, he discusses Frege's notion of analyticity in detail, as the conditional statements generated by the deduction theorem represent analytical judgements. What I find particularly interesting, besides the wealth of discussion of related concepts such as axiom, tautology, self-evidence, topic-neutrality and the like, is the fact that Sundholm, at least implicitly, alludes to the fact that in Frege there is a notion of proof from assumptions, in addition to proofs of hypothetical judgements, because otherwise talking of the deduction

Peter Schroeder-Heister
Department of Computer Science, University of Tübingen, Germany, e-mail: psh@uni-tuebingen.de

T. Piecha and K. F. Wehmeier (eds.), *Peter Schroeder-Heister on Proof-Theoretic Semantics*, Outstanding Contributions to Logic 29,
https://doi.org/10.1007/978-3-031-50981-0_17

theorem does not make much sense.[1] Though there are several places, where Frege speaks against assumptions as a special mode of judgement (e.g., Frege 1906; 1923), in the passage discussed by Sundholm he speaks of a deduction which starts from a *fact* ("Tatsache"), that is, something that comes from outside, and argues that the content of this deduction can be expressed by an (analytical) hypothetical judgement. The judgement we would call an assumption (as the starting point of the deduction), figures in Frege as the statement of an external fact. Formally, for the validity of the subsequent chain of deduction, it does not make any difference whether we start from an external fact or from an internally specified assumption: we have the idea of something from which a logical deduction starts, and which can then become ingredient of the deduction theorem.[2]

Tennant on Frege's class theory and the logic of sets. Tennant argues that Frege's fundamental assumptions would have led to a system corresponding to Core Logic, the foundational system for logic and mathematics Tennant has advocated and developed for several decades. Core Logic is itself very interesting from the standpoint of proof-theoretic semantics, as it relies on specific inferentialist assumptions concerning harmony, normal form etc. What makes it particularly significant is that is includes and accounts for crucial issues beyond standard proof-theoretic semantics, such as the notion of relevance as well as the non-insistance on transitivity (or cut in the sequent calculus). The latter is a point I have also put forward in several contexts, in particular in the development of definitional reflection and its application to paradoxes, a field where Tennant and myself have similar intuitions (Tennant, 1982, Schroeder-Heister, 2012b)[3]. What is also important about Core Logic is that it is a free logic, a topic underdeveloped in standard proof-theoretic semantics (or only discussed in connection with the denotation of proofs, but not in connection with singular terms), as the whole area of denotation is not given the attention it deserves. If we base our theorizing not on denotation and truth but on inference and proof as basic concepts, we must be able to say how we can incorporate the denotational concepts. Core Logic is perhaps the best worked-out system in this direction, apart from constructive type theories in the Martin-Löf tradition. Though itself set-theoretic in spirit, with the latter it has in common that it introduces the terms and predicates needed through particular rules. In standard approaches to set theory, the existence of certain sets is postulated axiomatically, but the terms denoting these sets are not primitive symbols of the language itself but have a status similar to eliminable definite descriptions. Tennant calls them "pasigraphs" as they are used throughout, even in

[1] When reading Frege-implications as sequents, an iterated implication with the antecedents ("Unterglieder") A and B and the succedent C ("Oberglied") cannot be distinguished from one with the antecedent A and succedent B-*implies*-C — it is just a different way of looking at the sequent, which means that a deduction theorem in this sense is built into Frege's notation (see Schroeder-Heister 1999; 2014).

[2] Frege obviously means *empirical facts* as he is making his remarks in the context of the discussion of induction, though his logical claim is completely independent of this context. His deduction theorem also holds when a deduction starts, for example, from a tautology or a contradiction.

[3] Even though concerning Ekman's paradox, our opinions diverge; see Schroeder-Heister and Tranchini (2017; 2021) and Tennant (2021).

the standard expositions, and insists that they be inferentially defined by introduction and elimination inferences.

Prawitz on validity of inference and argument. Proof-theoretic validity lies at the core of proof-theoretic semantics. Whether this programme will be successful depends in many respects on whether it can provide notions of validity that can compete with what is available in model-theoretic semantics. Defining a promising concept has been one of the central philosophical occupations of Prawitz's research after his groundbreaking work on natural deduction (Prawitz, 1965). The programme of defining validity has turned out more difficult than perhaps envisaged at the beginning of the 1970s. For a long time Prawitz has advocated a notion of validity of arguments or proofs, on which, as a secondary notion, a notion of validity of inferences could build (Prawitz 1971; 1973; 2006). I have myself tried to formally explicate this programme (Schroeder-Heister, 2006), though I have always tended to give the validity of inferences a conceptually primary status. For somewhat different reasons Prawitz now also favours such an approach. The big problem, however, is that the notions of validity of inference and validity of proof interact. If one wants to put validity of inferences first and that of proofs second — the validity of proofs as depending on that of the inferences involved in the proofs — there is the problem that there needs to be some property whose transmission from premisses to conclusion of an inference establishes its validity. This property would be some sort of validity of proofs (in the model-theoretic case it is the notion of truth in a model). Prawitz and myself are both working on hopefully resolving this prima facie circularity. His contribution to this volume can be seen as an intermediary step towards a solution. It characterizes the notions of inference, validity, argument, canonicality etc. together with their interdependencies, without yet trying to bring these concepts into a well-founded order. This is a step forward, as it sets up certain adequacy conditions these concepts have to obey. Currently it is not clear that in the end a well-founded order of concepts can be established. Perhaps the structural characterization of concepts and their interrelations is the best we can achieve: Proof-theoretic validity of inferences and arguments would then constitute a structure that can materialize in different ways. As this is work still under way, I leave the discussion here at this general level, and just stress its fundamental character.

De Campos Sanz on Kolmogorov and the theory of problems. Starting from Kolmogorov's interpretation of logical propositions as problems for which a solution is sought, de Campos Sanz arrives at a couple of distinctions that are directly relevant to proof-theoretic semantics. He presents a "reduction semantics", according to which the consequence from A to B is interpreted as the reduction of the problem B to the problem A, and proves its adequacy with respect to intuitionistic logic. Applying this apparatus to problem solving through constructions in elementary geometry leads him to consider additional logical constants, an example of which is what he calls "before-after conjunction", which is intended to capture the order in which constructions are carried out. According to de Campos Sanz it cannot be defined in terms of standard logical constants but should nevertheless be conceived as logical. He is even led to consider, besides the notion of *assumption* in the sense that a solution to a problem is supposed to be available, a more general notion of *hypothesis* in *reductio proofs* (which are abundant in geometry). From my point of view, Kolmogorov's *problem interpretation* of logic should be considered in connection with reductive approaches such as tableau systems and dialogical interpretations — in this sense de Campos Sanz's usage of the term "reduction semantics" is highly accurate. At a more general level, investigations such as de Campos Sanz's show us that other forms of argumentation together with other sorts of concepts come into play when we leave the narrow path of standard logical systems. This opens perspectives which are still way beyond the grasp of proof-theoretic semantics as it stands now.

Pereira, Haeusler and Nascimento on disjunctive syllogism and *ex falso*. The paper by Pereira, Haeusler and Nascimento presents two systems between minimal and intuitionistic logic in which *ex falso quodlibet* is not a consequence of disjunctive syllogism. This is a significant contribution to a discussion in logic since antiquity, as disjunctive syllogism is considered a *prima facie* plausible rule, whereas the *ex falso* principle does not have this degree of plausibility. The first system is a variant of Tennant's intuitionistic relevance logic, but without the relevance restriction concerning assumptions, the second is an adaptation of an intuitionistic multiple-conclusion system (developed by the authors in earlier work) to a single conclusion system, where the visibility of assumptions is restricted in a certain way. This is immediately relevant to proof-theoretic semantics as it gives us novel options of framing deductive reasoning which go beyond standard natural deduction (as the formal paradigm of proof-theoretic semantics). The first option, which the paper shares with Tennant, gives the absurdity constant a kind of structural role, when it occurs in generalized elimination inferences. The second option and its associated notion of "visibility" gives an intuitively more perspicuous rendering of what otherwise would be achieved by a not-so-perspicuous multiple conclusion system. *Visibility* might perhaps be considered to be an additional basic concept in the definition of a proof structure, which certainly can be extended to the general case of arbitrary n-ary connectives and even to general reasoning with inductively defined objects. The paper also shows that intuitionistic and classical logic are not the only logical systems to be discussed, and that an intermediate system may be more faithful to our semantical intuitions. This corresponds to observations also made in Piecha and Schroeder-Heister (2019) that intuitionistic logic is not the logic of standard proof-theoretic validity, and that a logic weaker than classical but stronger than intuitionistic logic might be taken into consideration.

Indrzejczak on the logicality of equality. In addition to being an encompassing overview of many treatments of equality in natural deduction and in the sequent calculus, this paper shows that equality can indeed be considered a logical constant in the sense of Došen's (1989) proposal using double line rules. However, Indrzejczak does not take over Došen's own proposal concerning equality rules, which for Indrzejczak is of a 'global' nature, since rephrasing it's rules in natural deduction would need a rewriting of a whole proof. He prefers a 'local' variant with 'free' [my terminology] predicate constants obeying the side condition that they do not occur otherwise in the sequent. I would consider this a hidden second-order treatment of equality which effectively expresses Leibniz equality. This shows the strength of Leibniz's concept if one wants to obtain a 'local' notion. It also points to the fact that the relationship between first- and second-order concepts is a critical topic of proof-theoretic semantics[4], which is far from being settled and which requires further research. At the same time Indrzejczak points to the unclear relationship between logicality and semantics. Even though many (such as Došen and also myself in several publications, e.g., Schroeder-Heister, 1984b) have tried to keep these two issues apart, the criteria for logicality, when formulated in terms of inference rules, are often of a kind that they can be read as meaning conferring and thus as semantical conditions.

Arndt on rules for implication elimination. Arndt discusses different ways of formulating the implication elimination rule in natural deduction. In an earlier publication (Arndt, 2019), he distinguished eight possible variants of implication rules in the sequent calculus. They differed in the way they could be derived, by means of the cut rule, from an axiomatic sequent expressing *modus ponens*. These variants included the standard implication-left rule as well as my proposal for a revised implication-left rule (Schroeder-Heister, 2011). Here this is carried over to natural deduction, where the role of cut is now taken by a rule of explicit composition (which is related to explicit substitution, see Abadi, Cardelli, Curien, and Lévy, 1991; Arndt and Tesconi, 2014). Furthermore, a notational device is added that forces certain premisses of rules to be assumptions (to "stand proud" in Tennant's (1992; 2002) terminology). The result is a congruence between natural deduction and the sequent calculus which puts them in much closer parallel than the usual translations between these types of systems. At the same time it gives a proper understanding of bidirectionality in natural deduction, which I had proposed (at least programmatically) in Schroeder-Heister (2009). Although these are observations at the syntactical level of formal systems, they are highly relevant to proof-theoretic semantics, as the way reasoning is framed depends on which options we have in unterstanding rules, assumptions, proof composition and the like. This is the first investigation I am aware of, which takes subtle differences in the formulations of inference rules into account by making explicit the possible differences in the status of their premisses.

Liang and Miller on focusing Gentzen's LK. The contribution of Liang and Miller gives, from the perspective of proof search, a presentation of the authors' "focused" sequent calculus which overcomes weaknesses of Gentzen's classical sequent calculus LK. It is well-known that the reduction procedures for LK, as applied, for example, in cut elimination proofs, are not very deterministic due to the fact that a great number of permutations between inferences are possible. This makes the sequent calculus differ from the calculus of natural deduction, which, however, has other deficiencies when it comes to proof search. In order to keep the advantages of the sequent calculus for computational purposes, it is here expanded to a system which is syntactically more involved by considering atoms and connectives of different polarities and, correspondingly, two different sequent arrows. One is compensated for this by a more streamlined and more deterministic structure of proofs which is not

[4] See also, in another context, the contribution by Pistone and Tranchini (2023, this volume).

only useful for computational purposes but also for demonstrations of metalogical results such as Herbrand's theorem. It generalizes systems of a related kind developed by Girard, Andreoli and others. This paper shows that the area of proof search, which at the time of the origin of proof-theoretic semantics was much neglected, is becoming an integral part of it in the sense that proof-search aspects are built into the semantically-understood inferences.

Pistone and Tranchini on intensional harmony as isomorphism. The concept of harmony is central in the proof-theoretic semantics of natural deduction. That the consequences of elimination rules of a logical sign have the same deductive power as the conditions of its introduction rules is often seen as a justification of these rules. Harmony also gives rise to identities between proofs in the sense that certain successions of introductions and eliminations can be "reduced" or "contracted", yielding a proof which is still considered the same as the original one. In this sense, harmony is the basis of an intensional proof-theoretic semantics based on the notion of identity of proofs. Because a general definition of harmony was a desideratum, in Schroeder-Heister (2015) I developed an approach which translated the conditions of introduction rules as well as consequences of eliminations rules into formulas of second-order logic with propositional quantification and defined harmony as logical equivalence of these translations. Pistone and Tranchini point out that equivalence is too weak a notion for an appropriate intensional notion of harmony and that some sort of isomorphism of these translations is needed. This is not available in the metatheory of second-order logic based on beta and eta reduction. As a solution, the authors propose an additional so-called "epsilon reduction" which is based on the idea that there is exactly one proof of polymorphic identity in second-order logic. This is a major step forward beyond what I had proposed, as for many cases it leads to a plausible notion of harmony. It also demonstrates the significance of second-order propositional logic for the proof-theoretic semantics of elementary propositional logic.

Wansing on synonymy. The intensional notion of proof identity induces a notion of isomorphism between propositions. One would consider A and B isomorphic to one another if there are proofs of B from A and of A from B, such that the composition of these proofs yields the identity proof of A from A and of B from B. "Yields" here means that the given composition of proofs is identical (in the sense of proof identity) to the identity proof. Wansing, who is working in a bilateral framework of proofs and disproofs, gives a different definition of synonymy (his term for isomorphism). For him A and B are synonymous, if there are identical proofs from A to B and from B to A, as well as identical disproofs between these propositions. His notion of proof identity does not require that the propositions proved (and assumed) are the same between identical proofs, which is against a principal *tenet* in standard intensional proof-theoretic semantics. Identity in Wansing's sense is defined by a structural correspondence of sequent-style proofs. This definitely gives novel incentives to the discussion on the identity of proofs, both from the structural point of view (identical proofs for different propositions), but also from the consideration of the sequent calculus and the appeal to bilateralism, which are very much neglected in current discussions of proof identity. So far, only the discussion of the principal type of a lambda term as representing the structure of a proof comes close to the idea that there can be identical proofs (proofs of identical structure) of different theorems (see Hindley, 1997; Rezende de Castro Alves, 2019). In the discussion of identity of proofs and thus in intensional proof-theoretic semantics (see Tranchini, 2023), there are many conceptual aspects still open, and Wansing is providing a fresh look at some of these.

Kahle and Santos on paradoxes. Kahle and Santos discuss the relationship between the conceptual constructions of logical, semantical and set-theoretic paradoxes and the logic used to derive a contradiction. It is the logic which renders these constructions paradoxical. On the other hand, it is extremely difficult to make specific logical features responsible for the paradoxical outcome — just passing to another logical system, for example from classical logic to a logic without excluded middle such as intuitionistic logic does not change the situation. What changes it are global considerations such as normalization requirements in the sense of Prawitz (1965, Appendix B) and Tennant (1982). Therefore Kahle and Santos plead to further scrutinize the conceptual constructions of the paradoxes, but from a consequentialist point of view, for which I have argued myself (based on ideas of Lars Hallnäs): not restricting conceptual definitions themselves, but classifying definitions according to possible consequences including the non-eliminability of cut etc. (Schroeder-Heister, 2012b). However, while Kahle and Santos discard the reference to substructural logics to avoid the paradoxes and criticize some attempts I made in this direction (Schroeder-Heister 2012a; 2016), I see an option that limits the rule of contraction. This is a limitation not in the global substructural, but in a local intensional sense (see Schroeder-Heister, 2022).

Hallnäs on the structure of proofs. As indicated in my autobiographical survey (Schroeder-Heister, 2023, Section 7), I owe to Lars Hallnäs many ideas I consider relevant for current proof-theoretic semantics, and even more relevant for its future development. His idea of definitional reflection, that is, the idea of a general principle to extract information from definitions, which can be partial and are not necessarily monotone, goes way beyond logic and has shaped my understanding of proof-theoretic semantics: not only because this approach represents a powerful extension of logic programming, as we presented it initially (Hallnäs and Schroeder-Heister, 1990), but, more importantly, because it constitutes a general reasoning principle from an intensional point of view. Hallnäs's contribution to this volume sketches the direction into which this might lead when we not only consider the function closure but the functional closure of definitions. Already the structure of natural deduction with its concept of assumption discharge and corresponding side conditions makes such an approach reasonable. The most original idea in this paper is the characterization of proofs in terms of their reductive behaviour, which allows Hallnäs to compare and identify proofs in different formal systems as this behaviour is independent of the formal system itself. I would look at it as an attempt to formalize the concept of 'proof idea', something that every mathematician is aware of and that is the driving force in defining the identity of proofs, but that has not really progressed so far. Hallnäs calls it a "structure theory of proofs" which is much more abstract than "structural proof theory" (Negri and von Plato, 2001) (which partly overlaps with what Prawitz (1971; 1972; 1973) called "general proof theory"). Using his general operators on proofs, he discusses in particular what I have called "Ekman's paradox", a topic of his former doctoral student Jan Ekman to which Hallnäs drew my attention in the early 1990s and which has fascinated me ever since, leading to recent work with Tranchini (Schroeder-Heister and Tranchini 2017; 2021). He can formally demonstrate that both Ekman's normalization paradox and Russell's set-theoretic paradox, though formulated in different formal systems, are based on the same idea, as they satisfy the same abstract proof equation. Hallnäs's analysis actually gives a structural rendering of Ekman's reduction of proof terms, which from the semantical perspective applied by Tranchini and myself remains invisible. The application of such general tools is a promising method in advanced intensional proof-theoretic semantics. This holds likewise for the advanced second-order tools used by Pistone and Tranchini (2023, this volume).

Francez and Kaminski on truth-value constants in multi-valued logics. The contribution by Francez and Kaminski can be seen as an application of proof-theoretic semantics to a system where formulas are signed with truth values. It is thus a generalization of bilateral systems, where one uses positively and negatively signed formulas, to the case of finitely many truth values. The elimination rule of the system formulated in sequent-style natural deduction corresponds to the general elimination rule proposed by Prawitz (1979) and myself (Schroeder-Heister, 1984a), but is now derived from introduction rules based on the truth functional meaning of the connective considered. The paper discusses in particular the case of the nullary constants truth and falsity and their generalizations to arbitrary truth values, and establishes that we have explosion rules for them corresponding to *ex falso quodlibet* in the case when the nullary constant is signed with a non-matching truth value. This shows that proof-theoretic semantics can be productively applied in the area of multi-valued logic and is not confined to intuitionistically inspired logics.

Więckowski on counterfactual assumptions and implications. Więckowski applies proof-theoretic semantics to causal and counterfactual reasoning, more precisely to reasoning from assumptions where assumptions are either *factual* ("since A is the case, B is the case", i.e., "B is the case, because A is the case") or *counterfactual* ("if A were the case, B would be the case"). He overcomes my general characterization of assumptions in natural deduction as "unspecific", which I used to distinguish assumptions in natural deduction from those in bidirectional sequent-based reasoning (Schroeder-Heister, 2004). His idea is to use two proof systems: a "reference system" which is used to infer the assumptions of the "modal system". When the reference system derives the assumption in a canonical way, the modal consequence is a factual or causal inference, while if this is not the case, the modal consequence is counterfactual. Thus the reference system allows one to distinguish between a derivation from an accepted assumption, from a non-accepted assumption, and from an unspecific assumption just laid down. As his reference system he chooses subatomic natural deduction as proposed by himself (Więckowski, 2011), which is particularly suited to deal with identity assumptions in the modal system. This approach is a further step (there are not so many yet[5]) to make proof-theoretic semantics fruitful for the investigation of intensional natural-language phenomena.

Bärtschi and Jäger on set-theoretic reduction principles. The contribution by Bärtschi and Jäger sits on the border between reductive and general proof theory. It investigates the strength of so-called separation principles in second-order arithmetic, which allow one to distinguish two disjoint unary formulas by means of a set containing instances of the first but no instance of the second. These principles play an important role in reverse mathematics. Under the name "reduction principles" (to distinguish them from set-theoretic "separation axioms") they are investigated here with respect to set-theoretic laws, in particular in Kripke-Platek set theory as compared to systems with transfinite recursion. This demonstrates how much remains to be done in proof-theoretic semantics to achieve significant results of mathematical proof theory, given that Kripke-Platek set theory is related to theories of inductive definitions. For me inductive definitions are a key topic in a proof-theoretic semantics with definitional reflection, in particular when functional closure as in Hallnäs's (2023) contribution (this volume) is taken into account.

To conclude, when I look at the breadth and depth of the content of these essays, I feel confirmed in my assessment that proof-theoretic semantics has a bright future.

Acknowledgements I am very grateful to the authors for their contributions, and to Thomas Piecha and Kai Wehmeier for compiling and editing this volume.

References

Abadi, M., L. Cardelli, P.-L. Curien, and J.-J. Lévy (1991). Explicit substitutions. *Journal of Functional Programming* 1, 375–416. DOI: 10.1017/S0956796800000186.

Arndt, M. (2019). Eight inference rules for implication. *Studia Logica* 107, 781–808. DOI: 10.1007/s11225-018-9821-9.

– (2023). Eight rules for implication elimination. In: *Peter Schroeder-Heister on Proof-Theoretic Semantics (this volume)*. Ed. by T. Piecha and K. F. Wehmeier. Berlin: Springer.

Arndt, M. and L. Tesconi (2014). Principles of explicit composition. In: *Second Pisa Colloquium in Logic, Language and Epistemology*. Ed. by E. Moriconi. Pisa: Edizioni ETS, 19–67.

Bärtschi, M. and G. Jäger (2023). Some set-theoretic reduction principles. In: *Peter Schroeder-Heister on Proof-Theoretic Semantics (this volume)*. Ed. by T. Piecha and K. F. Wehmeier. Berlin: Springer.

[5] One is the second half (Part II) of Francez's (2015) book on proof-theoretic semantics, which deals with applications in linguistics, as well as later work of this author such as Francez (2022).

de Campos Sanz, W. (2023). Kolmogorov and the general theory of problems. In: *Peter Schroeder-Heister on Proof-Theoretic Semantics (this volume)*. Ed. by T. Piecha and K. F. Wehmeier. Berlin: Springer.

Došen, K. (1989). Logical constants as punctuation marks. *Notre Dame Journal of Formal Logic* 30, 362–381. DOI: 10.1305/ndjfl/1093635154.

Francez, N. (2015). *Proof-Theoretic Semantics*. London: College Publications.

– (2022). Proof-theoretic semantics as a resource for expressing semantic variability. *Synthese* 200, article number 294. DOI: 10.1007/s11229-022-03780-1.

Francez, N. and M. Kaminski (2023). Truth-value constants in multi-valued logics. In: *Peter Schroeder-Heister on Proof-Theoretic Semantics (this volume)*. Ed. by T. Piecha and K. F. Wehmeier. Berlin: Springer.

Frege, G. (1884). *Die Grundlagen der Arithmetik: Eine logisch mathematische Untersuchung über den Begriff der Zahl*. Breslau: Wilhelm Koebner. Engl. *The Foundations of Arithmetic: A logico-mathematical enquiry into the concept of number*. Basil Blackwell: Oxford 1950, 2nd edition 1953.

– (1906). Über die Grundlagen der Geometrie. *Jahresbericht der Deutschen Mathematiker-Vereinigung* 15, 293–309, 377–403, 423–430. Engl. On the Foundations of Geometry (Second Series). In: G. Frege, On the Foundations of Geometry and Formal Theories of Arithmetic, trans. and ed. by E.-H. W. Kluge. New Haven: Yale University Press, 49–112.

– (1923). Logische Untersuchungen. Dritter Teil: Gedankengefüge. *Beiträge zur Philosophie des deutschen Idealismus* 3, 36–51. Engl. Logical Investigations. Third Part: Compound Thoughts. In: G. Frege, Logical Investigations, trans. by P. Geach and R. Stoothoff, ed. by P. Geach, Oxford: Blackwell 1977, 55–77.

Hallnäs, L. (2023). On the structure of proofs. In: *Peter Schroeder-Heister on Proof-Theoretic Semantics (this volume)*. Ed. by T. Piecha and K. F. Wehmeier. Berlin: Springer.

Hallnäs, L. and P. Schroeder-Heister (1990). A proof-theoretic approach to logic programming: II. Programs as definitions. *Journal of Logic and Computation* 1, 635–660. DOI: 10.1093/logcom/1.5.635.

Hilbert, D. and P. Bernays (1934). *Grundlagen der Mathematik I*. Berlin: Springer. Second edition 1968.

Hindley, J. R. (1997). *Basic Simple Type Theory*. Cambridge: Cambridge University Press.

Indrzejczak, A. (2023). Logicality of equality. In: *Peter Schroeder-Heister on Proof-Theoretic Semantics (this volume)*. Ed. by T. Piecha and K. F. Wehmeier. Berlin: Springer.

Kahle, R. and P. Santos (2023). Paradoxes, intuitionism, and proof-theoretic semantics. In: *Peter Schroeder-Heister on Proof-Theoretic Semantics (this volume)*. Ed. by T. Piecha and K. F. Wehmeier. Berlin: Springer.

Liang, C. and D. Miller (2023). Focusing Gentzen's LK proof system. In: *Peter Schroeder-Heister on Proof-Theoretic Semantics (this volume)*. Ed. by T. Piecha and K. F. Wehmeier. Berlin: Springer.

Negri, S. and J. von Plato (2001). *Structural Proof Theory*. Cambridge University Press.

Pereira, L. C., E. H. Haeusler, and V. Nascimento (2023). Disjunctive syllogism without *Ex falso*. In: *Peter Schroeder-Heister on Proof-Theoretic Semantics (this volume)*. Ed. by T. Piecha and K. F. Wehmeier. Berlin: Springer.

Piecha, T. and P. Schroeder-Heister (2019). Incompleteness of intuitionistic propositional logic with respect to proof-theoretic semantics. *Studia Logica* 107. Special issue *General Proof Theory*. Ed. by T. Piecha and P. Schroeder-Heister, 233–246. DOI: 10.1007/s11225-018-9823-7.

Pistone, P. and L. Tranchini (2023). Intensional harmony as isomorphism. In: *Peter Schroeder-Heister on Proof-Theoretic Semantics (this volume)*. Ed. by T. Piecha and K. F. Wehmeier. Berlin: Springer.

Prawitz, D. (1965). *Natural Deduction: A Proof-Theoretical Study*. Stockholm: Almqvist & Wiksell. Reprinted Mineola, NY: Dover Publ., 2006.

– (1971). Ideas and results in proof theory. In: *Proceedings of the Second Scandinavian Logic Symposium (Oslo 1970)*. Ed. by J. E. Fenstad. Amsterdam: North-Holland, 235–308. DOI: 10.1016/S0049-237X(08)70849-8.

– (1972). The philosophical position of proof theory. In: *Contemporary Philosophy in Scandinavia*. Ed. by R. E. Olson and A. M. Paul. Baltimore: Johns Hopkins Press, 123–134.

– (1973). Towards a foundation of a general proof theory. In: *Logic, Methodology and Philosophy of Science IV*. Ed. by P. Suppes et al. Amsterdam: North-Holland, 225–250.

– (1979). Proofs and the meaning and completeness of the logical constants. In: *Essays on Mathematical and Philosophical Logic: Proceedings of the Fourth Scandinavian Logic Symposium and the First Soviet-Finnish Logic Conference, Jyväskylä, Finland, June 29 – July 6, 1976*. Ed. by J. Hintikka, I. Niiniluoto, and E. Saarinen. Dordrecht: Kluwer, 25–40 [revised German translation 'Beweise und die Bedeutung und Vollständigkeit der logischen Konstanten', *Conceptus, 16*, 1982, 31–44]. DOI: 10.1007/978-94-009-9825-4_2.

– (2006). Meaning approached via proofs. *Synthese* 148. Special issue *Proof-Theoretic Semantics*, edited by R. Kahle and P. Schroeder-Heister., 507–524. DOI: 10.1007/s11229-004-6295-2.

– (2023). The validity of inference and argument. In: *Peter Schroeder-Heister on Proof-Theoretic Semantics (this volume)*. Ed. by T. Piecha and K. F. Wehmeier. Berlin: Springer.

Rezende de Castro Alves, T. (2019). *Synonymy and Identity of Proofs: A Philosophical Essay*. University of Tübingen: Doctoral Dissertation. DOI: 10.15496/publikation-32950.

Schroeder-Heister, P. (1984a). A natural extension of natural deduction. *Journal of Symbolic Logic* 49, 1284–1300. DOI: 10.2307/2274279.

– (1984b). Popper's theory of deductive inference and the concept of a logical constant. *History and Philosophy of Logic* 5, 79–110. DOI: 10.1080/01445348408837064.

– (1997). Frege and the resolution calculus. *History and Philosophy of Logic* 18, 95–108. DOI: 10.1080/01445349708837278.

Schroeder-Heister, P. (1999). Gentzen-style features in Frege. In: *Abstracts of the 11th International Congress of Logic, Methodology and Philosophy of Science, Cracow, Poland (August 1999)*, 449. DOI: 10.15496/publikation-72545.
– (2004). On the notion of *Assumption* in logical systems. In: *Selected Papers Contributed to the Sections of GAP5, Fifth International Congress of the Society for Analytical Philosophy, Bielefeld, 22-26 September 2003*. Ed. by R. Bluhm and C. Nimtz. Paderborn: Mentis, 27–48. URL: http://www.gap5.de/proceedings/. (Abstract and Extended Summary DOI: 10.15496/publikation-72545).
– (2006). Validity concepts in proof-theoretic semantics. *Synthese* 148. Special issue *Proof-Theoretic Semantics*. Ed. by R. Kahle and P. Schroeder-Heister, 525–571. DOI: 10.1007/s11229-004-6296-1.
– (2009). Sequent calculi and bidirectional natural deduction: on the proper basis of proof-theoretic semantics. In: *The Logica Yearbook 2008*. Ed. by M. Peliš. London: College Publications, 237–251. DOI: 10.15496/publikation-70817.
– (2011). Implications-as-rules vs. implications-as-links: an alternative implication-left schema for the sequent calculus. *Journal of Philosophical Logic* 40, 95–101. DOI: 10.1007/s10992-010-9149-z.
– (2012a). Paradoxes and structural rules. In: *Insolubles and Consequences: Essays in Honour of Stephen Read*. Ed. by C. Dutilh Novaes and O. Hjortland. London: College Publications, 203–211. DOI: 10.15496/publikation-70824.
– (2012b). Proof-theoretic semantics, self-contradiction, and the format of deductive reasoning. *Topoi* 31, 77–85. DOI: 10.1007/s11245-012-9119-x.
– (2014). Frege's sequent calculus. In: *Trends in Logic XIII: Gentzen's and Jaśkowski's Heritage – 80 Years of Natural Deduction and Sequent Calculi*. Ed. by A. Indrzejczak, J. Kaczmarek, and M. Zawidzki. Łódż University Press, 233–245. DOI: 10.15496/publikation-72324.
– (2015). Harmony in proof-theoretic semantics: a reductive analysis. In: *Dag Prawitz on Proofs and Meaning*. Ed. by H. Wansing. Berlin: Springer, 329–358. DOI: 10.1007/978-3-319-11041-7_15.
– (2016). Restricting initial sequents: the trade-offs between identity, contraction and cut. In: *Advances in Proof Theory*. Ed. by R. Kahle, T. Strahm, and T. Studer. Basel: Birkhäuser, 339–351. DOI: 10.1007/978-3-319-29198-7_10.
– (2022). Intensional proof-theoretic semantics and the rule of contraction. In: *Logically Speaking: A Festschrift for Marie Duží*. Ed. by P. Materna and B. Jespersen. London: College Publications, 5–21.
– (2023). Proof-theoretic semantics: an autobiographical survey. In: *Peter Schroeder-Heister on Proof-Theoretic Semantics (this volume)*. Ed. by T. Piecha and K. F. Wehmeier. Berlin: Springer.
Schroeder-Heister, P. and L. Tranchini (2017). Ekman's paradox. *Notre Dame Journal of Formal Logic* 58, 567–581. DOI: 10.1215/00294527-2017-0017.
– (2021). How to Ekman a Crabbé-Tennant. *Synthese* 199, 617–639. DOI: 10.1007/s11229-018-02018-3.
Sundholm, G. (2023). Grundlagen der Arithmetik, §17: Part 1. Frege's anticipation of the deduction theorem. In: *Peter Schroeder-Heister on Proof-Theoretic Semantics (this volume)*. Ed. by T. Piecha and K. F. Wehmeier. Berlin: Springer.

Tennant, N. (1982). Proof and paradox. *Dialectica* 36, 265–296. DOI: 10.1111/j.1746-8361.1982.tb00820.x.
– (1992). *Autologic*. Edinburgh: Edinburgh University Press.
– (2002). Ultimate normal forms for parallelized natural deductions. *Logic Journal of the IGPL* 10, 299–337. DOI: 10.1093/jigpal/10.3.299.
– (2021). Normalizability, cut eliminability and paradox. *Synthese* 199, 597–616. DOI: 10.1007/s11229-016-1119-8.
– (2023). Frege's class theory and the logic of sets. In: *Peter Schroeder-Heister on Proof-Theoretic Semantics (this volume)*. Ed. by T. Piecha and K. F. Wehmeier. Berlin: Springer.
Tranchini, L. (2023). *Harmony and Paradox: Intensional Aspects of Proof-Theoretic Semantics*. Berlin: Springer.
Wansing, H. (2023). A note on synonymy in proof-theoretic semantics. In: *Peter Schroeder-Heister on Proof-Theoretic Semantics (this volume)*. Ed. by T. Piecha and K. F. Wehmeier. Berlin: Springer.
Więckowski, B. (2011). Rules for subatomic derivation. *Review of Symbolic Logic* 4, 219–236. DOI: 10.1017/S175502031000033X.
– (2023). Counterfactual possibilities and might-counterfactuals. In: *Peter Schroeder-Heister on Proof-Theoretic Semantics (this volume)*. Ed. by T. Piecha and K. F. Wehmeier. Berlin: Springer.

Index

2Int, 357

Abadi, Martin, 447
Aczel, Peter, 9
admissible sets, 434
Adorno, Theodor W., 4
Ajjanagadde, Venkat, 21
Almukdad, Ahmad, 344
analytic, 62
Andreoli, Jean-Marc, 276, 310, 448
application of inference, 140
Aristotle, 13, 56, 142, 143
arithmetical, 427
arithmetical comprehension, 427
arithmetical transfinite recursion, 428, 429
Arndt, Michael, 14, 28, 29, 447
assertion, 136
assertive reasoning, 269
assumption, 136, 400
assumptive reasoning, 269
atomic systems, 27
Avron, Arnon, 28, 215
axiom of countability, 432

Béziau, Jean-Yves, 24–25
Baaz, Mathias, 222
bar induction, 428
basic set theory, 430
before-after conjunction, 187, 445
Behmann, Heinrich, 371
Bell, John, 220
Benedict XVI, 4
Bernays, Paul, 3, 366, 443
BHK interpretation, 343
BHK-interpretation, 144
BHK-proof, 145, 146
bidirectional natural deduction, 268–271
bilateralism, 24, 25, 343, 448
Binder, David, 11, 229
binding, 136
Boghossian, Paul, 136
Bolzano, Bernard, 5, 141
Boole, George, 13
Bourdeau, Michel, 29
Brandom, Robert, 2
Brouwer, L. E. J., 197, 367, 368
Brouwer, Luitzen Egbertus Jan, 152
Bubner, Rüdiger, 20
Buchholz, Wilfried, 3
Buhl, Günter, 5
Buldt, Bernd, 8
Bärtschi, Michael, 451

calculus of squares, 24
canonical identity statement, 86, 93
Carnap, Rudolf, 8, 317, 329, 341
characteristic aim, 136
Chateaubriand, Oswaldo, 31

T. Piecha and K. F. Wehmeier (eds.), *Peter Schroeder-Heister on Proof-Theoretic Semantics*, Outstanding Contributions to Logic 29,
https://doi.org/10.1007/978-3-031-50981-0

co-implication, 357
cognitive science, 20
composition, 148
comprehension principles, 426
conditional, 56
congruence relation, 353
constructive type theory, 9
contradiction, 393, 395
Contu, Patrizio, 22, 26
Coquand, Thierry, 14
Core Logic, 86, 88, 93, 94, 132, 444
counterfactual assumption, 409, 450
counterfactual implication, 408, 450
counterfactuals, 400, 450
counterpossibles, 417
Cozzo, Cesare, 137, 156, 159
Curry, Haskell, 365
Curry-Howard correspondence, 320
cut-elimination theorem, 288

de Campos Sanz, Wagner, 27, 344, 445
deduction theorem, 60, 443
definitional freedom, 370, 371
definitional reasoning, 16
definitional reflection, 16–17, 23–24
definitions, 376
Degtyarev, Anatoli, 219
dependency, 138
Devlin, Keith, 3
di Cosmo, Roberto, 341
Dialectica interpretation, 3
dialogical logic, 28
Diller, Justus, 3
Dingler, Hugo, 6
direct negation, 24
Dirichlet, Gustav Lejeune, 4
disjunctive syllogism, 193
disproof, 344
DLMPST, 31
Došen, Kosta, 17–19, 26–28, 30, 31, 211, 228, 318, 321, 341, 446, 447
double-barreled abstraction, 95
double-line rules, 17, 446
Duži, Marie, 30
Dummett, Michael, 5, 13, 17, 25, 29, 58, 145, 151–153, 316
Dunn, Michael, 18, 19
Dyckhoff, Roy, 18, 26
Däubler-Gmelin, Herta, 30

Egli, Urs, 7
Ehrenstein, Walter, 12
Ekman's paradox, 383
Ekman, Jan, 23, 450
encyclopedia, 6
equality, 212, 446
Eriksson, Lars-Henrik, 24
Euler, Leonhard, 6
evidence, 142
ex falso quodlibet, 193, 446, 450
explicit composition, 242, 246, 248, 253, 257, 258, 260, 263–265, 268, 270, 271
explosion, 392–395
exposed normal form, 255
expressive completeness, 9, 10
extensionality, 88, 101, 102, 107, 111
extensions of logic programming, 16–17

Fackeldey, Hubert, 4
factual assumption, 409, 450
factual implication, 408, 450
falsum (⊥), 391, 394, 395
Feferman, Solomon, 3
Felgner, Ulrich, 3
Felscher, Walter, 3
Feyerabend, Paul, 15
Fichot, Jean, 29
final conclusion, 140
focused proof system, 276
focusing, 447
Francez, Nissim, 1, 32, 450
free logic, 87, 91, 92, 100, 102, 104, 110, 112, 113, 116, 121, 128
Frege conditional, 54
Frege, Gottlob, 5, 8, 12–13, 54, 85–91, 96–100, 102, 103, 105, 106, 111, 136, 139, 443–445

Friedrichsdorf, Ulf, 8
functional closure, 376, 377
functional completeness, 9

Gabbay, Dov, 14, 19, 22
Gallier, Jean, 220
Gazzari, René, 29, 234
general elimination rule, 239, 242, 247, 248, 253, 254, 256–258, 261, 263, 265, 268
general logical laws, 64
general proof theory, 450, 451
generalized elimination rules, 9
generalized η-expansions, 321
generic, 137–139
Gentzen semantics, 1, 9
Gentzen, Gerhard, 1, 30, 59, 136, 140, 151, 152, 154, 276, 316, 447
Girard, Jean-Yves, 19, 27, 279, 309, 344, 448
Gratzl, Norbert, 229
Grelling, Kurt, 11, 363
Griffiths, Owen, 219
Grishin, Viacheslav, 19
ground, 142
Guenthner, Franz, 18–20
Gödel, Kurt, 2, 3, 10, 370

Hacking, Ian, 211, 224
Haeusler, Edward Hermann, 446
Hallnäs, Lars, 14, 16–17, 23, 368, 370, 449, 451
harmony, 26–27, 319, 395, 448
Hasenjaeger, Gisbert, 2–3, 9, 10
Heidegger, Martin, 5
Heinzmann, Gerhard, 31
Heister, Gabriele, 6, 12, 20
Henkin, Leon, 371
Herbrand's theorem, 307, 448
Herbrand, Jacques, 60
hereditarily countable, 432
Herre, Heinrich, 30
Hertz, Paul, 13, 30
Heyting, Arend, 144, 145
Heyting-Brouwer logic, 357
higher-level rules, 9
Hilbert's tenth problem, 2
Hilbert, David, 363, 364, 368, 370, 371, 443
Hindley, J. Roger, 449
Hintikka, Jaakko, 212
Hoering, Walter, 21, 30
Horkheimer, Max, 4
Hudelmaier, Jörg, 18, 21
Huet, Gérard, 14
Husserl, Edmund, 8
Hyland, Martin, 14
hypersequent, 17
hypo semantics, 168
hypothetical reasoning, 25

identity, 212
identity of proofs, 318, 328, 448
implication, 56
implication elimination, 447
implications-as-links, 245, 248
implications-as-rules, 243, 245
impredicative encodings (Russell-Prawitz translation), 318, 334
incompleteness, 27
inconsistency, 393
Indrzejczak, Andrzej, 234, 446
inference, 58
inference rule, 322
inherited identity, 351
intensional proof-theoretic semantics, 23–24, 448, 449
intuitionistic, 144, 155
inversion principle, 154
isomorphism, 318, 320, 327, 333, 448

Jaśkowski, Stanisław, 136, 234
Jensen, Ronald Björn, 3
Jäger, Gerhard, 30, 451
Jüngel, Eberhard, 4

Kahle, Reinhard, 22, 28–30, 212, 449
Kalish, Donald, 217
Kaminski, Michael, 450
Kanger, Stig, 219
Kant, Immanuel, 4

Kasper, Walter, 4
Keronen, Seppo, 21
Keuth, Herbert, 21
Kleene, Stephen C., 2, 365
Klev, Ansten, 219
Kolmogorov, Andrei N., 11, 161, 163, 165, 169, 170, 172, 176, 445
Koppelberg, Sabine, 3
Koslow, Arnold, 229
Kripke, Saul, 24
Kripke-Platek set theory, 430, 434, 451
Kuhn, Thomas S., 12
Käsemann, Ernst, 4
Küng, Hans, 4
Kürbis, Nils, 229

López-Escobar, Edgar, 26, 344
Lambek calculus, 19, 21
Lambek, Joachim, 17, 19
law of excluded middle, 100
laws of logic, 74
Leibniz Axiom, 215
Leibniz equality, 447
Leibniz, Gottfried Wilhelm, 13
Leitsch, Alex, 222
Leiß, Hans, 3, 7
Liang, Chuck, 447
linear logic, 276
LKF, focused version of LK, 276, 282
Lloyd, John W., 14
located constants, 394
located formula, 392
located sequent, 392
logic and cognition, 20–21
logic in philosophy, 22
logic programming, 14
logical framework, 14
logical ground inference, 261, 263, 264
logical ground sequent, 240, 249, 264
Logik der Forschung, 4
Loos, Rüdiger, 21
Lorenzen, Paul, 5
Lotze, Hermann, 13
Löwe, Benedikt, 28
Lüdecke, Rainer, 29

Maas, Wolfgang, 3
Machover, Moshe, 220
Manzano, Mara, 212
Martin-Löf type theory, 14, 15
Martin-Löf, Per, 9, 10, 12, 14, 29, 30, 142, 145, 146, 153, 218, 320, 370, 444
Maruyama, Yoshihiro, 229
Materna, Pavel, 30
Mates, Benson, 215
mathematical induction, 141
Matiyasevich, Yuri, 2
Maurer, Harald, 29
McClelland, James, 20
meaning, 141, 143
Merrill, Daniel, 13
Meyer, Paul Georg, 4
Meyer, Robert K., 19
Miller, Dale, 14, 17, 447
Miller, David, 11
minimal logic, 193
Minsky, Marvin, 15
Mints, Grigori, 26, 29, 31, 219
Mittelstraß, Jürgen, 6–8, 13, 15
modal proof system, 408
mode of assumption, 409
model-theoretic semantics, 1
Moltmann, Jürgen, 4
Montague grammar, 7
Montague, Richard, 217
Moreno, Manuel, 212
mountweazel, 7
multi-valued logics, 391, 395, 450
multifocused proof system, 310
Müller, Gert, 3

N4, 345, 346
Nagashima, Terui, 222
naive comprehension, 100
Nascimento, Victor, 446
natural deduction, 59, 151, 154, 158, 194, 393
negation by failure, 24

Negri, Sara, 220, 450
Nelson, David, 344
Nelson, Leonard, 363
Newell, Allen, 15
Nihilartikel, 7
non-creative, 157
non-primitive identity, 402

Oestermeier, Uwe, 21
Oliveira, Hermogenes, 29
Olkhovikov, Grigory, 26
Ono, Hiroakira, 19
Orlandelli, Eugenio, 229

paradox, 23–24, 449
  Curry's, 365–367
  Ekman's, 383, 444, 450
  Girard's, 370
  Grelling–Nelson, 364, 367
  Kleene–Rosser, 365
  Liar, 365–368, 370
  of inference, 58
  Richard, 365
  Russell's, 87, 88, 96, 98–100, 102, 111, 364–368, 382, 450
parametricity, 329
Parlamento, Franco, 215
pasigraph, 86, 112–122, 124–127, 130, 131, 133, 444
Paulson, Lawrence C., 14
Pearce, David, 19, 30
Peppinghaus, Benedikt, 3
Pereira, Luiz Carlos, 30, 446
perseity, 64
Petrić, Zoran, 341
Pfeiffer, Helmut, 3
Π^1_n reduction, 429
Piecha, Thomas, 10, 11, 27–30, 229, 344, 446
Pistone, Paolo, 27, 448, 450
Plotkin, Gordon, 14
Poggiolesi, Francesca, 214, 342
Pohlers, Wolfram, 3
polarization, 279
Popper, Karl R., 4, 7, 8, 11–13, 21, 228
practical proposition, 175
Prauss, Gerold, 5
Prawitz's completeness conjecture, 10
Prawitz, Dag, 1, 9–10, 15, 17, 25, 27, 29, 30, 32, 89, 206, 316, 319, 325, 340, 445, 449, 450
Prestel, Alexander, 2, 8, 9, 12, 15
Previale, Flavio, 215
probability, 15
problem, 175
problem interpretation, 161
PROLOG, 14, 16
proof equations, 381
proof search, 448
proof structure, 378
proof unfolders, 380
proof-theoretic semantics, 416
Proops, Ian, 59
proudness marker, 242, 252, 253, 255, 263, 264, 271
pure sets, 92, 96, 105, 111
Pym, David, 10, 22

quantum disjunction, 322, 334
Quine, Willard Van Orman, 216

Ratzinger, Joseph, 4
Read, Stephen, 18, 218, 369
reduction principles, 426
reduction procedure, 93, 94
Reduction Semantics, 166
reduction semantics, 168
Reeves, Steven, 222
reference proof system, 409
reflexivity of identity, 91, 99, 109
refutation, 344
Rejewski, Marian, 2
Restall, Greg, 234
reverse mathematics, 426
Reyle, Uwe, 14
Rezende de Castro Alves, Tiago, 29, 449
Robinson, Edmund, 22
Rosser, John Barkley, 365
rule of atomic denotation, 91, 93, 95, 97, 99, 105, 109, 113, 121

rule of functional denotation, 113, 116, 120
rules for set theory, 132
Rumelhart, David, 15, 20
Russell's paradox, 87, 88, 96, 98–100, 102, 111, 364–368, 382, 450
Russell, Bertrand, 364, 371

Sambin, Giovanni, 19, 28, 229
Sandqvist, Tor, 10, 27
Santos, Paulo Guilherme, 449
Scedrov, Andre, 19
Schaefer, Frank, 12
Schnorr, Claus Peter, 12
Scholz, Heinrich, 2
Schroeder-Heister, Peter, 54, 136, 145, 147, 154, 156, 159, 212, 219, 368–371, 399
Schröder, Ernst, 13
Schwabhäuser, Wolfram, 2, 3
Schwichtenberg, Helmut, 3, 9, 30, 221
Schütte, Kurt, 3
Scott, Dana, 17
second order arithmetic, 426
second-order propositional intuitionistic logic (System F), 316, 329
Seligman, Jerry, 219
semantical completeness, 10
separation principles, 426
sequent calculus, 212
set-abstraction operator, 86, 87, 96, 100, 102, 112, 115
set-theoretic reduction, 451
Shastri, Lokendra, 21
Σ^1_n reduction, 429
Simon, Herbert, 15
simple inference, 148
Simpson's set theory ATR^S_0, 430
single-barreled abstraction, 95, 110
Slaney, John, 18
Sneed, Joseph D., 12
Solovay, Robert M., 8
Soloviev, Sergei, 19
solution, 175
soundness, 142
speech act, 136
square of opposition, 24
Stålmarck, Gunnar, 18
stand proud, 249, 252–258, 261–266, 271
Stegmüller, Wolfgang, 2, 12
Sternefeld Wolfgang, 8
Stresius, Lothar, 4
structural proof theory, 450
structure of proofs, 449
structure theory of proofs, 375
subargument, 138
subatomic natural deduction, 450
subatomic system, 403
subformula property, 347
subformula sequent calculus, 347
substructural logics, 17, 19–20
Sundholm, Göran, 10, 19, 136, 140, 142, 144, 443
Suppes, Patrick, 218
support, 136
syllogism, 142
synonymy, 339, 340, 342, 354, 448
synthetic rules, 282

Tait, William, 29
Takeuti, Gaisi, 220
Tarski semantics, 1
Tarski, Alfred, 1, 60, 141, 217
Tennant, Neil, 18, 26, 201, 229, 369, 444, 446, 447, 449
term assumption, 403
Tesconi, Laura, 447
Textor, Mark, 234
Thiel, Christian, 5, 12, 13
Tichý, Pavel, 54, 136
tonk, 315
Tranchini, Luca, 23, 27, 29, 369, 370, 444, 448–450
Tristan chord, 15
Troelstra, Anne S., 3, 144, 145, 221, 343
truth-value constants, 391, 394, 395, 450

truthmaker, 86
truths of fact, 71
truths of reason, 71
Turing, Alan, 2
two-sorted typed λ-calculus, 358

Urelemente, 87, 88, 95, 96, 104–106, 108, 109, 111

validity
  immediate, 143
  mediate, 143
  proof-theoretic, 445
van Atten, Mark, 197
van Beethoven, Ludwig, 15
van Benthem, Johan, 19
van Dalen, Dirk, 3, 144, 145
variable-binding term-forming operator, 86
Veloso, Paulo, 166, 182
verum ($\top$), 391, 394, 395
virtual set theory, 132
Vogel, Martin, 6, 15
von Helmholtz, Hermann, 6
von Kutschera, Franz, 1, 9, 13, 54, 136
von Mises, Richard, 12
von Oettingen, Arthur, 6
von Plato, Jan, 26, 31, 220, 450
von Stechow, Arnim, 8
Voronkov, Andrei, 219

Wald, Abraham, 12
Wang, Hao, 219
Wansing, Heinrich, 1, 30, 32, 214, 234, 448
Wehmeier, Kai, 13, 29, 55, 212
Weingartner, Paul, 22
Welchman, Gordon, 2
Wette, Eduard, 3
Weyl, Hermann, 364, 371
Więckowski, Bartosz, 29, 30, 218, 450
Widebäck, Filip, 24
Wittgenstein, Ludwig, 5, 212
Wolters, Gereon, 6–7, 30
World Logic Day, 25, 32
Wundt, Wilhelm, 13

Zermelo, Ernst, 364
Zermelo-Fraenkel set theory, 434
Zimmermann, Ernst, 21, 30, 31
Zimmermann, Thomas E., 8

MIX
Papier aus verantwortungsvollen Quellen
Paper from responsible sources
FSC® C105338

If you have any concerns about our products,
you can contact us on
ProductSafety@springernature.com

In case Publisher is established outside the EU,
the EU authorized representative is:
Springer Nature Customer Service Center GmbH
Europaplatz 3, 69115 Heidelberg, Germany

Printed by Libri Plureos GmbH
in Hamburg, Germany